Basics of General, Organic, and Biological Chemistry

Basics of General, Organic, and Biological Chemistry

Editor:

Pam Wang, Ph.D.

Delve Publishing

www.delvepublishing.com

Basics of General, Organic, and Biological Chemistry

Editor: Pam Wang, Ph.D.

© 2017 Delve Publishing LLC
708 3rd Avenue, 6th Floor
New York, NY 10017
United States of America

www.delvepublishing.com

ISBN: 978-1-68095-901-7

Library of Congress Control Number: 2016951759

About the Editor

Pam Wang, Ph.D.

Pam Wang is a chemical biologist by training who has research experience in a variety of interdisciplinary fields. She received her PhD in 2014 studying artificial proteins and is now an industrial postdoc working on bacterial genome engineering. Her interests lie primarily in the application of biophysical, biochemical, and molecular engineering techniques to biotechnology and the pharmaceutical sciences.

Contents

List of Contributors

Stephen T. Skowron
School of Chemistry, University of Nottingham, University Park, Nottingham NG7 2RD, UK

Irina V. Lebedeva
Nano-Bio Spectroscopy Group and ETSF Scientific Development Centre, Departamento de Física de Materiales, Universidad del Pais Vasco UPV/EHU, San Sebastian E-20018, Spain

Andrey M. Popov
Institute for Spectroscopy of Russian Academy of Sciences, Troitsk, Moscow 142190, Russia

Elena Bichoutskaia
School of Chemistry, University of Nottingham, University Park, Nottingham NG7 2RD, UK

Peter Ertl
Novartis Institutes for BioMedical Research, Novartis Campus, CH-4056, Basel, Switzerland

Bernhard Rohde
Novartis Institutes for BioMedical Research, Novartis Campus, CH-4056, Basel, Switzerland

David Ferro-Costas
Departamento de Química Física, Universidade de Vigo, Facultade de Química, Lagoas-Marcosende s/n, 36310 Vigo, Spain

Ángel **Martín Pendás**
Departamento de Química Física y Analítica, Facultad de Química, Universidad de Oviedo, 33006-Oviedo, Spain

Leticia González
Institute of Theoretical Chemistry, University of Vienna, Währinger Str. 17, 1090 Vienna, Austria

Ricardo A. Mosquera
Departamento de Química Física, Universidade de Vigo, Facultade de Química, Lagoas-Marcosende s/n, 36310 Vigo, Spain

Richard P. Matthews
Department of Chemistry, Imperial College London, London, SW7 2AZ, UK

Ignacio J. Villar-Garcia
Department of Chemistry, Imperial College London, London, SW7 2AZ, UK
Department of Materials, Imperial College London, London, SW7 2AZ, UK

Cameron C. Weber
Department of Chemistry, Imperial College London, London, SW7 2AZ, UK

Jeraime Griffith
Department of Chemistry, Imperial College London, London, SW7 2AZ, UK

Fiona Cameron
Department of Chemistry, Imperial College London, London, SW7 2AZ, UK

Jason P. Hallett
Department of Chemistry, Imperial College London, London, SW7 2AZ, UK
Department of Chemical Engineering, Imperial College London, London, SW7 2AZ, UK

Patricia A. Hunt
Department of Chemistry, Imperial College London, London, SW7 2AZ, UK

Tom Welton
Department of Chemistry, Imperial College London, London, SW7 2AZ, UK

Leslie Glasser
Nanochemistry Research Institute, Department of Chemistry, Curtin University, Perth 6845, Western Australia, Australia

H. Donald Brooke Jenkins
Department of Chemistry, University of Warwick, Coventry, West Midlands CV4 7AL, UK

Dietmar Paschek
Universität Rostock, Institut für Chemie, Abteilung Physikalische und Theoretische Chemie, Albert-Einstein-Str. 21, D-18059 Rostock, Germany

Benjamin Golub
Universität Rostock, Institut für Chemie, Abteilung Physikalische und Theoretische Chemie, Albert-Einstein-Str. 21, D-18059 Rostock, Germany

Ralf Ludwig
Universität Rostock, Institut für Chemie, Abteilung Physikalische und Theoretische Chemie, Dr.-Lorenz-Weg 1, D-18059 Rostock, Germany
Leibniz Institut für Katalyse an der Universität Rostock e.V., Albert-Einstein-Str. 29a, D-18059 Rostock, Germany

Johann V. Pototschnig
Graz University of Technology, Institute of Experimental Physics, Petersgasse 16, 8010 Graz, Austria

Andreas W. Hauser
Graz University of Technology, Institute of Experimental Physics, Petersgasse 16, 8010 Graz, Austria

Wolfgang E. Ernst
Graz University of Technology, Institute of Experimental Physics, Petersgasse 16, 8010 Graz, Austria

Adam Wilson
Department of Organic Chemistry, University of Geneva, Geneva, Switzerland

Giulio Gasparini
Department of Organic Chemistry, University of Geneva, Geneva, Switzerland

Stefan Matile
Department of Organic Chemistry, University of Geneva, Geneva, Switzerland

Hongtao Zhao
Department of Biochemistry, University of Zurich, Zurich, Switzerland

Danzhi Huang
Department of Biochemistry, University of Zurich, Zurich, Switzerland

Anthony M. Reilly
The Cambridge Crystallographic Data Centre, 12 Union Road, Cambridge, CB2 1EZ, UK
Fritz-Haber-Institut der Max-Planck-Gesellschaft, Faradayweg 4-6, Berlin 14195, Germany

Alexandre Tkatchenko
Fritz-Haber-Institut der Max-Planck-Gesellschaft, Faradayweg 4-6, Berlin 14195, Germany

Andrew G. P. Maloney
EaStCHEM School of Chemistry and Centre for Science at Extreme Conditions, The University of Edinburgh, King's Buildings, West Mains Road, Edinburgh, Scotland, EH9 3FJ UK
Cambridge Crystallographic Data Centre, 12 Union Road, Cambridge, England, CB2 1EZ UK

Peter A. Wood
Cambridge Crystallographic Data Centre, 12 Union Road, Cambridge, England, CB2 1EZ UK

Simon Parsons
EaStCHEM School of Chemistry and Centre for Science at Extreme Conditions, The University of Edinburgh, King's Buildings, West Mains Road, Edinburgh, Scotland, EH9 3FJ UK

Matthew R. Davis
Division of Chemistry and Chemical Engineering, California Institute of Technology, 1200 E California Blvd, Pasadena CA 91125, USA

Dennis A. Dougherty
Division of Chemistry and Chemical Engineering, California Institute of Technology, 1200 E California Blvd, Pasadena CA 91125, USA

Paul A. Wender
Department of Chemistry, Department of Chemical and Systems Biology, Stanford University, CA 94305, USA

Victor Blanco
School of Chemistry, University of Manchester, Oxford Road, Manchester, M13 9PL, UK.
Departamento de Química Orgánica, Facultad de Ciencias, Universidad de Granada, Campus Fuentenueva S/N, 18071 Granada, Spain

David A. Leigh
School of Chemistry, University of Manchester, Oxford Road, Manchester, M13 9PL, UK

Vanesa Marcos
School of Chemistry, University of Manchester, Oxford Road, Manchester, M13 9PL, UK

Paul Wrede
Charité-Universitätsmedizin Berlin, Institut für Molekularbiologie und Bioinformatik, Arnimallee 22, D-14195 Berlin, Germany

Dominique Le´vesque
RNA Group/Groupe ARN, De´partement de Biochimie, Faculte´ de me´decine et des sciences de la sante´, Universite´ de Sherbrooke, Sherbrooke, Que´bec, Canada

Francis P. Brie`re
RNA Group/Groupe ARN, De´partement de Biochimie, Faculte´ de me´decine et des sciences de la sante´, Universite´ de Sherbrooke, Sherbrooke, Que´bec, Canada

Jean-Pierre Perreault
RNA Group/Groupe ARN, De´partement de Biochimie, Faculte´ de me´decine et des sciences de la sante´, Universite´ de Sherbrooke, Sherbrooke, Que´bec, Canada

Olga Romainczyk
Institute of Organic Chemistry and Chemical Biology, Max-von-Laue-Str. 7, 60438 Frankfurt am Main, Germany

Burkhard Endeward
Institute of Physical and Theoretical Chemistry and Center of Biomolecular Magnetic Resonance, Max-von-Laue-Str. 7, 60438 Frankfurt am Main, Germany

Thomas F. Prisner
Institute of Physical and Theoretical Chemistry and Center of Biomolecular Magnetic Resonance, Max-von-Laue-Str. 7, 60438 Frankfurt am Main, Germany

Joachim W. Engels
Institute of Organic Chemistry and Chemical Biology, Max-von-Laue-Str. 7, 60438 Frankfurt am Main, Germany

Ángel Catalá
Facultad de Ciencias Exactas, Instituto de Investigaciones Fisicoquímicas Teóricas y Aplicadas-Centro Científico Tecnológic La Plata-Consejo Nacional de Investigaciones Científicas y Técnicas, Universidad Nacional de La Plata, Argentina

Brandi L. Cantarel
Institute for Genome Sciences, University of Maryland, School of Medicine, Baltimore, Maryland, United States of America

Vincent Lombard
Architecture et Fonction des Macromole´cules Biologiques, Aix-Marseille Universite´, Centre National de la Recherche Scientifique, Marseille, France

Bernard Henrissat
Architecture et Fonction des Macromole´cules Biologiques, Aix-Marseille Universite´, Centre National de la Recherche Scientifique, Marseille, France

Preface

Chemistry is a physical science that studies the properties and activity of matter, as well as its interactions with other matter and energy. Classically, matter is found in one of three states—solid, liquid, or gas—and is made up of units known as atoms. Atoms are microscopic particles (typically between 100 and 500 pm in diameter) that consist of nucleons (protons and neutrons) and orbiting electrons; they can be neutral or ionized (ions), depending on the relative number of positively charged protons and negatively charged electrons. Elements in the Periodic Table are identified by the number of protons they possess in their atomic form. Under appropriate conditions, attractive forces can draw atoms towards each other, creating a chemical bond – the most well known types of which are ionic, covalent, dative, and metallic. Once these chemical bonds are formed, the resulting groups of atoms are referred to as molecules (covalent), complexes (dative), or lattices (ionic or metallic). Molecules can in turn be attracted to each other through intermolecular interactions, such as dipole-dipole, hydrogen bonding, Van der Waals, and cation-pi. The strength of these intermolecular interactions largely dictates the state (solid, liquid, or gas) of a compound at a given temperature, as well as its solubility in a given solvent.

The law of conservation of energy and the law of conservation of mass are two fundamental physical principles that lie at the heart of chemical reactivity. Reactions are commonly described by chemical equations, from which stoichiometry and thermodynamic parameters can be derived. Empirically measured reaction rates can also be used to obtain information on reaction kinetics, including rate constants and activation energies. For reversible reactions, equilibrium is defined as the point at which the forward and reverse reaction rates are equal. The effect of a change in conditions on a chemical equilibrium is predicted by Le Chatelier's principle, which states that when a change (e.g. in concentration or temperature) is introduced to a system, the system will adjust itself to counteract that change, thus creating a new equilibrium. Catalysts are agents that lower the activation energy of a reaction and increase the rates of both the forward and reverse reactions (hence do not change equilibrium positions). Organic chemistry is a branch of chemistry that studies the structure, properties, and reactions of organic compounds, which is defined as matter that contains carbon atoms, as well as other elements including, but not limited to, hydrogen, oxygen, nitrogen, sulfur, phosphorus, and the radiostable halogens. A systematic set of nomenclature for organic compounds has been established by IUPAC, although nonsystematic names are also frequently used. For complex molecules, structural formulas are the more common form of representation, allowing for clearer depictions of reaction mechanisms and synthetic routes. Organic synthesis is the construction of compounds from simpler building blocks, while reaction mechanisms are diagrammatic representations of proposed electron flows from starting materials to products. Organic compounds can generally be classified according to their functional groups, for example, alcohols, ketones, aldehydes, amines, sulfones, ethers, among many others. Concepts such as hybridization, electronegativity, stereochemistry, and conjugation lie at the core of organic reactions. Biological chemistry, or biochemistry, is a field that examines the chemical processes within living organisms. It encompasses all processes within and outside of the cell—the basic functional unit of life—ranging from the genetic code (DNA) to intercellular signaling events. Molecules in biology are largely organic compounds; the main classes of biomolecules are nucleic acids (carriers of genetic information), proteins (catalytic, structural, and signaling agents), carbohydrates (sources of energy), and lipids (cellular dividers for compartmentalization). Modern biochemical techniques include DNA sequencing, recombinant protein expression, Western blotting, biophysical binding assays, X-ray crystallography, electron microscopy, mass spectrometry, and flow cytometry. The integration of chemistry and biology has contributed significantly to medical advances in the past few decades.

Editor
Pam Wang, Ph.D.

INTRODUCTION

General chemistry

General chemistry (sometimes called "gen chem" for short) is a course often taught at the high school and introductory university level. It is intended to serve as a broad introduction to a variety of concepts in chemistry and is widely taught. At the university level, it is also sometimes used as a "weed out" course for disciplines (sometimes related, sometimes not) which are perceived to require a high level of intellectual rigor or large course loads. It is also one of the few chemistry courses in most universities that does not explicitly explore a particular discipline such as organic chemistry or analytical chemistry. General chemistry courses typically introduce concepts such as stoichiometry, prediction of reaction products, thermodynamics, nuclear chemistry, electrochemistry, chemical kinetics, and many of the rudiments of physical chemistry. Though the list of subjects covered is typically broad, leading some to criticize both the class and the discipline as encouraging memorization, most general chemistry courses are firmly grounded in several fundamental physical rules for which the primary challenge is understanding when the rules are applicable.

Branches of Chemistry

Chemistry itself has a number of branches:

- Analytical chemistry seeks to determine the composition of substances.
- Biochemistry is the study of chemicals found in living things (such as DNA and proteins).
- Inorganic Chemistry studies substances that do not contain carbon.
- Organic chemistry studies carbon-based substances. Carbon, as described in more detail in this book, has unique properties that allow it to make complex chemicals, including those of living organisms. An entire field of chemistry is devoted to substances with this element.
- Physical chemistry is the study of the physical properties of chemicals, which are characteristics that can be measured without changing the composition of the substance.

Chemistry as a discipline is based on a number of other fields. Because it is a measurement-based science, math plays an important role in its study and usage. A proficiency in high-school level algebra should be all that is needed in this text, and can be obtained from a number of sources. Chemistry itself is determined by the rules and principles of physics. Basic principles from physics may be introduced in this text when necessary.

Why Study Chemistry?

There are many reasons to study chemistry. It is one pillar of the natural sciences necessary for detailed studies in the physical sciences or engineering. The principles of biology and psychology are rooted in the biochemistry of the animal world, in ways that are only now beginning to be understood. Modern medicine is firmly rooted in the chemical nature of the human body. Even students without long-term aspirations in science find beauty in the infinite possibilities that originate from the small set of rules found in chemistry.

Chemistry has the power to explain everything in this world, from the ordinary to the bizarre. Why does iron rust? What makes propane such an efficient, clean burning fuel? How can soot and diamond be so different in appearance, yet so similar chemically? Chemistry has the answer to these questions, and so many more. Understanding chemistry is the key to understanding the world as we know it.

Organic chemistry

Organic chemistry is a branch of chemistry that involves the study of the structures, composition, and synthesis of carbon-containing compounds. In understanding this form of chemistry, it is important to note that all organic molecules include not only carbon, but also hydrogen. While it is true that organic compounds can contain other elements, the bond between carbon and hydrogen is what makes a compound organic.

Originally, this field was defined as the study of compounds created by living organisms, but its definition has been enlarged to include artificially synthesized substances as well. Before 1828, all organic compounds were obtained from living organisms. Scientists didn't believe it was possible to synthesize organic compounds from inorganic compounds, and many who attempted to do so, failed. In 1828, however, urea was made from inorganic substances, paving the way for a new definition of organic chemistry.

There are more than 6 million known organic compounds, and in addition to being plentiful, they are also unique. This is because carbon atoms have the ability to form strong bonds with many different elements. Carbon atoms are also able to bond covalently to other carbon atoms, while simultaneously forming strong bonds with other nonmetal atoms. When carbon atoms bond together, they can form chains consisting of thousands of atoms, as well as rings, spheres, and tubes.

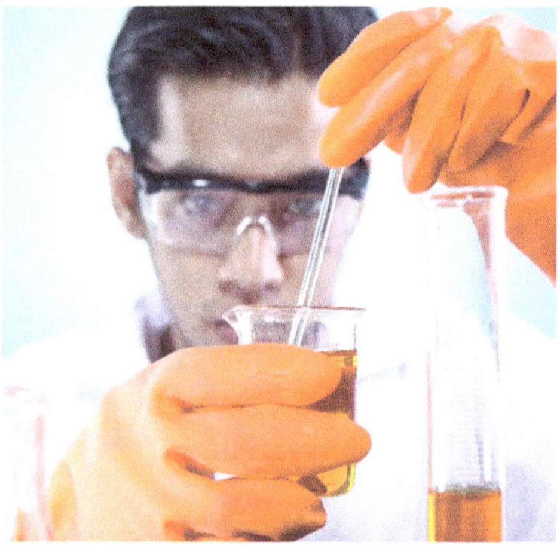

Many individuals consider organic chemistry to be very complicated and unrelated to daily life. Though the study of this discipline may be complex, it is very important to everyday life. In fact, organic compounds are a part of everything, from the foods people eat to the products they use. They are important in the creation of clothing, plastics, fibers, medications, insecticides, petroleum-derived chemicals, and a long list of products used to support life and to make it more convenient.

This field of study is important not only to those who are interested in science-related careers, but to every individual alive today and to those who will be born in the future. It is key in developing new products and improving those on which people have become dependent. Each year, organic chemists make discoveries that are helpful in improving medicines, aiding agricultural growth, understanding the human body, and performing countless tasks important to the average person.

Characterization

Since organic compounds often exist as mixtures, a variety of techniques have also been developed to assess purity, especially important being chromatography techniques such as HPLC and gas chromatography. Traditional methods of separation include distillation, crystallization, and solvent extraction.

Organic compounds were traditionally characterized by a variety of chemical tests, called "wet methods", but such tests have been largely displaced by spectroscopic or other computer-intensive methods of analysis. Listed in approximate order of utility, the chief analytical methods are:

- Nuclear magnetic resonance (NMR) spectroscopy is the most commonly used technique, often permitting complete assignment of atom connectivity and even stereochemistry using correlation spectroscopy. The principal constituent atoms of organic chemistry – hydrogen and carbon – exist naturally with NMR-responsive isotopes, respectively 1H and 13C.
- Elemental analysis: A destructive method used to determine the elemental composition of a molecule. See also mass spectrometry, below.
- Mass spectrometry indicates the molecular weight of a compound and, from the fragmentation patterns, its structure. High resolution mass spectrometry can usually identify the exact formula of a compound and is used in lieu of elemental analysis. In former times, mass spectrometry was restricted to neutral molecules exhibiting some volatility, but advanced ionization techniques allow one to obtain the "mass spec" of virtually any organic compound.
- Crystallography is an unambiguous method for determining molecular geometry, the proviso being that single crystals of the material must be available and the crystal must be representative of the sample. Highly automated software allows a structure to be determined within hours of obtaining a suitable crystal.

Traditional spectroscopic methods such as infrared spectroscopy, optical rotation, UV/VIS spectroscopy provide relatively nonspecific structural information but remain in use for specific classes of compounds. Traditionally refractive index and density were also important for substance identification.

Properties of Organic Compounds

Organic compounds are chemicals whose molecules contain carbon. There is no universally accepted, more precise definition of "organic," which makes defining the properties of organic compounds more difficult. Today, "organic compounds" most commonly refers to compounds that contain significant amounts of carbon, though they can be defined more loosely to include any molecules containing carbon or more strictly to include only molecules that contain carbon-carbon or carbon-hydrogen bonds.

The organic-inorganic distinction is somewhat arbitrary and is in large part derived from a theory called vitalism, which posited that the biochemical processes of living beings involve an additional energy or force that is not present in chemical processes occurring in nonliving matter. This suggested the idea that the substances involved in the biology of living creatures are therefore different from other substances in a way that goes beyond their physical and chemical properties and makes up a special category distinct from other matter. Vitalism is no longer a commonly accepted theory, and modern biochemistry explains biological processes and the properties of organic compounds according to the same physical laws that govern other chemical reactions, but the category "organic" remains in use, though today the term also encompasses a number of compounds that do not occur in living beings.

There is a vast number of different organic compounds with different properties, which are divided into smaller subclasses based on their composition or the composition of their functional groups and tend to have similar properties. Some organic categories have interesting properties that people often encounter in their daily lives. Hydrocarbons, a category that includes groups of molecules called alkanes, alkenes, and arenes according to their structure, are made up entirely of hydrogen and carbon and release a great deal of energy when burnt, making hydrocarbons such as propane, butane, and octane valuable fuel sources. Another type of organic molecules, called esters, make up vegetable and animal oils and fats.

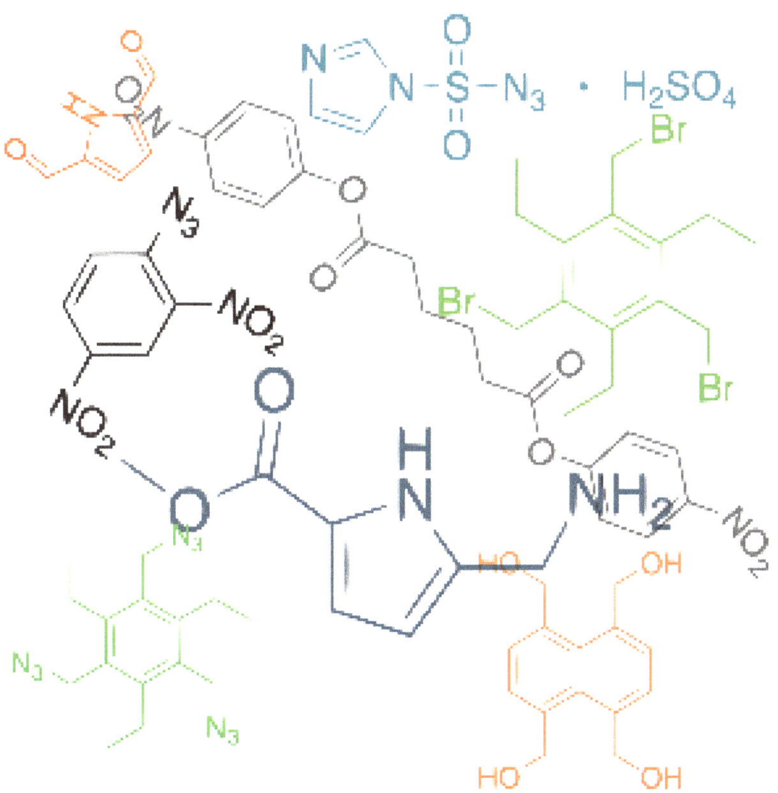

One of the important properties of organic compounds is the presence of structural units called functional groups. These are groups of atoms within a molecule that is joined to the rest of the molecule by a covalent bond between an atom of the functional group and a carbon atom in the molecule's main body. An organic molecule can have many functional groups, and these are largely responsible for the chemical properties of the molecule as a whole, because a given functional group will behave in more or less the same way regardless of the composition of the larger molecule it is attached to. Organic molecules are often classified and named according to the nature of their functional groups.

Molecules that are organic vary greatly in size. One of the important properties of organic compounds that makes them so important in biology comes from the fact that carbon atoms can bond to each other to form chains of carbon atoms in a row or loop, with additional atoms bonded to some of the carbon atoms. Many identical copies of individual structural units called monomers, which consist of a row of carbon atoms linked together with other atoms called side groups attached, can be linked together to form long repeating chains called polymers.

Organic compounds are not unique in their ability to do this, but carbon is particularly well suited to it. A single organic polymer molecule can have hundreds of repeated structural units, and more complex polymers can have separate branches splitting off from the central chain and form chemical bonds with other polymer chains to create a single molecule. Some organic compounds are made of vast structures called macromolecules.

Biological chemistry

Biological chemistry, or biochemistry, is the study of the chemical composition of living organisms at a cellular level. Included in this field of study are the structure of biological entities, the chemical properties of living organisms, and the changes in living cells caused by chemistry. It is a combination of life sciences and chemical sciences.

Someone seeking a career in biological chemistry will usually need at least a bachelor's degree. With a bachelor's degree, an individual can qualify for a job as a science teacher at the high school level, a research assistant, laboratory technician, or a scientist in a testing environment. Graduates may continue to study biological chemistry or proceed with graduate education in a healthcare setting or other profession. Advanced degrees generally open more doors for biochemists. Master's degree holders are qualified to work on simple research projects, among other jobs. Doctorate degrees qualify scientists to conduct independent research, teach college level courses, and perform administrative duties. Biochemistry plays a vital role in many industries. Pharmaceutical companies use biochemists to study disease and formulate drugs to treat and/or cure illness. The biochemist's comprehensive knowledge of chemicals at the molecular level is critical to understanding the relationships between a disease and possible drug treatments.

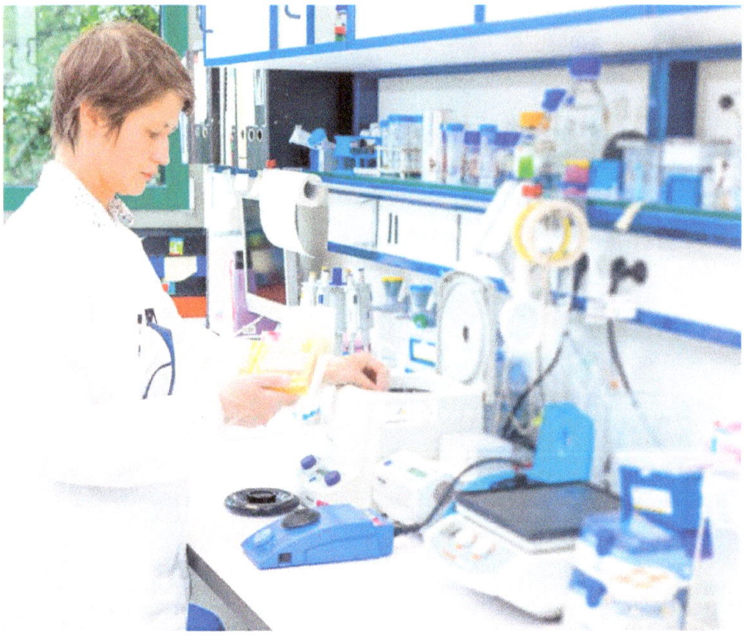

Biological chemistry is also significant in the industry of agriculture. It provides the basis for boosting the nutrient values of foods. Scientists also use biochemical science to develop methods for making plants insect and disease resistant. There are many uses for this science in the development of ingredients to preserve foods. Forensic science relies heavily on biochemistry. Scientists working in a forensic capacity use their specialized knowledge to assist in criminal investigations. Analysis of evidence, such as deoxyribonucleic acid (DNA) testing, is often vital to identifying and prosecuting a criminal. In a hospital setting, biochemistry tests assist in diagnosing disease. Analysis of kidney stones, various enzymes, and disease indicators aid doctors with the evaluation of a patient's condition, as well as its treatment. Vitamin levels may be assessed in relation to symptoms so that health care providers have all relevant data for treatment. Given that biological chemistry is the study of the effects of chemistry on living things at a cellular level, applications in cancer research are numerous. Cancer is the abnormal division and growth of cells. Examination of the biochemistry of the cells in question as well as the effect of various treatments upon those cells helps researchers develop increasingly effective treatments. Biological chemistry is essential to the understanding of living organisms in many settings. This field of science continues to evolve as technology advances, continually offering new applications. Sub-specialties such as biotechnology, neurochemistry, and genetics all benefit from the advances made possible by this evolving science.

In Chemistry, what is a Critical Point?

Critical point is a term used in thermodynamics to describe a pressure and temperature condition beyond which distinctions between phases, particularly between gas and liquid, cease to exist. Beyond such a point, a substance is neither completely liquid nor completely gaseous; it displays properties of both the liquid phase and the gas phase and is referred to as a supercritical fluid. There is also such a thing as a liquid-liquid critical point. Such a point denotes conditions of temperature, pressure, and composition beyond which a mixture will separate into two or more different liquid phases.

The critical point of a specific substance is often viewed on a phase diagram, a chart that shows the behavior of the substance at different temperature and pressure conditions. Such a chart shows, for example, a curve representing the equilibrium between the liquid and gaseous phases. On one side of the line, at higher temperature and lower pressure, the substance is in the gaseous phase. On the other side of the line, at lower temperature and higher pressure, the substance is in the liquid phase. At sufficiently high temperature and pressure, this line terminates in a single point beyond which the liquid-gas phase distinction disappears — the critical point.

Supercritical fluids, which exist beyond a substance's critical point, have many different properties that are reminiscent of the properties of both liquids and gases. Like liquids, supercritical fluids can serve as solvents; they can dissolve other substances. Like gases, however, they have no surface tension. The supercritical fluids that exist beyond the critical point are relatively rare in nature, but they do exist. Deep sea volcanoes, for instance, exist deep in the ocean and release extremely hot substances into the extremely high-pressure conditions at the ocean floor. This heats the water around the volcano, pushing it beyond its critical point and turning it into a supercritical fluid. Some planets, particularly gas giants, contain substances at very high temperature and pressure conditions at their cores. The high temperature and pressure conditions are far beyond the critical point for the substances involved, so they exist as supercritical fluids. There are many different applications of supercritical fluids in science, industry, and other fields. Their solubility properties makes them useful to chemists for dissolving certain substances. They can also be used in the production of biodiesel fuels. Supercritical methanol, for example, is put through a series of reactions that result in a quantity of usable fuel.

Biochemistry

Biochemistry is the study of the chemical processes and reactions that take place within living organisms. It can be considered a subdivision of both chemistry and biology, although the skills and techniques used within it place great emphasis on traditional chemistry.

For a very long time, it was thought that living and non-living matter were fundamentally different. Most scientists believed that only living beings could create special biological molecules, from other biological molecules obtained through food. These molecules were thought to be imbued with a "vital force" that made life possible. In 1828, the German chemist Freidrich Wöhler put an end to this by accidentally synthesizing the organic chemical urea — a major component of urine — from inorganic precursors. The field of biochemistry was born.

Since 1828, studies in the field have brought scientists knowledge of the way plants extract energy from the sun (photosynthesis); how animals convert glucose into the energy currency of the body, ATP (glycolysis); why the muscles burn when a person vigorously exercises (the production of lactic acid); how proteins are synthesized in the cell (protein expression); and much more. As living things tend to be the most useful and important arrangements of matter on Earth from many people›s perspective, knowledge of their inner workings has useful applications in many areas, including in medicine, agriculture, and molecular biology.

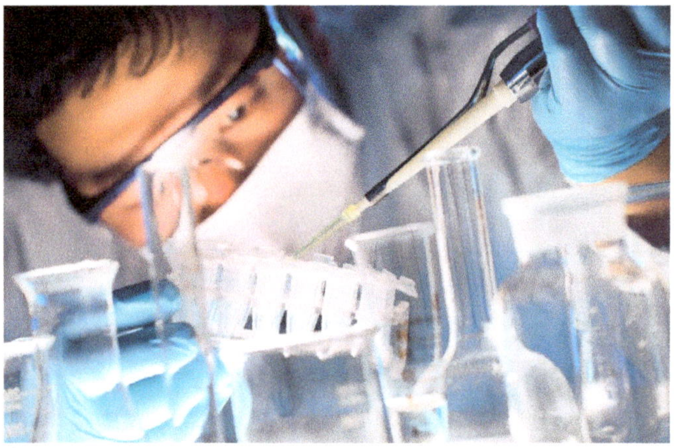

Some molecules studied by biochemists include carbohydrates, proteins, lipids, and nucleic acids. Most of these are organic polymers, meaning they primarily consist simple molecular patterns (monomers) repeated multiple times in a chain, sometimes thousands of times. The primary elements found in organic compounds are carbon, oxygen, hydrogen, nitrogen, calcium, and phosphorous, with trace amounts of chlorine, sulfur, potassium, sodium, magnesium, iron, and a few others.

Many of the molecules in the body, including the carbohydrates and proteins, serve in structural roles. Proteins are manufactured directly based on genetic instructions, and are among the most complex organic molecules. Nucleic acids are the building blocks of genetic instructions (DNA and RNA) found in all forms of life, from humans to viruses. The distinct pattern of nucleic acids found in the nuclei of a species' cells is called its genome. Then there are the lipids, the catch-all term for many non-water-soluble biomolecules. The fat in the body is made of lipids.

Biochemistry is studied by tens of thousands of professionals each day. They seek to understand better how life works and how people can both protect it and use it to make life better.

Biochemistry Research

Biochemistry research stands at the crossroads of chemistry and biology, seeking answers to questions about the chemistry of living beings and how chemical substances behave in the environment of a living person or animal or plant or other living being. It is also called both biochemical research and biological chemistry. According to the United States Bureau of Labor Statistics (BLS), most biochemists are involved in either academic or industrial biochemistry research.

Possible topics of study range from the complex chemical changes that occur at the cellular level to the processes within an individual living being, such as growth and brain function, to processes that take place across generations, such as heredity. Within any of these areas from the single cell to the family group—or even at larger levels—biochemists may study structures and their functions and processes, biochemical causes and effects, relationships between and among structures and organs. They may also attempt to synthesize or engineer products that would serve a role in medical science, for example.

With a focus on organic compounds, biochemistry research may focus on one of the four main types of organic matter found in a cell. One type is protein, the macromolecules made up of amino acids that form an essential component of the diet of people and animals. A second is carbohydrates, compounds composed entirely of oxygen, carbon, and hydrogen, which combine into foods such as sugars, cellulose, and starches. A third is called fats or lipids which are an important mechanism of providing energy reserves for organisms. Fourth is nucleic acids, which are present in all cells and which are key to protein synthesis and the transmission of genes.

Nutrition is another important area of biochemistry research and many biochemists are employed in the food industry. Biochemists study the biochemistry of food in and of itself as well as how it is used in the body, which involves the study of digestion. New food processes can be tested to see what, if any, impact they have on the nutritional quality of the food product, and new ideas for how to meet special nutritional needs can be devised. Other areas of biochemistry research include metabolism, hormones, the circulatory system, genetics, specific diseases, pharmacology, stem cell development, toxicology, and immunochemistry.

Biochemistry research was established by the late eighteenth century. The first time an organic compound was synthesized in the laboratory was when Friedrich Wöhler synthesized urea in 1828. Today, biochemistry research continues to look at some of the same areas that interested scientists several centuries ago, albeit using different equipment. Biochemistry researchers have some very specialized tools at their disposal. These include spectrophotometry, DNA gel electrophoresis, chromatography, and mass spectrometry.

CHAPTER 1

ENERGETICS OF ATOMIC SCALE STRUCTURE CHANGES IN GRAPHENE

Stephen T. Skowron, Irina V. Lebedeva, Andrey M. Popov and Elena Bichoutskaia

INTRODUCTION

Since the discovery of one- and two-dimensional carbon nanostructures, nanotubes[1] and graphene[2] respectively, a variety of interesting fundamental properties and applications of these nanostructures have been found. Contrary to bulk matter, one- and two-dimensional nanostructures do not correspond to the ground state of an infinite 3D system. Thus, studies of the kinetics of atomic scale processes are especially important for understanding atomistic formation mechanisms of these nanostructures. Such kinetics are determined by the energetics of bond realignment, emission and insertion of carbon atoms.

Transformation of carbon nanostructures under heating[3–6] or electron irradiation[3,4,7–14] leads to a production of entirely new species (see also ref. 15for a recent review on modelling transformation processes under electron irradiation). Among the most striking examples of such processes are the formation of graphene nanoribbons from different organic molecules inside carbon nanotubes under heating[3,4] and electron irradiation,[3,4] the formation of flat nanotubes from bilayer graphene nanoribbons,[7] fullerenes from initially flat graphene flakes[8] and carbon chains from graphene[13,14] under electron irradiation. Others include the transformation of polyhedral graphitic nanoparticles into quasi-spherical onions under heating[5] and electron irradiation,[9,10] the formation of double-walled nanotubes from single-walled nanotubes filled with fullerenes under heating[6] and electron irradiation,[11]and the formation of a trilobate structure from three La@C_{82} endofullerenes inside a carbon nanotube under electron irradiation.[12] Mesoscopic processes such as the shrinking of nanotube diameter under electron irradiation,[16] migration and merging of large holes in the outer wall of nanotubes at high temperature,[17] superelongation of carbon nanotubes at high temperature,[18,19] and the migration and annihilation of graphene grain boundaries under electron irradiation[20] have also been observed. These large transformations in nanostructures take place via bond realignments complemented by atom emission, and are determined by energetic characteristics of the corresponding atomic scale reactions.

The presence of defects in graphene structure[21–61] leads to new electronic,[33–44] magnetic,[21,24,35,45–55] thermal[56–61] and mechanical[22–32] properties. In addition to defect formation, self-healing of defects has been observed in carbon nanostructures, including vacancies and holes in nanotubes[62,63] and graphene[64–66] structure. The processes of defect formation, healing and changes of defect type due to bond realignment and atom emission reactions are determined by the activation energies (barriers) of thermally activated processes and the threshold energies of irradiation-induced processes.

The energetics of all atomic-scale reactions change upon approaching graphene edges,[52,56,67–71] and this is especially important for graphene nanoribbons. Edge reconstructions with the formation of topological defects are frequently observed under electron irradiation[13,72–75,270] and significantly deteriorate elastic[76] and fracture[77] properties of graphene nanoribbons. Creation of defects at the edges of graphene nanoribbons can also be used to tune electron[78–85] and thermal[86] transport. Atoms are less stable and much more reactive at the edge compared to the graphene interior, and so reactions at edges play a crucial role in transformations of carbon nanostructures. Notable examples include the transformation of a graphene flake to a fullerene,[8,70,71,87] the formation of nanotube caps,[88] sewing up graphene layers into a nanotube,[89,90] reconstruction of holes in graphene[91] and carbon nanotubes,[92–94] and graphene growth[95] and evaporation[96] at high temperatures.

Citation: Stephen T. Skowron, Irina V. Lebedeva, Andrey M. Popov and Elena Bichoutskaia, Energetics of atomic scale structure changes in graphene, DOI: 10.1039/c4cs00499j

There are currently several reviews devoted to the properties of carbon nanostructures with irradiation defects[97,98] and properties of graphene with defects.[99,100] A very recent review[101] covers the general study of graphene with TEM, while the various structures of interest in graphene under TEM are detailed in ref. 72. However, the energetics of rate determining bond reorganisations in graphene has not yet gained considerable attention within a single review. In the present review we provide an overview of the energetic characteristics of various thermally activated and irradiation-induced reactions in the graphene interior and at graphene edges. In addition to static characteristics, such as formation energies of defects relative to perfect structures, we consider activation energies and threshold energies under irradiation for various bond reorganisations and atom emission reactions. The relationship between atomic scale energetic characteristics and experimentally observable macroscopic processes is discussed.

As most energetic characteristics related to atomic scale structure changes in graphene cannot be determined from experiment, the majority of the data presented in the review is obtained by chemical calculations. However, the activation energies for thermally activated processes and the threshold energies for processes occurring under electron irradiation are rather sensitive to the level of theory used in calculations. Section 2 details some common computational approaches to calculating the key parameters of thermally- and irradiation-induced structure transformations. Section 3 presents the structure of defects in graphene as well as their formation energies and barriers for atomic scale reactions in thermally activated processes. The threshold energies for reactions under irradiation are described in Section 4. Conclusions based on published results are discussed in Section 5, together with some problems which remain unsolved.

COMPUTATIONAL APPROACHES

Methods for Calculating Interatomic Interactions in Carbon Nanostructures

Depending on the complexity of the problem under consideration, different methods should be chosen for the description of atomic interactions in carbon nanostructures. The most accurate results are obtained by quantum chemical methods that do not require a priori knowledge of any specific characteristics of the considered system. The optimal balance between accuracy and efficiency is achieved by density functional theory (DFT),[102,103] in which the total energy of the system is considered as a functional of the electron density and the description of the system is reduced to a set of self-consistent one-particle Schrödinger equations. In the simplest local density approximation (LDA),[104] the exchange–correlation energy depends only on the local electron density, which is sufficient for a description of periodic solids but can lead to significant errors for molecular systems and surfaces. Introduction of the dependence on the gradient of electron density in the generalized gradient approximation (GGA) improves results considerably. Perdew–Wang[105,106] (PW91) and Perdew–Burke–Ernzerhof[107,108] (PBE) GGA functionals are successfully used for calculations of defects in carbon nanostructures. Admixing of a fraction of exact exchange in hybrid functionals (PBE0,[109] B3LYP[110,111]) or inclusion of higher derivatives of the electron density in metaGGA functionals (such as M06-L[112,113]) further improves DFT performance, while dispersion corrections (vdW-DF,[114] DFT-D[115,116]) are important for accurate assessments of geometrical and energetic characteristics of defects in few-layer carbon nanostructures. The account of spin is critical for the analysis of defects with unpaired electrons, such as monovacancies, or graphene edges.

Though DFT offers an efficient framework for full-scale quantum modelling, the size of the considered systems is still limited to hundreds of atoms. Tight-binding models[117–124] represent a cheap alternative to first-principles methods, where quantum effects are captured by a direct, albeit simplified, description of electronic structure. Tight-binding parameters determining single-electron energies and a classical repulsive potential energy term are fitted to reference datasets or results of first-principles calculations. Tight-binding models can therefore be considered as a bridge between classical methods and full modelling of the quantum nature of chemical bonding. The incorporation of self-consistent calculations of the charge distribution[117,118] further improves the accuracy of tight-binding methods.

Classical force fields enable simulations with increased computational efficiency and system size, using analytical potentials to describe the interactions between atoms. These potentials are constructed and fitted to describe the particular behaviour of atoms in the system, observed in experiment or in accurate quantum mechanical calculations. Alongside simple analytical forms of pair potentials, such as the Morse potential,[125] more complex reactive potentials have been developed, capable of describing covalently bonded crystal structures such as graphene. In traditional forms of potentials all bonds are defined explicitly, which makes them unable to model chemical reactions due to the requirement of breaking and forming bonds. Reactive forms of potentials, however, deliberately avoid using explicit bonds in favour of bond orders, thus allowing for continuous bond formation and breaking. The development of reactive potentials is based on the decomposition of chemical bonding into individual contributions (angular, stretching, dihedral terms etc.) to the binding energy, while introducing cross-terms as a penalty for over- and/or under-coordination. Thirty years ago, Abell introduced a general expression for the binding energy as a sum of nearest neighbour pair interactions moderated by the local atomic environment.[126] In the 1990s, Tersoff[127] and Brenner[128] successfully applied the Abell formalism to various solid-state carbon structures including hydrocarbons. Subsequently, Brenner and Stuart et al. extended the original Brenner empirical potential into a second-generation reactive empirical bond order (REBO) potential energy expression for hydrocarbons[129] and non-bonded and dihedral-angle interactions were included (AIREBO).[130]Extensions of these potentials to metals[87,131–133] make simulations of complex catalytically activated processes possible. Van Duin, Goddard and co-workers developed a new formulation of the reactive potential, ReaxFF, based on accurate benchmarking density functional theory studies and extended its use to various materials, including hydrocarbon reactions,[134] transition metals,[135] silicon[136] and other materials such as polymers and ceramics.

In spite of this significant progress in the development of reactive empirical potentials for carbon systems, they are still not sufficiently accurate for studies of some problems related to graphene energetics. As mentioned below, the reactive bond-order potentials (initially developed for hydrocarbons) fail to reproduce the adequate ground state for monovacancies in graphene. Further improvement of existing potentials for the description of specific problems is possible, by fitting their parameters to certain properties that are important for the phenomena under consideration. For example, the parameters of the Brenner potential have recently been fitted to energies of zigzag and armchair graphene edges and elastic energies of the C_{60} and C_{70} fullerenes.[133] This modification of the potential improved the description of the balance between the fullerene elastic energy and graphene edge energies, which is important for the energetics of graphene formation or its transformation to other carbon nanostructures.

Analysis of Thermally Activated Reactions of Defects

A significant contribution to the understanding of the formation and transformation of defects in carbon nanostructures is provided by the analysis of their free energy surface. To reduce computational cost, these reactions are considered in simulation cells with periodic boundary conditions (PBC) or for finite-size flake models. The reliability of such calculations is determined not only by the accuracy of the description of the interatomic interactions, but also by the convergence of the results with respect to the size of the simulation cell (so that periodic images of the defects do not interact) or size of the finite model (to eliminate edge effects) and other simulation parameters. The saddle point in reactions of defects can be sought by restricting some degrees of freedom of the system or by modified geometry optimisation methods.[137,138] A more efficient nudged elastic band method,[139,140] which has been widely used in recent years, is based on the optimisation of the total energy of a string of system images connecting known initial and final states. Another useful method for searching for saddle points is action-derived molecular dynamics, which generates dynamical trajectories with fixed pre-assigned initial and final boundary conditions.[141] Such a method was used, for instance, to find a path with a set of saddle points for vacancies approaching and coalescing in graphene.[142] Using a simple Arrhenius formula, the reaction barriers obtained by these methods can be used to estimate the reaction rates at experimental conditions. The lifetime t_s of a state can be calculated in this way as

$$t_s = \nu^{-1} \exp(\Delta E_t/kT) \tag{1}$$

where v is the characteristic frequency, ΔE_t is the barrier between states, k is the Boltzmann constant, and T is the temperature.

Molecular dynamics (MD)[143,144] and Monte Carlo[143,145] methods allow computational modelling of the evolution of carbon nanostructures with time. In Monte Carlo methods, one of the possible elementary processes at each simulation step is chosen on the basis of their probabilities, obtained according to their energetics. A list of possible elementary processes and their parameters can be known in advance or determined "on-the-fly" by probing the free energy surface. Molecular dynamics, in which trajectories of atoms are obtained by the numerical integration of Newton's equations of motion for nuclei, with the forces evaluated according to the classical force fields or ab initio methods, does not require any information on possible elementary processes or analysis of the free energy surface. However, even for classical systems, this method is typically limited by the time scales of microseconds. Limitations of ab initio molecular dynamics methods are much more severe and the corresponding numerical experiments are three orders of magnitude shorter in time. Practically all processes of formation and transformation of defects are activated and would not fit into this timeframe under experimental conditions (e.g., at room temperature). A common approach to overcome this difficulty is to accelerate kinetics by increasing the simulation temperature. In this way, one can access valuable information on the process mechanism and gain insight into the kinetics of the process. However, extraction of kinetic parameters requires extensive statistics. Furthermore, care should be taken in the analysis of the results. Overheating the system in the numerical experiment distorts the free energy surface and affects the kinetics of various reactions in different ways, which may complicate the interpolation of the data to experimental conditions. Therefore, static analysis of the potential energy surface at zero temperature is still helpful for supplementing and verifying molecular dynamics results.

Calculations of Threshold Energies

The threshold energy of an irradiation-induced process is the minimum energy that must be transferred from an electron of the electron beam (e-beam) to an atom in graphene in order for the process to take place without immediate reversal. Although this can in principle be determined by experiment, due to the development of atomic resolution microscopy, currently there are few examples of experimentally determined thresholds.[146,267] The threshold energy of a process of interest is therefore usually obtained computationally, via a series of MD simulations. At the beginning of each simulation, a kinetic energy corresponding to the energy transferred from the beam electron is assigned to the impacted atom. The structural evolution of the entire system over time is then computed using MD in order to determine whether or not the process occurs. This is repeated over a range of initial kinetic energies, and occasionally over a range of impact angles, in order to determine the threshold energy to the desired accuracy. The binary search algorithm has previously been used[225] to choose the value of kinetic energy to apply in order to determine the threshold energy. However, if the computational cost of an individual MD simulation is high while the number of threshold energies to be determined is low (for example when calculating a single threshold energy at a very high level of theory), manual inspection of the MD trajectories and estimation of a suitable kinetic energy based on the results can be more efficient.

In addition to computationally inexpensive classical MD, density functional tight-binding (DFTB) has also been widely used to calculate threshold energies in graphene.[147,151,225,265,268] Its low computational cost relative to DFT enables calculations on systems with large numbers of atoms, or allows a large series of MD simulations to be run in order to obtain precise threshold energies. Apart from calculating threshold energies, DFTB can be generally very useful when modelling the effect of electron impacts on graphene structures, for example in the case of Stone–Wales rearrangements at a variety of impact angles,[147] in which the 27 000 DFTB-MD simulations performed would have been unfeasible at a higher level of theory. Despite the low computational cost, these calculations typically give comparable results to DFT for carbon atoms in graphene, although caution must be used when employing DFTB; emission threshold energies of zigzag edges in graphene for example have been shown to be overestimated by approximately 2.5 eV compared to the DFT case, ascribed to stronger local bonding in the DFTB relaxed structures (shown in Fig. 13).[148] Disparities for other systems have also been shown, such as in hexagonal boron nitride (h-BN), for which DFTB simulations predicted similar threshold energies for the emission of boron and nitrogen[149] while DFT simulations[150] later agreed with the experimentally shown asymmetry. This discrepancy was assigned to an inadequate description of charge transfer in the DFTB model.

DFT-MD provides an increase in accuracy and versatility compared to DFTB, at the cost of increased calculation time. Calculations using LDA and GGA exchange–correlation functionals are commonly employed in calculations of threshold energies.[148,150,266,269,271] Higher levels of theory, such as hybrid exchange–correlation functionals that include a degree of exact exchange (most notably B3LYP), can be used to obtain more accurate values, but the computational cost of these methods tends to limit the possible system size.

Infinite graphene can be emulated computationally in one of two ways: by using PBC and by using large graphene flakes or nanoribbons. Supercells used in calculations using PBC must be large enough that the atom hit by the electron beam is not influenced by its periodic image, while graphene flakes must be large enough to negate edge effects, shown to extend at least 10 Å from the edge.[67] Simulations using PBC are far more commonly used for calculating threshold energies than large flakes, due to the requirement of a much larger number of atoms to neutralize the large effects of the extended edges compared to a point defect. The atomic processes of structure change studied in graphene in this way typically take of the order of 100 fs to occur when the energies close to the threshold energy are applied, and a 1 fs time step is usually used. Zero-point velocities or thermal vibrations derived from the Maxwell–Boltzmann distribution are used for the initial velocities of other atoms in the system.

The CompuTEM Algorithm

The CompuTEM algorithm[15,151] is a method for linking molecular dynamics and high-resolution transmission electron microscopy (HRTEM) image simulations. The link is established by incorporating structural information from the MD simulations into multislice image simulations, as well as using realistic estimations of the signal-to-noise ratio from the electron dose rate and detector limitations. The effect of the e-beam on a sample is described by incorporating structural changes caused by the electron irradiation from the MD simulations, at realistic rates determined from the electron dose rate and threshold energies. Fig. 1 illustrates this algorithm, in which a random irradiation-induced event occurring in a sequence of such events is described as follows: (1) the nanostructure is equilibrated at a temperature corresponding to experimental conditions in HRTEM, (2) each atom in the nanostructure is classified with respect to the number and strength of its chemical bonds, (3) the probability of an irradiation-induced event (such as atom removal and/or changes to the local atomic structure) is assigned to each atom in accordance with the atom type determined at step 2, so that the sum of the probabilities over all atoms and all considered types of events for each atom is equal to unity, (4) a single random irradiation-induced event is introduced, (5) MD simulation at a temperature corresponding to experimental conditions for a duration sufficient for bond reorganisation or atom removal, (6) MD simulation at the elevated temperature taking into account the structure relaxation after the irradiation-induced event. The introduction of the irradiation-induced event can be described by one of two approaches, an in depth discussion of which can be found in ref. 15.

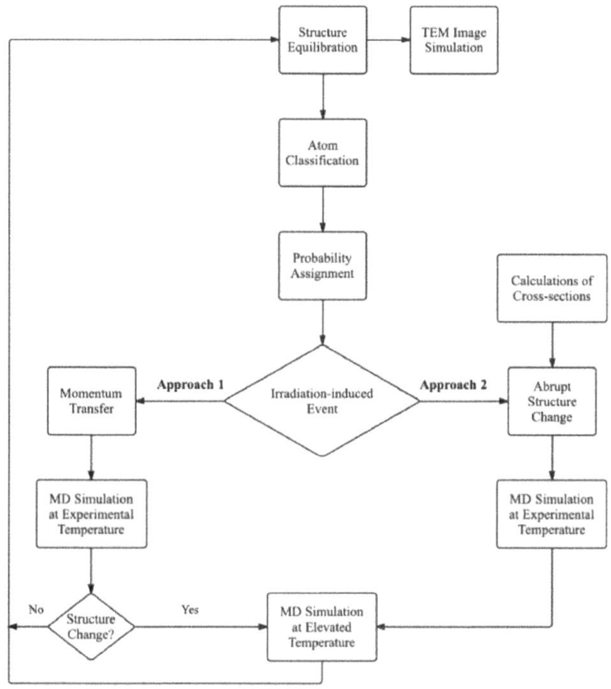

Figure. 1 CompuTEM algorithm for including irradiation-induced structure changes in TEM image simulations. Reprinted (adapted) from ref. 15.

STRUCTURE AND ENERGETICS OF DEFECTS IN GRAPHENE

Vacancies

Effects of Vacancies on Mechanical, Electronic and Magnetic Properties. The influence of vacancies on the physical properties of graphene has been widely studied. The Young modulus,[23–28] Poisson ratio,[25] and tensile strength[22,27,31] of graphene have been shown to decrease with an increase in vacancy concentration. The dependence of fracture strain on the concentration is non-monotonic; there is a decrease in strain at low concentrations and an increase at higher concentrations.[22] A set of calculations demonstrated that the presence of vacancies induces magnetism in graphene.[21,24,35,45–54] According to other calculations the presence of divacancies leads to a decrease of graphene conductivity.[33,34] A decrease in the conductivity of graphene nanoribbons is also observed upon the incorporation of vacancies.[41] Vacancies, as all local defects, do not affect the universal quantization of low-temperature thermal conductance, while they lead to a significant decrease in the thermal conductivity of graphene[26,27,58] and graphene nanoribbons[56,59,60] at room temperature. The thermal conductance of narrow graphene nanoribbons is found to be very sensitive to the position of the vacancy relative to the edge.[56]

Structure and Energetics of Monovacancies

Although pioneer works consider the energetics of the monovacancy with D_{3h} symmetry and dangling bonds,[119,152] subsequent calculations showed that the local three-fold symmetry breaks down to C_{2v} symmetry due to Jahn–Teller distortion.[21,25,47,52,153–162] Most of these calculations show that the reconstructed vacancy has a 5/9 structure[21,47,52,67,119,152–162] where a new bond with length b_v arises between a pair of the three atoms with dangling bonds, forming a pentagon with displacement d_v of the third atom out of graphene plane (Fig. 2a). The structure of a reconstructed monovacancy with two new equivalent bonds and without any out of plane displacement of atoms has also been reported.[48] The formation energy of a defect is defined as $E_f = E_d + n\mu - E_p$, where E_d and E_p are total energy of defective and perfect systems, respectively, μ is the chemical potential estimated by calculations as the total energy per atom in graphene, and n is number of removed atoms for vacancies (positive value) or added adatoms (negative value). The values of formation energies E_v and structural parameters b_v and d_v of the monovacancy in graphene calculated using different methods[21,25,35,47,48,52,54,67,119,152–164] are listed in Table 1. To summarise the results presented inTable 1, recent DFT calculations for infinite graphene give vacancy formation energies in the range E_v = 7.6–7.9 eV and the length of the newly formed bond b_v = 1.8–2.0 Å. However, the out of plane distortions caused by defects in graphene are not fully understood. A reliable value for the out of plane displacement d_v has not yet been obtained, as these deformations are highly sensitive to the system size and edge structure (for flake geometries). Note that the calculated value for the length of the newly formed bond coincides with the experimental one b_v = 1.9 Å, measured by an aberration-corrected TEM study.[159] TEM studies cannot directly measure out of plane displacement, however the observation of the projections of two bonds of the third under-coordinated atom of the vacancy being compressed to 1.37 and 1.21 Å (ref. 159) can be considered as an indirect argument that out of plane displacement d_v takes place.

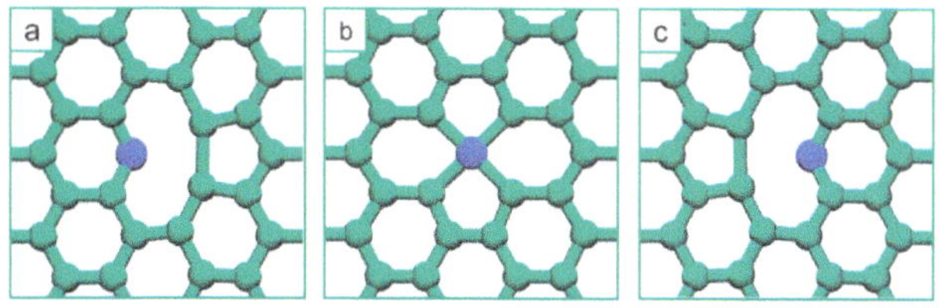

Figure. 2 Initial ground state (a), transition state (b), and final ground state (c) for one migration step of the reconstructed 5/9 monovacancy in graphene. Reprinted (adapted) from ref. 161.

Table 1 Energetic and structural parameters of the reconstructed monovacancy in graphene (see Fig. 2), graphite surface and bulk graphite: formation energy E_v (in eV), barrier ΔE_m for vacancy migration (in eV), bond length b_v of new formed bond and displacement d_v of the atom colored in dark blue out of graphene plane (both in Å). The system considered (periodic boundary conditions (PBC) or flake), number of atoms in the computational cell per layer without the vacancy and calculation method (SP and NSP stand for "spin-polarised" and "non-spin-polarised" calculations, respectively) are indicated.

Ref.	System	Number of atoms	Method	E_v	ΔE_m	b_v	d_v
	Single-layer graphene						
21	PBC	128	DFT-PW91, SP	7.7		2.02	0.18
23, 24	PBC	128	DFT-PBE, SP		1.17		
35	PBC	56	DFT-LDA, SP	8.09			
			DFT-PBE, SP	7.72			
			DFT-PBE-D2, SP	7.77			
48	PBC	100	DFT-PW91, SP	7.80		2.40[a]	0[a]
49	PBC	72	DFT-PBE, SP	7.665		1.95	0.184
52	PBC	960	DFT-PBE, SP			1.96	
54	PBC	50	DFT-LDA/DZP, SP	7.83			
		32	DFT-M06-L/6-31G*, SP	7.78			
		50	DFT-vdW-DF/DZP, SP	7.40			
		72	DFT-vdW-DF/DZP, SP	7.44			
154			DFTB[120]	7.38			
155			DFTB[120]	7.6	1.4	~2	
			DFT-PW91	7.7	1.3	~2	
156	PBC	98	DFT-LDA	7.65		2.15	0.43
		200		7.3			
157	PBC	128	DFT-PW91, SP	7.85	1.37	2.37	0.29
158	PBC	72	LAPW[b]-PBE, SP	7.87		2.4	0
159	PBC	200	DFT-PBE			1.90	

160	PBC	288	DFT-LDA(PW91), SP	7.91	1.24	1.83	0
	PBC	288	DFT-PBE, SP	7.36	1.25	1.80	0
162	PBC	128	DFT-PBE/DNP, SP	7.73			
163	PBC	144	DFT-PW91, SP	7.63			
164	PBC		DFT-LDA, SP	7.40		2.10	
			DFTB, SP	7.51	1.29	2.20	
173	PBC	128	DFT-LDA, NSP		0.91		
			DFT-LDA, SP		1.01		
67	Flake C_{116}		DFT-B3LYP/6-31G*, SP	8.0		1.73	
153	Flake $C_{120}H_{17}$		DFT, NSP	7.4	1.6	2.1 ± 0.1	
25	PBC	968	TB model[119]	6.89		1.7	0.1
119	PBC	112	TB model[119]	7.3	1.0	—[c]	—[c]
	Graphite surface						
35	PBC[d]	56	DFT-LDA, SP	8.03			
			DFT-PBE, SP	7.71			
			DFT-PBE-D2, SP	7.72			
47	PBC[d]	256	DFT-LDA, SP			1.79	
54	PBC[d]	32	DFT-LDA/DZP, SP	8.01			
			DFT-M06-L/6-31G*, SP	7.90			
			DFT-vdW-DF/DZP, SP	7.50			
		50	DFT-vdW-DF/DZP, SP	7.42			
165	PBC[e]	96	cDFT[f]-PW91, SP	7.87	0.99		
	Bulk graphite						
152	PBC[g]	18	DFT-LDA	7.6	1.6	—[c]	—[c]
160	PBC[h]	288	DFT-LDA(PW91), SP	7.96	1.07	1.89	0
			DFT-PBE, SP	7.35	1.15	1.86	0
165	PBC[h]	64	cDFT[f]-PW91, SP	8.58	1.4	2.11	0
166	PBC[i]	18	DFT-PW91, NSP	7.6			

[a] The structure of reconstructed vacancy not is not 5/9 state, see text.

[b] All-electron density-functional linear augmented plane wave method (LAPW).

[c] Optimisation of structure is restricted by consideration of symmetrical vacancy.

[d] Bilayer graphene in AB stacking.

[e] Upper layer of three-layer graphene.

[f] DFT with empirical van der Waals correction.

[g] Three layers of graphite in ABC stacking.

[h] Two layers of graphite in AB stacking.

[i] Three layers of graphite in ABA stacking.

In graphite, the ground state of a single vacancy occurs when it is positioned over the hexagon centre of a neighbouring layer. This configuration has a lower formation energy than a vacancy positioned over an atom of a neighbour layer, by 0.03–0.04 eV for bulk graphite[47,160,165] and 0.03 eV for a graphite surface.[165] The newly formed bonds of this vacancy in graphite are 0.01 Å (ref. 47) and 0.03–0.04 Å (ref. 160 and 165) longer than when positioned over an atom of a neighbour layer. The formation energy of the vacancy on a graphite surface is found to be 0.7 eV lower than for bulk graphite.[165] Analogously to graphite, in AB-stacked bilayer graphene, a monovacancy positioned over a hexagon centre is preferred by up to 0.04 eV compared to a vacancy over a carbon atom.[35,54] The formation energy of a vacancy in twisted bilayer graphene is 0.01–0.07 eV larger than in AB-stacked bilayer graphene, depending on the relative positions of the layers.[35] The energetic and structural parameters of the vacancy in bulk graphite, on a graphite surface and in bilayer graphene, calculated using different methods,[35,47,54,152,160,165,166] are also listed in Table 1. The map of bond lengths around the vacancy in graphite at its ground and transition states has also been presented.[165]

According to DFT calculations, the barrier for transition between the three possible equivalent states of the 5/9 vacancy is only ΔE_t = 0.13 eV. This leads to a small lifetime of t_s = 15 ps and t_s = 32 μs at room and liquid helium temperature, respectively,[153] estimated by the Arrhenius formula (eqn (1)) with a characteristic frequency of $\nu \sim 10^{-13}$ s^{-1}. In the case of such a small lifetime, the superposition of three equivalent states should be observed in TEM observations, which have imaging timescales of the order of one second. However, both the reconstructed 5/9 vacancy[159,167,168](observed for at least 90 s, ref. 159) and the symmetrical vacancy[159] have been found by HRTEM studies. The authors of this study proposed that the under-coordinated carbon atom is functionalised by a low mass contaminant such as hydrogen, precluding the oscillations between three equivalent states. However, the 80 keV electron beam used in this study should be able to easily remove hydrogen atoms. Note also that these rather old calculations giving the small barrier for transition between three equivalent states of the 5/9 vacancy were performed on a small graphene flake $C_{120}H_{17}$ where the considered vacancy is close to the edge even in the middle of the flake.[153] However as discussed below, the proximity of the edge has drastic effects on the energetic characteristics of the vacancy.[56,67,68] The barrier ΔE_t for transitions between equivalent states of the vacancy should correlate with values of the structural parameters b_v and d_v of the vacancy, which characterise the distortion of the symmetrical vacancy. Table 1 shows the wide spread of these parameters calculated using different methods. Note also the barrier ΔE_t cannot exceed the relaxation energy (the energy difference between vacancies with reconstructed 5/9 structure and the vacancy with all bond lengths as in pristine graphene and three-fold symmetry). The relaxation energy of the vacancy with D_{3h} symmetry into a 5/9 vacancy can be considered as upper estimate of the barrier ΔE_t. DFT-calculated values of this energy are also considerably scattered: 0.9 eV (ref. 156) and 0.29 eV (ref. 53) for graphene, 0.11–0.15 eV for bulk graphite,[165] and 0.14–0.16 eV for graphite surface.[165] Recent studies using density-functional methods show that the vacancy in graphene forms a dynamical JT centre in graphene (where the nuclei tunnel between the minima) owing to the small quantum mechanical barrier for nuclear tunnelling.[169] Note also that calculations for infinite graphene based on the tight-binding (TB) model give a considerably greater barrier for the transition between equivalent states of vacancy, ΔE_t = 0.78 eV,[142] than the DFT calculations for the small graphene flake. Thus further experimental and theoretical studies are necessary to clarify the ground state structure of the vacancy in graphene.

Structure and Energetics of Divacancies

Three different reconstructed states of the divacancy in graphene are commonly observed on graphene under electron irradiation in HRTEM:[65,147,167,170] the V_2(5-8-5), V_2(555-777) and V_2(5555-6-7777) states, see Fig. 3. Divacancies with the 5-8-5 structure were also found by scanning tunnelling microscopy in Ar^+ irradiated graphene.[171] 60 kV HRTEM studies of monolayer graphene following bombardment with energetic gold particles allowed observation of the unreconstructed divacancy.[172] The transitions between the three reconstructed divacancy states under electron irradiation were observed in HRTEM.[65,147,167] The formation energies of the V_2(5-8-5) and V_2(555-777) divacancies and one structural parameter of V_2(5-8-5) divacancy in graphene,[25,33,34,49,50,153,154,156,157,160,162–164,171,173] calculated using TB,[119–121] and DFT methods, are listed in Table 2. To summarise the results presented in Table 2, recent DFT calculations for infinite graphene give formation energies of the V_2(5-8-5) divacancy in the range E_{2v} = 7.5–8.0 eV and energy gains for the transition from the V_2(5-8-5) to the V_2(555-777) state in the range ΔE_2 = 0.8–1.0 eV. The energy gain due to the transition from the V_2(5-8-5) to V_2(5555-6-7777) state was calculated to be 0.33 eV.[162] Maps of bond length changes around the divacancy have been presented.[25,161] The calculations of divacancy energetics in graphene and molecular dynamics simulations of its dynamical behaviour at high temperature (3000–4000 K) have also been performed[174] using the AIREBO potential. These calculations show that this potential gives the correct ground state of the divacancy but the wrong ground state of the monovacancy. Although 18 other possible states of the divacancy have been found during these simulations, these can only be considered as a qualitative description of possible divacancy metastable states. The barriers and transition states for transformation between the V_2(5-8-5) and V_2(555-777) states and between the V_2(555-777) and V_2(5555-6-7777) states have been found by DFT calculations[162] and are shown in Fig. 3. Close values of the barrier for the V_2(5-8-5) to V_2(555-777) transformation were found by other DFT studies, 5.27 eV,[170] 5.17 eV (ref. 173) and 5.1 eV.[160] All three states of divacancy are therefore stable at room temperature.

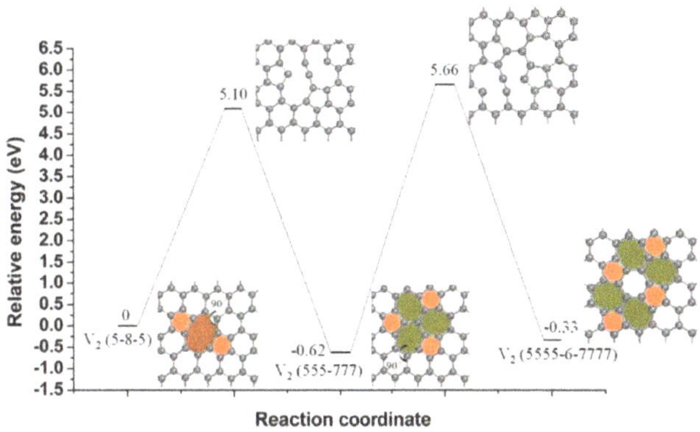

Figure. 3 Schematic representation of the structure and energetics for three states of a divacancy in graphene: V_2(5-8-5) state, V_2(555-777) state and V_2(5555-6-7777) state, and transition states for transition between V_2(5-8-5) and V_2(555-777) states and between V_2(555-777) and V_2(5555-6-7777) states. Reprinted (adapted) with permission from ref. 162. Copyright (2013) American Chemical Society.

Table 2 Energetic and structural parameters of reconstructed divacancies in graphene (see Fig. 3): formation energy of the V_2(5-8-5) divacancy E_{2v} (in eV), energy difference between the V_2(5-8-5) and V_2(555-777) states of divacancy ΔE_2 (in eV) and bond length d_v of newly formed bond of pentagon of V_2(5-8-5) divacancy (in Å). The system considered (periodic boundary conditions (PBC) or flake), number of atoms in the computational cell per layer without the vacancy and calculation method (SP and NSP stand for "spin-polarised" and "non-spin-polarised" calculations, respectively) are indicated.

Ref.	System	Number of atoms	Method	E_{2v}	ΔE_2	d_v
49	PBC	72	DFT-PBE, SP	8.998		1.78
50	PBC	72	DFT-PBE, SP	6.37		
154			DFTB[120]	7.52		
162	PBC	128	DFT-PBE/DNP, SP	6.79	0.62	
157	PBC	128	DFT-PW91, SP	7.53		1.82
171	PBC	128	DFT-vdW-DF/DZP	7.8	0.8	
160	PBC	288	DFT-LDA(PW91), SP	7.81	0.92	
		288	DFT-PBE, SP	7.11	1.07	
156	PBC	98	DFT-LDA	7.59		1.77
163	PBC	144	DFT-PW91, SP	8.08		
173	PBC	128	DFT-LDA, SP, NSP		0.91	
			DFT-PW91		0.90	
33, 34	PBC	98	DFT-LDA/DZP, NSP		0.9	
164	PBC		DFT-LDA, SP	8.25		1.72
			DFTB, SP	8.19		1.69
153	Flake $C_{120}H_{17}$		DFT, NSP	8.7		
25	PBC	968	TB model[119]	7.52	0.47	1.66

Structure and Energetics of Multi-vacancies

A set of studies is devoted to the structure and energetics of trivacancies and multi-vacancies in graphene, obtained by structure reconstruction after the removal of neighbour atoms without bond rearrangements.[49,50,156] The trivacancy with this structure was obtained by bombardment with energetic gold particles.[172] For the vacancies with structures in which an odd number of carbon atoms are missing, at least one dangling bond persists, whereas for structures missing an even number of atoms (2, 4 and 6 missing atoms) more stable structures lacking dangling bonds are possible.[49,156] Other interesting types of multi-vacancies form in graphene as a result of vacancy coalescence and then subsequent bond rearrangement under electron irradiation in HRTEM.[65,66,167] In the case of multi-vacancies with an even number of missing atoms, the formation of lines of two or three V_2(5-8-5) vacancies is observed in the graphene membrane under electron irradiation in HRTEM.[65,66,167]In such pairs or lines, adjacent divacancies have a common pentagon when aligned in the zigzag direction, while a tetragon is formed when aligned in the armchair direction, due to the overlap of two pentagons. The DFT-calculated energy gain is 1.32 and 2.01 eV per divacancy pair, as compared to isolated divacancies, for alignment in the zigzag and armchair directions, respectively.[167] The multi-vacancy created by combining two and three V_2(5555-6-7777) divacancies has been considered.[167] The DFT-calculation-based comparison of formation energies for several types of multi-vacancies in graphene with even missing atoms has also been performed.[50,175,176] For the case of multi-vacancies with odd numbers of missing atoms, the formation of a bridging atom stabilizing the structure is observed in HRTEM.[66] The DFT-calculated energy gain is 1.55 eV for a trivacancy with a bridging atom, as compared to a trivacancy with a dangling bond.[66]

Vacancy Migration

Recent DFT calculations of formation energies of mono- and divacancies in infinite graphene[50,156,157,160–163] show that the V_2(5-8-5) divacancy has a 7.2–8.7 eV lower energy than a pair of separated vacancies. This means that the coalescence of two monovacancies into one divacancy is an exothermic process, which can take place via the thermally activated diffusion of vacancies or under electron and ion irradiation. In a similar manner, trivacancies have a lower energy than a separated monovacancy and divacancy, and so on.[49,156] The coalescence of large numbers of vacancies can lead to the formation of amorphous graphene structure.[146] The directed motion of a vacancy to the edge, due to a decrease in total energy at each step towards the edge, has been proposed based on DFT calculations[67,68] and molecular dynamics simulations.[67] The rates of the above processes are determined by the barriers for vacancy motion. The structures of the initial, transition and final states corresponding to one step of monovacancy motion is shown in Fig. 2. The transition state is a so-called spiro structure, where the moving atom is of equal distance to four nearest neighbours.[161,165] Note that due to the high symmetry of the spiro structure it should correspond to the minimum, maximum or saddle point of the potential energy of the system.

The described step of vacancy migration was directly observed by a TEM study.[159] It is interesting to note that during the time between two subsequent image exposures, this TEM study observed the migration of a vacancy by two steps significantly more frequently than the migration by one step. Calculated values of the barrier ΔE_m of migration of a vacancy in graphene,[23,24,152,153,155,157,160,164,173] bulk graphite[152,160,165] and on the graphite surface[165] are listed in Table 1. Some DFT calculations give the value of the barrier ΔE_m of migration of a vacancy in graphite as being 0.15–0.26 eV lower than in graphene,[160] whereas DFT calculations with an empirical van der Waals correction give the opposite result; the barrier ΔE_m is greater for graphite by 0.4 eV compared with the graphite surface.[165] Only two experimental estimates of the barrier ΔE_m are currently available for bulk graphite and its surface. The first, $\Delta E_m = 1.8 \pm 0.3$ eV, was obtained by Raman measurements of disorder relaxation in He^+ irradiated graphite.[177] Recent STM studies of vacancy aggregation on vacancy-decorated graphite surfaces at different temperatures give the value $\Delta E_m = 0.9–1$ eV.[178] The calculated values of ΔE_m listed in Table 1 range between these two experimental estimates and are only in tentative agreement with few experiments. Further studies are necessary to obtain accurate values of the barriers for migration of vacancy in graphene, bulk graphite and graphite surface.

Contrary to monovacancy migration, the barrier for divacancy migration is considerably large. The estimation of this barrier for a small graphene flake ($C_{120}H_{17}$) gives the value of about 7 eV.[153] An estimation based on migration in two steps, via the dissociation and merging of two vacancies, gives the value 7.49 eV.[157] HRTEM studies reveal the migration of the V_2(5-8-5) divacancy under the action of electron irradiation through the formation of an intermediate metastable dislocation dipole (the V_2(55-77) state).[167,170] Divacancy migration under electron irradiation in HRTEM has also been observed as a result of transitions between the V_2(5-8-5), V_2(555-777) and V_2(5555-6-7777) divacancy states.[147,170] It has been found using DFT calculations in combination with careful analysis of HRTEM images that one step of the migration of the V_2(5-8-5) divacancy occurs through two intermediate V_2(55-77) states which have a higher total energy by 3.44 eV (other DFT calculations 3.72 eV (ref. 167)). The barrier for transformation from the V_2(5-8-5) to V_2(55-77) state is 5.27 eV, whereas the migration of the V_2(555-777) divacancy occurs through a transformation to the V_2(5-8-5) state with a barrier of 6.20 eV.[170] Thus, thermally activated migration of divacancies at room temperature can be excluded.

Effects of Edges and Deformation on the Energetics of Vacancies

The structure and energetics of the monovacancy near the edge of the graphene layer have been studied on examples of narrow graphene nanoribbons with the vacancy in the middle,[52,56] graphene flakes with dangling bonds at the edge[67] and graphene flakes with the edge terminated by hydrogen atoms[68,69] with different positions of the vacancy. Note that the case of an edge with dangling bonds is realised under the action of electron irradiation in TEM and is important when considering irradiation-induced processes such as the graphene flake to fullerene transformation.[8] It is evident that the structural parameters and formation energy of vacancies should change near the edge. For armchair graphene nanoribbons with widths of 7 and 5 hexagons, a stronger new bond is formed for the reconstructed 5/9 vacancy with the length b_v of 1.9 and 1.8 Å, respectively, slightly less than the value $b_v = 1.96$ Å obtained for bulk graphene using the same parameters of spin-polarised DFT calculations.[52] DFT-calculated formation energies of reconstructed 5/9 vacancies in the middle of graphene flakes (ranging in size from C_{52} to C_{116}) with dangling bonds at the edge[67] are listed in Table 3. Table 3 also presents the difference ΔE in the total energy of the flake with the vacancy in the middle and at the edge. A considerable difference of $\Delta E = 6.6–8.9$ eV in the total energies has been found for all flakes. The total energies corresponding to the subsequent steps of vacancy motion from the middle to the edge have been calculated for the C_{116}[67] and $C_{41}H_{16}$[68] graphene flakes.

Table 3 The DFT-calculated energy characteristics of flakes with a monovacancy located at distance d (in Å) from the edge: E_v (in eV) is the energy of vacancy formation, and ΔE (in eV) is the difference in the total energy between a flake with the vacancy in the middle and at the edge. Reprinted (adapted) from ref. 67

Flake	E_v	ΔE	d
C_{52}	6.8	7.3	4.3
C_{69}	7.3	6.6	7.1
C_{88}	7.3	7.3	7.1
C_{103}	7.7	8.9	8.6
C_{116}	8.0	7.6	8.6

Moreover, upon decreasing the distance between the vacancy and the graphene layer edge, a change of energetic and structural parameters of the reconstructed 5/9 vacancy is predicted, as well as a drastic change of the whole vacancy structure.[56,67,68] DFT calculations show that the vacancy in the middle of an armchair graphene nanoribbon with the width of 5 hexagons has a spiro ground state with an sp^3 atom with four equivalent bonds 1.64 Å in length[56] in the centre of the vacancy (the same bond lengths as in the spiro state monovacancy in the middle of the C_{116} graphene flake[67]). The coexistence of the 5/9 and spiro states of the monovacancy near the graphene edge has been obtained by DFT calculations for graphene flakes with both dangling bonds[67] and hydrogen atoms[68] at the edge. It was also found that the spiro state of the vacancy is more stable than the 5/9 state in the middle of a graphene flake with hydrogen atoms at the edge when the distance between the vacancy and the edge is less than 7 Å.[68] Furthermore, these calculations give the stable spiro state of the vacancy even at a distance of 15.6 Å from the edge,[68] contrary to in infinite graphene where it corresponds to a saddle point.[161] Note that calculations using the old version of the Brenner potential gives the 5/9 structure for the ground state of the vacancy whereas the new version of this potential gives the spiro structure.[154] The influence of the vacancy on the surrounding structure is more significant near the edge than in infinite graphene. For flakes consisting of approximately one hundred carbon atoms, the presence of a vacancy in the centre resulted in considerable bending of the entire structure, for both flakes with dangling bonds at the edge[67] and flakes terminated at the edge by hydrogen atoms.[69,153] Vacancies in the spiro state in the middle of narrow graphene nanoribbons causes them to coil into spiral helices.[68]

The vicinity of the edge has an even more drastic effect on vacancy migration than on its structure and energetics.[67,68] As both 5/9 and spiro states of the vacancy near the graphene edge correspond to local potential energy minima, the only vacancy migration step near the edge is the transition between the 5/9 and spiro states.[67] The scheme of possible vacancy migration steps near the edge is presented in Fig. 4. A transition from the initial 5/9A state to two spiro states SA and SB is possible. Four transitions are possible from each spiro state SA and SB, which can be achieved by breaking one of the four bonds: a return to the initial 5/9A state, transition to another 5/9 state of the same vacancy (5/9B and 5/9F) and migration of the vacancy (5/9C, 5/9D, 5/9E, and 5/9G).[67] The energies of local minima and transition states for vacancy migration in the middle of the C_{116}[67] and from the middle to the edge of the $C_{41}H_{16}$[68] graphene flakes have been calculated. These calculations reveal that the barrier for vacancy motion towards the edge is considerably lower than the barrier for motion away from the edge. Thus, both the difference in the total energy of graphene with the vacancy at the edge and at distance d from the edge (see Table 3) and the barrier to vacancy migration decrease with the decrease of distance d. This leads to the conclusion that thermally activated directional motion of the vacancy towards the edge should take place,[67,68] as confirmed by molecular dynamics simulations.[67] As these barriers for the vacancy motion towards the edge are smaller than barriers for vacancy motion in infinite graphene, the motion to the edge will occur too fast to be observed by STM or TEM studies at room temperature.

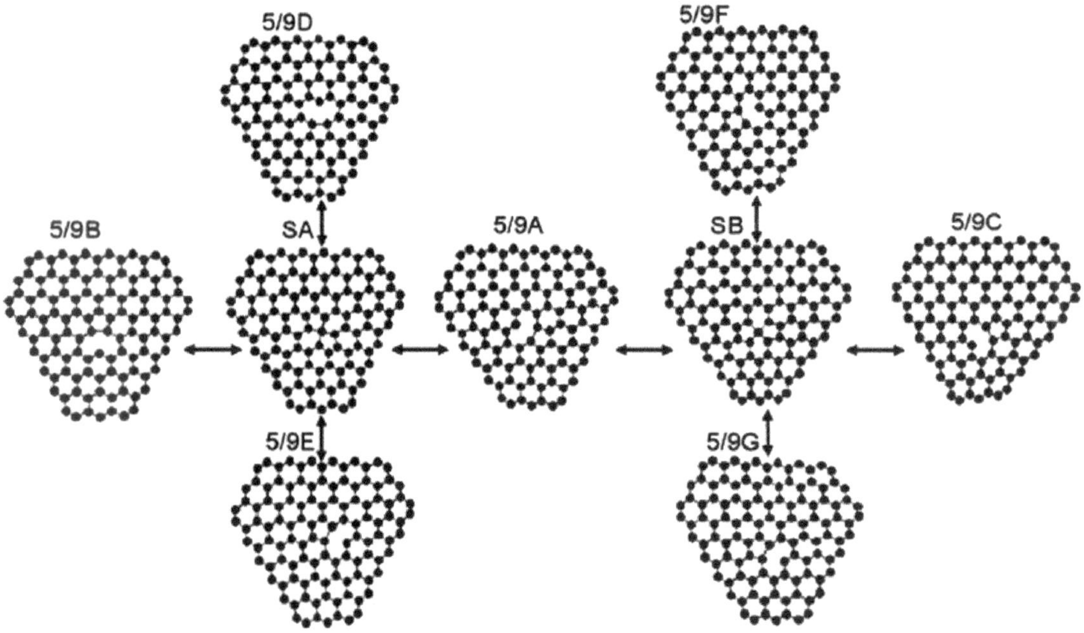

Figure. 4 Scheme of possible transitions between the neighbouring 5/9 and spiro vacancy states in the middle of the C_{116} flake. Reprinted (adapted) from ref. 67.

According to DFT calculations, a 3% compression of graphene leads to a change in the ground state of the vacancy to the spiro state.[51] The deformation of graphene also exerts considerable influence on the barriers ΔE_m for vacancy migration. DFT calculations show that the value for pristine graphene ΔE_m = 1.17 eV changes to 3.15 and 0.75 eV for elongation by 5% in the armchair and zigzag directions, respectively, and to 0.12 and 1.79 eV for compression by 5% in armchair and zigzag directions, respectively.[23] Thus the possibility of directional motion of vacancies caused by applied graphene layer deformation has been proposed.[23]

Interaction and Coalescence of Vacancies

A set of studies is devoted to the interaction of vacancies and their coalescence into divacancies.[50,142,157,160,162,173] This set includes a TB molecular dynamics study of the convergence and coalescence of vacancies,[173] action-derived molecular dynamics for energetics of certain paths of vacancies migrating towards each other,[142] DFT calculations of energies of the subsequent states at the convergence of vacancies and barriers between these states,[160,162] and barriers of the final step of coalescence of pairs of vacancies.[157,160,162,173]The formation energy of vacancies in graphene as a function of their concentration has also been studied.[50] Thorough DFT calculations give the energies of pairs of vacancies for different neighbour locations, as well as barriers to transition between these locations.[160] The barriers for the final step of coalescence of a pair of vacancies into a V_2(5-8-5) divacancy are found in this work to be 2.10 and 0.82 eV, for different final neighbour locations[160] (in agreement with previous DFT calculations: 1.52 eV (ref. 173) and 2.17 eV (ref. 157)). Another DFT study revealed two different transition paths for the final step of coalescence into the V_2(5-8-5) divacancy, with barriers of 1.17 and 1.92 eV.[162] Two examples of energy changes for the migration towards and coalescence of vacancies, based on a TB model, are shown in Fig. 5.[142] These calculations give the values of the barrier for the final step of coalescence as 1.3 and 1.9 eV for different approaching pathways with V_2(555-777) divacancy formation. Detailed DFT-calculated maps of energies of vacancies located near the V_2(5-8-5) divacancy and a trivacancy, as well as transition barriers between neighbouring vacancy positions, have been used to study the dynamical evolution of the coalescence of monovacancies with divacancies and trivacancies by kinetic

Monte Carlo simulations.[161] Analogously to the coalescence of a pair of monovacancies, the barriers for the final stage of coalescence depend on the pathway of the vacancy approaching the di- and trivacancies. These barriers lie in the range 0.9 to 4.7 eV and 0.9 to 1.5 eV for different neighbour positions for the V_2(5-8-5) divacancy and trivacancy, respectively.[161] Examples of the final steps of coalescence of the monovacancy and V_2(5-8-5) divacancy are shown in Fig. 6.

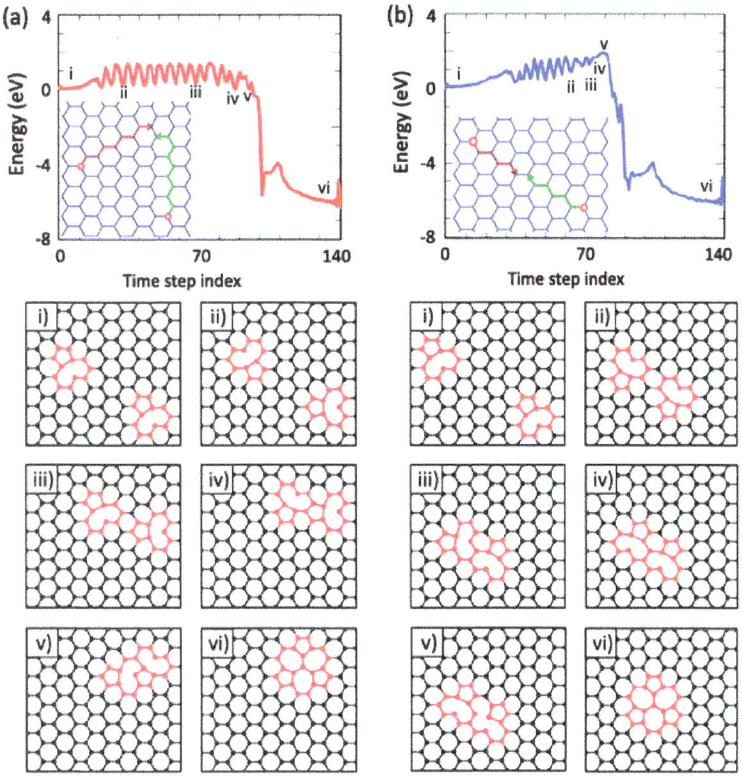

Figure. 5 The process of merging monovacancies into a V_2(555-777) divacancy for different pathways for the approaching migration of vacancies. The potential energy of system relative to the initial state along migration pathways is shown as function of time step index in action-derived molecular dynamics simulation. Reprinted (adapted) from ref. 142. Reproduced by permission of IOP Publishing. All rights reserved.

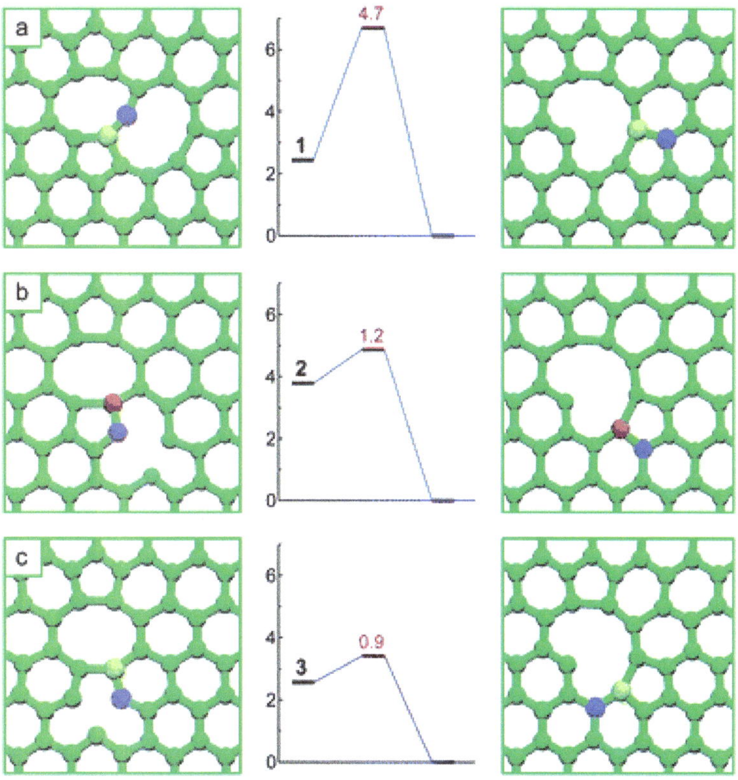

Figure. 6 The final steps of the merging of a 5/9 monovacancy with a V_2(5-8-5) divacancy, from the three non-equivalent 3rd nearest neighbour positions (a), (b), and (c), where the two atoms involved in each transition have been coloured to track their movement. The energies of initial and transition states relative the final state are indicated. Reprinted (adapted) from ref. 161.

Adatoms

The healing of various defects of graphene by the migration of adatoms has been observed by HRTEM.[64,65] Possible magnetism of adatoms on graphene has been discussed,[179,180] and the structure and energetics of the adatom on graphene have been considered in a set of papers.[157,163,179–183] A DFT study gives three stable positions of the adatom on graphene: a bridge position over the centre of a bond, a dumbbell arrangement where the adatom has bonds with three atoms of graphene, and an off-top structure where the adatom and one atom of graphene each have four bonds and the same position relative the graphene plane.[181] The bridge position is found to be a ground state,[166,179,181] with the dumbbell arrangement and off-top structure having 0.37 and 0.22 eV higher energies, respectively.[181] The bridge position of an adatom on graphene was observed by HRTEM[184] and scanning transmission electron microscopy (STEM).[185] DFT calculations give the adsorption energy of an adatom as 1.40 eV (ref. 179) and 1.7 eV (ref. 53 and 166) and low values for the barrier to adatom migration on graphene: 0.35 eV,[53] 0.45 eV,[183] 0.47 eV,[179] 0.52 eV,[163,180] and 0.53 eV.[157] The DFT-based TB method gives the same ground state bridge position of the adatom, and close values of the adsorption energy (2 ± 0.3 eV) and barrier for migration (0.4 ± 0.1 eV).[182,183] The migration of adatoms can therefore easily take place at room temperature. DFT calculations predict an energy gain of up to 0.25 eV for close positions of neighbour adatoms, with a decrease of the barrier to migration to only 0.225 eV,[180] meaning that the agglomeration of adatoms at low temperatures is possible. The structure and energetics of adatom pairs on graphene (i.e. the case where chemical bonds exist between adatoms) have been also considered by DFT.[181] The most stable pair is found to be the so-called 7-5-5-7 defect that is comprised of two pentagons and two heptagons and introduces a local elevation of 2 Å out of the graphene plane.

As monovacancies can easily migrate and merge into divacancies, whereas thermally activated migration of divacancies is not possible, the interaction of adatoms with divacancies is an important process observed by HRTEM.[184] DFT energy calculations of states of an adatom and divacancy approaching one another, as well as the barriers between these states, have been performed.[162,163] For the V_2(5-8-5) divacancy, the coalescence of the divacancy and adatom takes place with three barriers during the approach of the adatom (0.91, 0.73 and 1.36 eV) before the coalescence and a total energy release of 5.79 eV,[162] whereas a different study gives even higher barriers of 1.55 and 2.49 eV between neighbouring bridge positions near the V_2(5-8-5) divacancy.[163] Thus a state in which an adatom near a V_2(5-8-5) divacancy is stable at room temperature is possible, in agreement with HRTEM observations.[184] However, the adatom does not coalesce with the V_2(555-777) divacancy – a dumbbell-like configuration is formed where the adatom is adsorbed over the central atom of the V_2(555-777) divacancy.[162] The motion of the adatom to this final location occurs with the subsequent barriers 0.43, 1.00, 0.78, 0.39 and 0.56 eV and the total energy release is 1.8 eV. The creation of an inverse Stone–Wales defect on the periphery of a divacancy, by one adatom hopping within the vicinity of another, was proposed.[163]

Topological Defects and Bond-realignment Reactions

An important class of reactions in carbon nanostructures is related to bond rotations that lead to the formation or transformation of topological defects, while not being accompanied by a loss or addition of atoms. The simplest example of such a reaction is the 90 degree rotation of a bond in a perfect hexagonal network, leading to the formation of a Stone–Wales (SW) defect[72,168] comprised of two pentagons and two heptagons (Fig. 7a).

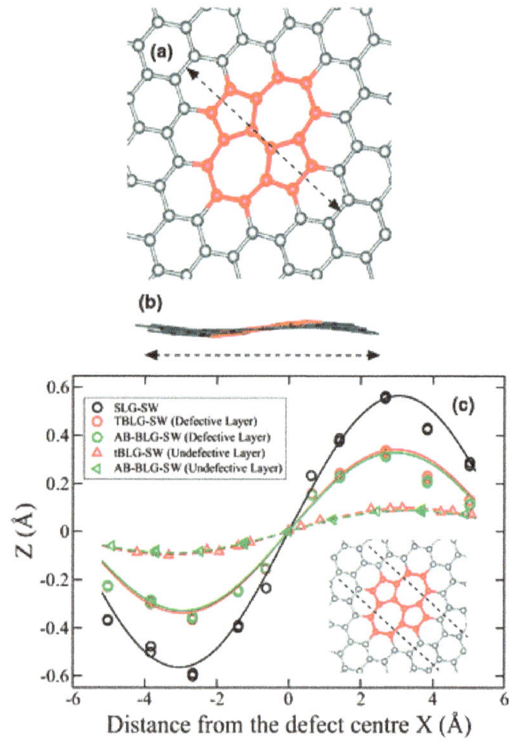

Figure. 7 Structure of the SW defect in single-layer graphene (SLG): (a) top view and (b) side view showing the distortion of the defective graphene sheet. The side view in (b) is taken by viewing perpendicular to the dashed line in (a). (c) The distortion pattern for the SW defects in SLG, twisted bilayer graphene (TBLG) and bilayer with the AB stacking (AB-BLG). A distortion amplitude of zero corresponds to the centre of the SW defect. The dashed line in the inset encloses the atoms over which the deviations from zero are measured. Reprinted (adapted) from ref. 35. Copyright (2014) by the American Physical Society.

Effects of SW Defects on Mechanical, Electronic and Magnetic Properties

SW defects only very weakly affect the elastic properties of graphene such as the Young modulus[26–30] and Poisson ratio,[30] although this effect becomes more pronounced with increased defect concentrations.[26–28] Numerous studies show that pre-existing SW defects serve as nucleation centres for fracture[22,27] and substantially decrease the failure strain and tensile strength of graphene.[22,27,29,31,32] The accumulation of SW defects in graphene degrades the ultimate tensile strength and failure strain to saturated levels that are 30–50% lower than in pristine graphene.[22,27] The SW transformations are also direct participants of plastic deformation in graphene, which is discussed below.

The analysis of the density of states in graphene showed that p_z orbitals of the carbon atoms of the rotated bond in SW defects give rise to a defect band ~0.5 eV above the Fermi level, with the width and height dependent on the defect concentration.[33–40] The conductivity of defected graphene exhibits systematic degradations around defect resonance energies.[33,34] Electron backscattering regions are also introduced in transmission spectra of graphene nanoribbons with SW defects.[41–44] SW defects generally do not carry a magnetic moment[33–35,40] and are neutral,[40] although some charge transfer was detected between layers of defected bilayer graphene[35] and asymmetric arrangements of SW defects in graphene nanoribbons can affect their magnetic moment.[55] The effect of SW defects on the thermal conductivity of graphene[26,27,57,58] and graphene nanoribbons[59–61] is similar to that of other local defects, such as vacancies.

Structure and Energetics of SW Defects

The combined DFT[35,36,54,70,164,166,186–188] and TB data[164,189] show that the formation energy of a SW defect in a graphene layer is 4.5–5.0 eV (Table 4). A slightly larger formation energy (by 0.3 eV) is predicted on the basis of quantum Monte Carlo (QMC) simulations.[186] Significant variations[35,36,53,54,70,91,119,152,164,166,173,186–194] are observed depending on the supercell used in the calculations (Table 4), due to interactions of periodic images of SW defects through generated long-range stress and strain fields.[190,191] In the calculations for finite graphene flakes, the formation energy[195–198] of SW defects might be affected by the vicinity of the edges. While the results of TB methods[119,121–123] are very close to those of DFT,[188] the semiempirical potentials[127–129] and continuum models[199] are not very accurate in exact numerical values for energies.[70,71,138,200–202] However, they give the correct trends and qualitative behaviour for bond realignment reactions.[70,91,188,202–204]

Table 4 Formation energy E_{SW}, activation energy $E_{a,SW}$, energy gain from out-of-plane buckling δE_b (in eV) and maximum difference in out-of-plane coordinate of atoms h (in Å) for SW defects in single-layer and bilayer graphene and graphite. The system considered (periodic boundary conditions (PBC) or flake), number of atoms in the computational cell per layer and calculation method are indicated

Ref.	System	Number of atoms in cell	Method	E_{SW} (eV)	$E_{a,SW}$ (eV)	δE_b (eV)	h (Å)
	Single-layer graphene						
35	PBC	56	DFT-LDA	5.25		0.22	1.2
			DFT-PBE	5.14		0.21	
			DFT-PBE-D2	5.09		0.23	
36	PBC	98	DFT-LDA	5.16		0.28	1.3
		Infinite-size limit	DFT-LDA	4.66			
53	PBC	128	DFT-PBE	5.70			
54	PBC	72	DFT-LDA/DZP	4.43			
	PBC	50	DFT-LDA/DZP	4.29			
			DFT-M06-L/6-31G*	4.31			
			DFT-vdW-DF/DZP	4.26			
70	PBC		DFT	5.16	9.91		
91	PBC		DFT-PBE	5.4	9.4		
166	PBC	50	DFT-PW91	4.8	9.2		
			DFT-LDA	5.2	9.4		
186	PBC	242	DFT-PBE	4.59		0.231	1.61
	PBC	50	DFT-PBE	5.26		0.269	1.16
			DFT-LDA	5.42		0.210	1.07
			DFT-PBE0	5.63		0.061	0.79
			QMC	5.82 ± 0.03		0.096 ± 0.020	
164	PBC		DFT-LDA	4.86			
			DFTB[121]	4.85	10.4		
187	PBC	98	DFT-GGA	3.83			
190, 191	PBC	32–192	DFT-LDA	4.20 to 5.90			
188	PBC		DFT-LDA	5	9.9		
192	PBC	60	DFT-LDA	5.4, 4.8[a]			

193	PBC	128	DFT-LDA/DZ, PW	5.9	11.0, 10.3		
	Adatom-catalysed			2.6	4		
194	PBC	48	DFT-PBE	4.82	9.3	0.07	
195	Flake $C_{248}H_{42}$		DFT-PBE/3-21G		10.0		
196	Flake $C_{62}H_{20}$		DFT-LDA	3.4	9.0		
	Adatom-catalysed			1.6	2.3		
198	Flake, $C_{42}H_{16}$		DFT -B3LYP/6-31G(d)		6.2		
	Adatom-catalysed				0.86		
197	Flake, 6 shells		Hückel		6.02		
173	PBC	128	TB model[123]		10.2		
189	PBC	572	TB model[122]	4.43			
119	PBC	112	TB model[119]	5.8	9.8		
200	PBC		Tersoff–Brenner[128]		5.9		
138	PBC		Tersoff–Brenner[128]	3	6		
202	PBC		Tersoff–Brenner[128]	2.54	6.16		
70, 71	PBC		Tersoff–Brenner[128]	4.43	7.88		
199	PBC		Classical continuum model	3.7			
	PBC		Gauge field theory	3.2			
	Bilayer graphene[b]						
35	PBC	56	DFT-LDA	5.39		0.08	
			DFT-PBE	5.15		0.23	
			DFT-PBE-D2	5.27		0.04	
54	PBC	72	DFT-LDA/DZP	4.66			
	PBC	50	DFT-LDA/DZP	4.93			
			DFT-M06-L/6-31G*	5.15			
			DFT-vdW-DF/DZP	4.83			
	Bulk graphite[c]						
152	PBC	18	DFT-LDA	10.4	13.7		

[a]Dependent on orientation.

[b]With AB stacking.

[c]Graphite with ABC stacking.

It is seen that although all atoms in a SW defect have the same number of bonds as in the perfect hexagonal network, it has a significant formation energy. The reason is that upon the SW rotation, many bonds get compressed or stretch ed[36,152,164,166,186,198,205,206] (Table 5), and the angles between the bonds deviate from 120°. The angle of the pentagons at the atoms forming the rotated bond was calculated to be 115.5°,[36] 115.0° [206] and 118.2°.[205]The most deformed bond is the central rotated bond (separating two heptagons in the SW defect), which experiences compression by almost 10% (Table 5 and Fig. 8a).[152,164,166,186,198,205,206] Some compression is also observed for the bonds between pentagons and hexagons, while the bonds between heptagons and hexagons and the bonds between heptagons and pentagons are stretched[152,164,166,205,206] (Fig. 8a). This behaviour is in agreement with general observations that pentagons and heptagons are centres of compression and tensile stress, respectively.[207–210]

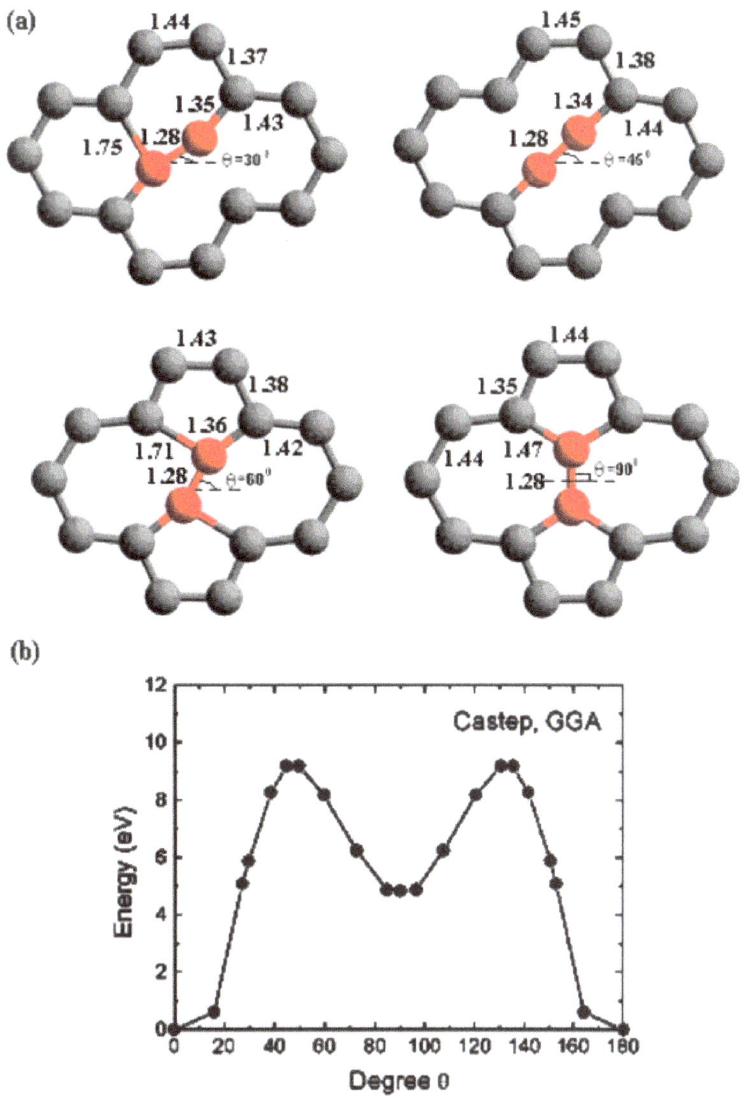

Figure. 8 The Stone–Wales transformation: (a) Intermediate structures corresponding to different angles of bond rotation θ (in degrees) as indicated in the panels. The atoms at the defect core are shown in red. All the bond lengths are in Å. (b) The corresponding formation energy as a function of the rotation angle θ. Reprinted (adapted) from ref. 166. Copyright (2005) by the American Physical Society.

Table 5 Geometry of SW defect in single-layer graphene: lengths (in Å) of bonds between two heptagons (r_{77}), between a heptagon and a pentagon (r_{57}), between a pentagon and a hexagon (r_{56}) and between a heptagon and a hexagon (r_{76}). The system considered (periodic boundary conditions (PBC) or flake), cell size, number of atoms in the computational cell and calculation method are indicated

Ref.	**System**	**Number of atoms**	**Method**	r_{77}	r_{57}	r_{56}	r_{76}
36	PBC	98	DFT-LDA	1.31			
152	PBC	18	DFT-LDA	1.28	1.50	1.43–1.45	1.43–1.45
186	PBC	242	DFT-PBE	1.32			
164			DFT	1.32	1.46		1.45
			DFTB[121]	1.32	1.45		1.44
166	PBC	50	DFT-PW91	1.28	1.47	1.35, 1.44	1.44
206	PBC		DFTB[121]	1.338	1.466		1.450, 1.459
198	Flake, $C_{42}H_{16}$		DFT-B3LYP/6-31G	1.35			
205	PBC		Tersoff–Brenner[128]	1.38	1.43	1.39	1.43

Geometrical optimisation of a SW defect starting with all atoms in the same plane yields a flat SW defect structure.[119,152,166,190,191,205] However, this structure has two imaginary frequencies, and the true minimum is characterised by the carbon atoms forming the rotated bond at the defect core buckling out-of-plane above and below the graphene plane[35,36,186,194] (Fig. 7b). The energy gain from out-of-plane buckling of the core atoms is 60 – 300 meV depending on the method used and the supercell considered,[35,36,186,194] and comes from slight elongation of the compressed rotated bond at the defect core.[186] The maximum displacement between atoms perpendicular to the graphene plane in the buckled energy minimum configuration is 0.8–1.6 Å (ref. 35, 36 and 186) (Fig. 7b).

The analysis of correlation effects in the formation energy of two SW defects has revealed that both repulsion and attraction between two SW defects is possible, depending on their relative position and orientation.[138,201] Attraction valleys correspond to a 'diagonal' location of defects with respect to each other for both aligned[138] and misaligned defects.[201] The most energetically favourable configuration is immediately adjacent but not overlapping.

The formation of SW defects in bilayer graphene with the AB stacking of graphene layers (AB-BLG)[35,54] and with twisted graphene layers (TBLG)[35] was investigated. It was shown that in bilayer graphene, SW defects are also stabilised by out-of-plane buckling, but the magnitude of this distortion is reduced to 0.7 Å compared to 1.2 Å in single-layer graphene (SLG),[35]i.e. the presence of the second layer inhibits the distortion. The stabilization energy from buckling is also decreased from 220 meV in SLG to 40 meV in AB-BLG and 60 meV in TBLG.[35] The defect formation energy in TBLG was found to be almost the same as in AB-BLG, while there was only slight variation (by tens of meV) for different positions of the SW defect due to the differences in the interlayer coupling.[35] Compared to in SLG, creation of a SW defect requires an additional 140–180 meV (ref. 35) in TBLG and 180 meV (ref. 35) (230 meV (ref. 54)) in AB-BLG. This difference was mostly attributed to the inefficient energy relaxation by out-of-plane buckling in the presence of the second layer.

Typical DFT values for the activation energy of the SW transformation lie in the range 9–11 eV [70,91,119,164,166,188,193–196] (Table 4 and Fig. 8b). The barriers calculated with semiempirical potentials can be substantially underestimated,[70,71,138,197,200,202] although they are still large enough to reflect that two carbon bonds are broken at the same time in this transformation. The transition state for the SW transformation corresponds to a rotation of the carbon bond by 45–55° (Fig. 8b).[119,138,152,166,194,195,200] As in the final SW defect, compressive stress in the transition state caused by the shortened rotating bond (1.29 Å)[194] and other bonds of the central carbon atoms (1.36 Å)[194] is relieved by out-of-plane buckling[138,194,195,200] with an energy gain of about 0.9 eV,[194] 0.4 eV (ref. 138) or 0.3 eV.[200] Different modes of the transition state (Fig. 9) in which (1) both of the central atoms are displaced in the same direction out of the plane (S_{++}), (2) both are displaced in different directions out of the plane (S_{-+}), (3) only one of the atoms is displaced out of the plane (S_{+}), and (4) both of the atoms lie in plane (S_0) have been compared. According to the Tersoff–Brenner potential,[128] the reaction path $S_{-+}$is the most probable, followed by S_+, S_{++} and finally S_0.[138,200] However, DFT studies[194,195] predict that the true transition state with only one imaginary frequency corresponds to the S_+ structure with the central bond tilted by 15° [195] or one of the central atoms buckling out by 0.95 Å.[194] The maximum displacement between atoms perpendicular to the graphene plane in the buckled transition state configuration is of the order of that in the final SW defect, at 1.42 Å.[194]

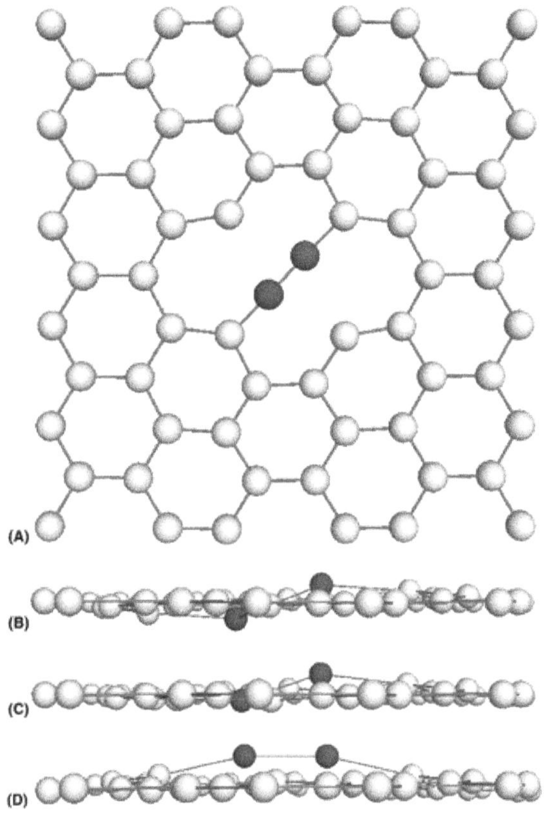

Figure. 9 Atomic configurations for the four modes of the SW transition state. All the modes correspond to bond orientations about 45° (A): (B) S_{-+}, (C) S_+, (D) S_{++}. Mode S_0 is not shown and corresponds to all atoms in the plane. Reprinted (adapted) from ref. 138.

It has been shown that the barrier to the SW transformation can be considerably reduced in the presence of carbon adatoms,[193,196,198] hydrogen and hydroxyl groups[194] or transition metals.[91] Participation of a nickel atom reduces the activation energy of the SW transformation in graphene by 1.8 eV.[91] One hydrogen atom, two hydrogen atoms and two hydroxyl groups reduce the activation energy by 2.5 eV, 5.6 eV and 3.4 eV respectively.[194]The effect of a carbon adatom is even more pronounced as the corresponding mechanism involves only one bond breaking at a time and the activation energy is decreased by 5–7 eV (Table 4).[193,196,198]

Modification of SW Energetics by Strain

It is seen that the activation energy for a SW transformation is very high, both for the formation of non-hexagonal rings and their annihilation. Therefore, SW defects are unlikely to be formed in pristine graphene even at high temperatures. However, additional manipulations such as electron irradiation or mechanical deformation can induce the formation of SW defects. As the rotated bond of the SW defect is highly compressed, stretching a graphene layer along the direction of the rotated bond leads to a decrease in the formation energy of SW defects[138,188,193,195,199–204,206,211] and activation energy of SW transformations.[31,138,188,193,195,200,202,211] The formation of SW defects can therefore be facilitated by strain, and it has been considered as the first step in the plastic deformation of carbon nanotubes and graphene. The energy gain associated with the rotation of a bond depends on its orientation, and increases with a reduction in the angle of the rotated bond with respect to the loading direction.[31,138,188,195,199–201,206,211] This implies that relieving lateral strain in graphene by SW rotations is most efficient when the layer is stretched along the zigzag direction. The critical strain at which formation of SW defects in graphene becomes thermodynamically favourable was calculated to be 6–9%[188,193,199,202] and 12–17%[188,199] when the strain is applied in the zigzag and armchair directions, respectively. The annihilation of unfavourably oriented SW defects under strain has been demonstrated with MD simulations.[212]

Although both the formation energy of SW defects and activation energy of SW transformations are reduced by several eV at the critical strain, such transformations are still kinetically limited[138,188,193,195,200,202,211] and greater strains should be applied to observe them experimentally. The Arrhenius equation was used to estimate that the yield strain at which the formation of SW defects should take place during experiments at room temperature is on the order of 17–22%[195] for different loading directions in graphene or nanotubes of different chirality, in agreement with the experimentally measured yield strains for carbon nanotubes up to 10%[213] and 17%.[214] The minimum activation energy was reached for an intermediate angle of the rotated bond between 0° and 30° with respect to the loading direction. Calculations[200] on the basis of the Tersoff–Brenner potential[128] predicted that this angle is 11°, while DFT calculations[195] for nanotubes gave an estimate of 22°. The approximate formation energy of SW defects[138,201] and activation energy for SW transformations[195,200] have been proposed as functions of strain magnitude and angle of the rotated bond with respect to the loading direction.

Dislocations

Nucleation of a SW defect "unlocks" carbon nanostructures for further plastic relaxation under strain:[215] (1) plastic flow[138,203,204,211,216,217] and (2) the generation of multiple SW defects[138,203,204] and large rings, such as octagons.[138,216] The first scenario that explains experimentally observed superelongation of carbon nanotubes under strain[18,19] is realised by a separation of pentagon–heptagon (57) pairs[184,218,219,220](Fig. 10) that represent crystal dislocations[138,190,191,203,204,211,216] (a SW defect is a dislocation dipole in which two edge dislocations with opposite Burgers vectors are displaced by one lattice unit). Separated 57 pairs introduce long-ranged stress fields and consequently high local strain energies. These strain energies can, nevertheless, be partially relieved by the buckling of graphene and formation of protrusions of 2–3 Å (ref. 221 and 222) height near the defects, providing a finite formation energy of isolated 57 pairs of 7.5 eV,[221] 6.2 eV (ref. 203) and 5 eV (ref. 223) according to different calculations.

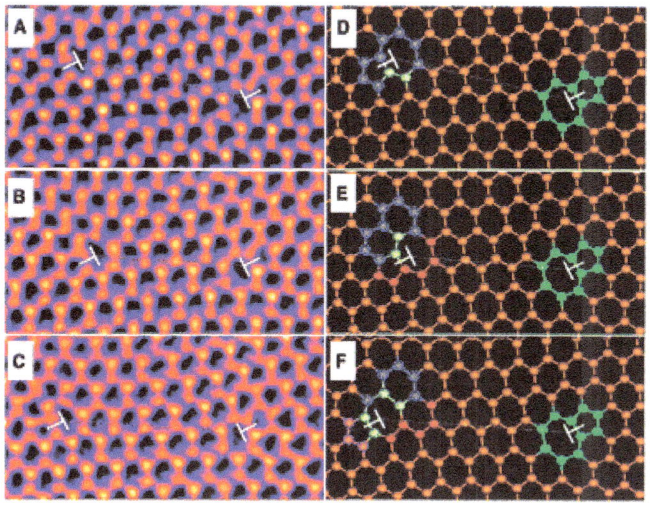

Figure. 10 Real-time dislocation dynamics. HRTEM images showing changes in the position of an edge glide dislocation with time under continuous electron beam irradiation. Time (A) 0 s. (B) 141 s. (C) 321 s. (D to F) Atomic models illustrating the structures inferred from (A) to (C), respectively. The white T indicates the position of the dislocation. The transition from structures (A, D) to (B, E) corresponds to a glide of the dislocation core, while the transition from (B, E) to (C, F) corresponds to a climb step, associated with the emission of a pair of atoms. Reprinted (adapted) from ref. 218 with permission from AAAS.

The glide of 57 pairs, i.e. displacement by one Burgers vector at a time, is possible by a SW rotation of the bond between the heptagon and the hexagon adjacent to the pentagon (Fig. 10A, B, D, E). In unstrained graphene,[138,190,191,211,218] separating 57 pairs with one intervening hexagon costs 1–6 eV, although the energy required decreases with further separation of the 57 pairs[211,216,218] (Fig. 11). As with the initial formation of a SW defect, separation of 57 pairs is stabilised by the application of an appropriate load.[138,203,204,211,216] Therefore, the long range strain fields favour gliding of 57 dislocation cores in opposite directions, while the local curvature energy of two oppositely directed 57 pairs is minimised when they are adjacent and form a SW defect.[138,190,191,203,204,211,216,218] The glide of 57 pairs can also be facilitated by the formation of chains of dislocation dipoles (so-called "dislocation worms"), which effectively screen stress fields of the separated 57 pairs.[203,204]

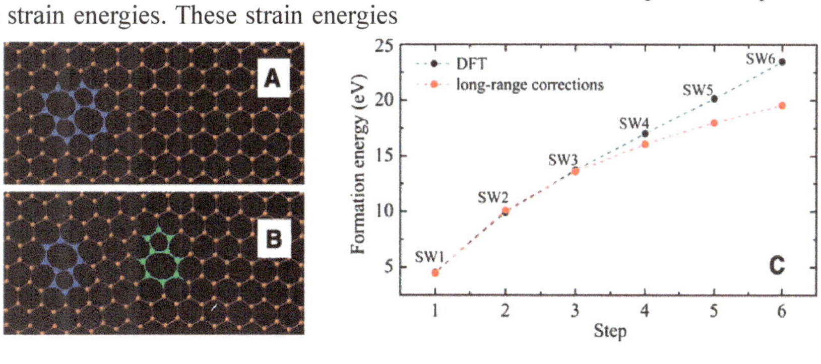

Figure. 11 (A and B) Atomistic models obtained within DFT for one and four Stone–Wales bond rotations (steps 1 and 4). (C) Formation energy at each step (one dislocation is stationary and the other one is moving along the glide direction). The black dots represent the DFT data. The

red dots represent the DFT data corrected for long-range strain-field interactions. Reprinted (adapted) from ref. 218 with permission from AAAS.

In addition to the glide of dislocation cores by SW rotations, another type of motion, a climb step, is possible. Emission of a pair of carbon atoms from the edge between the pentagon and the hexagon adjacent to the heptagon[217] (Fig. 10B, C, E, F) results in motion along the defect axis in the direction from the heptagon to the pentagon. This was shown to be responsible for nanotube sublimation that preserves perfect hexagonal structure even upon significant loss of mass.[224] A combination of the two mechanisms, glide and climb, explains the motion of 57 pairs along any arbitrary trajectory.[217] A combined motion of 57 pairs in graphene by both glide and climb[218,225] (Fig. 10) as well as by glide, assisted by the formation of "dislocation worms",[225] was demonstrated for graphene experimentally and is in agreement with HRTEM observations[219,226] for carbon nanotubes.

Unlike in nanotubes, in graphene the formation of a second SW defect is preferable to the dissociation of the first SW defect into 57 pairs, both under strain and in its absence.[138,203,204] Simple dissociation of SW defects into 57 pairs in graphene is therefore thermodynamically unfavourable, and in most cases the plastic deformation of a graphene layer should start through the formation of additional defects, multiple SW defects or larger rings such as octagons. Once the defects are present in graphene, subsequent SW rotations become much easier. For example, the activation energy of the transformation of the V_2(5-8-5) structure to V_2(555-777) was found to be 5.74 eV, according to the TB calculation[173] (5.17 eV according to LDA calculation), which is almost half the 10.2 eV activation energy for the formation of a SW defect in perfect graphene, obtained in the same paper. It is important to note, however, that plastic deformation through the creation of multiple defects can also finally lead to the formation of isolated 57 pairs. For instance, chains of SW defects are known to rearrange giving 57 pairs screened by "dislocation worms".[203,204]

Deformation of graphene under strain, nevertheless, is hardly the main source of lone separated 57 pairs so frequently observed in HRTEM.[218,227,228] Creation of dislocations under HRTEM conditions can be explained by the reconstruction of vacancies and adatoms introduced by electron irradiation. Tight-binding molecular dynamics simulations for carbon nanotubes with multi-vacancies demonstrate that such nanotubes are sewed up almost perfectly with only two separated 57 pairs left and a reduced diameter.[92,93] The transformation of graphene with a chain of missing atoms to a nearly ideal layer with only two separated 57 pairs was shown to be preferred over the formation of a local haeckelite structure composed of collective 555-777 defects when the multivacancy size exceeds ten,[176] in agreement with the experimental data.[225] The formation of separated 57 pairs was found to be favourable over other possible reconstructions for multivacancy sizes above 4 in narrow graphene nanoribbons of width within 1 nm, and above 26 in wide graphene nanoribbons of width greater than 10 nm.[175]

According to the experimental data,[225,228] however, even scattered vacancies tend to reconstruct and form a nearly ideal hexagonal network with only two separated 57 pairs. Divacancies can dissociate into 57 pairs by glide and pseudoclimb similar to SW defects.[217,227] However, 57 pairs formed in this way turn out to be unstable, as seen in tight-binding molecular dynamics simulations.[228] Stable dislocations are rather produced by agglomeration and collective reconstruction of several vacancies assisted by carbon adatoms helping to convert non-hexagonal rings into hexagons, as follows both from computer simulations[228,229] and experimental observations.[228]

The geometry of stable dislocations depends critically on whether they are formed from adatoms or vacancies.[225,228] The dislocation cores formed from adatoms (i.e. pointing towards each other) have only shallow minima at nanometre distances associated with the antisymmetric hillock–basin configuration, while at short distances they prefer the symmetric hillock–hillock configuration and tend to migrate towards one another by SW rotations.[225] The dislocation cores formed by multi-vacancy reconstruction, on the other hand, repel at short distances and have only one minimum at the distance determined by the number of vacancy units, in which the antisymmetric hillock–basin configuration is slightly preferred over the symmetric hillock–hillock one.[228]

In addition to 57 pairs, double pentagon (octagon) rings have been recently identified as dislocation cores in graphene.[230] As with separated 57 pairs, these defects are formed by the reconstruction of vacancy constellations and move by glides involving one bond rotation at each step.[230]

Graphene Grain Boundaries

The non-hexagonal rings in polycrystalline graphene play an important structural role as grain boundaries[20,231–236] between arbitrarily oriented graphene flakes. Dislocations characterised by any Burgers vector as well as grain boundaries, covering the whole range of possible misorientation angles, can be constructed on the basis of coupled ((1,0) dislocation) and dissociated ((1,1) dislocation) 57 pairs separated by hexagons[221–223,237] (Fig. 12). The stress fields of isolated dislocation cores in the limit of misorientation angles $\theta \rightarrow 0°$ and $\theta \rightarrow 60°$, which correspond to small-angle armchair and zigzag regimes, respectively, induce strong buckling of graphene with protrusion heights on the order of 2–3 Å.[221,222] The possibility of such buckling in free-standing graphene results in a considerable reduction of the grain boundary energy.[203,204,221,238,239] These theoretical results are supported by experimental STM images of graphene on Ir [232] and SiC,[233] where hillocks with large separations are observed at small-angle grain boundaries. Buckling is also predicted to take place in disordered grain boundaries.[240]

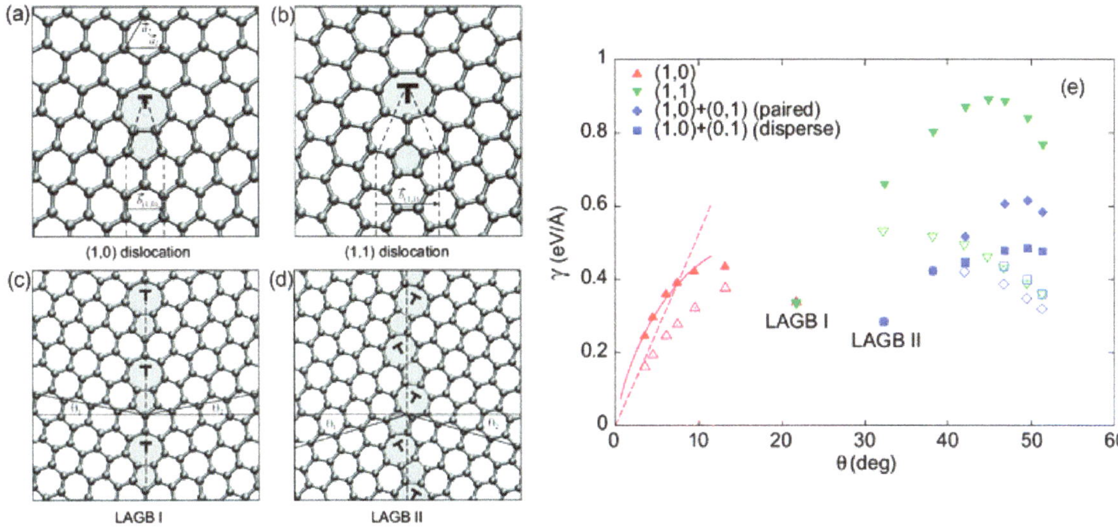

Figure. 12 (a, b) Atomic structures of (1,0) and (1,1) dislocations. The dashed lines delimit the semi-infinite strips of graphene introduced at the dislocation core. Non-hexagonal rings are shaded. (c, d) Atomic structures of the $\theta = 21.8°$ (LAGB I) and the $\theta = 32.2°$ (LAGB II) symmetric large-angle grain boundaries, respectively. The dashed lines show the boundary lines and the solid lines definite angles θ_1 and θ_2 ($\theta = \theta_1 + \theta_2$) defining orientations of the graphene grains. (e) Grain-boundary energy per unit length γ as a function of misorientation angle θ for various flat (filled symbols) and buckled (open symbols) grain-boundary structures. The two energetically favorable large-angle grain boundaries, LAGB I and LAGB II, are labeled. The dashed curve shows the asymptotic linear dependence of γ for the buckled small-angle armchair grain boundaries ($E_f = 7.5$ eV). Reprinted figures with permission from ref. 221. Copyright (2010) by the American Physical Society.

The energies of the most favourable straight grain boundaries covering all angle ranges lie within 0.5 eV Å^{-1} (Fig. 12e),[221–223,238,239,241]i.e. they are smaller than typical values for bare graphene edges (1.0–1.2 eV Å^{-1}).[96,164,242–248] Especially stable large-angle grain boundaries are formed by close packing of (1,0) (LAGBI) and (1,1) (LAGBII) dislocation cores (Fig. 12c–e).[221] These grain-boundaries with misorientation angles 21.8° and 32.2° have formation energies of only 0.338 eV Å^{-1} and 0.284 eV Å^{-1}, respectively.[221] Though they contain the maximum number of dislocation cores, their stress fields mutually cancel each other providing the lowest formation energy and maximum mechanical strength.[208,221,249,250] Unlike in small-angle grain boundaries, these structures are flat,[220,221] as confirmed by the experimental data.[233] The formation energy of LAGBII per 57 pair is almost half that of SW defects (1.3 eV vs. 2.5 eV according to ref. 251), leading to its abundance in experimental observations[233–236] and realisation of grain boundaries at other large angles as pieces of LAGB II separated by kinked sites.[235] Similar results were obtained for rotational grain boundaries, where the structure formed by a loop of six alternating pentagons and heptagons (the flower defect) was identified as the topological defect with the lowest formation energy per dislocation core (1.2 eV),[251] explaining why its formation is frequently observed via coalescence of mobile dislocations or SW defects.[20,225]

The glide and climb of 57 pairs constituting grain boundaries[222,252] make their migration possible.[20] The driving force for this migration depends only on the local in-plane boundary curvature and does not depend on the atomistic structure, providing that curved grain boundaries tend to become straight, while grain boundary loops annihilate upon annealing.[20] Tight-binding molecular dynamics simulations[252] of grain boundary motion showed that C_2 dimer emission, corresponding to a climb of the 57 pairs, is preceded by the formation of an adatom.

Effects of Edges on the Formation of SW Defects

The close vicinity of graphene edges allows for a more efficient relaxation of SW defects, providing a decrease in both the formation energy of SW defects and the activation energy for this process.[70,71] The formation energy of SW defects was shown to reduce from 4.4 eV inside a large graphene flake to 2.3 eV at the zigzag flake edge[70,71] according to the Tersoff–Brenner potential[128] and from 5.2 eV to 3.1 eV according to DFT calculations.[70] The corresponding changes in the activation energy were from 7.9 eV to 6.7 eV according to the semiempirical potential[70,71] and from 9.9 eV to 6.8 eV according to DFT calculations.[70] The formation energy and barrier for SW defects at the armchair edge of the flake were calculated to be 2.8 eV and 7.4 eV, respectively.[71]

Edge Reconstructions

In addition to the effect on the energetics of reactions inside graphene layers, edges enable new reactions with the participation of under-coordinated atoms at the very edge. As these atoms are already destabilised compared to ones in the graphene interior, the corresponding reactions have considerably reduced barriers and reaction energies[70,71] This means that such processes can occur at experimentally accessible times at high temperatures even without the assistance of catalysts, strain or irradiation, as opposed to SW transformations inside the layers.

In particular, reconstruction of the zigzag (ZZ) edge via the transformation of pairs of adjacent hexagons to pentagon–heptagon pairs (Fig. 13a) is energetically favourable[77,96,148,164,242–248,253] and is often observed experimentally.[13,75,254,255] This reconstruction (ZZ(57)) decreases the zigzag edge energy by 0.1–0.4 eV Å^{-1}, making it even slightly more stable than the armchair (AC) edge (Table 6). These results can be understood by considering edge geometries. The pristine zigzag edge is known to be much more expensive energetically than the armchair one, as two-coordinated atoms at the zigzag edge have dangling bonds, while two-coordinated atoms at the armchair edge form triple bonds, as seen by short bond lengths between such atoms (1.22–1.25 Å,[164,242,243,245] close to the 1.20 Å triple bond in acetylene (Table 7)). The same triple bonds of 1.23–1.26 Å length[78,164,242,243,245,247,253] are observed between two-coordinated atoms in the reconstructed ZZ(57) zigzag edge, guaranteeing its stability (Table 7). The initial compressive stress of the pristine zigzag edge of −0.5 eV $Å^{-1}$,[247] −2.05 eV $Å^{-1}$,[76] and −2.09 eV $Å^{-1}$ [248] is completely reversed to a tensile stress of 1.5 eV $Å^{-1}$,[247] 0.24 eV $Å^{-1}$,[76] and 0.14 eV $Å^{-1}$ [248] in the reconstructed ZZ(57) edge that provides its planarity. The 100% reconstructed edge also becomes non-spin-polarised.[73,247,248]

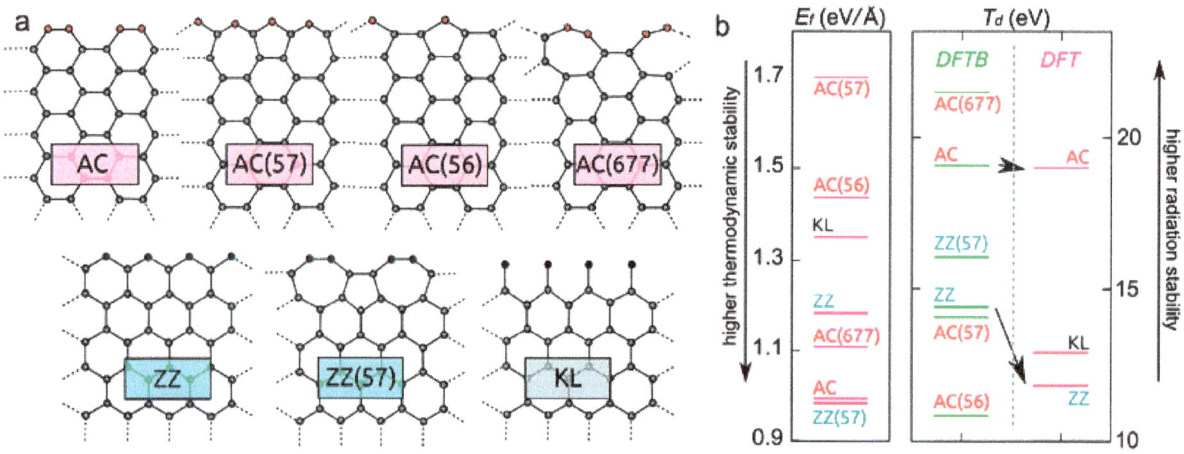

Figure. 13 (a) Different edge reconstructions as optimised by DFT and (b) their formation energies E_f and displacement thresholds T_d. Displacement thresholds as calculated with both DFT and DFTB are presented, showing the disagreement between DFT and DFTB for ZZ edges. Uncertainties in the T_d are of the order of 0.1 eV. Reprinted (adapted) with permission from ref. 148. Copyright (2014) American Chemical Society.

Table 6 Energies (in eV $Å^{-1}$) of pristine zigzag (ZZ), pristine armchair (AC), reconstructed zigzag (ZZ(57)) and reconstructed armchair (AC(677)) edges of graphene. The calculation method (SP and NSP stand for "spin-polarised" and "non-spin-polarised" calculations, respectively) is indicated

Ref.	Method	ZZ(57)	AC	AC(677)	ZZ
96	DFT-LDA	1.06	1.09		1.43
	TB model[124]	1.10	1.12		1.41
164	DFTB[121]	1.07	1.08		1.21
242	DFT-LDA, SP	1.09	1.10		1.34
243	DFT-PBE	0.96	0.98	1.11	1.31
244	DFT-PW91, SP	0.97			1.21
245	DFT-PBE/DZP, SP	0.98	1.02		1.15
	DFT-PBE/DZP, NSP	0.98	1.02		1.34
	Modified LCBOPII[245]	0.81	0.75		1.05
246	DFT-PBE, SP	0.965	1.008		1.145
247	DFT, SP	0.97	1.00	1.08	1.17
248	DFT-LDA/DZP, NSP	1.147	1.202		1.543
	DFT-LDA/DZP, SP	1.147	1.202		1.391

Table 7 Geometry of graphene edges: lengths (in Å) of triple (central) edge bonds of heptagons (r_{7c}), other edge bonds of heptagons (r_7), edge bonds of pentagons (r_5), between heptagons and pentagons (r_{57}), between heptagons and hexagons (r_{76}), between pentagons and hexagons (r_{56}) in the reconstructed zigzag edge, edge bonds in the pristine zigzag edge (r_z), triple (central) edge bonds of hexagons (r_{a1}) and other edge bonds (r_{a2}) of the pristine armchair edge. The calculation method is indicated

Ref.	Method	r_{7c}	r_7	r_5	r_{57}	r_{76}	r_{56}	r_z	r_{a1}	r_{a2}
78	DFT-LDA	1.24	1.39	1.43	1.45	1.43	1.40			
164	DFTB[121]	1.24	1.43	1.43				1.40	1.25	1.42
242	DFT-LDA	1.24	1.38	1.45				1.42	1.22	1.38

243	DFT-PBE	1.24	1.41	1.44	1.47			1.39	1.24	1.39
247	DFT	1.23								
248	DFT-LDA/DZP							1.40	1.26	
253	DFT-PBE/DZP	1.26								
245	Modified LCBOPII[245]	1.28	1.41	1.41	1.47			1.40	1.28	1.40

Reconstruction of the armchair edge through the transformation of three adjacent hexagons to two heptagons and a pentagon (an incomplete SW defect, Fig. 13a, AC(677)) was shown to be energetically unfavourable although still preferred over the unreconstructed zigzag edge (Table 6).[243,247]Nevertheless, it was also predicted that partial reconstruction of the armchair edge can further stabilise it.[247] The minimum edge energy of 0.98 eV $Å^{-1}$, very close to the edge energy of the reconstructed ZZ(57) zigzag edge, was calculated for about 25% of hexagons transformed into 6757 fragments.[247] Such a structure was also shown to have a small edge stress, indicating that it is stable against deformations, while the fully reconstructed armchair edge has a considerable tensile stress of 4 eV $Å^{-1}$.[247]

It should be emphasised, however, that these considerations are only valid at very low hydrogen pressures, while at ambient conditions the edges are passivated and no dangling bonds are left.[246] In this case, reconstruction of the edges has a considerable energy cost, while the edges consisting only of hexagons and with two-coordinated atoms bonded to hydrogen are the most preferred.[192,246,253,256] The energetics and warping of various hydrogen-passivated reconstructed edges was analysed.[192,246,256] The reconstruction of the zigzag edge into a sequence of alternating pentagons and heptagons is also unfavourable on transition metal surfaces, while other new types of reconstructions with an increased number of low-coordinated carbon atoms become possible.[257] Decoration of graphene edges with silicon[258] and other adatoms[259] is also known to affect their structure.

The activation energy for simultaneous and complete reconstruction of the zigzag edge is proportional to the edge length, and was calculated to be 0.6 eV per hexagon pair.[243] The simulations using a semiempirical potential[245] gave a rather close value of 0.7 eV at the low temperature limit.[245]Though it was shown to have non-monotonic behaviour at high temperatures (increasing up until 700 K and slowly decreasing at higher temperatures[245]), the cost for this simultaneous transformation remains too high and the transformation actually occurs step-by-step via the formation of pentagon–heptagon pairs. Therefore, sequential steps of the process have to be analysed.

Bond Rotation Reactions at Graphene Edges

The formation energy of a 57 pair at the zigzag edge of a 12-atom square hole in graphene was found to be 1.35 eV by first-principles calculations, much smaller than 3.83 eV for the SW transformation inside the layer.[187] The formation energy of a 57 pair at the pristine zigzag edge of a graphene flake was found to be about 1 eV according to the semiempirical potential,[128] 3.4 eV smaller than for a SW defect inside the graphene flake.[70,71] The activation energy of this reaction was found to be around 3 eV, which is less than half that of the SW transformation inside the layer.[70,71] Such a decrease was explained by the fact that bond switching at the edge requires only one bond to be broken at a time, compared to two in the perfect graphene layer.

A consecutive energy decrease by 1.51 eV and 1.78 eV was observed in successive steps of the generation of 57 pairs[96] according to TB and unpolarised LDA calculations, which are known to underestimate the stability of the pristine zigzag edge.[77,245,247,248] Activation energies of just 0.69 eV and 0.44 eV for each respective step were obtained, indicating that such a transformation should take place spontaneously at room temperature.[96] A similar free energy barrier of 0.83 eV was obtained at room temperature using a semiempirical potential.[245] The barrier was shown to increase up to 1 eV with increasing temperature.[245] Recent spin-polarised DFT calculations have, however, refined these values. The energy increase by about 0.3 eV was found for the formation of the first 57 pair, followed by a total energy gain of 0.6 eV from the generation of the second one[77] (Fig. 14). An activation energy of 1.80 eV (ref. 77) or 1.12 eV (ref. 73) was obtained for the first step, and 1.11 eV for the second one[77] (Fig. 14), demonstrating that the pristine zigzag edge is stable at room temperature, as confirmed by experiments.[72,74,260] An energy decrease of 0.21 eV was predicted for the formation of the first 57 pair in ref. 79 and the activation energy for this reaction was found to be 1.61 eV. It was also suggested that the formation of these pairs can occur through the recombination of an adatom and a vacancy, with an activation energy of just 0.89 eV.[79] The barriers for the migration of an adatom and vacancy on the zigzag edge were calculated in the same paper to be 0.76 eV and 2.12 eV, respectively.

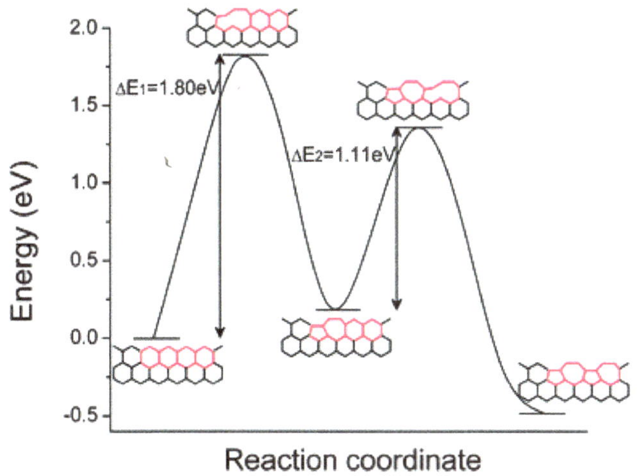

Figure. 14 The minimum energy path of the first and second step transitions from a hexagon–hexagon pair to a pentagon–heptagon pair. Reprinted (adapted) from ref. 77.

Tensile stress was shown to increase energies of the pristine and defected zigzag edges, decreasing the activation energy for the first-step transformation with 0.21 eV per %

rate.[73] Therefore, a uniaxial strain of 5% is sufficient to provoke reconstruction of the zigzag edge at room temperature.[73,77] Fracture of a monolayer graphene was shown to be governed by the competition between bond breaking and bond rotation at a crack tip.[261] Using atomistic reaction pathway calculations, a kinetically favourable fracture path was identified as an alternating sequence of bond rotation at the graphene edge and bond breaking.[261] The generation of 57 defects was also shown to play a crucial role in graphene evaporation, as the emission of carbon atoms was found to proceed preferentially from heptagons.[96] In particular, the energy gain from the formation of several 57 pairs at the zigzag edge was revealed to be sufficient to compensate for atom evaporation.[96]

Similar energetics are observed for other defects and graphene edges. The formation of a 757 fragment on the armchair edge of a hole in graphene was calculated to be 2.01 eV, which is significantly reduced as compared to the formation energy of a SW defect.[187] The energetics of various defects at zigzag and armchair edges of graphene flakes were investigated.[70,71] Similar barriers on the order of 3 eV were obtained for all reactions proceeding through the breaking of only one bond, including diffusion of pentagons and heptagons along the edges. Reactions forming carbon chains by breaking a bond between adjacent rings were recognised as important for all types of edges at high temperatures. Though the potential energy change in such reactions is relatively high, on the order of 2 eV, the formation of carbon chains becomes favourable at high temperatures due to an increase in entropy, related to their low-frequency vibrations.[70,71] The formation of carbon chains and their reconnection followed by edge reconstruction was shown to be the principal mechanism of transformation of a graphene flake to a fullerene[70,71,87] and formation of nanotube caps from open nanotube ends.[88] The transformation of narrow graphene nanoribbons into carbon chains was observed experimentally under electron irradiation.[13,14] The formation of carbon chains also took place in simulations of strained graphene nanoribbons at high temperature.[262,263]

REACTIONS UNDER IRRADIATION

While the important values when discussing thermally activated reactions are the activation and formation energies, threshold energies are key for irradiation-induced reactions. The threshold energy of a process is the minimum energy required for that process to occur without immediate reversal, considering a specific direction of atom motion induced by the electron beam. The difference between the activation energy and threshold energy for the same reaction can be explained as follows. Firstly, as the reaction pathway for an irradiation-induced reaction is determined by the direction of momentum transferred to a given atom from the incident electron, it does not necessarily coincide with the optimal reaction way for the thermally activated reaction. Secondly, even for optimal directions of transferred momentum (at the minimum threshold energy), the collective motion of multiple atoms will usually not coincide with the optimal reaction way for thermally activated reactions. The threshold energy of an irradiation-induced process therefore tends to be higher than the activation energy of the equivalent thermally induced process.

Under the electron beam, graphene is constantly exposed to an external stimulus. In these non-equilibrium conditions, low formation energies do not therefore imply the structures most likely to form and survive. In fact, the presence of possible routes induced by the beam towards a structure and the subsequent stability of the structure against the electron irradiation are key. The preferred configuration of graphene edges in TEM is a good example of the importance of these considerations. Two main edge configurations are observed: zigzag (ZZ) and armchair (AC). Although early observations and calculations[260] concluded that the ZZ configuration would be the most stable, the consideration of dynamic effects following electron impacts demonstrated the very high stability of AC edges with respect to irradiation.[148]

For another demonstration of this effect consider the structure of the most commonly observed tetravacancy, the extended linear structure.[264] It was proposed that the high occurrence of the most common V_4 structures in TEM images is due to the specifics of the irradiation-induced pathways towards their creation, but an explanation for the long lifetime of the linear structure over the more frequently modelled structures is perhaps due to the shared structural features between this vacancy and the V_2(5555-6-7777) divacancy (or 'butterfly' vacancy). It has been shown that the central atoms of this divacancy have an even larger emission threshold than pristine graphene,[147] implying a very strong stability against electron irradiation. It would therefore be expected that this is also true of the very similar tetravacancy structure.

Atom Emission

From Pristine Graphene

The emission of carbon atoms from pristine graphene in TEM has been extensively studied, both theoretically and experimentally. It has been observed that in order to eliminate atom emission and the creation of holes and defects, an accelerating voltage of 80 kV or lower must be used. This corresponds to a maximum transmittable energy to a stationary carbon atom of 15.8 eV, implying that the emission threshold energy of carbon must be slightly above this. However, Meyer et al.[146,267] demonstrated the importance of including the effects of the thermal motions of the graphene lattice when considering the transfer of momentum from beam electron to sample atom, explaining the apparent overestimation of theoretical emission thresholds compared to experiment. This inconsistency had previously been attributed to the absence of any beam-induced electronic excitations in calculations,[150,265] but was revealed to be an effect of the increased transmittable energy from the beam due to thermal motion of the atoms; the threshold energies are generally accurate when taking this into consideration (Table 8). An experimentally derived rate of atom emission from graphene was found at various accelerating voltages and dose rates, and used to calculate the threshold energy of 23.6 eV including these lattice vibration effects.[146,267] It is worth noting that this is higher than any published theoretical value, by at least 0.6 eV. A comparison[266] of theoretical threshold energies calculated by DFT to excited state time-dependent DFT (TDDFT) calculations that explicitly included the electron dynamics concluded that the ground state Born–Oppenheimer dynamics universally used when calculating threshold energies is indeed justified. Qualitative trends were very similar between DFT and TDDFT, despite small quantitative differences; DFT resulted in the transfer of slightly more energy to the surrounding graphene lattice, while TDDFT showed similar kinetic behaviour of the emitted atom and the graphene lattice at a delay of 10 fs.

Table 8 Threshold energies E_d (in eV) calculated for various processes in graphene. The system considered (periodic boundary conditions (PBC), flake or nanoribbon), number of atoms in the computational cell and calculation method are indicated

Process	Ref.	System	Size of supercell	Method	E_d (eV)
Emission from pristine graphene	146,267	Experimental measurement	—	—	23.6
	268			DFTB	22
	265	PBC	7 × 7 supercell	DFTB	23
	150	PBC	4 × 6 supercell	DFT-PBE	22.03
	147		120–200 atoms	DFTB	22.50
	266	PBC	478 atoms (converged to 1342)	DFT-LDA	23
	151	Graphene flake		DFTB	23.0
	225 (ESI)	PBC	152 atoms	DFTB	21.34
	269	PBC	160 atoms	DFT-PW91, SP	22
Emission from V_1(5-9)	148	PBC		DFT-PBE	14.7
Emission from V_2(5-8-5)	148	PBC		DFT-PBE	16.2
Emission from pentagon site	151	Graphene flake		DFTB	16.9
Emission from 2-coordinated edge	151	Graphene flake		DFTB	13.4
Emission from AC edge	148	Graphene nanoribbon, PBC	6–7 graphene unit cells wide	DFT-PBE	19.0
Emission from ZZ edge	148	Graphene nanoribbon, PBC	6–7 graphene unit cells wide	DFT-PBE	12.0
Emission from KL edge	148	Graphene nanoribbon, PBC	6–7 graphene unit cells wide	DFT-PBE	12.9
SW formation	266	PBC	478 atoms (converged to 1342)	DFT-LDA	19
	271	PBC	72 atoms	DFT-LDA/DZP	22.5
	271	PBC graphene with h-BN substrate	144 atoms	DFT-LDA/DZP	31
SW healing	266	PBC	478 atoms (converged to 1342)	DFT-LDA	13
	271	PBC	72 atoms	DFT-LDA/DZP	14
	271	PBC graphene with h-BN substrate	144 atoms	DFT-LDA/DZP	17

Although areas of pristine graphene are generally considered to be protected from irradiation damage at or below 80 kV, the use of very high electron dose rates ($>10^8$ e nm^{-2} s^{-1}) at 80 kV has been shown to create defects.[65] Similar defect creation is not witnessed under lower dose rates ($<10^6$e nm^{-2} s^{-1}) and long irradiation times, where large total numbers of electrons are passed through the sample ($\sim10^{10}$ e nm^{-2}),[146,267] meaning that a dose-rate dependent effect is either lowering the threshold energy or increasing the energy that can be transferred by increasing the out-of-plane atomic vibrations. This lowered threshold energy is calculated to be 19.7 eV, and weakening of the bonds due to ionization or plasmon excitations was suggested as a possible cause. Alternatively, while beam-induced heating effects were shown to be not responsible, a flexural phonon mode induced by inelastic collisions could result in the required increase in out-of-plane vibrations. This electron dose rate dependent effect has been exploited with the ability to create defects at a high dose rate and then observe their dynamics without creating additional defects at a lower dose rate, all at an 80 kV accelerating voltage.[65,159,264]

The threshold energies given for atom emission only correspond to the case when the transmission of energy from beam electron to graphene atom occurs perfectly perpendicular to the graphene plane. This is the angle at which a minimum energy will be required to emit the atom, while more energy will be needed if transferred at a shallower angle. When energy is transferred at non-perpendicular angles, there will also be a variation in threshold energy with the azimuthal angle φ; the in-plane direction of energy transfer. In the typical setup of the electron beam being perpendicular to the graphene sheet, this minimum threshold energy coincides with the angle at which most energy can be transferred ($\theta = \pi$, where θ is the electron scattering angle). This means that for emission processes, the single value of the threshold energy is highly informative, despite the anisotropic dependence of threshold energy on angles of momentum transfer. For other processes, such as the SW rearrangement discussed later, this overlap of the minimum threshold energy E_d with the angle at which the maximum energy can be transferred from the beam does not hold, and so mapping of the anisotropy is vital. Mapping of the anisotropy of the threshold energy of emission from pristine graphene, shown in Fig. 15,[265] confirmed the very sharp increase in E_d at $\theta < \pi$, rising from 23 eV to 43–780 eV at $\theta = \frac{\pi}{2}$, depending on φ.

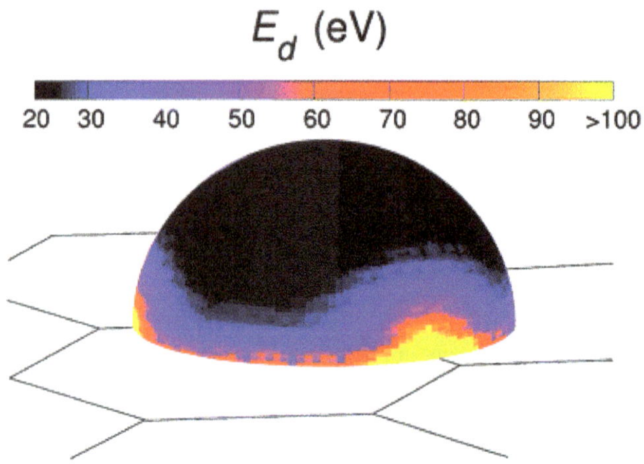

Figure. 15 Mapping of the threshold energy for emission of an atom from pristine graphene as a function of the angle of transferred momentum from the e-beam. The threshold energy of 23 eV rises very rapidly to large energies at angles parallel to the graphene sheet. Reprinted figure with permission from ref. 265. Copyright (2007) by the American Physical Society.

From Vacancy Structures

Following the emission of a single atom from pristine graphene, a symmetrical vacancy or reconstructed 5-9 structure is formed. In any structure with an odd number of missing atoms, there will always be at least one under-coordinated atom bonded to fewer than three atoms. These atoms will be more easily lost via collisions with electrons from the e-beam due to a decreased emission threshold, calculated as 12 eV for the unreconstructed vacancy[266] and 14.7 eV in the case of the reconstructed monovacancy.[148] As this energy is easily accessible to even an 80 kV electron beam, correlated sputtering of two atoms from areas of pristine graphene is normally observed. Even-numbered vacancies are fully coordinated with three bonds to each atom and subsequently have larger threshold energies, although the value varies depending on the ring size and to a smaller extent on the specific local structure. For example, atoms in pentagons typically have threshold energies around 16.9 eV,[151] depending on the local atomic structure. This generally leads to the formation of separate divacancies on the graphene lattice, as once a divacancy is formed there is a much smaller preference as to which atom will be emitted next, and the typically low concentration of defect sites means that additional emissions are likely to occur on pristine graphene.

This effect is increased with the existence of the two commonly observed reconstructed divacancy structures shown in Fig. 3. As the central atoms of these structures have been shown to have a larger emission threshold than pristine graphene,[147] the probability of direct trivacancy formation is further reduced.

In dislocation dipoles, the emission of two atoms leads to a climb step, increasing the distance between the two dislocations (shown in Fig. 10).[218]Emission of atoms from the area surrounding the dislocation has been observed to occur more frequently than in pristine graphene, confirmed by calculations of the threshold energies for all surrounding atoms.[225] The threshold energies ranged from 18.64 to 24.14 eV, compared to 21.34 eV calculated for emission from pristine graphene using the same method. This study also demonstrated the importance of the beam direction in defects which introduce out-of-plane topology. As the buckled structure of the dislocation is not flat, there are two distinct directions in which it can be oriented with respect to the beam, up or down. Depending on the direction of the e-beam in the z-axis, the emission threshold energy of an individual atom varied by up to 2.9 eV (Fig. 16).

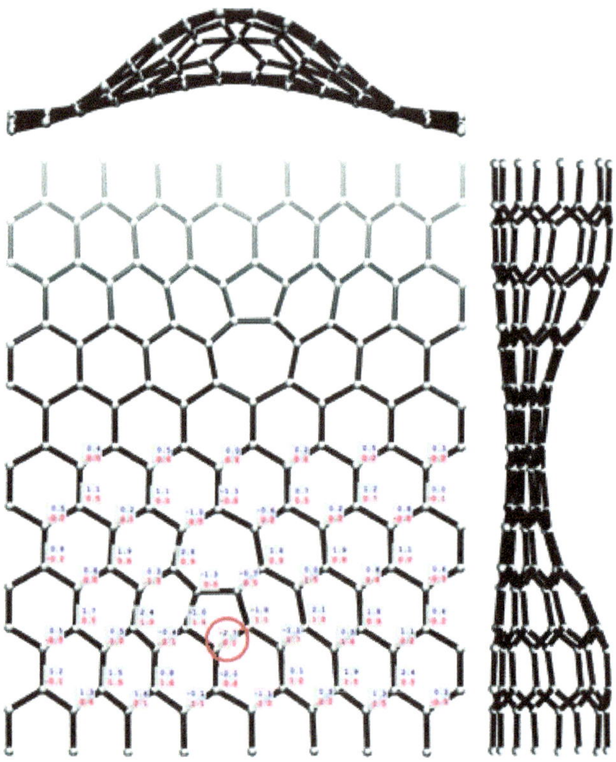

Figure. 16 Emission threshold energies calculated for atoms surrounding a dislocation core. Values are given as deviations from the pristine value calculated using the same method (21.34 eV). As the structure is buckled out-of-plane (side views), it is not symmetrical, resulting in different threshold energies when the momentum is transferred upwards (blue numbers) and downwards (red numbers). Reprinted (adapted) from ref. 225 (ESI).

From Graphene Edges

A study of the emission threshold energies of various simple and reconstructed edge configurations used graphene nanoribbons of various widths.[148] DFTB calculations were undertaken for all edge configurations shown in Fig. 13. Termination of various edge configurations by hydrogen atoms was shown to have no effect on the emission threshold energies of edge atoms. The threshold energies calculated by this method are shown, ranging from ~11 eV for the AC(56) edge to ~22 eV for the AC(677) edge. Also shown are the results of more computationally demanding DFT simulations, carried out for the unreconstructed AC (19.0 eV), ZZ (12.0 eV) and KL (12.9 eV) edges. This set of calculations showed that the specific local structure of edges has a very large impact on the irradiation stability, and demonstrated the lack of correlation between irradiation stability and low formation energy; the AC and KL edges exhibit very flexible behaviour following impacts from the electron beam, which increases their resistance to atom emission. The discrepancy between these predictions of the stability of AC configurations over ZZ, compared to experimental observations of preferential ZZ formation in holes in graphene,[260] was attributed to the differences between an approximately circular hole and extended edges. The predicted stability of the KL edge under the e-beam was very recently confirmed by experimental HRTEM images showing their existence in graphene nanoribbons and at the edge of bulk graphene.[270]

Bond Rotations

In Pristine Graphene

Other than emission, the only major process of irradiation-induced atomic structure change in graphene is the Stone–Wales (SW) rearrangement, a 90° rotation of a single carbon–carbon bond. In pristine graphene a SW rearrangement leads to a SW defect, composed of two 5- and two 7-membered rings. This has been observed in TEM and is seen to heal quickly, on the order of the electron dose per image used (~10^7 e nm^{-2}).[147] Theoretical calculations confirmed the creation of the SW defect via impacts with electrons from the e-beam, revealing two mechanisms of formation. The energy required for SW defect formation was calculated as being typically 1 eV lower than the emission threshold (calculated as 22.5 eV), with 2 eV below this resulting in SW formation at certain space angles of transferred momentum.

The threshold energies for the formation and healing of a SW defect were calculated as being 22.5 eV and 14 eV respectively, confirming the reversible nature of the defect.[271] The same study revealed the ability of substrates to alter the energy required for each process; placing the graphene on hexagonal boron-nitride increased them to 31 eV and 17 eV respectively. An additional study[266] (using a much larger, converged supercell) gives comparable numbers for the pristine graphene case, although giving slightly lower values of 19 eV and 13 eV for formation and healing of the SW defect. The anisotropy of the energy required to heal the SW defect at space angles (θ,φ) was mapped,[266] revealing the strong dependence on the direction of momentum transfer, as previously discussed for atom emission. In addition, this paper revealed the ability of a SW defect to migrate due to a single electron impact, provided one of two combinations of space angles and energy (18 or 20 eV) are achieved.

Vacancy Structure Rearrangements

Divacancy structures have been seen to convert between different stable states under electron irradiation.[147] These transformations can be understood in terms of individual or multiple SW rearrangements on bonds in the vacancy structures. A very in-depth study[266] mapped the threshold energies required for single-impact structure changes at different space angles for each of the V_2(5-8-5), V_2(555-777), V_2(555-6-777) and V_2(55-77) divacancy structures. It showed that the amount of energy required for a SW rearrangement is highly dependent on the angles of transferred momentum, ranging from 14 to 22 eV, and that different structures can result from impacts on the same atom, depending on the direction of momentum transfer. Fig. 17 shows a summary of these findings, giving the threshold energy for each SW rearrangement indicated by an arrow. It is worth noting the inclusion of the reconstructed V_2(55-77) divacancy, as typically only the previous three structures are discussed.[99] This structure can be achieved via the rotation of a different bond in the V_2(5-8-5) structure, however with a threshold energy of 21 eV compared to the 16 eV required for the V_2(5-8-5) to V_2(555-777) transformation. In addition, the reverse process of V_2(55-77) to V_2(5-8-5) has the lowest threshold energy of any divacancy rearrangement at 14 eV, and so while this structure should be able to form under irradiation, it would be expected to have a shorter lifetime compared to the other reconstructed divacancies. Indeed, while it is very rarely seen in TEM, an example of this structure forming from a V_2(5-8-5) vacancy, before quickly undergoing the reverse transformation, has been observed.[167] This structure also provides a low energy route to the formation of trivacancies, with only 14 eV required to remove an atom and form the V_3(5-10-5) structure.

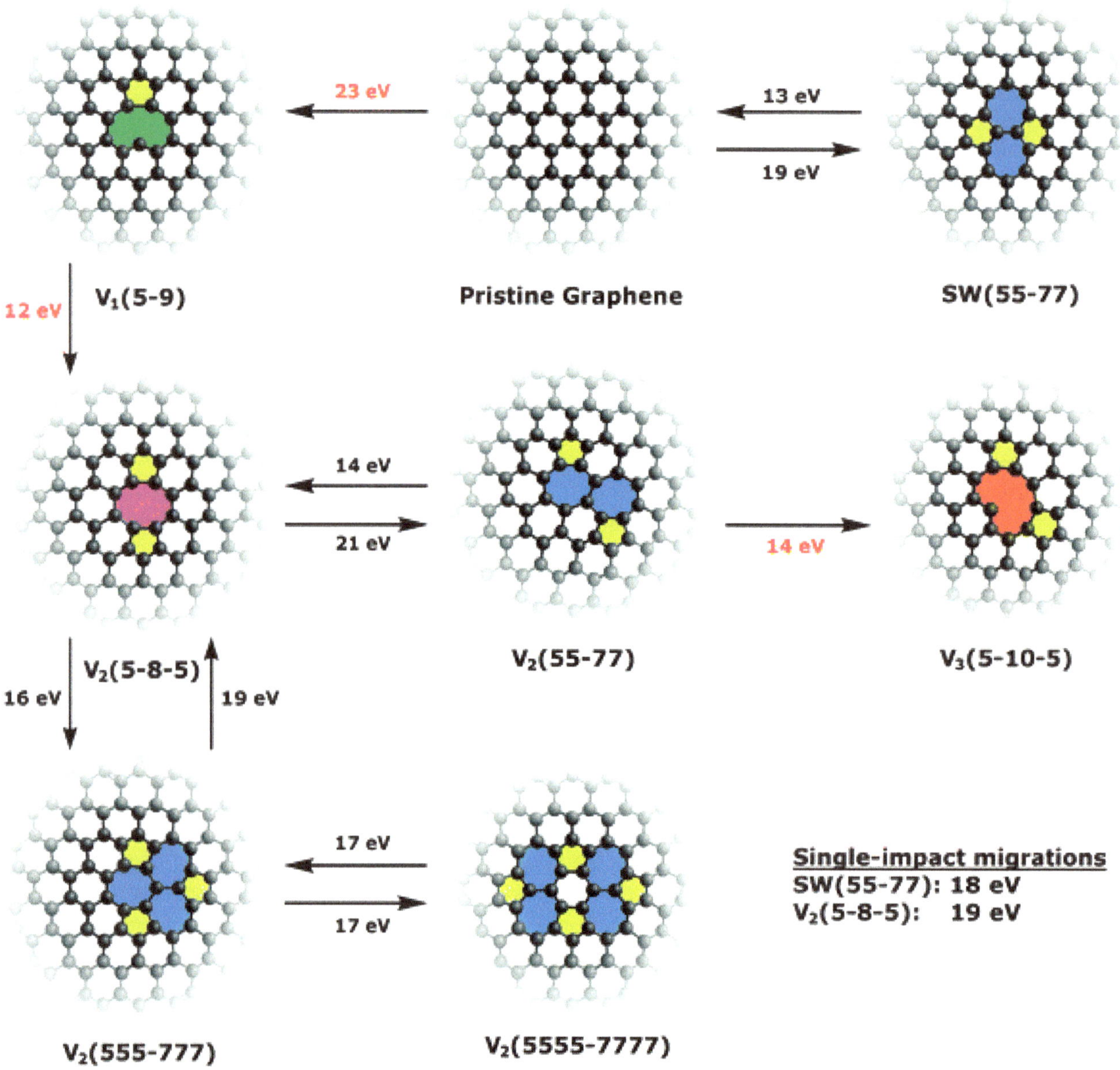

Figure. 17 A summary of the threshold energies calculated for irradiation-induced processes in ref. 266. Emission processes are labelled in red and bond rotations (SW rearrangements) in black. Rings are coloured according to their size: 5, yellow; 6, white; 7, blue; 8, purple; 9, green; 10, red. Threshold energies are also given for the direct migration of structures across the graphene lattice due to a single impact from the e-beam, only witnessed for a small number of combinations of angle space and transferred energy.

Although the SW rearrangement has been shown to be responsible for a wide range of dynamical behaviour in larger vacancy structures, the need to map the anisotropy of the threshold energy, combined with an increasing number of inequivalent atoms and potential structures in larger vacancies, makes the calculation of threshold energies prohibitive. While several key structures of particular stability were found for the tetravacancy,[264] a large number of other structures were observed experimentally, each living for seconds at a time. Fig. 18 shows the 17 most frequently observed tetravacancy structures together with the SW rearrangements required to convert between them. Tight-binding MD simulations were carried out in order to demonstrate the evolution of the tetravacancy under electron irradiation in this manner,[264] however a full search of the impact angle space in order to determine threshold energies for defect structures larger than divacancies has not been performed.

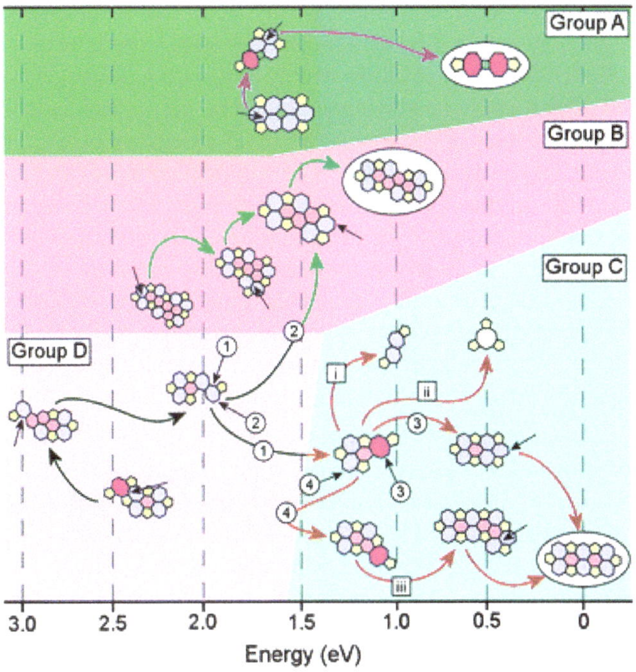

Figure. 18 SW rotations that allow for the conversion between different experimentally observed tetravacancy structures, the most frequently observed of which are circled. Each coloured arrow indicates a single SW rotation, except those numbered with Roman numerals which require two bond rotations. The bonds which rotate are indicated by black arrows. Configurations which can develop in one of two ways have numbered pathways corresponding to the respective numbered bond rotation. Reprinted (adapted) from ref. 264. Copyright (2014) American Chemical Society.

Vacancy Migration

In addition to conversion between different stable structures, vacancy structures have been observed to migrate across the graphene lattice.[167] As the barrier to the thermal migration of monovacancies is in the region of 1–1.5 eV (Table 1), it is assumed that the monovacancy migration is primarily thermally driven. Impacts from the electron beam will typically transfer much more energy than this (at 80 kV the maximum transferrable energy to a carbon atom is 15.8 eV), and so electron impacts on monovacancies are generally assumed to result in the emission of the under-coordinated carbon atom, as discussed above. However, the barriers for thermal migration of larger vacancy structures are unreasonably large, and so the effects of the e-beam are assumed to be responsible for the experimentally witnessed migration. Two consecutive SW rearrangements are proposed to be responsible for a migration of the V_2(5-8-5) vacancy by one lattice unit via the V_2(55-77) structure.[167] The theoretical study described above[266] provided confirmation of the possibility of this proposed two-step mechanism; it confirms the ability of SW rearrangements to convert between the two divacancy structures, and the low irradiation stability of the V_2(55-77) structure (with the lowest threshold energy calculated, 14 eV) provides an explanation for both rearrangements happening in quick succession. An additional two step mechanism was proposed[167] for the V_2(5-8-5) vacancy rotating 60° around a pentagon, via an intermediate V_2(555-777) structure. The calculated threshold energies of these two steps, 16 eV and 19 eV, suggest that while this mechanism will be less likely to occur than the migration via the V_2(55-77) structure, it will be possible given an electron beam of sufficient energy. During the mapping of the threshold energies of various divacancy structures,[266] a previously unpredicted single-impact mechanism of V_2(5-8-5) migration was discovered, via bond breaking and reforming rather than through a direct bond rotation. The threshold energy of this process is 19 eV, and due to the requirement of only one electron impact it would be expected that this mechanism of migration would be especially prevalent at sufficiently high e-beam energies.

At Graphene Edges

The threshold energies for the bond rotation between the zigzag (ZZ) and reconstructed zigzag pentagon–heptagon ZZ(57) edges in graphene were calculated from experimental TEM images of torn graphene.[255] These calculations assumed that every scattering event that transferred energy above the threshold energy would result in a transformation; they ignored the anisotropic dependence of the threshold energy. By observing the rates of conversion between the two structures, it was shown that the threshold for the ZZ to ZZ(57) rotation is 1.3 eV, while the ZZ(57) to ZZ rotation is 2.5 eV. These values are far lower than theoretical calculations of thresholds for SW rearrangements in the graphene interior, consistent with the smaller number of bonds that must be broken at the edge in order for a SW rearrangement to take place. It would be expected that in general, bond rotations will be far more prevalent at the edge than in the graphene interior.

In addition to conversions between the ZZ and ZZ(57) structures via a bond rotation, other SW rearrangements have been observed occurring during simulations of electron impacts below the emission threshold at various edge configurations.[148] Conversions between the AC and AC(57) edges and the AC and AC(667) edges were witnessed, while the AC to AC(56) transformation was always seen to occur following the emission of an atom. Threshold energies of these and other similar processes have not yet been calculated, but the knowledge of rates of conversion between structures could provide valuable insights into the dynamical behaviour of graphene edges under electron irradiation.

CONCLUSIONS

In the present paper we review the recent studies devoted to the energetics of formation, transformation, migration, coalescence and healing of defects in graphene. The cases of various vacancy-type and topological defects created due to bond realignment reactions in the graphene interior and at the graphene edge are considered, with both thermally activated and irradiation-induced processes taken into account.

The structure and formation energies for the monovacancy and many types of multi-vacancies in the graphene interior have been studied in detail. It has also been found that adatoms and single vacancies have low barriers for migration, about 0.5 eV and 1.0–1.4 eV, respectively, whereas barriers for divacancy migration exceed 5 eV. Thermally activated processes related to multi-vacancy formation and healing at room temperature will therefore be determined by the migration of adatoms and single vacancies. However, contradictory results have been obtained for the values of out-of-plane displacement, the barrier for transition between equivalent states of the 5/9 vacancy, and influence of interlayer interaction on the barrier for 5/9 vacancy motion in graphite. DFT calculations demonstrate a drastic change in the structure and energetics of a single vacancy near the graphene edge, and predict the fast directional motion of the vacancy to the edge at room temperature.[67,68] However these results have only been obtained on a few examples of rather small graphene flakes and narrow nanoribbons. Further detailed studies of the energetics of the vacancy near armchair and zigzag edges of semi-infinite graphene, both with dangling bonds and terminated by hydrogen and other atoms are necessary. A set of multi-vacancies with various structures formed due to vacancy coalescence have been found for free standing graphene. As strain has a drastic effect on the barrier to vacancy migration,[23,161] local strain could be used to control vacancy coalescence. Studies investigating the possibility of such control remain an open problem.

First-principles calculations clearly demonstrate that bond realignment reactions in the graphene interior have large barriers and should not take place even at high temperatures. The effect of strain on these reactions has been extensively studied in recent years, shedding light on the mechanisms of plastic deformations of graphene and carbon nanotubes. The catalytic effects of metal, carbon and hydrogen adatoms have been considered. Similar to vacancies, topological defects have been shown to result in out of plane buckling of graphene layers thus providing stress relaxation. However, the magnitude and energetics of the out of plane distortions are very sensitive to the calculation method used and the size of the periodic cell (or a finite system) considered. Therefore, more accurate studies are still required to get reliable data for the out of plane deformations caused by defects in graphene.

Graphene edges are found to be much more reactive than the graphene interior, and in the absence of strain or catalysing agents high-temperature transformations in graphene are concentrated at the edges. Significant efforts have been made to describe the reconstruction of zigzag edges. However, other types of bond realignment reactions at the edge have mostly been studied using semiempirical potentials, which are only qualitatively correct. Moreover, the investigation of sequential steps of defect generation has been focused on free (unfunctionalised) edges. More detailed first-principles analysis of various bond realignment reactions and the inclusion of functionalised edges into consideration are therefore of interest. In particular, the formation of carbon chains at edges has been shown to play a leading role in high-temperature transformations of graphene flakes[70,71,87] and open-ended nanotubes,[88] and has been observed experimentally for graphene nanoribbons under electron irradiation.[13] More detailed investigation into chain formation reactions is required to understand these processes.

Quantitative studies of irradiation-induced processes of structure change in graphene overwhelmingly deal with the case of atom emission, predominantly from pristine graphene. The threshold energy for this process is very well known, with good agreement between experiment[146,267] and theoretical studies (Table 8), but similar studies for interesting defective types of graphene structure are lacking. While simulations considering edges[148] and small vacancy structures (up to divacancies)[266] have produced useful insights into the dynamics of these areas under electron irradiation, these remain the minority of cases. Even among the various divacancy structures, the V_2(55-77) dislocation dipole typically receives little attention, presumably due to it being a much rarer sight experimentally. However, the presence of low energy routes from this structure mean that it could be playing an important part in irradiation-induced vacancy dynamics. A threshold energy of 14 eV for the rearrangement back to the V_2(5-8-5) structure provides an accessible mechanism of divacancy migration, and a threshold energy of only 14 eV for the emission of an atom provides a low energy pathway to trivacancies and therefore to the creation of larger vacancy structures. Indeed, the low number of experimental observations of this structure may merely indicate that it typically has a short lifetime under the e-beam due to the presence of these low threshold processes.

Larger vacancy structures and commonly observed structures remain to be examined at all in terms of threshold energies of irradiation-induced processes. An especially important example of this is the stability and migration of grain boundaries in graphene. Emission threshold energies of different grain boundary configurations could help to explain experimentally seen structures, while the HRTEM observation of the atom-by-atom migration of a grain boundary has been reported,[20] mediated by individual bond rotations.

Bond rotations have been shown to be responsible for a wide range of atomic structure transformations in graphene, providing accessible routes between different stable structures. However, the theoretical estimation of threshold energies for this kind of process remains difficult due to the requirement of mapping the energy over the range of impact angles. This is reflected in the severe lack of data available for SW rearrangements in non-pristine graphene, with only several key studies providing quantitative understanding of these processes. For example, despite knowledge of the irradiation stabilities of a variety of edge configurations with respect to emission,[148] one of the very few examples of a quantitative study of bond rotations at the edge only deals with the conversion between a single pair of edge configurations,[255] and the threshold energies are calculated from the experimentally observed rates of conversion. As this makes the large assumption that any impact above the threshold energy would result in a bond rotation, theoretical calculations are necessary for comparison; mapping the anisotropy of this case as has been performed for divacancies[266] could provide accurate values of the threshold energy. Simulations have shown that similar conversions can occur between a variety of other edge reconstructions,[148] however even the relative threshold energies of these processes, and therefore the relative likelihood of their occurrence under the e-beam, are unknown.

Knowledge of threshold energies of irradiation-induced processes are key to understanding the dynamical behaviour of graphene under the e-beam, with multiple examples showing that this behaviour cannot be explained by simply considering the equilibrium energetics. Despite this, and with HRTEM increasingly becoming the de facto tool for the experimental study of the atomic structure of graphene, theoretical calculations of threshold energies and resulting cross-sections remain scarce compared to the wealth of data on thermal activation and formation energies.

In summary, there are several areas in which further calculations of the energetic characteristics of atomic scale structure changes in graphene are necessary: (1) the presence and magnitude of out of plane displacements in 5/9 monovacancies, SW defects, dislocation cores and other defects, (2) the barrier for the transition between equivalent states of the 5/9 vacancy, (3) detailed energetics for the vacancy migration to graphene edges, (4) the control of vacancy migration and coalescence by local strain, (5) detailed energetics for bond realignment reactions at graphene edges, beyond the formation of pentagon–heptagon pairs at the zigzag edge, (6) detailed energetics of bond realignment reactions at hydrogenated and other types of functionalised edges, (7) detailed energetics of the formation of carbon chains at different types of edges, (8) threshold energies of structure changes (emission and migration by bond rotation) at grain boundaries, (9) both qualitative (relative rates) and quantitative (threshold energies) details of conversions between different edge configurations, (10) threshold energies of processes involving vacancy structures larger than divacancies.

ACKNOWLEDGEMENTS

EB acknowledges ERC Consolidator Grant for financial support. AP and IL acknowledge Russian Foundation of Basic Research (14-02-00739-a). IL acknowledges support by the Marie Curie International Incoming Fellowship within the 7th European Community Framework Programme (Grant Agreement PIIF-GA-2012-326435 RespSpatDisp), Grupos Consolidados del Gobierno Vasco (IT-578-13).

NOTES AND REFERENCES

1. S. Iijima, Nature, 1991, **354**, 56–58, DOI:10.1038/354056a0.
2. K. S. Novoselov, A. K. Geim, S. V. Morozov, D. Jiang, Y. Zhang, S. V. Dubonos, I. V. Grigorieva and A. A. Firsov, Science, 2004, **306**, 666–669, DOI:10.1126/science.110. 2896.
3. A. Chuvilin, E. Bichoutskaia, M. C. Gimenez-Lopez, T. W. Chamberlain, G. A. Rance, N. Kuganathan, J. Biskupek, U. Kaiser and A. N. Khlobystov, Nat. Mater., 2011, **10**, 687–692, DOI:10.1038/nmat3082.
4. T. W. Chamberlain, J. Biskupek, G. A. Rance, A. Chuvilin, T. J. Alexander, E. Bichoutskaia, U. Kaiser and A. N. Khlobystov, ACS Nano, 2012,**6**, 3943–3953, DOI:10.1021/nn300137j.
5. W. A. De Heer and D. Ugarte, Chem. Phys. Lett., 1993, **207**, 480–486, DOI:10.1016/0009-2614(93)89033-E.
6. S. Bandow, M. Takizawa, K. Hirahara, M. Yadasako and S. Iijima, Chem. Phys. Lett., 2001, **337**, 48–54, DOI:10.1016/S0009-2614(01)00192-0.
7. B. Song, G. F. Schneider, Q. Xu, G. Pandraud, C. Dekker and H. Zandbergen, Nano Lett., 2011, **11**, 2247–2250, DOI:10.1021/nl200369r.
8. A. Chuvilin, U. Kaiser, E. Bichoutskaia, N. A. Besley and A. N. Khlobystov, Nat. Chem., 2010, **2**, 450–453, DOI:10.1038/nchem.644.
9. D. Ugarte, Nature, 1992, **359**, 707–709, DOI:10.1038/359707a0.
10. D. Ugarte, Chem. Phys. Lett., 1993, **207**, 473–479, DOI:10.1016/0009-2614(93)89032-D.
11. J. Sloan, R. E. Dunin-Borkowski, J. L. Hutchison, K. S. Coleman, V. C. Williams, J. B. Claridge, A. P. E. York, C. G. Xu, S. R. Bailey, G. Brown, S. Friedrichs and M. L. H. Green, Chem. Phys. Lett., 2000, **316**, 191–198, DOI:10.1016/S0009-2614(99)01250-6.
12. C. S. Allen, Y. Ito, A. W. Robertson, H. Shinohara and J. H. Warner, ACS Nano, 2011, **5**, 10084–10089, DOI:10.1021/nn204003h.
13. A. Chuvilin, J. C. Meyer, G. Algara-Siller and U. Kaiser, New J. Phys., 2009, **11**, 083019, DOI:10.1088/1367-2630/11/8/083019.
14. C. Jin, H. Lan, L. Peng, K. Suenaga and S. Iijima, Phys. Rev. Lett., 2009, **102**, 205501, DOI:10.1103/PhysRevLett.102.205501.
15. S. T. Skowron, I. V. Lebedeva, A. M. Popov and E. Bichoutskaia, Nanoscale, 2013, **5**, 6677–6692, **10.1039/c3nr02130k**.
16. P. M. Ajayan, V. Ravikumar and J.-C. Charlier, Phys. Rev. Lett., 1998, **81**, 1437–1441, DOI:10.1103/PhysRevLett.81.1437.
17. C. Jin, K. Suenaga and S. Iijima, Nano Lett., 2008, **8**, 1127–1130, DOI:10.1021/nl0732676.
18. J. Y. Huang, S. Chen, Z. Q. Wang, K. Kempa, Y. M. Wang, S. H. Jo, G. Chen, M. S. Dresselhaus and Z. F. Ren, Nature, 2006, **439**, 281, DOI:10.1038/439281a.
19. J. Y. Huang, S. Chen, Z. F. Ren, Z. Wang, K. Kempa, M. J. Naughton, G. Chen and M. S. Dresselhaus, Phys. Rev. Lett., 2007, **98**, 185501, DOI:10.1103/PhysRevLett.98.185501.
20. S. Kurasch, J. Kotakoski, O. Lehtinen, V. Skákalová, J. Smet, C. E. Krill, A. V. Krasheninnikov and U. Kaiser, Nano Lett., 2012, **12**, 3168–3173, DOI:10.1021/nl301141g.
21. Y. Ma, P. O. Lehtinen, A. S. Foster and R. M. Nieminen, New J. Phys., 2004, **6**, 68, DOI:10.1088/1367-2630/6/1/068.
22. L. Xu, N. Wei and Y. Zheng, Nanotechnology, 2013, **24**, 505703, DOI:10.1088/0957-4484/24/50/505703.
23. A. S. Fedorov, D. A. Fedorov, Z. I. Popov, Y. E. Anan'eva, N. S. Eliseeva and A. A. Kuzubov, J. Exp. Theor. Phys., 2011, **112**, 820–824, DOI:10.1134/S1063776111040042.
24. A. S. Fedorov, Z. I. Popov, D. A. Fedorov, N. S. Eliseeva, M. V. Serjantova and A. A. Kuzubov, Phys. Status Solidi B, 2012, **249**, 2549–2552, DOI:10.1002/pssb.201200105.
25. R. Dettori, E. Cadelano and L. Colombo, J. Phys.: Condens. Matter, 2012, **24**, 104020, DOI:10.1088/0953-8984/24/10/104020.
26. F. Hao, D. Fang and Z. Xu, Appl. Phys. Lett., 2011, **99**, 041901, DOI:10.1063/1.3615290.
27. B. Mortazavi and S. Ahzi, Carbon, 2013, **63**, 460–470, DOI:10.1016/j.carbon.2013.07.017.
28. N. Jing, Q. Xue, C. Ling, M. Shan, T. Zhang, X. Zhou and Z. Jiao Z, RSC Adv., 2012, **2**, 9124–9129, **10.1039/c2ra21228e**.
29. R. Ansari, S. Ajori and B. Motevalli, Superlattices Microstruct., 2012, **51**, 274–289, DOI:10.1016/j.spmi.2011.11.019.

30. S. P. Wang, J. G. Guo and L. J. Zhou, Physica E, 2013, **48**, 29–35, DOI:10.1016/j.physe.2012.11.002.

31. M. C. Wang, C. Yan, L. Ma, N. Hu and M. W. Chen, Comput. Mater. Sci., 2012, **54**, 236–239, DOI:10.1016/j.commatsci.2011.10.032.

32. C. Baykasoglu and A. Mugan, Eng. Fract. Mech., 2012, **96**, 241–250, DOI:10.1016/j.engfracmech.2012.08.010.

33. A. Lherbier, S. M.-M. Dubois, X. Declerck, S. Roche, Y.-M. Niquet and J.-C. Charlier, Phys. Rev. Lett., 2011, **106**, 046803, DOI:10.1103/PhysRevLett.106.046803.

34. A. Lherbier, S. M.-M. Dubois, X. Declerck, Y.-M. Niquet, S. Roche and J.-C. Charlier, Phys. Rev. B: Condens. Matter Mater. Phys., 2012, **86**, 075402, DOI:10.1103/PhysRevB.86.075402.

35. K. Ulman and S. Narasimhan, Phys. Rev. B: Condens. Matter Mater. Phys., 2014, **89**, 245429, DOI:10.1103/PhysRevB.89.245429.

36. S. N. Shirodkar and U. V. Waghmare, Phys. Rev. B: Condens. Matter Mater. Phys., 2012, **86**, 165401, DOI:10.1103/PhysRevB.86.165401.

37. J. Kang, J. Bang, B. Ryu and K. J. Chang, Phys. Rev. B: Condens. Matter Mater. Phys., 2008, **77**, 115453, DOI:10.1103/PhysRevB.77.115453.

38. S. Chowdhury, S. Baidya, D. Nafday, S. Halder, M. Kabir, B. Sanyal, T. Saha-Dasgupta, D. Jana and A. Mookerjee, Physica E, 2014, **61**, 191–197, DOI:10.1016/j.physe.2014.04.002.

39. V. H. Crespi, M. L. Cohen and A. Rubio, Phys. Rev. Lett., 1997, **79**, 2093–2096, DOI:10.1103/PhysRevLett.79.2093.

40. J. C. Carlsson and M. Scheffler, Phys. Rev. Lett., 2006, **96**, 046806, DOI:10.1103/PhysRevLett.96.046806.

41. D. Selli and F. Mercuri, Carbon, 2014, **75**, 190–200, DOI:10.1016/j.carbon.2014.03.052.

42. G. Wang, Phys. Chem. Chem. Phys., 2011, **13**, 11939–11945, **10.1039/c1cp20541b**.

43. H. Zeng, J.-P. Leburton, Y. Xu and J. Wei, Nanoscale Res. Lett., 2011, **6**, 254, DOI:10.1186/1556-276X-6-254.

44. Y. Ren and K.-Q. Chen, J. Appl. Phys., 2010, **107**, 044514, DOI:10.1063/1.3309775.

45. P. O. Lehtinen, A. F. Foster, Y. Ma, A. V. Krasheninnikov and R. M. Nieminen, Phys. Rev. Lett., 2004, **93**, 187202, DOI:10.1103/PhysRevLett.93.187202.

46. O. V. Yazyev and L. Helm, Phys. Rev. B: Condens. Matter Mater. Phys., 2007, **75**, 125408, DOI:10.1103/PhysRevB.75.125408.

47. S. Choi, B. W. Jeong, S. Kim and G. Kim, J. Phys.: Condens. Matter, 2008, **20**, 235220, DOI:10.1088/0953-8984/20/23/235220.

48. R. Singh and P. Kroll, J. Phys.: Condens. Matter, 2009, **21**, 196002, DOI:10.1088/0953-8984/21/19/196002.

49. X. Q. Dai, J. H. Zhao, M. H. Xie, Y. N. Tang, Y. H. Li and B. Zhao, Eur. Phys. J. B, 2011, **80**, 343–349, DOI:10.1140/epjb/e2011-10955-x.

50. R. Faccio and A. W. Mombrú, J. Phys.: Condens. Matter, 2012, **24**, 375304, DOI:10.1088/0953-8984/24/37/375304.

51. E. J. G. Santos, S. Riikonen, D. Sánchez-Portal and A. Ayuela, J. Phys. Chem. C, 2012, **116**, 7602–7606, DOI:10.1021/jp300861m.

52. B. Wang and S. T. Pantelides, Phys. Rev. B: Condens. Matter Mater. Phys., 2012, **86**, 165438, DOI:10.1103/PhysRevB.86.165438.

53. V. O. Özçelik, H. H. Gurel and S. Ciraci, Phys. Rev. B: Condens. Matter Mater. Phys., 2013, **88**, 045440, DOI:10.1103/PhysRevB.88.045440.

54. P. A. Denis, R. Faccio and F. Iribarne, Comput. Theor. Chem., 2012, **995**, 1–7, DOI:10.1016/j.comptc.2012.06.014.

55. P. Lu, Z. Zhang and W. Guo, Phys. Lett. A, 2009, **373**, 3354–3358, DOI:10.1016/j.physleta.2009.07.031.

56. J.-W. Jiang, B.-S. Wang and J.-S. Wang, Appl. Phys. Lett., 2011, **98**, 113114, DOI:10.1063/1.3567768.

57. Z. G. Fthenakis, Z. Zhu and D. Tománek, Phys. Rev. B: Condens. Matter Mater. Phys., 2014, **89**, 125421, DOI:10.1103/PhysRevB.89.125421.

58. Y. Y. Zhang, Y. Cheng, Q. X. Pei, C. M. Wang and Y. Xiang, Phys. Lett. A, 2012, **376**, 3668–3672, DOI:10.1016/j.physleta.2012.10.048.

59. Z. X. Xie, K.-Q. Chen and W. Duan, J. Phys.: Condens. Matter, 2011, **23**, 315302, DOI:10.1088/0953-8984/23/31/315302.

60. J. Haskins, A. Kınacı, C. Sevik, H. Sevinçli, G. Cuniberti and T. Çağın, ACS Nano, 2011, **5**, 3779–3787, DOI:10.1021/nn200114p.

61. D. Yang, F. Ma, Y. Sun, T. Hu and K. Xu, Appl. Surf. Sci., 2012, **258**, 9926–9931, DOI:10.1016/j.apsusc.2012.06.052.

62. A. V. Krasheninnikov, K. Nordlund and J. Keinonen, Phys. Rev. B: Condens. Matter Mater. Phys., 2002, **65**, 165423, DOI:10.1103/PhysRevB.65.165423.

63. J.-W. Jiang and J.-S. Wang, J. Appl. Phys., 2010, **108**, 054303, DOI:10.1063/1.3481004.

64. B. Song, G. F. Schneider, Q. Xu, G. Pandraud, C. Dekker and H. Zandbergen, Nano Lett., 2011, **11**, 2247–2250, DOI:10.1021/nl200369r.

65. A. W. Robertson, C. S. Allen, Y. A. Wu, K. He, J. Olivier, J. Neethling, A. I. Kirkland and J. H. Warner, Nat. Commun., 2012, **3**, 1144–1151, DOI:10.1038/ncomms2141.

66. A. W. Robertson, G.-D. Lee, K. He, E. Yoon, A. I. Kirkland and J. H. Warner, Nano Lett., 2014, **14**, 3972–3980, DOI:10.1021/nl501320a.

67. A. Santana, A. M. Popov and E. Bichoutskaia, Chem. Phys. Lett., 2013, **557**, 80–87, DOI:10.1016/j.cplett.2012.11.077.

68. X. Gao, L. Liu, S. Irle and S. Nagase, Angew. Chem., Int. Ed., 2010, **49**, 3200–3202, DOI:10.1002/anie.200907347.

69. M. W. C. Dharma-wardana and M. Z. Zgierski, Physica E, 2008, **41**, 80–83, DOI:10.1016/j.physe.2008.06.007.

70. I. V. Lebedeva, A. A. Knizhnik, A. A. Bagatur'yants and B. V. Potapkin, Physica E, 2008, **40**, 2589–2595, DOI:10.1016/j.physe.2007.09.155.

71. I. V. Lebedeva, A. A. Knizhnik and B. V. Potapkin, Russ. J. Phys. Chem. B, 2007, **1**, 675–684, DOI:10.1134/S1990793107060255.
72. A. W. Robertson and J. H. Warner, Nanoscale, 2013, **5**, 4079–4093, **10.1039/c3nr00934c**.
73. Y. C. Cheng, H. T. Wang, Z. Y. Zhu, Y. H. Zhu, Y. Han, X. X. Zhang and U. Schwingenschlögl, Phys. Rev. B: Condens. Matter Mater. Phys., 2012, **85**, 073406, DOI:10.1103/PhysRevB.85.073406.
74. Z. Liu, Y.-C. Lin, C.-C. Lu, C.-H. Yeh, P.-W. Chiu, S. Iijima and K. Suenaga K, Nat. Commun., 2014, **5**, 4055, DOI:10.1038/ncomms5055.
75. P. Koskinen, S. Malola and H. Häkkinen, Phys. Rev. B: Condens. Matter Mater. Phys., 2009, **80**, 073401, DOI:10.1103/PhysRevB.80.073401.
76. C. D. Reddy, A. Ramasubramaniam, V. B. Shenoy and Y.-W. Zhang, Appl. Phys. Lett., 2009, **94**, 101904, DOI:10.1063/1.3094878.
77. Y. C. Cheng, Z. Y. Zhu and U. Schwingenschlögl, J. Mater. Chem., 2012, **22**, 24676–24680, **10.1039/c2jm34068b**.
78. J. N. B. Rodrigues, P. A. D. Gonçalves, N. F. G. Rodrigues, R. M. Ribeiro, J. M. B. Lopes dos Santos and N. M. R. Peres, Phys. Rev. B: Condens. Matter Mater. Phys., 2011, **84**, 155435, DOI:10.1103/PhysRevB.84.155435.
79. J. Li, Z. Li, G. Zhou, Z. Liu, J. Wu, B.-L. Gu, J. Ihm and W. Duan, Phys. Rev. B: Condens. Matter Mater. Phys., 2010, **82**, 115410, DOI:10.1103/PhysRevB.82.115410.
80. J. Bhattacharjee, J. Chem. Phys., 2012, **137**, 094705, DOI:10.1063/1.4747547.
81. S. M.-M. Dubois, A. Lopez-Bezanilla, A. Cresti, F. Triozon, B. Biel, J.-C. Charlier and S. Roche, ACS Nano, 2010, **4**, 1971–1976, DOI:10.1021/nn100028q.
82. P. Hawkins, M. Begliarbekov, M. Zivkovic, S. Strauf and C. P. Search, J. Phys. Chem. C, 2012, **116**, 18382–18387, DOI:10.1021/jp304676h.
83. S. Ihnatsenka and G. Kirczenow, Phys. Rev. B: Condens. Matter Mater. Phys., 2013, **88**, 125430, DOI:10.1103/PhysRevB.88.125430.
84. L. T. Nguyen, C. H. Pham and V. L. Nguyen, J. Phys.: Condens. Matter, 2011, **23**, 295503, DOI:10.1088/0953-8984/23/29/295503.
85. H. Zheng, J. Zhao, D. Xu, J. Wei and H. Zhang, Eur. Phys. J. B, 2013, **86**, 80, DOI:10.1140/epjb/e2013-30754-7.
86. J. Lan, Y. Cai, G. Zhang, J.-S. Wang and Y.-W. Zhang, J. Phys. D: Appl. Phys., 2014, **47**, 265303, DOI:10.1088/0022-3727/47/26/265303.
87. I. V. Lebedeva, A. A. Knizhnik, A. M. Popov and B. V. Potapkin, J. Phys. Chem. C, 2012, **116**, 6572–6584, DOI:10.1021/jp212165g.
88. G. Zheng, S. Irle, M. Elstner and K. Morokuma, J. Phys. Chem. A, 2004, **108**, 3182–3194, DOI:10.1021/jp0373090.
89. M. Volpe and F. Cleri, J. Chem. Phys., 2001, **115**, 3308–3314, DOI:10.1063/1.1387448.
90. S. S. Han, K. S. Lee and H. M. Lee, Chem. Phys. Lett., 2004, **383**, 321–325, DOI:10.1016/j.cplett.2003.11.033.
91. L. Meng, J. Jiang, J. Wang and F. Ding, J. Phys. Chem. C, 2014, **118**, 720–724, DOI:10.1021/jp409471a.
92. G.-D. Lee, C. Z. Wang, E. Yoon, N. M. Hwang and K. M. Ho, Appl. Phys. Lett., 2008, **92**, 043104, DOI:10.1063/1.2837632.
93. G.-D. Lee, C. Z. Wang, E. Yoon, N. M. Hwang and K. M. Ho, Phys. Rev. B: Condens. Matter Mater. Phys., 2007, **76**, 165413, DOI:10.1103/PhysRevB.76.165413.
94. I. V. Lebedeva, T. W. Chamberlain, A. M. Popov, A. A. Knizhnik, T. Zoberbier, J. Biskupek, U. Kaiser and A. N. Khlobystov, Nanoscale, 2014,**6**, 14877, **10.1039/c4nr05006a**.
95. S. Haghighatpanah, A. Börjesson, H. Amara, C. Bichara and K. Bolton, Phys. Rev. B: Condens. Matter Mater. Phys., 2012, **85**, 205448, DOI:10.1103/PhysRevB.85.205448.
96. G.-D. Lee, C. Z. Wang, E. Yoon, N.-M. Hwang and K. M. Ho, Phys. Rev. B: Condens. Matter Mater. Phys., 2010, **81**, 195419, DOI:10.1103/PhysRevB.81.195419.
97. A. V. Krasheninnikov and F. Banhart, Nat. Mater., 2007, **6**, 723–733, DOI:10.1038/nmat1996.
98. A. V. Krasheninnikov and K. Nordlund, J. Appl. Phys., 2010, **107**, 071301, DOI:10.1063/1.3318261.
99. F. Banhart, J. Kotakoski and A. V. Krasheninnikov, ACS Nano, 2011, **5**, 26–41, DOI:10.1021/nn102598m.
100. H. Terrones, R. Lu, M. Terrones and M. S. Dresselhaus, Rep. Prog. Phys., 2012, **75**, 062501, DOI:10.1088/0034-4885/75/6/062501.
101. A. Bachmatiuk, J. Zhao, S. M. Gorantla, I. G. G. Martinez, J. Wiedermann, C. Lee, J. Eckert and M. H. Rummeli, Small, 2015, **5**, 515–542, DOI:10.1002/smll.201401804.
102. P. Hohenberg and W. Kohn, Phys. Rev., 1964, **136**, B864–B871, DOI:10.1103/PhysRev.136.B864.
103. W. Kohn and L. J. Sham, Phys. Rev., 1965, **140**, A1133–A1138, DOI:10.1103/PhysRev.140.A1133.
104. J. P. Perdew and A. Zunger, Phys. Rev. B: Condens. Matter Mater. Phys., 1981, **23**, 5048–5079, DOI:10.1103/PhysRevB.23.5048.
105. J. P. Perdew, in Electronic Structure of Solids '91, ed. P. Ziesche and H. Eschrig, Akademie Verlag, Berlin, 1991, pp. 11–20 Search .
106. J. P. Perdew, J. A. Chevary, S. H. Vosko, K. A. Jackson, M. R. Pederson, D. J. Singh and C. Fiolhais, Phys. Rev. B: Condens. Matter Mater. Phys., 1992, **46**, 6671–6687, DOI:10.1103/PhysRevB.46.6671.
107. J. P. Perdew, K. Burke and M. Ernzerhof, Phys. Rev. Lett., 1996, **77**, 3865–3868, DOI:10.1103/PhysRevLett.77.3865.
108. J. P. Perdew, K. Burke and M. Ernzerhof, Phys. Rev. Lett., 1997, **78**, 1396, DOI:10.1103/PhysRevLett.78.1396.
109. C. Adamo and V. Barone, J. Chem. Phys., 1999, **110**, 6158–6170, DOI:10.1063/1.478522.
110. S. F. Boys and F. Bernardi, Mol. Phys., 1970, **19**, 553–566, DOI:10.1080/00268977000101561.
111. E. Ruiz, A. Rodríguez-Fortea, J. Tercero, T. Cauchy and C. Massobrio, J. Chem. Phys., 2005, **123**, 074102, DOI:10.1063/1.1999631.

112. Y. Zhao and D. G. Truhlar, J. Chem. Phys., 2006, **125**, 194101, DOI:10.1063/1.2370993.

113. Y. Zhao and D. G. Truhlar, Acc. Chem. Res., 2008, **41**, 157–167, DOI:10.1021/ar700111a.

114. M. Dion, H. Rydberg, E. Schröder, D. C. Langreth and B. I. Lundqvist, Phys. Rev. Lett., 2004, **92**, 246401, DOI:10.1103/PhysRevLett.92.246401.

115. S. Grimme, J. Comput. Chem., 2004, **25**, 1463–1473, DOI:10.1002/jcc.20078.

116. S. Grimme, J. Comput. Chem., 2006, **27**, 1787–1799, DOI:10.1002/jcc.20495.

117. T. Frauenheim, G. Seifert, M. Elstner, Z. Hajnal, G. Jungnickel, D. Porezag, S. Suhai and S. Scholz, Phys. Status Solidi B, 2000, **217**, 41–62, DOI:10.1002/(SICI)1521-3951(200001)217:1<41::AID-PSSB41>3.0.CO;2-V.

118. M. Elstner, D. Porezag, G. Jungnickel, J. Elsner, M. Haugk, T. Frauenheim, S. Suhai and G. Seifert, Phys. Rev. B: Condens. Matter Mater. Phys., 1998, **58**, 7260–7268, DOI:10.1103/PhysRevB.58.7260.

119. C. H. Xu, C. L. Fu and D. F. Pedraza, Phys. Rev. B: Condens. Matter Mater. Phys., 1993, **48**, 13273–13279, DOI:10.1103/PhysRevB.48.13273.

120. T. Frauenheim, G. Seifert, M. Elstner, T. Niehaus, C. Köhler, M. Amkreutz, M. Sternberg, Z. Hajnal, A. Di Carlo and S. Suhai, J. Phys.: Condens. Matter, 2002, **14**, 3015–3047, DOI:10.1088/0953-8984/14/11/313.

121. http://www.dftb-plus.info/ .

122. M. S. Tang, C. Z. Wang, C. T. Chan and K. M. Ho, Phys. Rev. B: Condens. Matter Mater. Phys., 1996, **53**, 979–982, DOI:10.1103/PhysRevB.53.979.

123. A. J. Lu and B. C. Pan, Phys. Rev. Lett., 2004, **92**, 105504, DOI:10.1103/PhysRevLett.92.105504.

124. G.-D. Lee, C. Z. Wang, E. Yoon, N.-M. Hwang and K. M. Ho, Phys. Rev. B: Condens. Matter Mater. Phys., 2006, **74**, 245411, DOI:10.1103/PhysRevB.74.245411.

125. P. M. Morse, Phys. Rev., 1929, **34**, 57–64, DOI:10.1103/PhysRev.34.57.

126. G. C. Abell, Phys. Rev. B: Condens. Matter Mater. Phys., 1985, **31**, 6184–6196, DOI:10.1103/PhysRevB.31.6184.

127. J. Tersoff, Phys. Rev. Lett., 1988, **61**, 2879–2882, DOI:10.1103/PhysRevLett.61.2879.

128. D. W. Brenner, Phys. Rev. B: Condens. Matter Mater. Phys., 1990, **42**, 9458–9471, DOI:10.1103/PhysRevB.42.9458.

129. D. W. Brenner, O. A. Shenderova, J. A. Harrison, S. J. Stuart, B. Ni and S. B. Sinnott, J. Phys.: Condens. Matter, 2002, **14**, 783–802, DOI:10.1088/0953-8984/14/4/312.

130. S. J. Stuart, A. B. Tutein and J. A. Harrison, J. Chem. Phys., 2000, **112**, 6472–6486, DOI:10.1063/1.481208.

131. K. Albe, K. Nordlund and R. S. Averback, Phys. Rev. B: Condens. Matter Mater. Phys., 2002, **65**, 195124, DOI:10.1103/PhysRevB.65.195124.

132. K. O. E. Henriksson and K. Nordlund, Phys. Rev. B: Condens. Matter Mater. Phys., 2009, **79**, 144107, DOI:10.1103/PhysRevB.79.144107.

133. A. S. Sinitsa, I. V. Lebedeva, A. A. Knizhnik, A. M. Popov, S. T. Skowron and E. Bichoutskaia, Dalton Trans., 2014, **43**, 7499–7513, **10.1039/c3dt53385a**.

134. A. C. T. van Duin, S. Dasgupta, F. Lorant and W. A. Goddard III, J. Phys. Chem. A, 2001, **105**, 9396–9409, DOI:10.1021/jp004368u.

135. K. D. Nielson, A. C. T. van Duin, J. Oxgaard, W. Q. Deng and W. A. Goddard III, J. Phys. Chem. A, 2005, **109**, 493–499, DOI:10.1021/jp046244d.

136. M. J. Buehler, A. C. T. van Duin and W. A. Goddard III, Phys. Rev. Lett., 2006, **96**, 095505, DOI:10.1103/PhysRevLett.96.095505.

137. C. Peng and H. B. Schlegel, Isr. J. Chem., 1993, **33**, 449–454, DOI:10.1002/ijch.199300051.

138. Ge. G. Samsonidze, G. G. Samsonidze and B. I. Yakobson, Comput. Mater. Sci., 2002, **23**, 62–72, DOI:10.1016/S0927-0256(01)00220-8.

139. G. Henkelman, B. P. Uberuaga and H. A. Jonsson, J. Chem. Phys., 2000, **113**, 9901–9904, DOI:10.1063/1.1329672.

140. E. Bitzek, P. Koskinen, F. Gähler, M. Moseler and P. Gumbsh, Phys. Rev. Lett., 2006, **97**, 170201, DOI:10.1103/PhysRevLett.97.170201.

141. D. Passerone and M. Parrinello, Phys. Rev. Lett., 2001, **87**, 108302, DOI:10.1103/PhysRevLett.87.108302.

142. A. T. Lee, B. Ryu, I.-H. Lee and K. J. Chang, J. Phys.: Condens. Matter, 2014, **26**, 115303, DOI:10.1088/0953-8984/26/11/115303.

143. M. P. Allen and D. J. Tildesley, Computer simulation of liquids, Clarendon Press, Oxford, 1987 Search .

144. M. J. Field, A practical introduction to the simulation of molecular systems, Cambridge University Press, Cambridge, 2007 Search .

145. K. Binder and D. Heermann, Monte Carlo simulation in statistical physics - an introduction, Springer-Verlag, Berlin Heidelberg New York, 2010 Search .

146. J. C. Meyer, F. Eder, S. Kurasch, V. Skakalova, J. Kotakoski, H. J. Park, S. Roth, A. Chuvilin, S. Eyhusen, G. Benner, A. V. Krasheninnikov and U. Kaiser, Phys. Rev. Lett., 2012, **108**, 196102, DOI:10.1103/PhysRevLett.108.196102.

147. J. Kotakoski, J. C. Meyer, S. Kurasch, D. Santos-Cottin, U. Kaiser and A. V. Krasheninnikov, Phys. Rev. B: Condens. Matter Mater. Phys., 2011,**83**, 245420, DOI:10.1103/PhysRevB.83.245420.

148. J. Kotakoski, D. Santos-Cottin and A. V. Krasheninnikov, ACS Nano, 2012, **6**, 671–676, DOI:10.1021/nn204148h.

149. A. Zobelli, A. Gloter, C. P. Ewels, G. Seifert and C. Colliex, Phys. Rev. B: Condens. Matter Mater. Phys., 2007, **75**, 245402, DOI:10.1103/PhysRevB.75.245402.

150. J. Kotakoski, C. H. Jin, O. Lehtinen, K. Suenaga and A. V. Krasheninnikov, Phys. Rev. B: Condens. Matter Mater. Phys., 2010, **82**, 113404, DOI:10.1103/PhysRevB.82.113404.

151. A. Santana, A. Zobelli, J. Kotakoski, A. Chuvilin and E. Bichoutskaia, Phys. Rev. B: Condens. Matter Mater. Phys., 2013, **87**, 094110, DOI:10.1103/PhysRevB.87.094110.

152. E. Kaxiras and K. C. Pandey, Phys. Rev. Lett., 1988, **61**, 2693–2697, DOI:10.1103/PhysRevLett.61.2693.

153. A. A. El-Barbary, R. H. Telling, C. P. Ewels, M. I. Heggie and P. R. Briddon, Phys. Rev. B: Condens. Matter Mater. Phys., 2003, **68**, 144107, DOI:10.1103/PhysRevB.68.144107.

154. J. Kotakoski, A. V. Krasheninnikov and K. Nordlund, Phys. Rev. B: Condens. Matter Mater. Phys., 2006, **74**, 245420, DOI:10.1103/PhysRevB.74.245420.

155. A. V. Krasheninnikov, P. O. Lehtinen, A. S. Foster and R. M. Nieminen, Chem. Phys. Lett., 2006, **418**, 132–136, DOI:10.1016/j.cplett.2005.10.106.

156. M. Saito, K. Yamashita and T. Oda, Jpn. J. Appl. Phys., 2007, **46**, L1185–1187, DOI:10.1143/JJAP.46.L1185.

157. H. Zhang, M. Zhao, X. Yang, H. Xia, X. Liu and Y. Xia, Diamond Relat. Mater., 2010, **19**, 1240–1244, DOI:10.1016/j.diamond.2010.06.010.

158. B. R. K. Nanda, M. Sherafati, Z. S. Popović and S. Satpathy, New J. Phys., 2012, **14**, 083004, DOI:10.1088/1367-2630/14/8/083004.

159. A. W. Robertson, B. Montanari, K. He, C. S. Allen, Y. A. Wu, N. M. Harrison, A. I. Kirkland and J. H. Warner, ACS Nano, 2013, **7**, 4495–4502, DOI:10.1021/nn401113r.

160. C. D. Latham, M. I. Heggie, M. Alatalo, S. O. Öberg and P. R. Briddon, J. Phys.: Condens. Matter, 2013, **25**, 135403, DOI:10.1088/0953-8984/25/13/135403.

161. T. Trevethan, C. D. Latham, M. I. Heggie, P. R. Briddon and M. J. Rayson, Nanoscale, 2014, **6**, 2978–2986, **10.1039/c3nr06222h**.

162. L. Wu, T. Hou, Y. Li, K. S. Chan and S.-T. Lee, J. Phys. Chem. C, 2013, **117**, 17066–17072, DOI:10.1021/jp405130c.

163. M. T. Lusk and L. D. Carr, Phys. Rev. Lett., 2008, **100**, 175503, DOI:10.1103/PhysRevLett.100.175503.

164. A. Zobelli, V. Ivanovskaya, P. Wagner, I. Suarez-Martinez, A. Yaya and C. P. Ewels, Phys. Status Solidi B, 2012, **249**, 276–282, DOI:10.1002/pssb.201100630.

165. G. Teobaldi, H. Ohnishi, K. Tanimura and A. L. Shluger, Carbon, 2010, **48**, 4145–4161, DOI:10.1016/j.carbon.2010.07.029.

166. L. Li, S. Reich and J. Robertson, Phys. Rev. B: Condens. Matter Mater. Phys., 2005, **72**, 184109, DOI:10.1103/PhysRevB.72.184109.

167. J. Kotakoski, A. V. Krasheninnikov, U. Kaiser and J. C. Meyer, Phys. Rev. Lett., 2011, **106**, 105505, DOI:10.1103/PhysRevLett.106.105505.

168. J. C. Meyer, C. Kisielowski, R. Erni, M. D. Rossell, M. F. Crommie and A. Zettl, Nano Lett., 2008, **8**, 3582–3586, DOI:10.1021/nl801386m.

169. Z. S. Popović, B. R. K. Nanda and S. Satpathy, Phys. Rev. B: Condens. Matter Mater. Phys., 2012, **86**, 85458, DOI:10.1103/PhysRevB.86.085458.

170. Y. Kim, J. Ihm, E. Yoon and G.-D. Lee, Phys. Rev. B: Condens. Matter Mater. Phys., 2011, **84**, 075445, DOI:10.1103/PhysRevB.84.075445.

171. M. M. Ugeda, I. Brihuega, F. Hiebel, P. Mallet, J.-Y. Veuillen, J. M. Gómez-Rodríguez and F. Ynduráin, Phys. Rev. B: Condens. Matter Mater. Phys., 2012, **85**, 121402, DOI:10.1103/PhysRevB.85.121402.

172. H. Wang, Q. Wang, Y. Cheng, K. Li, Y. Yao, Q. Zhang, C. Dong, P. Wang, U. Schwingenschlögl, W. Yang and X. X. Zhang, Nano Lett., 2012,**12**, 141–144, DOI:10.1021/nl2031629.

173. G.-D. Lee, C. Z. Wang, E. Yoon, N.-M. Hwang, D.-Y. Kim and K. M. Ho, Phys. Rev. Lett., 2005, **95**, 205501, DOI:10.1103/PhysRevLett.95.205501.

174. J.-M. Leyssale and G. L. Vignoles, J. Phys. Chem. C, 2014, **118**, 8200–8216, DOI:10.1021/jp501028n.

175. L. Wang, F. Yan, H. L. W. Chan and F. Ding, Nanoscale, 2012, **4**, 7489–7493 **RSC**.

176. B. W. Jeong, J. Ihm and G.-D. Lee, Phys. Rev. B: Condens. Matter Mater. Phys., 2008, **78**, 165403, DOI:10.1103/PhysRevB.78.165403.

177. E. Asari, M. Kitajima, K. G. Nakamura and T. Kawabe, Phys. Rev. B: Condens. Matter Mater. Phys., 1993, **47**, 11143, DOI:10.1103/PhysRevB.47.11143.

178. J. I. Paredes, P. Solís-Fernández, A. Martínez-Alonso and J. M. D. Tascón, J. Phys. Chem. C, 2009, **113**, 10249–10255, DOI:10.1021/jp901578c.

179. P. O. Lehtinen, A. S. Foster, A. Ayuela, A. V. Krasheninnikov, K. Nordlund and R. M. Nieminen, Phys. Rev. Lett., 2003, **91**, 017202, DOI:10.1103/PhysRevLett.91.017202.

180. I. C. Gerber, A. V. Krasheninnikov, A. S. Foster and R. M. Nieminen, New J. Phys., 2010, **12**, 113021, DOI:10.1088/1367-2630/12/11/113021.

181. L. Tsetserisa and S. T. Pantelides, Carbon, 2009, **47**, 901–908, DOI:10.1016/j.carbon.2008.12.002.

182. A. V. Krasheninnikov, K. Nordlund, P. O. Lehtinen, A. S. Foster, A. Ayuela and R. M. Nieminen, Carbon, 2004, **42**, 1021–1025, DOI:10.1016/j.carbon.2003.12.025.

183. A. V. Krasheninnikov, K. Nordlund, P. O. Lehtinen, A. S. Foster, A. Ayuela and R. M. Nieminen, Phys. Rev. B: Condens. Matter Mater. Phys., 2004, **69**, 073402, DOI:10.1103/PhysRevB.69.073402.

184. A. Hashimoto, K. Suenaga, A. Gloter, K. Urita and S. Iijima, Nature, 2004, **430**, 870–873, DOI:10.1038/nature02817.

185. U. Bangert, M. H. Gass, A. L. Bleloch, R. R. Nair and J. Eccles, Phys. Status Solidi A, 2009, **206**, 2115–2119, DOI:10.1002/pssa.200982207.

186. J. Ma, D. Alfè, A. Michaelides and E. Wang, Phys. Rev. B: Condens. Matter Mater. Phys., 2009, **80**, 033407, DOI:10.1103/PhysRevB.80.033407.

187. L. Chen, Y. Ouyang, Y. Wang, Y. Sun and H. Pan, Physica E, 2010, **43**, 593–597, DOI:10.1016/j.physe.2010.10.003.

188. Q. Zhao, M. Buongiorno Nardelli and J. Bernholc, Phys. Rev. B: Condens. Matter Mater. Phys., 2002, **65**, 144105, DOI:10.1103/PhysRevB.65.144105.

189. B. C. Pan, W. S. Yang and J. Yang, Phys. Rev. B: Condens. Matter Mater. Phys., 2000, **62**, 12652–12655, DOI:10.1103/PhysRevB.62.12652.

190. E. Ertekin, M. S. Daw and D. C. Chrzan, Philos. Mag. Lett., 2008, **88**, 159–167, DOI:10.1080/09500830701822104.

191. E. Ertekin, D. C. Chrzan and M. S. Daw, Phys. Rev.

B: Condens. Matter Mater. Phys., 2009, **79**, 155421, DOI:10.1103/PhysRevB.79.155421.

192. S. Bhowmick and U. V. Waghmare, Phys. Rev. B: Condens. Matter Mater. Phys., 2010, **81**, 155416, DOI:10.1103/PhysRevB.81.155416.
193. P. Jensen, J. Gale and X. Blase, Phys. Rev. B: Condens. Matter Mater. Phys., 2002, **66**, 193403, DOI:10.1103/PhysRevB.66.193403.
194. A. J. M. Nascimento and R. W. Nunes, Nanotechnology, 2013, **24**, 435707, DOI:10.1088/0957-4484/24/43/435707.
195. T. Dumitrica and B. I. Yakobson, Appl. Phys. Lett., 2004, **84**, 2775–2777, DOI:10.1063/1.1695630.
196. C. P. Ewels, M. I. Heggie and P. R. Briddon, Chem. Phys. Lett., 2002, **351**, 178–182, DOI:10.1016/S0009-2614(01)01371-9.
197. L. G. Zhou and S.-Q. Shi, Appl. Phys. Lett., 2003, **83**, 1222–1224, DOI:10.1063/1.1599961.
198. C. Wang and Y. H. Ding, J. Mater. Chem. A, 2013, **1**, 1885–1891, **10.1039/c2ta00736c**.
199. Z. Li, P. Dharap, P. Sharma, S. Nagarajaiah and B. I. Yakobson, J. Appl. Phys., 2005, **97**, 074303, DOI:10.1063/1.1870102.
200. Ge. G. Samsonidze, G. G. Samsonidze and B. I. Yakobson, Phys. Rev. Lett., 2002, **88**, 065501, DOI:10.1103/PhysRevLett.88.065501.
201. B. I. Yakobson, G. Samsonidze and G. G. Samsonidze, Carbon, 2000, **38**, 1675–1680, DOI:10.1016/S0008-6223(00)00093-2.
202. M. Buongiorno Nardelli, B. I. Yakobson and J. Bernholc, Phys. Rev. B: Condens. Matter Mater. Phys., 1998, **57**, R4277–R4280, DOI:10.1103/PhysRevB.57.R4277.
203. S. Chen, Buckling and topological defects in graphene and carbon nanotubes, PhD thesis, University of California, Berkeley, 2012 Search .
204. S. Chen, E. Ertekin and D. C. Chrzan, Phys. Rev. B: Condens. Matter Mater. Phys., 2010, **81**, 155417, DOI:10.1103/PhysRevB.81.155417.
205. I. Zsoldos, Nanotechnol., Sci. Appl., 2010, **3**, 101–106, DOI:10.2147/NSA.S13905.
206. B. B. Fan, X. B. Yang and R. Zhang, Phys. Lett. A, 2010, **374**, 2781–2784, DOI:10.1016/j.physleta.2010.04.066.
207. Z. Song, V. I. Artyukhov, B. I. Yakobson and Z. Xu, Nano Lett., 2013, **13**, 1829–1833, DOI:10.1021/nl400542n.
208. R. Grantab, V. B. Shenoy and R. S. Ruoff, Science, 2010, **330**, 946–948, DOI:10.1126/science.1196893.
209. L. Yi, Z. Yin, Y. Zhang and T. Chang, Carbon, 2013, **51**, 373–380, DOI:10.1016/j.carbon.2012.08.069.
210. Y. J. Wei, J. Wu, H. Yin, X. Shi, R. Yang and M. S. Dresselhaus, Nat. Mater., 2012, **11**, 759–763, DOI:10.1038/nmat3370.
211. P. Zhang, P. E. Lammert and V. H. Crespi, Phys. Rev. Lett., 1998, **81**, 5346–5349, DOI:10.1103/PhysRevLett.81.5346.
212. Y. J. Sun, F. Ma, D. Y. Ma, K. W. Xu and P. K. Chu, J. Phys. D: Appl. Phys., 2012, **45**, 305303, DOI:10.1088/0022-3727/45/30/305303.
213. D. Bozovic, M. Bockrath, J. H. Hafner, C. M. Lieber, H. Park and M. Tinkham, Phys. Rev. B: Condens. Matter Mater. Phys., 2003, **67**, 033407, DOI:10.1103/PhysRevB.67.033407.
214. B. G. Demczyk, Y. M. Wang, J. Cumings, M. Hetman, W. Han, A. Zettl and R. O. Ritchie, Mater. Sci. Eng., A, 2002, **334**, 173–178 CrossRef.
215. B. I. Yakobson, Appl. Phys. Lett., 1998, **72**, 918–920, DOI:10.1063/1.120873.
216. M. Buongiorno Nardelli, B. I. Yakobson and J. Bernholc, Phys. Rev. Lett., 1998, **81**, 4656–4659, DOI:10.1103/PhysRevLett.81.4656.
217. F. Ding, K. Jiao, M. Wu and B. I. Yakobson, Phys. Rev. Lett., 2007, **98**, 075503, DOI:10.1103/PhysRevLett.98.075503.
218. J. H. Warner, E. R. Margine, M. Mukai, A. W. Robertson, F. Giustino and A. I. Kirkland, Science, 2012, **337**, 209–212, DOI:10.1126/science.1217529.
219. K. Suenaga, H. Wakabayashi, M. Koshino, Y. Sato, K. Urita and S. Iijima, Nat. Nanotechnol., 2007, **2**, 358–360, DOI:10.1038/nnano.2007.141.
220. J. Guo, J. R. Morris, Y. Ihm, C. I. Contescu, N. C. Gallego, G. Duscher, S. J. Pennycook and M. F. Chisholm, Small, 2012, **8**, 3283–3288, DOI:10.1002/smll.201200894.
221. O. V. Yazyev and S. G. Louie, Phys. Rev. B: Condens. Matter Mater. Phys., 2010, **81**, 195420, DOI:10.1103/PhysRevB.81.195420.
222. T.-H. Liu, G. Gajewski, C.-W. Pao and C.-C. Chang, Carbon, 2011, **49**, 2306–2317, DOI:10.1016/j.carbon.2011.01.063.
223. Y. Liu and B. I. Yakobson, Nano Lett., 2010, **10**, 2178–2183, DOI:10.1021/nl100988r.
224. F. Ding, K. Jiao, Y. Lin and B. I. Yakobson, Nano Lett., 2007, **7**, 681–684, DOI:10.1021/nl0627543.
225. O. Lehtinen, S. Kurasch, A. V. Krasheninnikov and U. Kaiser, Nat. Commun., 2013, **4**, 2089, DOI:10.1038/ncomms3098.
226. J. Y. Huang, S. Chen, S. H. Jo, Z. Wang, D. X. Han, G. Chen, M. S. Dresselhaus and Z. F. Ren, Phys. Rev. Lett., 2005, **94**, 236802, DOI:10.1103/PhysRevLett.94.236802.
227. C. Gómez-Navarro, J. C. Meyer, R. S. Sundaram, A. Chuvilin, S. Kurasch, M. Burghard, K. Kern and U. Kaiser, Nano Lett., 2010, **10**, 1144–1148, DOI:10.1021/nl9031617.
228. G.-D. Lee, E. Yoon, K. He, A. W. Robertson and J. H. Warner, Nanoscale, 2014, **6**, 14836–14844, **10.1039/c4nr04718d**.
229. G.-D. Lee, E. Yoon, N.-M. Hwang, C.-Z. Wang and K.-M. Ho, Appl. Phys. Lett., 2013, **102**, 021603, DOI:10.1063/1.4775671.
230. M. H. Gass, U. Bangert, A. L. Bleloch, P. Wang, R. R. Nair and A. K. Geim, Nat. Nanotechnol., 2008, **3**, 676–681, DOI:10.1038/nnano.2008.280.
231. P. Simonis, C. Goffaux, P. A. Thiry, L. P. Biro, Ph. Lambin and V. Meunier, Surf. Sci., 2002, **511**, 319–322, DOI:10.1016/S0039-6028(02)01511-X.

232. J. Coraux, A. T. N'Diaye, C. Busse and T. Michely, Nano Lett., 2008, **8**, 565–570, DOI:10.1021/nl0728874.

233. Y. Tison, J. Lagoute, V. Repain, C. Chacon, Y. Girard, F. Joucken, R. Sporken, F. Gargiulo, O. V. Yazyev and S. Rousset, Nano Lett., 2014, **14**, 6382–6386, DOI:10.1021/nl502854w.

234. P. Y. Huang, C. S. Ruiz-Vargas, A. M. van der Zande, W. S. Whitney, M. P. Levendorf, J. W. Kevek, S. Garg, J. S. Alden, C. J. Hustedt, Y. Zhu, J. Park, P. L. McEuen and D. A. Muller, Nature, 2011, **469**, 389–393, DOI:10.1038/nature09718.

235. B. Yang, H. Xu, J. Lu and K. P. Loh, J. Am. Chem. Soc., 2014, **136**, 12041–12046, DOI:10.1021/ja5054847.

236. K. Kim, Z. Lee, W. Regan, C. Kisielowski, M. F. Crommie and A. Zettl, ACS Nano, 2011, **5**, 2142–2146, DOI:10.1021/nn1033423.

237. O. V. Yazyev, Solid State Commun., 2012, **152**, 1431–1436, DOI:10.1016/j.ssc.2012.04.045.

238. B. I. Yakobson and F. Ding, ACS Nano, 2011, **5**, 1569–1574, DOI:10.1021/nn200832y.

239. J. Zhang, J. Zhao and J. Lu, ACS Nano, 2012, **6**, 2704–2711, DOI:10.1021/nn3001356.

240. S. Malola, H. Häkkinen and P. Koskinen, Phys. Rev. B: Condens. Matter Mater. Phys., 2010, **81**, 165447, DOI:10.1103/PhysRevB.81.165447.

241. A. Cao and Y. Yuan, Appl. Phys. Lett., 2012, **100**, 211912, DOI:10.1063/1.4722786.

242. V. V. Ivanovskaya, A. Zobelli, P. Wagner, M. I. Heggie, P. R. Briddon, M. J. Rayson and C. P. Ewels, Phys. Rev. Lett., 2011, **107**, 065502, DOI:10.1103/PhysRevLett.107.065502.

243. P. Koskinen, S. Malola and H. Häkkinen, Phys. Rev. Lett., 2008, **101**, 115502, DOI:10.1103/PhysRevLett.101.115502.

244. J. Kunstmann, C. Özdoğan, A. Quandt and H. Fehske, Phys. Rev. B: Condens. Matter Mater. Phys., 2011, **83**, 045414, DOI:10.1103/PhysRevB.83.045414.

245. J. M. H. Kroes, M. A. Akhukov, J. H. Los, N. Pineau and A. Fasolino, Phys. Rev. B: Condens. Matter Mater. Phys., 2011, **83**, 165411, DOI:10.1103/PhysRevB.83.165411.

246. T. Wassmann, A. P. Seitsonen, A. M. Saitta, M. Lazzeri and F. Mauri, Phys. Rev. Lett., 2008, **101**, 096402, DOI:10.1103/PhysRevLett.101.096402.

247. B. Huang, M. Liu, N. Su, J. Wu, W. Duan, B. Gu and F. Liu, Phys. Rev. Lett., 2009, **102**, 166404, DOI:10.1103/PhysRevLett.102.166404.

248. C. K. Gan and D. J. Srolovitz, Phys. Rev. B: Condens. Matter Mater. Phys., 2010, **81**, 125445, DOI:10.1103/PhysRevB.81.125445.

249. J. M. Carlsson, L. M. Ghiringhelli and A. Fasolino, Phys. Rev. B: Condens. Matter Mater. Phys., 2011, **84**, 165423, DOI:10.1103/PhysRevB.84.165423.

250. Y. Wei, J. Wu, H. Yin, X. Shi, R. Yang and M. Dresselhaus, Nat. Mater., 2012, **11**, 759–763, DOI:10.1038/nmat3370.

251. E. Cockayne, G. M. Rutter, N. P. Guisinger, J. N. Crain, P. N. First and J. A. Stroscio, Phys. Rev. B: Condens. Matter Mater. Phys., 2011, **83**, 195425, DOI:10.1103/PhysRevB.83.195425.

252. G. D. Lee, E. Yoon, C.-Z. Wang and K.-M. Ho, J. Phys.: Condens. Matter, 2013, **25**, 155301, DOI:10.1088/0953-8984/25/15/155301.

253. L. L. Song, X. H. Zheng, R. L. Wang and Z. Zeng, J. Phys. Chem. C, 2010, **114**, 12145–12150, DOI:10.1021/jp1040025.

254. J. H. Warner, Y.-C. Lin, K. He, M. Koshino and K. Suenaga, Nano Lett., 2014, **14**, 6155–6159, DOI:10.1021/nl5023095.

255. K. Kim, S. Coh, C. Kisielowski, M. F. Crommie, S. G. Louie, M. L. Cohen and A. Zettl, Nat. Commun., 2013, **4**, 2723, DOI:10.1038/ncomms3723.

256. C.-I. Chia and V. H. Crespi, Phys. Rev. Lett., 2012, **109**, 076802, DOI:10.1103/PhysRevLett.109.076802.

257. V. I. Artyukhov, Y. Liu and B. I. Yakobson, Proc. Natl. Acad. Sci. U. S. A., 2012, **109**, 15136–15140, DOI:10.1073/pnas.1207519109.

258. J. Lee, Z. Yang, W. Zhou, S. J. Pennycook, S. T. Pantelides and M. F. Chisholm, Proc. Natl. Acad. Sci. U. S. A., 2014, **111**, 7522–7526, DOI:10.1073/pnas.1400767111.

259. J. H. Warner, Z. Liu, K. He, A. W. Robertson and K. Suenaga, Nano Lett., 2013, **13**, 4820–4826, DOI:10.1021/nl402514c.

260. Ç. Ö. Girit, J. C. Meyer, R. Erni, M. D. Rossell, C. Kisielowski, L. Yang, C.-H. Park, M. F. Crommie, M. L. Cohen, S. G. Louie and A. Zettl,Science, 2009, **323**, 1705–1708, DOI:10.1126/science.1166999.

261. S. S. Terdalkar, S. Huang, H. Yuan, J. J. Rencis, T. Zhu and S. Zhang, Chem. Phys. Lett., 2010, **494**, 218–222, DOI:10.1016/j.cplett.2010.05.090.

262. J. Zhu and D. Shi, J. Appl. Phys., 2011, **110**, 104311, DOI:10.1063/1.3662183.

263. E. Erdogan, I. Popov, C. G. Rocha, G. Cuniberti, S. Roche and G. Seifert, Phys. Rev. B: Condens. Matter Mater. Phys., 2011, **83**, 041401, DOI:10.1103/PhysRevB.83.041401.

264. A. W. Robertson, G. D. Lee, K. He, E. Yoon, A. I. Kirland and J. H. Warner, Nano Lett., 2014, **14**, 1634–1642, DOI:10.1021/nl500119p.

265. A. Zobelli, A. Gloter, C. P. Ewels, G. Seifert and C. Colliex, Phys. Rev. B: Condens. Matter Mater. Phys., 2007, **75**, 245402, DOI:10.1103/PhysRevB.75.245402.

266. Z. Wang, Y. G. Zhou, J. Bang, M. P. Prange, S. B. Zhang and F. Gao, J. Phys. Chem. C, 2012, **116**, 16070–16079, DOI:10.1021/jp303905u.

267. J. C. Meyer, F. Eder, S. Kurasch, V. Skakalova, J. Kotakoski, H. J. Park, S. Roth, A. Chuvilin, S. Eyhusen, G. Benner, A. V. Krasheninnikov and U. Kaiser, Phys. Rev. Lett., 2013, **110**, 239902, DOI:10.1103/PhysRevLett.110.239902.

268. A. Krasheninnikov, F. Banhart, J. Li, A. Foster and R. Nieminen, Phys. Rev. B: Condens. Matter Mater. Phys., 2005, **72**, 125428, DOI:10.1103/PhysRevB.72.125428.

269. L. Tsetseris, B. Wang and S. T. Pantelides, Phys. Rev. B: Condens. Matter Mater. Phys., 2014, **89**, 035411, DOI:10.1103/PhysRevB.89.035411.

270. K. He, A. W. Robertson, S. Lee, E. Yoon, G. D. Lee and J. H. Warner, ACS Nano, 2014, **8**, 12272–12279, DOI:10.1021/nn504471m.

271. K. Jin, H. Y. Xiao, Y. Zhang and W. J. Weber, Appl. Phys. Lett., 2014, **104**, 203106, DOI:10.1063/1.4879258.

CHAPTER 2

THE MOLECULE CLOUD - COMPACT VISUALIZATION OF LARGE COLLECTIONS OF MOLECULES

Peter Ertl and Bernhard Rohde

Keywords: Molecule cloud, Visualization, Scaffold analysis, Chemical databases, Open source

BACKGROUND

One of the most typical tasks that a cheminformatics expert in pharmaceutical or agrochemical industry performs practically daily is analyzing and visualizing large collections of molecules. Typical areas, where this is needed are the analysis of the company compound archive and its enhancement by purchasing additional molecules from commercial compound providers, analysis of high-throughput screening results, design of combinatorial libraries, chemogenomics analysis of bioactivity data and many others [1]. But also researchers in academia are facing similar challenges when they need to process and visualize large molecular databases that have become freely available in the last few years [2] or were even generated in silico [3]. A number of methods to analyze and visualize large collections of molecules exist [4]. Among the most commonly used ones are various clustering techniques, dimensionality reduction approaches [5] or methods based on substructure analysis, particularly the analysis of molecule scaffolds [6]. But even if such sophisticated methods are applied, at the end it is necessary to visually inspect large tables of molecular structure diagrams.

The necessity to analyze and visualize large data sets, is of course, nothing specific to chemistry. Practically all areas of science need to cope with this problem. A simple, but very efficient method to analyze large data sets has been introduced in text processing. A method called "text cloud" or "word cloud" identifies the most common words in a text and displays them in form of a cloud diagram, where the frequency of a word is indicated by the size of the font used to render it. The most common words in the text like "and", "or" or "the" do not contain useful information and are therefore not displayed. This type of diagram allows very efficient and compact visualization of the most important "message" contained in the text (Figure 1). The word cloud graphs summarizing the content of web pages become one of the symbols of Web 2.0. Several types of word cloud diagrams exist, differing in their layout and the way how the text is rendered.

Figure 1 Word cloud diagram created from the text of this article by interactive service [16].

The word cloud diagrams inspired us to develop a method for compact visualization of large molecular data sets. The method is based on the same principle, but instead of words molecular structures are displayed in the graph. The size of molecules corresponds to their frequency in the particular data sets. The largest structures catch the eyes of the observer first and therefore a single look at the diagram provides basic, condensed information about the content of the parent data set. In analogy with the "word cloud" we call this method "Molecule Cloud".

RESULTS AND DISCUSSION

Molecule Cloud Methodology

The principle of the Molecule Cloud method is very simple. The most common substructure features present in the analyzed data set are identified and then displayed in such a way that their size corresponds to their frequency. The most common

Citation: Ertl and Rohde: The Molecule Cloud - compact visualization of large collections of molecules. Journal of Cheminformatics 2012 4:12. doi:10.1186/1758-2946-4-12.

substructures are rendered the largest and therefore immediately catch the eye of an observer. Molecules are represented by their scaffolds, i.e. cores remaining when all non-ring substituents are removed. The concept of scaffold as the central part of a molecule is one of the basic concepts of medicinal chemistry and scaffolds play an important role in several drug discovery techniques like combinatorial chemistry and scaffold hopping [7]. Molecules without any rings are represented by their major chain, i.e. the longest chain, containing the largest number of heteroatoms. Reduction of molecules to scaffolds simplifies the analysis considerably. For example 35 million molecules from the PubChem database are represented by about 3.9 million scaffolds (50% of which are singletons, present only once in the database) and about 150 thousand chains.

Once the most common scaffolds and chains are identified, they need to be visualized with their size scaled according to the frequency in the parent database. It is well known that the frequency of various substructure features like scaffolds, substituents or linkers in molecular databases follows the power law (so called "long tail" distribution) [8]. This means that only few scaffolds in a database are very common, while there are many rare scaffolds including large number of singletons. Before using scaffold frequencies as a scaling factor, they therefore need to be transformed into the logarithmic scale. Benzene is a special case, practically in all large data sets the benzene is clearly the most frequent scaffold. In many cases it is therefore advisable not to display it. Even after logarithmic transformation of frequencies benzene would be disproportionally large, and it would not contribute any useful information. Removal of benzene is similar to the removal of the common stop words in classical text clouds.

According to our experience the optimal number of substructures to be displayed in the Molecule Cloud is between 100 and 250. This number usually contains 30 - 50 large structures, easily recognizable, the rest are smaller structures that optically fill the image. Of course, when displaying the graph in a larger area (for example as a poster) the number of structures that can be displayed is proportionally larger.

The greatest challenge in generating the Molecule Cloud is the esthetically pleasing layout of molecules in the graph. This is done by a two-pass layout algorithm developed specially for this purpose. In the first step molecules are placed in the display area in such a way that their overlap is minimal. The process starts by sorting molecule images according to their size and placing the largest one in the center. Then other molecules, one by one are placed in a loop on a dense grid of predefined layout points and an "overlap score" is calculated for each placement. At the end the molecule is placed at the position with the best score (i.e. position where the overlap with other structures is minimal) and the next molecule is processed. This procedure is illustrated in the Figure 2, where one can see the layout after placement of 10, 25, 50 and 200 molecules. The "overlap score" used to identify the best position to place a molecule is calculated as the sum of overlap areas between molecular frames and the sum of the distances between molecule centers.

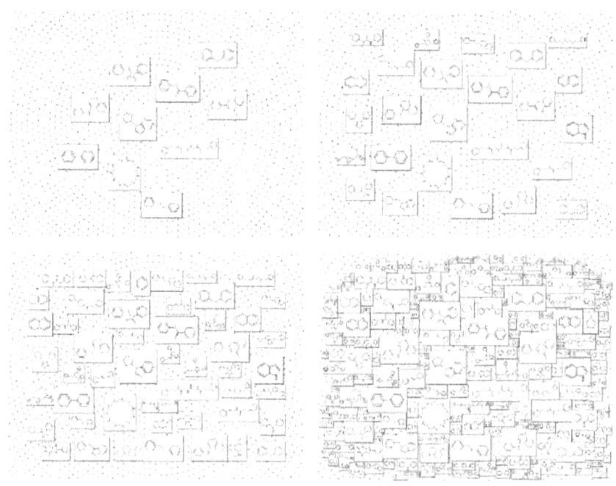

Figure 2 Illustration of the layout procedure.

In most cases already this initial layout provides quite good placement of molecules. To further improve it, a second layout step is performed, namely an iterative optimization loop. In this step molecules, one by one are slightly moved in the direction that improves the total "overlap score". The convergence is fast and after few seconds the final layout is achieved. During the optimization slight repulsive forces are also placed in the corners of the drawing area to provide aesthetically more pleasing "oval" display instead of completely filling the available image rectangle.

The Molecule Cloud layout algorithm has been implemented in Java. The layout itself does not require any "chemical intelligence", it operates simply on the rectangles representing molecules. The required molecular processing capabilities, particularly parsing of SMILES and molecule depiction are defined by a Java interface class and may be implemented by using any cheminformatics toolkit. The program requires as input only a list of SMILES codes of structures to display with their frequencies and desired size of the final image as input. We tested the algorithm using two cheminformatics engines, the depiction engine from Molinspiration [9] and the recently released Novartis open source Avalon Cheminformatics Toolkit [10]. To interested parties the Java source code of the Molecule Cloud layout algorithm is available from the corresponding author under the terms of the BSD license. The distribution provides also instructions how to interface the program with the Avalon Cheminformatics Toolkit.

Application Examples

In this section Molecule Cloud diagrams are presented for several popular publicly available data sets. PubChem [11] is the largest publicly available molecular structure database. In June 2012 it contained nearly 33 million unique structures. The Molecule Cloud of PubChem is shown in Figure 3. In this image, scaffolds of bioactive molecules are indicated by magenta background, where the color intensity is proportional to the ratio between bioactive and all molecules containing this scaffold. Bioactive molecules were identified by the PubChem advanced search as molecules having activity better than 10 μm in any PubChem assay.

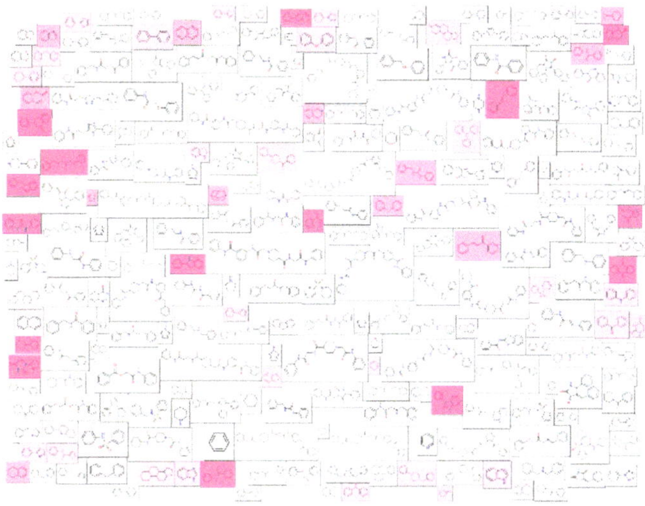

Figure 3 Molecule Cloud for molecules in the PubChem database. Magenta background indicates presence of biological activity.

A very useful resource for drug discovery, particularly for researchers in academia is the ZINC database [12]. ZINC, created and maintained by John Irwin from the University of California, San Francisco is a collection of commercially available compounds that may be used in virtual screening. The Molecule Cloud graph with the most common scaffolds present in about 12 million ZINC structures is shown in Figure 4.

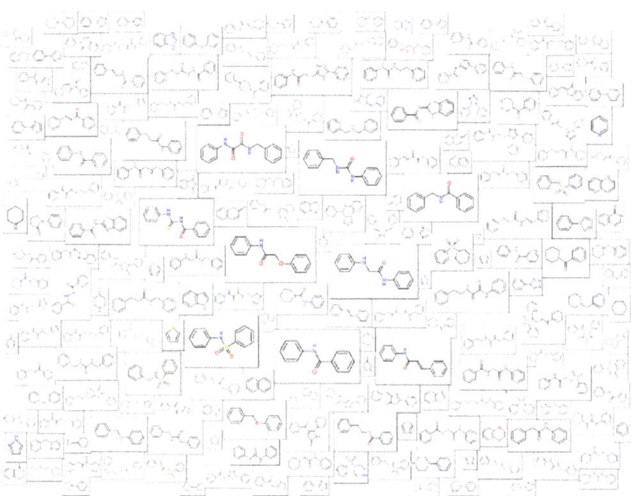

Figure 4 Molecule Cloud of the ZINC database.

The last example shows results of analyses of molecules in the ChEMBL database [13]. ChEMBL is a database of molecules extracted from medicinal chemistry journals and other sources. It also contains biological activities and information about the respective targets. This database is an extremely useful source of information particularly for scientists in academia, providing the type of information that was before available only to researchers in pharmaceutical industry. Molecule Cloud visualizing the most common scaffolds of more than 350,000 bioactive molecules (having activity below 10 μm) from ChEMBL is shown in Figure 5. In this figure, information about targets is also displayed by using color. The following six target classes were considered: GPCRs, ion channels, nuclear receptors, kinases, proteases and other enzymes. The scaffold box is colored if at least 70% of molecules containing this scaffold show activity on particular target class. Scaffolds exhibiting multiple types of activities are not colored. Among these, one can see several known "privileged scaffolds" [14] including biphenyl, indole, quinoline and purine.

Figure 5 Molecule Cloud of the ChEMBL database. Preference of scaffold for particular target class is indicated by color (magenta - GPCRs, blue – kinases, red – proteases, green – other enzymes, brown – nuclear receptors, yellow – ion channels), non-colored scaffolds show activity on multiple targets.

The Molecule Cloud allows, of course, visualization of also other structural elements than scaffolds. This is illustrated in Figure 6, where the most common substituents (up to 15 atoms) from the ChEMBL database are shown. Although majority of the substituents are the same as in the other databases [15], the Molecule Cloud visualization provides clear advantage in comparison with classical molecule grid display.

Figure 6 The most common substituents in the ChEMBL database.

CONCLUSIONS

Molecule Cloud - a method for compact visualization of the typical substructures present in large collections of molecules was described here. The Molecule Cloud graphs allow to recognize scaffolds and other substructure features that are typical for particular data set by a single look. Additional information about molecules may be included by using colors. Further enhancement of generated images, for example by adding textual information, or possibility to click on particular scaffold to see the molecules containing it, should be with help of HTML5 technology easy.

The Java source code of the Molecule Cloud layout algorithm is available from the corresponding author on request. When combined with the recently released Avalon Cheminformatics Toolkit the software allows generation of Molecule Cloud diagrams (Figure 7) with a completely open source solution.

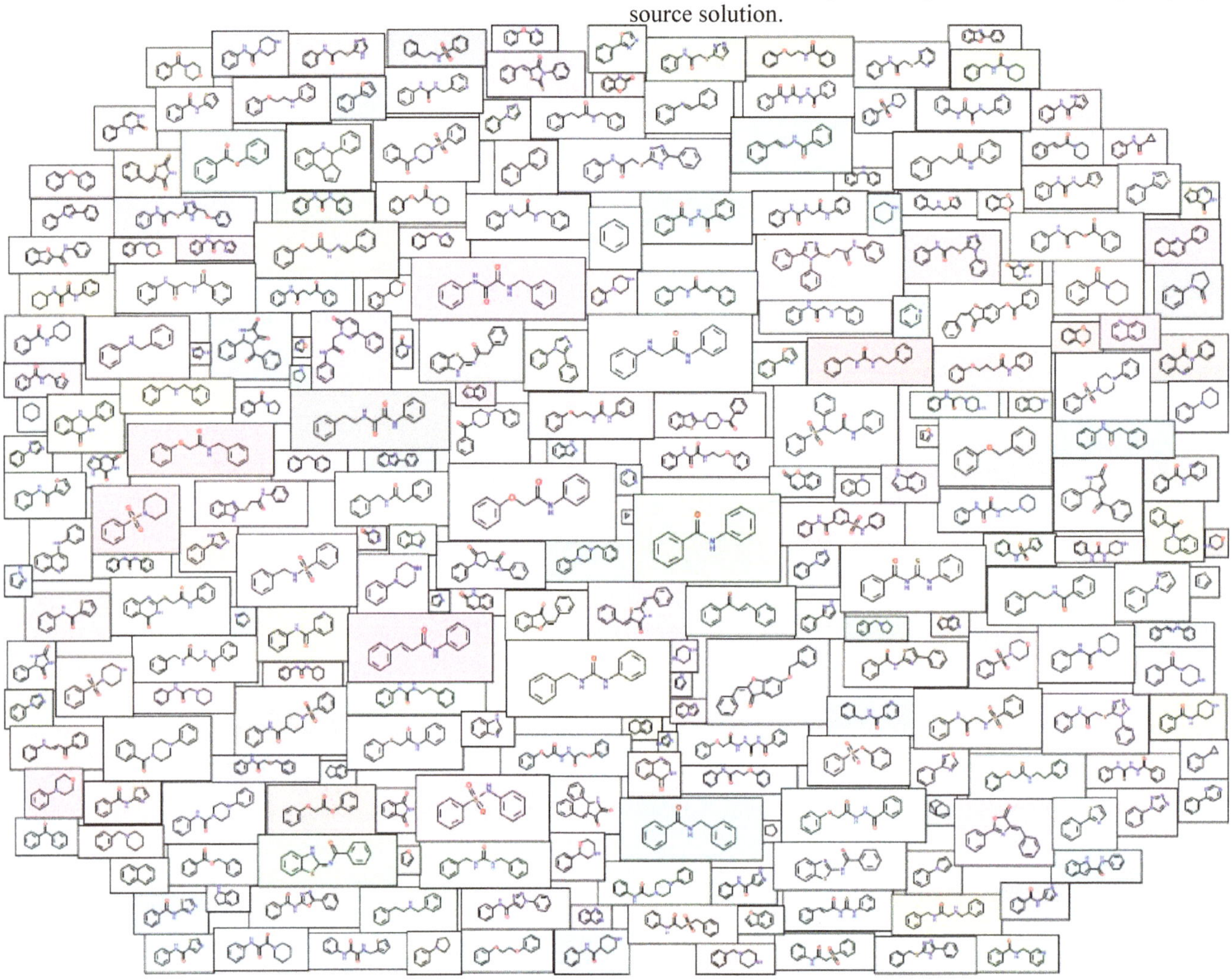

Figure 7 Molecule Cloud diagram of the ZINC database rendered by the Avalon Cheminformatics Toolkit.

AUTHORS' CONTRIBUTIONS

PE developed the idea to display set of molecules as a Molecule Cloud in analogy to the visualization of text information as a word cloud and implemented the layout algorithm in Java. BR is author of the Avalon Cheminformatics Toolkit and he interfaced this toolkit with the Molecule Cloud. Both authors read and approved the final manuscript.

REFERENCES

1. Martin E, Ertl P, Hunt P, Duca J, Lewis R: Gazing into the crystal ball; the future of computer-aided drug design. J Comp-Aided Mol Des. 2011, 26: 77-79.
2. Langdon SR, Brown N, Blagg J: Scaffold diversity of exemplified medicinal chemistry space. J Chem Inf Model. 2011, 26: 2174-2185.
3. Blum LC, Reymond J-C: 970 Million druglike small molecules for virtual screening in the chemical universe database GDB-13. J Am Chem Soc. 2009, 131: 8732-8733. 10.1021/ja902302h.
4. Dubois J, Bourg S, Vrain C, Morin-Allory L: Collections of compounds - how to deal with them?. Cur Comp-Aided Drug Des. 2008, 4: 156-168. 10.2174/157340908785747410.
5. Medina-Franco JL, Martinez-Mayorga K, Giulianotti MA, Houghten RA, Pinilla C: Visualization of the chemical space in drug discovery. Cur Comp-Aided Drug Des. 2008, 4: 322-333. 10.2174/157340908786786010.
6. Schuffenhauer A, Ertl P, Roggo S, Wetzel S, Koch MA, Waldmann H: The Scaffold Tree - visualization of the scaffold universe by hierarchical scaffold classification. J Chem Inf Model. 2007, 47: 47-58. 10.1021/ci600338x.
7. Langdon S, Ertl P, Brown N: Bioisosteric replacement and scaffold hopping in lead generation and optimization. Mol Inf. 2010, 29: 366-385. 10.1002/minf.201000019.
8. Lipkus AH, Yuan Q, Lucas KA, Funk SA, Bartelt WF, Schenck RJ, Trippe AJ: Structural diversity of organic chemistry. A scaffold analysis of the CAS Registry. J Org Chem. 2008, 73: 4443-4451. 10.1021/jo8001276.

9. mib 2010.10, Molinspiration Cheminformatics: http://www.molinspiration.com,

10. Bernhard R: Avalon Cheminformatics Toolkit. http://sourceforge.net/projects/avalontoolkit/,

11. Wang Y, Bolton E, Dracheva S, Karapetyan K, Shoemaker BA, Suzek TO, Wang J, Xiao J, Zhang J, Bryant SH: An overview of the PubChem BioAssay resource. Nucleic Acids Res. 2009, 38: D255-D266.

12. Irwin JJ, Shoichet BK: ZINC−a free database of commercially available compounds for virtual screening. J Chem Inf Model. 2004, 45: 177-182.

13. Gaulton A, Bellis LJ, Bento AP, Chambers J, Davies M, Hersey A, Light Y, McGlinchey S, Michalovich D, Al-Lazikani B, Overington JP: ChEMBL: a large-scale bioactivity database for drug discovery. Nucleic Acids Res. 2012, 40: D1100-D1107. 10.1093/nar/gkr777.

14. Welsch ME, Snyder SA, Stockwell BR: Privileged scaffolds for library design and drug discovery. Curr Opin Chem Biol. 2010, 14: 347-361. 10.1016/j.cbpa.2010.02.018.

15. Ertl P: Cheminformatics analysis of organic substituents: Identification of the most common substituents, calculation of substituent properties, and automatic identification of drug-like bioisosteric groups. J Chem Inf Comp Sci. 2003, 43: 374-380. 10.1021/ci0255782.

16. TagCrowd: http://tagcrowd.com

CHAPTER 3

QUANTUM THEORY OF ATOMS IN MOLECULES

David Ferro-Costas, Ángel Martín Pendás, Leticia González and Ricardo A. Mosquera

INTRODUCTION

Understanding the nature of electronically excited states is fundamental to disciplines ranging from basic synthetic organic and inorganic chemistry to nanotechnology and biology. Photoinduced processes are basic to life itself through photosynthesis, and it has become increasingly clear that many biologically important molecules have been subjected to strong natural selection favoring their photostability.[1] A theoretical approach to these problems has traditionally been hindered due to difficulties in solving the time-independent Schrödinger equation for states different from the ground state. However, computational advances in the last decade have started reverting the situation and, with the introduction of multireference methods and the developments in linear scaling techniques,[2]ab initio predictability in excited states for moderately sized systems can be reached.[3]

Conspicuously, despite the increasing wide black-box availability of computational methods to aid practicing chemists, most experiments are designed and/or interpreted in terms of a set of rules that were introduced to understand chemical bonding in electronic ground states. Some of these rules are based on the one-electron approximation inherent to one-determinant descriptions as, for example, the use of molecular orbital (MO) arguments to understand reactivity pathways. Unfortunately, this approximation is sometimes poor, especially in the case of electronic excited states. It is for this reason that a working framework to understand chemical bonding in excited states still needs to be appropriately developed.

Any chemical bonding paradigm for excited states must be able to tackle multireference systems and, as a consequence, it cannot rely on a fixed set of one-electron functions (orbitals). Mainstream ground state chemical bonding theory is still dominated by MO thinking (an intrinsically one-electron scheme). As such, more than often, it offers results strongly dependent on the method used to obtain the approximated wave function because MOs are not invariant under orbital transformations. One textbook example is the description of the C=C double bond in ethylene. While the Pipek–Mezey localization method[4] provides a representation based on one σ MO and one π orbital, the Boys localization method[5,6] describes the bond by two equivalent banana bond orbitals. Such a situation could be, in a certain way, confusing for some practicing chemists. A possible way out to this ambivalence is to rewrite chemical bonding theory in terms of orbital invariant quantities. These objects can be obtained in terms of reduced density matrices (RDMs), either in position or momentum space. As the chemist's language is developed in the physical space, most of the implementations proposed so far are based on **r** dependent RDMs, and these methods are collectively known as real space theories of chemical bonding.

RDMs depend on spin-spatial coordinates and lack atomic information intrinsic to orbitals. Fortunately, Richard F. W. Bader paved the road by showing how atomic, and in general, chemically meaningful regions may be extracted from the RDMs[7,8] by examining the topology induced by the electron density, ρ. This, so-called Quantum Theory of Atoms in Molecules (QTAIM), has provided deep insights into the nature of the chemical bond for over thirty years, and its methods have been widely generalized to develop what is called Quantum Chemical Topology (QCT).[9] QTAIM thinking satisfies all of the requirements previously commented: it is invariant under orbital transformations, all of its descriptors are equally defined and obtained for one determinant or correlated wave functions (thus providing a smooth procedure to understand how correlation affects bonding), and it includes an exact energy partitioning into atomic and interatomic contributions.[10]

Citation: David Ferro-Costas, Ángel Martín Pendás, Leticia González and Ricardo A. Mosquera, Quantum theory of atoms in molecules. DOI: 10.1039/c4cp00431k

Traditionally, QTAIM is thought of as a theory of ground state molecules. This historical prejudice roots to the difficulties of the topological procedure to properly define atoms in simple excited states, like the 2s H atom, for which each attraction basin of ρ, i.e. each of its local maxima, defines a quantum atom in the theory. A similar situation is found in the ground state of some molecules, e.g. in Li_2, which displays a non-nuclear maximum of the electron density at the internuclear midpoint.[11] However, the controversy on ground state non-nuclear maxima is now over, their existence related to important concepts like F-centers in ionic solids.[12] Despite that a large class of chemically important molecules do not show excited states with non-nuclear attractors, the use of QTAIM to aid understanding of excited state processes is scarce.[13–17]

To our knowledge, this paper is the first in performing an energetic partition scheme based on QTAIM regions that allows us to understand physically the evolution of the charge density and geometry of a molecule upon excitation. This type of analysis can be extremely helpful to explain, in an orbital invariant language, photophysical and photochemical phenomena. As a practical example, a paradigmatic process in photochemistry, the n $\rightarrow \pi^*$ electronic transition, has been selected. A simple but very well studied molecule, formaldehyde, is chosen to illustrate how QTAIM can explain the molecular changes that take place after the electronic excitation. Three electronic states will be analyzed: the ground state S_0 and the first singlet S_1 and triplet T_1 states. In the traditional MO language, S_1 and T_1 are understood as one-electron excitations from a non-bonding MO centered at the oxygen atom (the n lone pair) to the C–O antibonding π^* orbital. According to textbook MO machinery, populating the π^* MO weakens the C–O double bond and this manifests as a lengthening of the bond. Concurrently, the hybridization of the C atom changes from sp^2-like to sp^3 leading to a C pyramidalization in any of the two S_1 and T_1 excited states.[18] In this work we shall show the ability of the QTAIM to cope with excited states and we will carefully monitor the QTAIM interpretation against the traditional MO explanation, to correlate both views as much as possible.

METHODOLOGY

Basic Descriptors of the Quantum Theory of Atoms in Molecules

Bader and coworkers showed that the topology induced by the electron density, $\rho(\mathbf{r})$, comes from generalizing the quantum mechanics to subsystems in $\mathbb{R}^3$. These subsystems (Ω) are associated with the atoms in the molecule and they are referred as "atomic basins". In the following, the QTAIM descriptors employed in this paper are briefly revised. Further information can be found in ref. 7, 8 and 19.

Some elements of chemical structure can be identified with critical points of $\rho(\mathbf{r})$. For example, critical points exhibiting three negative curvatures are called nuclear critical points (NCPs), as they are normally placed at, or close to, the position of the nuclei. Critical points with two negative curvatures (and one positive) are called bond critical points (BCPs). The line of locally maximum density linking two NCPs define what is known as the bond path and its presence implies that the corresponding atoms are bonded to one another. The interaction between a given pair of atoms bonded is characterized according to the properties of the electron and energy densities at the BCP.

Two interesting descriptors are defined at the BCP. One is the electron density at that point, ρ^{BCP}, and it is related to the strength of the bonding. The second is the ellipticity, ε, which measures the extent to which density is preferentially accumulated in a given plane containing the bond path, and it is defined as follows

$$\varepsilon^{BCP} = \frac{\lambda_1}{\lambda_2} - 1, \quad \text{with} \quad |\lambda_1| \geq |\lambda_2| \tag{1}$$

where λ_1 and λ_2 are the eigenvalues of the Hessian matrix associated with the eigenvectors which form a plane perpendicular to the bond path. If $\lambda_1 = \lambda_2$ then $\varepsilon = 0$ and the bond is cylindrically symmetrical – a situation that can be found e.g. at the C–C bonds of ethane and ethine. As such, ε^{BCP} has been related to the π character, providing reasonable results for most of single and double bonds between carbons.

The average number of electrons associated with the atomic basin Ω, $N(\Omega)$, can be obtained by

$$N(\Omega) = \int_{\Omega} \rho(\mathbf{r})d\mathbf{r} \tag{2}$$

These atomic electron populations are usually transformed into net atomic charges by

$$q(\Omega) = Z - N(\Omega) \tag{3}$$

where Z is the nuclear charge of the corresponding atom. Moreover, it can be shown that $N(\Omega)$ can be split into two kinds of terms:

$$N(\Omega) = \lambda_{\Omega} + \frac{1}{2}\sum_{\Omega' \neq \Omega} \delta_{\Omega,\Omega'} \tag{4}$$

The term λ_{Ω} is associated exclusively to Ω and is called localization index. The terms related to the pair of basins, $\delta_{\Omega,\Omega'}$, are called delocalization indices and measure the number of electrons shared between Ω and Ω'.

Energy Partition

Within the Born–Oppenheimer approximation, the expected value for the molecular electronic energy, E, is given by:

$$E = \langle \Psi | \hat{H} | \Psi \rangle = T + V_{ne} + V_{ee} + V_{nn} \tag{5}$$

where $\hat{H}$ is the Hamiltonian of the system, T is the kinetic energy, V_{ne} is the nucleus–electron attraction, and V_{ee} and V_{nn} are, respectively, the electron–electron and the nucleus–nucleus repulsions.

All these quantities can be written in terms of the electron density of the system, $\rho(\mathbf{r})$:

$$V_{ne} = -\sum_{\alpha=1}^{M} \int \frac{Z_\alpha \rho(\mathbf{r})}{|\mathbf{R}_\alpha - \mathbf{r}|} d\mathbf{r} \tag{6}$$

$$V_{coul} = \frac{1}{2}\int \frac{\rho(\mathbf{r}_1)\rho(\mathbf{r}_2)}{|\mathbf{r}_2 - \mathbf{r}_1|} d\mathbf{r}_1 d\mathbf{r}_2 \tag{7}$$

the spin-free first order RDM, $\gamma(\mathbf{r},\mathbf{r}')$:

$$T = -\frac{1}{2}\int_{\mathbf{r}' \to \mathbf{r}} \nabla^2 \gamma(\mathbf{r},\mathbf{r}') d\mathbf{r} \tag{8}$$

and the exchange–correlation component of the diagonal second order RDM, $\Gamma_{xc}(\mathbf{r}_1,\mathbf{r}_2)$:

$$V_{xc} = \frac{1}{2}\int \frac{\Gamma_{xc}(\mathbf{r}_1,\mathbf{r}_2)}{|\mathbf{r}_2 - \mathbf{r}_1|} d\mathbf{r}_1 d\mathbf{r}_2 \tag{9}$$

where $\mathbf{r}$, $\mathbf{r}_1$ and $\mathbf{r}_2$ represent electron coordinates; M is the number of nuclei; and Z_α and $\mathbf{R}_\alpha$ are the charge and the position vector of the α nucleus. The total electron–electron repulsion, V_{ee}, is split into a classic Coulomb term (V_{coul}, eqn (7)) and another term associated with the exchange–correlation of the system (V_{xc}, eqn (9)):

$$V_{ee} = V_{coul} + V_{xc} \tag{10}$$

As QTAIM provides a partition of the three-dimensional space into atomic basins, the electron density is also susceptible of partitioning into atomic contributions:

$$\rho(\mathbf{r}) = \sum_{\Omega} \rho_\Omega(\mathbf{r}) = \sum_{\Omega} w_\Omega(\mathbf{r})\rho(\mathbf{r}) \tag{11}$$

where Ω represents an atomic basin and the function w_Ω is defined, for the QTAIM partition, by

$$w_\Omega(\mathbf{r}) = \begin{cases} 1 & \text{if } \mathbf{r} \in \Omega \\ 0 & \text{elsewhere} \end{cases} \tag{12}$$

Likewise, the RDMs can be also partitioned as

$$\gamma(\mathbf{r},\mathbf{r}') = \gamma(\mathbf{r},\mathbf{r}')\sum_{\Omega} w_\Omega(\mathbf{r}') = \sum_{\Omega} \gamma_\Omega(\mathbf{r},\mathbf{r}') \tag{13}$$

$$\begin{aligned} \Gamma_{xc}(\mathbf{r}_1,\mathbf{r}_2) &= \sum_{\Omega}\sum_{\Omega'} w_\Omega(\mathbf{r}_1) w_{\Omega'}(\mathbf{r}_2)\Gamma_{xc}(\mathbf{r}_1,\mathbf{r}_2) \\ &= \sum_{\Omega}\sum_{\Omega'} \Gamma_{xc}^{\Omega\Omega'}(\mathbf{r}_1,\mathbf{r}_2) \end{aligned} \tag{14}$$

allowing for a whole division of the molecular energy into atomic and diatomic terms.[10] For instance, using eqn (13) in (8), we obtain:

$$T = \sum_{\Omega} -\frac{1}{2}\int_{\mathbf{r}'\to\mathbf{r}} \nabla^2 \gamma(\mathbf{r},\mathbf{r}') w_\Omega(\mathbf{r}') d\mathbf{r} = \sum_{\Omega} T(\Omega) \tag{15}$$

Similarly, through eqn (11), V_{ne} is given by

$$V_{ne} = \sum_{\Omega}\sum_{\alpha=1}^{M} \int_\Omega \frac{-Z_\alpha \rho(\mathbf{r})}{|\mathbf{R}_\alpha - \mathbf{r}|} d\mathbf{r} = \sum_{\Omega}\sum_{\alpha=1}^{M} V_{ne}(\alpha,\Omega) \tag{16}$$

Finally, the electron–electron repulsion results in:

$$\begin{aligned} V_{ee} &= \sum_{\Omega} \underbrace{\{V_{coul}(\Omega,\Omega) + V_{xc}(\Omega,\Omega)\}}_{V_{ee}(\Omega,\Omega)} \\ &+ \frac{1}{2}\sum_{\Omega}\sum_{\Omega'\neq\Omega} \underbrace{\{V_{coul}(\Omega,\Omega') + V_{xc}(\Omega,\Omega')\}}_{V_{ee}(\Omega,\Omega')} \end{aligned} \tag{17}$$

where the Coulomb components are:

$$V_{coul}(\Omega,\Omega) = \frac{1}{2}\int_\Omega\int_\Omega \frac{\rho(\mathbf{r}_1)\rho(\mathbf{r}_2)}{|\mathbf{r}_2 - \mathbf{r}_1|} d\mathbf{r}_1 d\mathbf{r}_2 \tag{18}$$

$$V_{coul}(\Omega,\Omega') = \int_\Omega\int_{\Omega'} \frac{\rho(\mathbf{r}_1)\rho(\mathbf{r}_2)}{|\mathbf{r}_2 - \mathbf{r}_1|} d\mathbf{r}_1 d\mathbf{r}_2 \tag{19}$$

and the exchange–correlation:

$$V_{xc}(\Omega,\Omega) = \frac{1}{2}\int_\Omega\int_\Omega \frac{\Gamma_{xc}(\mathbf{r}_1,\mathbf{r}_2)}{|\mathbf{r}_2 - \mathbf{r}_1|} d\mathbf{r}_1 d\mathbf{r}_2 \tag{20}$$

$$V_{xc}(\Omega,\Omega') = \int_\Omega\int_{\Omega'} \frac{\Gamma_{xc}(\mathbf{r}_1,\mathbf{r}_2)}{|\mathbf{r}_2 - \mathbf{r}_1|} d\mathbf{r}_1 d\mathbf{r}_2 \tag{21}$$

Once the partition of all the energetic terms associated with the operators comprising the Hamiltonian of the system is done, different terms can be grouped. The collection of all the terms that takes place exclusively in one atomic basin results in the definition of the net energy for that atom, E_{net},

$$E_{net}(\Omega) = T(\Omega) + V_{ne}(\omega,\Omega) + V_{ee}(\Omega,\Omega) \tag{22}$$

where ω represents the nucleus that belongs to the Ω basin. Similarly, the interaction energy (V_{int}) between 2 atoms can be defined according to the equation,

$$V_{int}(\Omega,\Omega') = V_{nn}(\omega,\omega') + V_{ne}(\omega,\Omega') + V_{ne}(\omega',\Omega) + V_{ee}(\Omega,\Omega') \tag{23}$$

for $\Omega \neq \Omega'$ and $\omega \neq \omega'$. The introduction of both kinds of energies (net and interaction) into eqn (5) yields the chief equation of the interacting quantum atoms (IQA) energy partition scheme:

$$E = \sum_{\Omega} E_{net}(\Omega) + \frac{1}{2}\sum_{\Omega}\sum_{\Omega'\neq\Omega} V_{int}(\Omega,\Omega') \tag{24}$$

According to this equation, a molecular system can be analysed in terms of its constituent elements (atoms or even groups of them) and their interactions. This point of view resembles the traditional conception of chemistry before the introduction of quantum mechanics and it does not invoke orbital definitions.

Computational Details

All the geometries have been optimized using the Complete Active Space Self Consistent Field (CASSCF) procedure[20] with an active space of 12 electrons distributed in 10 MOs. The active space is depicted in Fig. 1 and comprises three σ, σ* pairs, the π_{CO}, π_{CO}* orbital pair, and two lone pairs of electrons of the oxygen atom. For computational sake, the IQA scheme[10] has been performed using the wave function corresponding to a more modest CASSCF(4,3) method (restricted to the most relevant lone pair of O and the π, π* pair of MOs – see Fig. 1). The energies of each state are computed state-specific. An analysis of the electronic wavefunctions obtained with the (12,10) and (4,3) active spaces reveals that the main configurations contributing to the state are the same. In all cases, the basis set employed is the cc-pVTZ.[21] All the geometries and energies were obtained using the MOLCAS 7.8 program.[22]

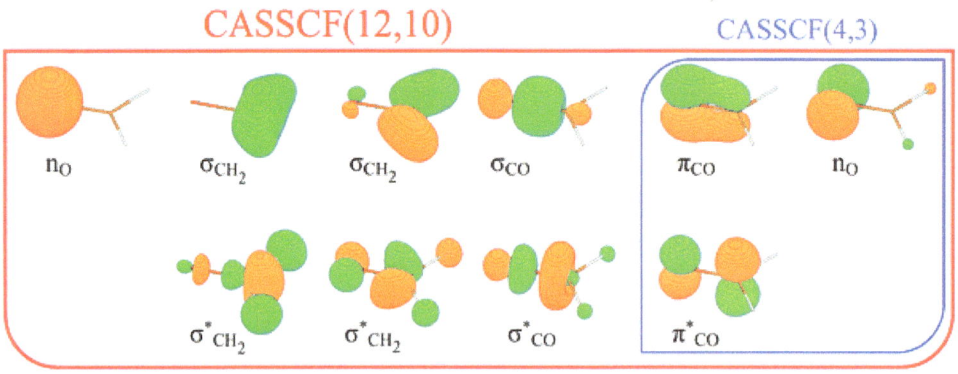

Figure. 1 Natural molecular orbitals of formaldehyde included in the CASSCF(12,10) active space; here, exemplary for the S_0 state. The subset selected for the subsequent CASSCF(4,3) single point calculations is in the blue box.

The Molden2AIM[23] software has been used for the generation of wfn files, needed to carry out the QTAIM analysis of the electron density. This analysis has been done using the AIMPAC package of programs,[24] as well as using the PROMOLDEN code,[10] developed by the Quantum Chemistry group of Oviedo University. Additionally, the MULTIWFN software[25] was used to obtain the Laplacian of the ρ function ($\nabla^2\rho$) and the Electron Localization Function (ELF).[26]

RESULTS

Geometries

In order to study the evolution of the S_1 and T_1 excited states step by step, a total of five geometries, associated with critical points in different potential energy surfaces (PES), have been considered. Fig. 2 shows the equilibrium geometry in the electronic ground state S_0 (denoted as min S_0) and the optimized S_1 (min S_1) and T_1 (min T_1) minima. As it can be seen, the optimized S_1 and T_1 minima show pyramidalized C atoms and C–O bonds elongated with respect to the ground state by ca. 0.14 Å. To separate the effect of elongation and pyramidalization, constrained planar optimizations have also been performed in each electronic excited state (planar S_1 and planar T_1). The latter C_{2v} geometries correspond to structures with an imaginary vibration frequency associated with the loss of planarity. The relative energy of each critical point in each PES is collected in Table 1. As expected the energy of T_1 is smaller than that of S_1 for each geometry.

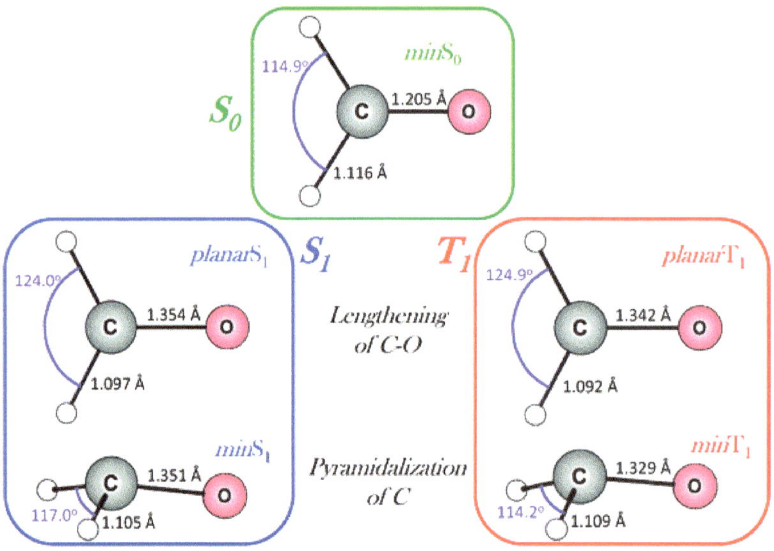

Figure. 2 Equilibrium geometries of formaldehyde in the S_0, S_1 and T_1electronic states, as well as planar structures in the S_1 and T_1 excited states. Bond distances in Å and the H–C–H angle in degrees.

Table 1 Relative energies (eV) of formaldehyde obtained at the CASSCF(12,10)/cc-pVTZ level of theory

	S_0	S_1	T_1
min S_0	0.00[a]	4.52	4.26
planar S_1	0.66	4.00	3.79
min S_1	1.07	3.94	3.67

planar T_1	0.58	4.01	3.78
min T_1	1.12	3.96	3.66

[a]Energy of −114.04552 au.

QTAIM ANALYSIS IN THE EXCITED STATE

To understand which processes are taking place upon excitation, it is illustrative to analyse the charge density separately in (a) the vertical electronic excitation, (b) the lengthening of the C–O and (c) C pyramidalization steps.

Vertical Transition to the Excited State

In Fig. 3 the variation of the electron density function, $\Delta\rho(\mathbf{r})$, associated with the $S_0 \rightarrow S_1$ excitation process is depicted. The electron density is mainly depleted (shown in red) from the surroundings of the oxygen atom. This depletion is basically equivalent to the $\rho(\mathbf{r})$ associated with a non-bonding orbital. This electron density is then gained (green) by both the C and O atoms, but there is no electron density between the atoms—a fact that can be associated with the idea of populating a $\pi_{C-O}{}^*$ MO. As one can see, the analysis of $\Delta\rho(\mathbf{r})$, which is invariant with regard to MO rotations, naturally provides the same understanding about the charge distribution upon excitation, as the MO theory.

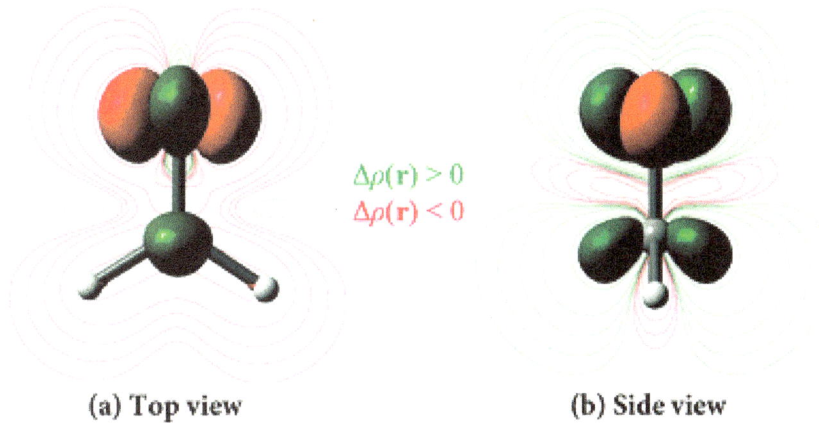

Figure. 3(a) Top view and (b) side view of the electron density variation, $\Delta\rho(\mathbf{r})$, in the vertical electronic excitation $S_0 \rightarrow S_1$. Isosurfaces of ±0.025 au and isolines of ±0.001, ±0.002, ±0.004, ±0.008, and ±0.020 au are shown (positive in green, negative in red).

Fig. 4a shows the reorganization of the atomic electron populations in the vertical excitation. As expected, the most important change takes place in the C–O moiety, where the oxygen transfers 0.293 electrons to the carbon basin.

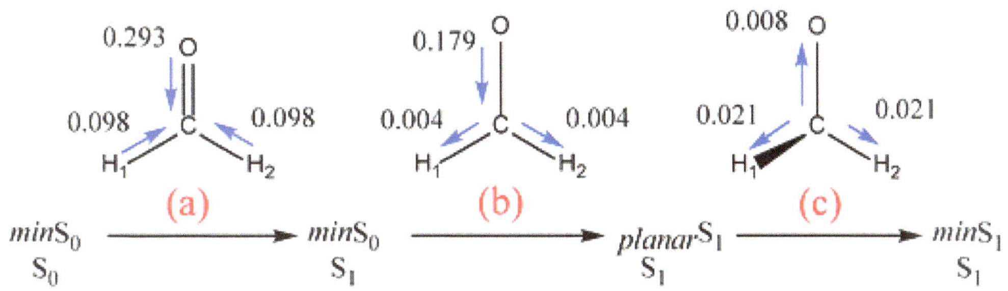

Figure. 4 Reorganization of the electron density (au) from min S_0 (in S_0 state) to min S_1 (in S_1) in (a) the vertical electronic excitation $S_0 \rightarrow S_1$ in the minS_0 geometry, (b) the lengthening of the C–O bond in the excited state, and (c) the pyramidalization of the carbonyl carbon.

The electron density can be also expanded in terms of the natural MO χ_i:

$$\rho(\mathbf{r}) = \sum_i n_i |\chi_i(\mathbf{r})|^2 = \sum_{a\in\sigma} n_a |\chi_a(\mathbf{r})|^2 + \sum_{b\in\pi} n_b |\chi_b(\mathbf{r})|^2 \tag{25}$$

where n_i is the occupation number of the natural MO χ_i. As also shown in eqn (25), the natural MOs can be grouped according to their symmetry (σ or π) and, consequently, the atomic electron population can be split into σ and π contributions:

$$N^{\sigma}(\Omega) = \sum_{a\in\sigma} n_a \int_{\Omega} |\chi_a(\mathbf{r})|^2 \mathrm{d}\mathbf{r} \tag{26}$$

$$N^{\pi}(\Omega) = \sum_{b\in\pi} n_b \int_{\Omega} |\chi_b(\mathbf{r})|^2 \mathrm{d}\mathbf{r} \tag{27}$$

It should be emphasized that N^{σ} and N^{π} come from the MO framework and thus they are not physical observables. However, their analysis may still be useful in chemical bonding theory. Moreover, it is interesting to analyze them if we want to find the parallelism between QTAIM results and MO concepts. The σ–π partitioning of the atomic population in the S_0 state is shown in Table 2, together with their variations, ΔN, upon vertical excitation. Overall, one "σ electron" is converted into one "π electron" in the electronic excitation. While O loses 0.759 au from its σ distribution, both C and O increase their π electronic population to 0.484 and 0.466, respectively (i.e. larger at the C, as expected from the antibonding character of the MO). This means that around 60% of the oxygen electronic population involved in the electronic transition still belongs to its basin. Similarly, it can be seen that around 40% of the electronic population gained by the C comes from the H atoms, indicating that the transition is not completely described by an $n_O \rightarrow \pi_{CO}^*$ transition.

Table 2 Atomic populations for the S_0 state (total as well as σ and π components) and their change after the vertical $S_0 \rightarrow S_1$ electronic excitation. All values are in au and referred to the min S_0 geometry

	N_{S_0}	$N_{S_0}^{\sigma}$	$N_{S_0}^{\pi}$	**Δ*N***	**Δ*N*$^{\sigma}$**	**Δ*N*$^{\pi}$**
C	4.850	4.365	0.484	0.489	0.006	0.484
O	9.118	7.629	1.489	−0.293	−0.759	0.466
H1/H2	1.016	1.001	0.015	−0.098	−0.122	0.023
Σ	16.000	13.997	2.003	0.000	−0.996	0.996

Also interesting is to follow the change in the electron density and the ellipticity at the bond critical points[7] (Table 3). The $n_O \rightarrow \pi^*$ electronic transition produces an increase of the electron density in the C basin, implying a larger electronic repulsion to its environment. As a consequence, the electron density at the BCP of different bonds in which the carbon is involved decreases. Important to note is that although ρ^{BCP} decreases in the three bonds, only the C–O bond distance increases after excitation (see Fig. 2), clearly illustrating that the $\Delta\rho^{BCP}$ between two electronic states is not enough to predict the geometric reorganization taking place upon excitation. In passing we note that correlations between ρ^{BCP} and the bond length should only be done within the same kind of chemical bonds and an electronic excitation does not assure that bonds in different electronic states share the same chemical nature.

Table 3 Electron density ρ (au) and ellipticity ε at the bond critical point (BCP) of C–O and C–H bonds. The absolute values refer to the S_0 state at the min S_0 geometry while variations are calculated for (a) the vertical excitation $S_0 \rightarrow S_1$, (b) the C–O bond elongation, and (c) the C pyramidalization

	$\rho_{S_0}^{BCP}$	$\varepsilon_{S_0}^{BCP}$	**(a)**		**(b)**		**(c)**	
			Δρ^{BCP}	**Δε^{BCP}**	**Δρ^{BCP}**	**Δε^{BCP}**	**Δρ^{BCP}**	**Δε^{BCP}**
C–O	0.41953	0.124	−0.01196	0.015	−0.10505	0.023	0.00227	−0.064
C–H	0.28318	0.021	−0.02414	0.067	0.01912	−0.004	−0.00239	−0.030

The ellipticity increases in both C–O and C–H bonds, in agreement with the fact that the electron density is accumulated at both sides of the molecular plane over the C atom, implying that the electron density distribution along the bonds is less cylindrical. This is also a clear example where the association of the ellipticity (or, in general, properties emerging from the electron density) with the bond order (or concepts related to the orbital conception) can be dangerous.

The results of the energy partition following the IQA scheme are collected in Table 4. The first column refers to the change in energies for the vertical excitation. Accordingly, in the S_1 excited state the net energies of C and O decrease whereas the interaction between them, ΔV_{int}(C,O), becomes very much repulsive (1131 kJ mol^{-1}). As O is negatively charged in the ground state (charge $q_O = -1.118$ au), the withdrawal of electron density from its basins upon excitation stabilizes the atom through the reduction of the electron–electron repulsion in its basin. At the same time, the C in S_0 is positively charged ($q_C = +1.150$ au) and the increment of its atomic electron population increases its nucleus–electron attraction.‡ However, this redistribution of the electron density also intensifies the electron–electron repulsion between both basins (C–O), destabilizing the interaction between them. It can be also of interest to sum up all the inter-atomic interaction energies depending on ρ(**r**) in what we call a classical potential, V_{clas}, defined as $V_{int} - V_{xc}$ (recall eqn (21) and (23)). The variation of V_{clas} for the CO pair (998.9 kJ mol^{-1}) is pretty close to the one obtained by employing a point-charge model (around 877 kJ mol^{-1}). The same agreement is obtained for the C–H bond (35.1 kJ mol^{-1}vs. ca. 65 kJ mol^{-1}). Moreover, a ΔV_{xc}(C,O) of 132.5 kJ mol^{-1} after vertical excitation (see Table S1 in the ESI†) indicates that an important part of the C=O double bond breaks. In summary, it can be seen that the stabilization of the individual atoms comprised in the bond and destabilization of the interaction, making the bond strength weaker, is in line with the idea of populating an antibonding MO.

Table 4 Changes in the intra-atomic net energies and the inter-atomic interaction energies (kJ mol^{-1}) for the processes (a), (b), and (c) considered in Fig. 4. The data correspond to the CASSCF(4,3)/cc-pVTZ level of theory calculation

	(a)	(b)	(c)
ΔE_{net}(C)	−606.0	−290.2	57.8
ΔE_{net}(O)	−203.1	−398.4	10.5
ΔE_{net}(H1)[a]	37.7	−9.7	−10.1
ΔV_{int}(C,O)	1131.4	654.5	−62.2
ΔV_{int}(C,H1)[a]	36.8	−32.0	−5.9
ΔV_{int}(O,H1)[a]	−43.8	28.1	11.8
ΔV_{int}(H1,H2)	16.1	0.0	−6.6

[a] Same as H1 for H2.

Using the CASSCF(4,3) wave function employed in the IQA scheme, localization and delocalization indices[7] can be obtained (Table 5).§ In the ground state, the most important differences, comparing the expected values of the ideal Lewis structure for the formaldehyde, are found in C and O atoms (λ_C, λ_O, and $\delta_{C,O}$). These differences can be basically justified considering the other resonance form of the C=O bond (C^+–O^-). Upon excitation, the localization indices (λ) follow the total basin electron populations (cf.Table 2). Since oxygen suffers a loss of electron density, it becomes less prone to share it, decreasing $\delta_{C,O}$. This behavior is again the one expected according to the orbital interpretation because the π* MO should entail a reduction in the bond order. Indeed, $\delta_{C,O}$ = 1.098 au for S_1 is close to the expected value for a single bond. Therefore, the evolution of λ and δ values could be taken as an indication of the partial heterolytic cleavage of the π bond upon vertical electronic excitation. As shown in Table 5, the electrons delocalized in the bond now become localized in one of the two atoms (C).

Table 5 Expected values for the localization (λ, diagonal values) and the delocalization (δ) indices for the Lewis standard structure of formaldehyde at S_0, the calculated values for S_0 at min S_0 geometry, and the change of these indices for the $S_0 \rightarrow S_1$ excitation, C–O bond elongation and C pyramidalization. All values are in au and calculated at the CASSCF(4,3)/cc-pVTZ level of theory

	λ_{Ω} & $\delta_{\Omega,\Omega'}$	C	O	H1	H2
S_0 Lewis	C	4.000			
	O	2.000	7.000		
	H1	1.000	0.000	0.500	
	H2	1.000	0.000	0.000	0.500
S_0 at min S_0	C	3.305			
	O	1.297	8.389		
	H1	0.908	0.112	0.458	
	H2	0.908	0.112	0.051	0.458
$S_0 \rightarrow S_1$ at min S_0	C	0.559			
	O	−0.199	−0.222		
	H1	0.022	0.007	−0.088	
	H2	0.022	0.007	−0.019	−0.088
min $S_0 \rightarrow$ planar S_1 in the S_1 state	C	0.147			
	O	−0.103	−0.033		
	H1	0.029	−0.036	0.001	
	H2	0.029	−0.036	0.000	0.001

planar $S_1 \rightarrow$ min S_1 in the S_1 state	C	−0.064			
	O	−0.016	0.006		
	H1	−0.006	0.010	0.032	
	H2	−0.006	0.010	0.002	0.032

As it has been previously reported,[28] a change in $\delta_{\Omega,\Omega'}$ accompanying vertical excitation may be used, in general, to predict the evolution of the corresponding bond length. Thus, $\delta_{C,O}$ decreases (−0.199 au) and $\delta_{C,H}$ increases (0.022 au) upon electronic excitation, facts which can be directly related to a stretch and a shrink of the C–O and C–H bonds, respectively (Fig. 2).

Lengthening of the C–O bond in the S_1 Excited State

After vertical excitation, the nuclei are subjected to new forces and, as a consequence, a geometrical reorganization occurs (in this case the C–O bond elongates and the C pyramidalizes, see min S_1 in Fig. 2). In this section we shall analyse energies and populations obtained within the bond elongation only, i.e. at the planar S_1 geometry.

The atomic electron population reorganizations that occur in the C–O elongation process are shown in Fig. 4b. The oxygen transfers the electron density to the C basin (as expected when the C–O bond is breaking, since at an infinite C–O distance O should be neutral). A detailed analysis of the σ and π populations reveals that the nature of the electronic reorganization is practically σ (Table 6). Delocalization indices (Table 5) indicate that, in this process, the C–O bond order is reduced to a value of 0.995 au. Thus, elongation breaks the double bond, homolytically in the π distribution, as no significant changes are observed in π populations, and partially heterolytically in the σ one.

Table 6 Atomic populations for the S_1 state (total as well as σ and π components) at the min S_0 geometry and their change in the C–O bond elongation. All values are in au

	N_{S_1}	$N^{\sigma}_{S_1}$	$N^{\pi}_{S_1}$	ΔN	ΔN^{σ}	ΔN^{π}
C	5.339	4.371	0.968	0.170	0.174	−0.004
O	8.826	6.871	1.955	−0.179	−0.190	0.011
H1/H2	0.918	0.880	0.038	0.004	0.008	−0.004
Σ	16.000	13.001	2.999	0.000	0.000	0.000

During the C–O bond elongation process, important changes in the values of ρ^{BCP} and ε^{BCP} are not observed, with the exception of the electron density in the C–O bond, which obviously decreases as a consequence of the C–O bond cleavage (Table 3).

In the IQA scheme (Table 4) the same trends found in the excitation process are observed: stabilization of C and O net energies and destabilization in their interaction energies. This is again in agreement with the fact that the antibonding character of the π* MO is responsible for the elongation. However, our analysis indicates that the lengthening of the C–O bond is not accompanied by a transfer of the π electron density, but a σ one.

Pyramidalization of the C Atom in the S_1 Excited State

The electron density reorganization accompanying the pyramidalization (see Fig. 4c) is negligible in comparison with the other two processes. Since the total atomic populations remain almost invariable, the localization and delocalization indices (Table 5) as well as the ellipticity and the electron density at the BCPs (Table 3) hardly change. Despite being small, it is interesting to note that the variation of the ellipticity at the BCP in the C–O bond (from 0.162 au at planar S_1 to 0.097 at min S_1) is associated with a larger σ character, in agreement with the idea that the C atom acquires sp^3-like hybridisation.

The IQA energy partition (Table 4) for pyramidalization does not show any remarkable stabilization. The most important term arises from the C–O interaction, which behaves oppositely to the excitation and C–O bond elongation processes. This stabilization, however, is in line with a deformation of the C electron density to reach a sp^3-like hybridisation that occurs to make the bond stronger. In terms of IQA energies this is reflected as a destabilization in E_{net} and a stabilization in V_{int}.

In order to understand why the C pyramidalises, it is useful to look at the terms that define V_{int} (recall eqn (17) and (23)) which are collected in Table S3 of the ESI.† As it can be seen, the origin of the C–O stabilization is due to the reduction in the electron–electron repulsion between both atomic basins. Thus, the driving force of the pyramidalization seems to be the reduction of the initial electron–electron repulsion introduced by the vertical excitation (Table S1, ESI†). This repulsion has its origin (see above) from the π contribution of the electron density (see Fig. 3).

For the sake of simplicity, let us consider the ideal spherical out-of-plane π electron densities shown in Fig. 5. Taking into consideration only the large interatomic interactions (that is, interactions between spheres in the same side of the plane) depicted as double red arrows, we find that: (i) the ρwith π-symmetry in the planar conformation (II) gives rise to $V^{\pi}_{ee} \propto 2nm$ between C and O bonds (being n and m the electronic charge of the spheres); (ii) after distortion of the π distribution, upon transferring certain a and b amounts of electron density (with a,b ≥ 0) across the plane (as indicated in the imaginary state III), repulsion decreases, since $V^{\pi}_{ee} \propto 2(nm - ab)$;¶ and (iii) due to the distortion, the local geometric environment of the atoms is modified, causing the pyramidalization of the atom (IV). In plain English, a symmetry-breaking in the electron density reduces the electron–electron repulsion.

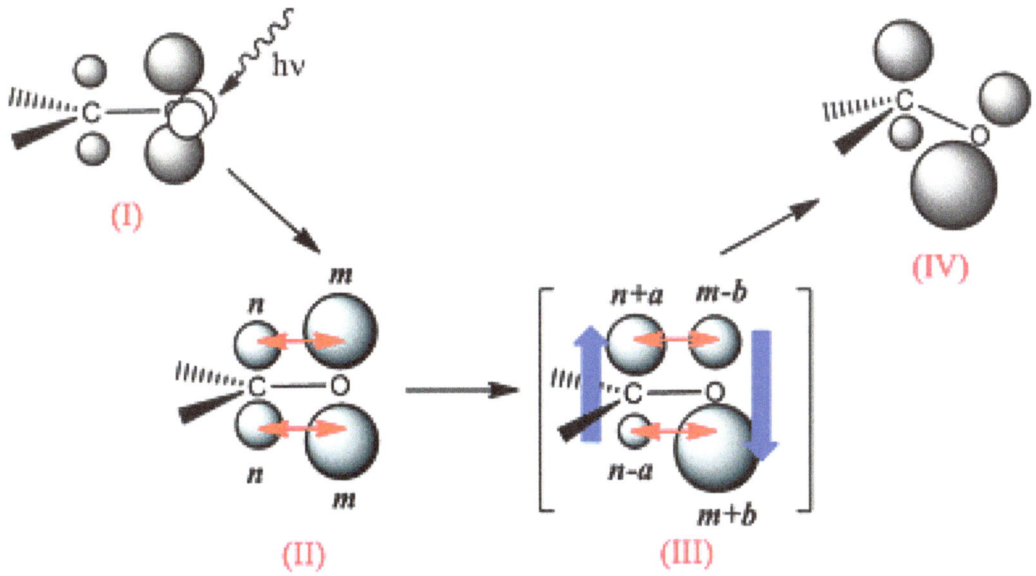

Figure. 5 Scheme of the pyramidalization of the carbonyl carbon. This process takes place in order to reduce the repulsion between the out-of-plane electron densities at C and O.

In addition, V_{clas} and V_{xc} values for the C–O pair (−71 kJ mol^{-1} and 8.8 kJ mol^{-1}, respectively) indicate that this process is governed by classical energy terms.

Comparison between S_1 and T_1

In the previous section, an analysis of the changes undergone by formaldehyde after excitation from S_0 to S_1 has been done. A similar study can be done for the T_1 state, at the planar T_1 and min T_1 geometries, and the results obtained are very similar (Table S4, ESI†).

In Fig. 6, the $\Delta\rho(\mathbf{r})$ function between the T_1 and S_1 states is shown at planar S_1 geometry. The triplet state is characterized by a larger population at the C basin, which has mainly π character (Table 7). This means that the π population of C and O basins are more "alike" in the triplet state than in the singlet (0.964 and 1.967 au vs. 1.006 and 1.920 au, for C and O basins, Table 7). According to the mechanism described in Fig. 5, larger repulsions between C and O are expected in T_1 and the pyramidalization in the triplet state should be slightly larger. Indeed, the HHOC dihedral angle for min S_1 is 22.8 degrees while it is 26.9 degrees for min T_1.

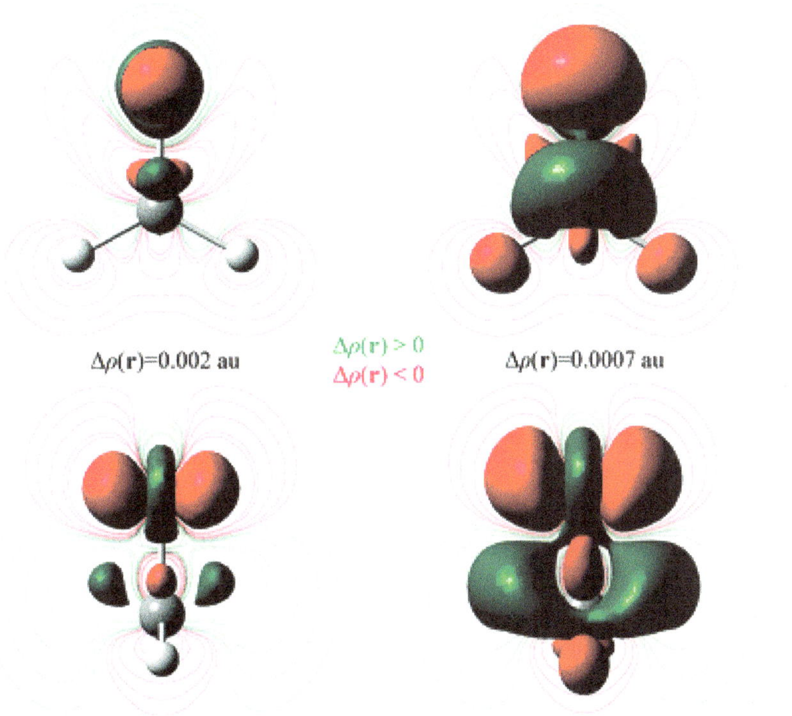

Figure. 6 $\rho_{T1} - \rho_{S1}$ at the planar S_1 geometry. Isosurfaces of ±0.002 and ±0.0007 au and isolines of ±0.0001, ±0.0002, ±0.0004, ±0.0008, and ±0.001 au are shown (positive in green and negative in red.)

Table 7 Atomic populations for S_1 and T_1 states (total as well as σ and π components) at the planar S_1 geometry and their difference (T_1 minus S_1). All values are in au

	N_{S_1}	$N^{\sigma}_{S_1}$	$N^{\pi}_{S_1}$	N_{T_1}	$N^{\sigma}_{T_1}$	$N^{\pi}_{T_1}$	ΔN	ΔN^{σ}	ΔN^{π}
C	5.509	4.545	0.964	5.556	4.551	1.006	0.048	0.005	0.042
O	8.647	6.680	1.967	8.617	6.697	1.920	−0.030	0.017	−0.047
H 1 / H2	0.922	0.888	0.035	0.913	0.877	0.037	−0.009	−0.011	0.002
Σ	16.000	13.001	2.999	16.000	13.001	2.999			

Differences in IQA energetic terms (at planar S_1 geometry) are all smaller than 5.0 kJ mol^{-1} in absolute value (Table S5, ESI), with the exception of the E_{net}(C), whose change is −18.1 kJ mol^{-1}. This is due to the increase in the nucleus–electron attraction in the C basin.

Laplacian of the Electron Density in the Excited States

In QTAIM, the Laplacian of the electron density, $\nabla^2\rho(\mathbf{r})$, is typically used to recover the main features of the Lewis and Valence-Shell Electron-Pair Repulsion (VSEPR) models.[29] Similarly, the electron localization function (ELF)[26] is known for its ability to analyze the distribution of electron pairs in the molecule. However, the usual formulation of the ELF is only well defined for monodeterminatal wave functions, while $\nabla^2\rho(\mathbf{r})$ can be obtained at any level of theory.

Isosurfaces of the Laplacian of the electron density of S_0 at min S_0 and of S_1 at min S_0 and min S_1 (Fig. 7) show two interesting facts. First, upon vertical electronic excitation, the $\nabla^2\rho < 0$ region typically associated with the O lone pairs rotates 90 degrees around the C–O bond. And second, during the pyramidalization, a $\nabla^2\rho < 0$ region (not associated with any bond) appears in the surroundings of the C atom. The isosurfaces of T_1 at minS_0 and min T_1 are identical to those of S_1 and therefore not shown.

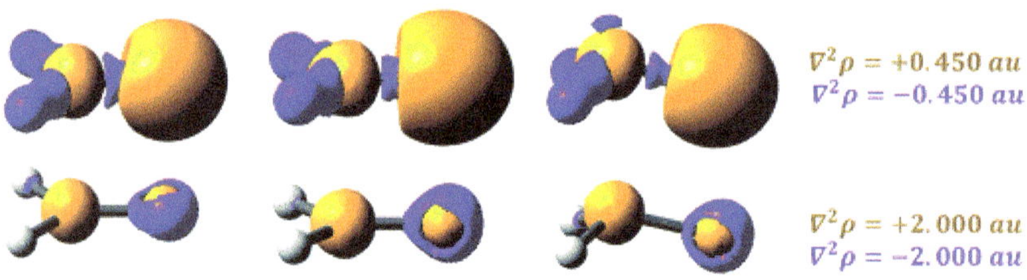

Figure. 7 $\nabla^2\rho$ isosurfaces (±0.4 au and ±2.0 au; positive in yellow, negative in violet) obtained from the electron density of the S_0 (at min S_0) and S_1(at min S_0 and min S_1) states. The electron density was obtained at a CASSCF(12,10)/cc-pVTZ level of theory.

In order to visualize the ELF not only in the S_0 state but also in the excited state of formaldehyde, we analyzed the T_1 state, which can be described with one Slater determinant in an unrestricted Hartree–Fock (UHF) formulation. This is in this case justified because the S_1 and T_1 states exhibit similar properties and thus, the ELF analysis may be extended "mutatis mutandis" to describe also the S_1 state. Interestingly, the two facts observed with $\nabla^2\rho(\mathbf{r})$ are also reproduced by the topology of the ELF,[30] as observed in Fig. 8.

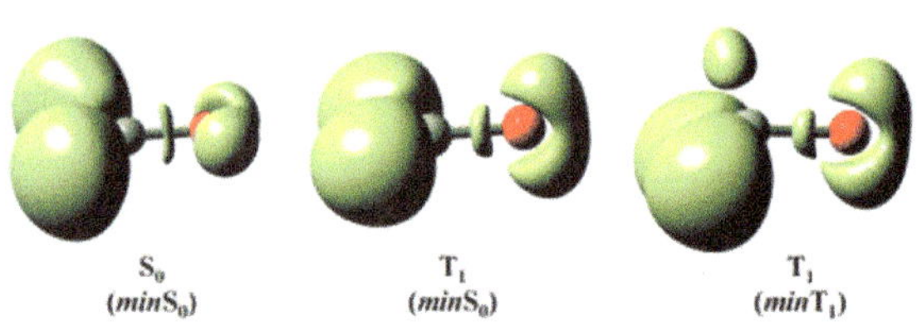

Figure. 8 0.85 au ELF isosurface for the the S_0 (at minS_0) and T_1 (at min S_0 and min T_1) states. The ELF was obtained at the UHF/cc-pVTZ level of theory.

It is of importance to point that according to the traditional nomenclature of the $S_0 \rightarrow S_1$ transition ($n \rightarrow \pi^*$), it is expected that the electron density of the lone pairs is transformed into that of a π* MO. As it is shown here, the topological analysis of ρ (both QTAIM or ELF) enriches this description, showing that this conversion is caused by a rotation of the ρ associated with the O lone pairs, what could be interpreted as that the "σ" lone pairs become "π" lone pairs and this triggers a concomitant weakening of the C–O double bond.

The rotation of the O lone pairs is very relevant, as it appears in a huge number of photophysical processes. It seems reasonable that the O lone pairs are in a σ disposition due to the presence of the C–O double bond in the S_0 state, even if it maximizes the repulsion between C–H bonding pairs and the O lone pairs. In the S_1 electronic arrangement, the repulsion between the O lone pairs and the C–H bonding pairs is minimized.

CONCLUSIONS

The main conclusions may be summarized as follows:

- The QTAIM has proved to be versatile enough to be applied to systems in their electronic excited states, generating results that are in line with firmly-established chemical facts and interpretations. Moreover, QTAIM results have not only been useful as those coming from

MO concepts, but even capable of adding details of electronic rearrangements that are not easily deduced from the MO framework.

- Excitation to the singlet $n\pi^*$ state is accompanied by the stabilization of the individual atoms of the carbonyl bond but with a destabilization of the interaction between them. This fact is in accordance with the idea of populating a π^* MO, being also an indication that the nature of the C–O bond has changed.
- The energy partitions associated with the vertical excitation and the lengthening of the carbonyl bond show the same trend, which could be also ascribed to the antibonding character of the π_{C-O}^* MO. However, the C–O bond lengthening does not involve a change in the π contribution of the atomic populations, but in the σ one.
- The pyramidalization of the C atom can be understood as a breaking of the symmetry of the electron density in order to reduce the electron–electron repulsion between C and O basins.
- The analysis of the electron density (through QTAIM and ELF) reveals that the $n \rightarrow \pi^*$ transition may be also understood as a rotation of the oxygen lone pairs to a π disposition, accompanied by a synchronous weakening of the C–O double bond.

ACKNOWLEDGEMENTS

D.F-C. and R.A.M. thank Spanish Ministry of Economy for the project CTQ2010-21500 and A.M.P. for project CTQ2012-31174. D. F-C. also thanks Spanish Ministry of Education for an FPU fellowship and the University of Vigo for the "Estadía en centros de investigación" grant.

REFERENCES

1. M. Klessinger and J. Michl, Excited States and Photochemistry of Organic Molecules, VH Publishers, Inc., Toronto, 1995.
2. F. Weigend, M. Kattannek and R. Ahlrichs, J. Chem. Phys., 2009, **130**, 164106.
3. L. González, D. Escudero and L. Serrano-Andrés, ChemPhysChem, 2012, **13**, 28–51
4. J. Pipek and P. G. Mezey, J. Chem. Phys., 1989, **90**, 4916–4926
5. J. M. Foster and S. F. Boys, Rev. Mod. Phys., 1960, **32**, 300–302
6. D. A. Kleier, T. A. Halgren, J. H. Hall Jr and W. N. Lipscomb, J. Chem. Phys., 1974, **61**, 3905–3919
7. R. F. W. Bader, Atoms in Molecules: A Quantum Theory, Clarendon Press, Oxford, 1995.
8. R. F. W. Bader, Chem. Rev., 1991, **91**, 893–928
9. P. L. A. Popelier, Struct. Bonding, 2005, **115**, 1–56
10. M. A. Blanco, A. M. Pendás and E. Francisco, J. Chem. Theory Comput., 2005, **1**, 1096–1109
11. J. Cioslowski, J. Phys. Chem., 1990, **94**, 5496–5498
12. P. Mori-Sanchez, J. Recio, B. Silvi, C. Sousa, A. M. Pendas, V. Luaña and F. Illas, Phys. Rev. B: Condens. Matter Mater. Phys., 2002, **66**, 075103.
13. R. F. W. Bader, D. Bayles and G. L. Heard, J. Chem. Phys., 2000, **112**, 10095
14. Y.-G. Wang, K. B. Wiberg and N. H. Werstiuk, J. Phys. Chem. A, 2007, **111**, 3592
15. P. B. Coto, D. Roca-Sanjuán, L. Serrano-Andrés, A. M. Pendás, S. Martí and J. Andrés, J. Chem. Theory Comput., 2009, **5**, 3032
16. R. Chávez-Calvillo and J. Hernández-Trujillo, J. Phys. Chem. A, 2011, **115**, 13036–13044
17. V. Tognetti and L. Joubert, Chem. Phys. Lett., 2013, **557**, 150–153
18. N. J. Turro, Modern molecular photochemistry, The Benjamin Cummings, Menlo Park, California, 1978.
19. C. F. Matta and R. J. Boyd, The Quantum Theory of Atoms in Molecules: From Solids State to DNA and Drug Design, Wiley-VCH, 2007.
20. B. O. Roos, P. R. Taylor and P. E. M. Siegbahn, Chem. Phys., 1980, **48**, 157–173
21. T. H. Dunning, J. Chem. Phys., 1989, **90**, 1007–1023
22. F. Aquilante, L. De Vico, N. Ferré, G. Ghigo, P. Malmqvist, P. Neogrády, T. B. Pedersen, M. Pitoňák, M. Reiher, B. O. Roos, L. Serrano-Andrés, M. Urban, V. Veryazov and R. Lindh, J. Comput. Chem., 2010, **31**, 224–247.
23. W. Zou, Molden2AIM, http://people.smu.edu/wzou/program/index.html.
24. R. Biegler-Knig, R. Bader and T.-H. Tang, J. Comput. Chem., 1982, **13**, 317
25. T. Lu and F. Chen, J. Comput. Chem., 2012, **33**, 580–592
26. A. D. Becke and K. E. Edgecombe, J. Chem. Phys., 1990, **92**, 5397–5403
27. A. M. Pendás, M. A. Blanco and E. Francisco, J. Comput. Chem., 2006, **28**, 161–184.
28. A. M. Pendás, M. Blanco and E. Francisco, J. Comput. Chem., 2009, **30**, 98–109.
29. R. J. Gillespie and P. L. A. Popelier, Chemical Bonding and Molecular Geometry, Oxford University Press, 2001
30. I. Fourré, B. Silvi, P. Chaquin and A. Sevin, J. Comput. Chem., 1999, **20**, 897–910.

CHAPTER 4

IONIC LIQUID MIXTURES

Richard P. Matthews, Ignacio J. Villar-Garcia, Cameron C. Weber, Jeraime Griffith, Fiona Cameron, Jason P. Hallett, Patricia A. Hunt and Tom Welton

INTRODUCTION

Ionic liquids (ILs) are low melting salts that have found application in areas as diverse as organic and inorganic synthesis,[1,2] catalysis,[3] the preparation of materials,[4–6] energy applications,[7]separations[8] and tribology.[9] One of the ideas central to many of these applications has been the ability to modify IL properties through appropriate ion selection, behaviour that has led to these liquids being referred to as 'designer solvents'.[10] An extension of this concept is the fine tuning of IL properties through the use of mixtures.[11] To accomplish this in a rational fashion, the impact of mixing ILs on their structure and the effect this has on their physical and chemical properties needs to be fully understood.

The structures of ILs are governed by an array of different intermolecular interactions, including Coulombic forces, dispersion interactions, H-bonding and π-interactions.[12] While a substantial amount of work has been conducted examining the structure of simple ILs, the impact of mixing on solution structure has received considerably less attention.[12] Nonetheless, a major finding consistent across a range of IL mixtures is that the structure of IL mixtures is dominated by the random distribution of ions driven by Coulombic interactions.[13–17] Within the ionic environment, more subtle structural perturbations arise as a result of the other (weaker) intermolecular forces present within these complex fluids.

A number of studies have looked at some of the effects of H-bonding interactions on IL mixtures. Brüssel et al. examined $[C_2C_1im]Cl_x[SCN]_{1-x}$ mixtures using ab initio molecular dynamics (MD) simulations and NMR experiments.[18,19] They found that the $[SCN]^-$ anion in the mixture is displaced from interacting with H^2 on the imidazolium cation (see Fig. 1 for the labelling of imidazolium cation atoms) by the stronger coordination of the Cl^- anion. This illustrates that the cation can preferentially H-bond with one particular anion over another. Similar H-bonding effects have been observed by Rebelo and coworkers for mixtures of $[NH_4][SCN]$ with ILs $[C_4C_1im][EtSO_3]$, $[C_4C_1im][EtSO_4]$ and $[C_4C_1im][OAc]$. Utilising NMR spectroscopy it was found that the presence of the $[NH_4]^+$ cation reduced the strength of H-bonding interaction between the imidazolium ring hydrogens and the stronger H-bond accepting anion, which was inferred to arise from competitive H-bonding interactions between the $[NH_4]^+$ cation and this anion.[20,21] MD simulations indicated that the preferential H-bonds between these ions did not affect the random distribution of ions in the network. Wang et al. and D'Anna et al. observed preferential H-bonding interactions between imidazolium cations and specific anions that give rise to non-linear NMR behaviour, particularly for H^2, in $[C_2C_1im][OAc]_x[NTf_2]_{1-x}$ ($[NTf_2]$ = bis(trifluoromethanesulfonyl)imide) and 1-benzyl-3-butylimidazolium ($[BzC_4im][BF_4]_x[NTf_2]_{1-x}$) mixtures respectively.[22,23]

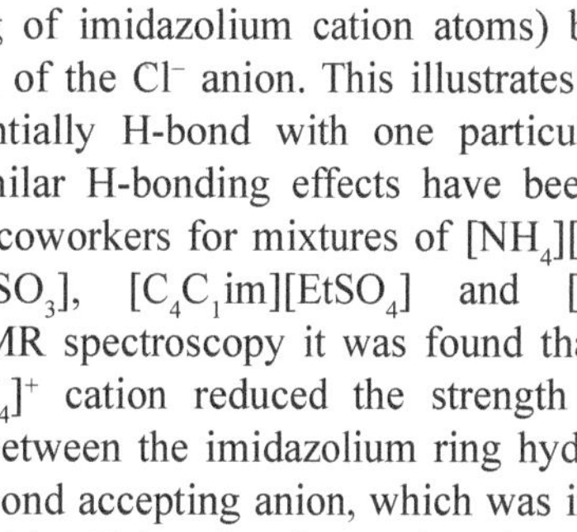

Figure. 1 The $[C_4C_1im]^+$ cation with atomic sites labelled as they will be referred to within this article.

Other forms of spectroscopy have also observed preferential H-bonding behaviour. Optical Kerr effect (OKE) spectroscopy was used to study $[C_5C_1im]Br_x[NTf_2]_{1-x}$ and $[C_5C_1im][PF_6]_x[TFA]_{1-x}$ (TFA = trifluoroacetate).[24,25] $[C_5C_1im][PF_6]_x[TFA]_{1-x}$ was found to exhibit non-additivity in the

---

Citation: Richard P. Matthews, Ignacio J. Villar-Garcia, Cameron C. Weber, Jeraime Griffith, Fiona Cameron, Jason P. Hallett, Patricia A. Hunt and Tom Welton, ionic liquid mixtures, DOI: 10.1039/c6cp00156d

OKE spectra which was interpreted as random ion association without preferential interactions, however, $[C_5C_1im]Br_x[NTf_2]_{1-x}$ yielded additive behaviour. This was inferred as evidence for the formation of a nanostructured random block network, consistent with the development of a H-bonding network based on the more effective H-bond accepting Br^- anion. Far-infrared spectroscopy illustrated the formation of preferential H-bonds between the NH group on the protic triethylammonium cation and $[OAc]^-$ anions relative to $[OTf]^-$ anions for $[C_2C_2C_2N][OAc]_x[OTf]_{1-x}$ mixtures.[26] Furthermore, the preferential association of the Cl^- anion with H^2 has been predicted computationally for $[C_4C_1im]Cl_x[PF_6]_{1-x}$ with no strong preference for either anion determined for $[C_4C_1im][BF_4]_x[PF_6]_{1-x}$ due to the similar size and H-bond accepting capacity of these anions.[13] Collectively these results suggest that preferential H-bonding interactions do occur within IL mixtures. In imidazolium ILs the interaction of the stronger H-bond accepting anion with H^2 appears to be favourable although the orientation of the weaker H-bond accepting anion and an understanding of the magnitude of this preference has not been thoroughly established.

A less studied feature of imidazolium ILs is their ability to engage in π–π stacking interactions.[27] The term π–π stacking is taken here to include the π^+–π^+ interactions between cationic aromatic imidazolium rings within an IL. For these interactions, the positive charge acts to both reduce dispersive interactions through the contraction of the π electron cloud, and to introduce a repulsive cation–cation Coulombic contribution. In this context π^+–π^+ interactions are only observed in the presence of additional molecules (solvent or counter anions) which facilitate and stabilise the interaction.[27] Moreover the term "stacking" suggests rings which are positioned directly one above the other, however this is not the lowest energy configuration and possible stacking conformers (depending on the relative orientation of the two imidazolium rings) include stacked-parallel, stacked-rotated, stacked-antiparallel, stacked-displaced and T-shaped conformers.[27]

Evidence for π^+–π^+ interactions between imidazolium rings has been observed in the solid state for ILs including $[C_2C_1im][NO_3]$, $[C_2C_1im][NO_2]$, $[C_2C_1im]_2[SO_4]{\cdot}H_2O$, $[C_1C_1im][OTf]$ and $[C_1C_1im][NTf_2]$.[28–31] Despite evidence for π^+–π^+ stacking in the solid state, $[C_1C_1im][NTf_2]$ was not found to possess these interactions in the liquid state.[28,31] Moreover, π^+–π^+ motifs have been assigned based on NOE data for $[C_2C_1im]Cl$ in dichloromethane and in neat $[C_4C_1im][BF_4]$ and $[C_4C_1C_1{}^2im][BF_4]$.[32,33] π^+–π^+ stacking interactions have been examined computationally for imidazolium salts and found to occur in ILs including $[C_1C_1im]Cl$ and $[C_1C_1im][NO_3]$, but not in ILs with more weakly interacting anions.[27,34] This phenomenon was attributed to the stabilisation of the cationic π^+–π^+ interactions through both Coulombic and H-bonding interactions between the IL cation and anion.[35] Further examples of π^+–π^+interactions have been determined computationally. These include the observation of parallel and anti-parallel orientated cation rings for $[C_2C_1im]Br$[36] and further relationships between anion size/type and π^+–π^+ interactions have been shown,[18,37] with a general finding being that π^+–π^+ interactions are favoured by smaller anions as larger anions induce a greater distance between imidazolium rings, preventing their interaction.

The effect of forming an IL mixture on π^+–π^+ stacking has not been widely examined, although MD simulations performed by Brüssel et al. discovered that the $[SCN]^-$ anion in $[C_2C_1im]Cl_x[SCN]_{1-x}$ is capable of disrupting these interactions as the $[SCN]^-$ anion preferentially occupies a location above the plane of the imidazolium ring.[18] The preference of larger more diffuse anions to locate above the imidazolium ring and engage in anion–π^+ interactions may be competitive with π^+–π^+ stacking, thus limiting the range of ILs that exhibit π^+–π^+phenomena, particularly in mixtures.

Several physicochemical studies have demonstrated that mixtures of ILs often show close to ideal mixing behaviours.[38–42] A mixture is considered ideal if the chemical potential of its components can be directly related to the chemical potential of the pure component and its mole fraction in the mixture.[11] A consequence of this definition is that for ideal mixtures the enthalpy of mixing is zero and properties thermodynamically related to the chemical potential (such as molar volume) scale with the mole fraction of the component. There is no direct relationship between the excess enthalpy or excess molar volume in terms of sign or magnitude although non-zero values of either component are indicative of a non-ideal mixture.

We have recently reported the physical properties of an extensive range of IL mixtures.[43] Most of the IL mixtures studied possessed negligibly small excess molar volumes, indicating that they display close to ideal thermodynamic behaviour. Non-ideality was observed for mixtures with a substantial difference in the H-bond accepting abilities of the anions, however, the magnitude of non-ideality was small with excess molar volumes below 0.5% of the total molar volume being observed.

Nevertheless, not all mixtures of ILs will be ideal or nearly ideal and exceptions to the ideality of IL mixtures have been observed. Mixtures based on bulky phosphonium cations, such as trihexyltetradecylphosphonium chloride ($[C_6C_6C_6C_{14}P]Cl$), which contain large alkyl groups that effectively shield the central phosphorous, exhibit non-ideal behaviour at concentrations where both ILs are miscible.[44,45] In such IL mixtures it is anticipated that the steric bulk of the alkyl groups interferes with the ability of the anion to associate closely with the cationic core, resulting in an increased association of anions with the smaller cation, which has a more accessible charge centre. ILs composed of one ion possessing large non-polar groups (typically the cation) have been discovered to be immiscible with ILs containing smaller highly interacting ions, moreover large excess enthalpies of mixing have been calculated for these systems.[46]

Despite the prevalence of specific interactions such as preferential H-bonding between different ions, close to ideal mixing of ILs has been shown to be remarkably commonplace, warranting a systematic study of the structures of IL mixtures. It is clear from previous studies that perturbations to H-bonding interactions can arise when anions of different H-bond accepting capacity are employed in mixtures. The effect of the anion on phenomena such as the π^+–π^+ stacking of imidazolium rings within mixtures has also not been studied experimentally to date.

In order to address these issues, six different IL mixtures based on a common $[C_4C_1im]^+$ cation were studied experimentally using NMR techniques and theoretically through the use of MD simulations. IL mixtures were chosen to give a variety of anion sizes and H-bond accepting differences;

$[C_4C_1im][Me_2PO_4]_x[NTf_2]_{1-x}$, $[C_4C_1im][OAc]_x[NTf_2]_{1-x}$, $[C_4C_1im][OTf]_x[NTf_2]_{1-x}$, $[C_4C_1im]Cl_x[NTf_2]_{1-x}$, $[C_4C_1im]Cl_x[SCN]_{1-x}$ and $[C_4C_1im]Cl_x[OTf]_{1-x}$. Furthermore, our previous physicochemical investigations found that $[C_4C_1im][Me_2PO_4]_x[NTf_2]_{1-x}$ formed non-ideal mixtures whereas $[C_4C_1im][OTf]_x[NTf_2]_{1-x}$ mixtures were very close to ideal, with $[C_4C_1im]Cl_x[NTf_2]_{1-x}$ being intermediate between these two extremes. Thus, these mixtures were chosen to enable a direct comparison between mixtures of similar ILs that exhibit an increase in non-ideality.[43] $[C_4C_1im]Cl_x[SCN]_{1-x}$ was included to build upon the computational work of Brüssel et al.[18]

RESULTS AND DISCUSSION

NMR of Undiluted Ionic Liquids and Their Mixtures

The ^1H and ^{13}C NMR spectra of the undiluted ILs and their mixtures were obtained to gain an understanding of any preferential interactions, such as H-bonding, between specification sites and the anions, as well as the magnitude of these associations. While the absolute chemical shifts obtained for each chemical environment do not provide direct information on the structure of the mixture, the examination of different chemical environments for each mole fraction of a component within the mixture allows for the presence of specific interactions to be inferred (vide infra). For the NMR spectra, a capillary filled with DMSO-d_6 was used as a lock and chemical shift reference. As $[C_4C_1im]Cl$ is solid at room temperature, these investigations were performed at 80 °C to enable all compositions to be explored.

Fig. 2 depicts the H^2, H^4 and H^5 signals for $[C_4C_1im][Me_2PO_4]_x[NTf_2]_{1-x}$, $[C_4C_1im]Cl_x[SCN]_{1-x}$ and $[C_4C_1im][OTf]_x[NTf_2]_{1-x}$. All graphs are depicted with the concentration of the most H-bond accepting anion in the mixture on the x-axis to enable a more facile comparison between mixtures. ^1H and ^{13}C NMR chemical shifts are provided in ESI,† Tables S28–S50 and Fig. S24–S42, and details of the fitted model are provided in the ESI,† (Section D) and below.

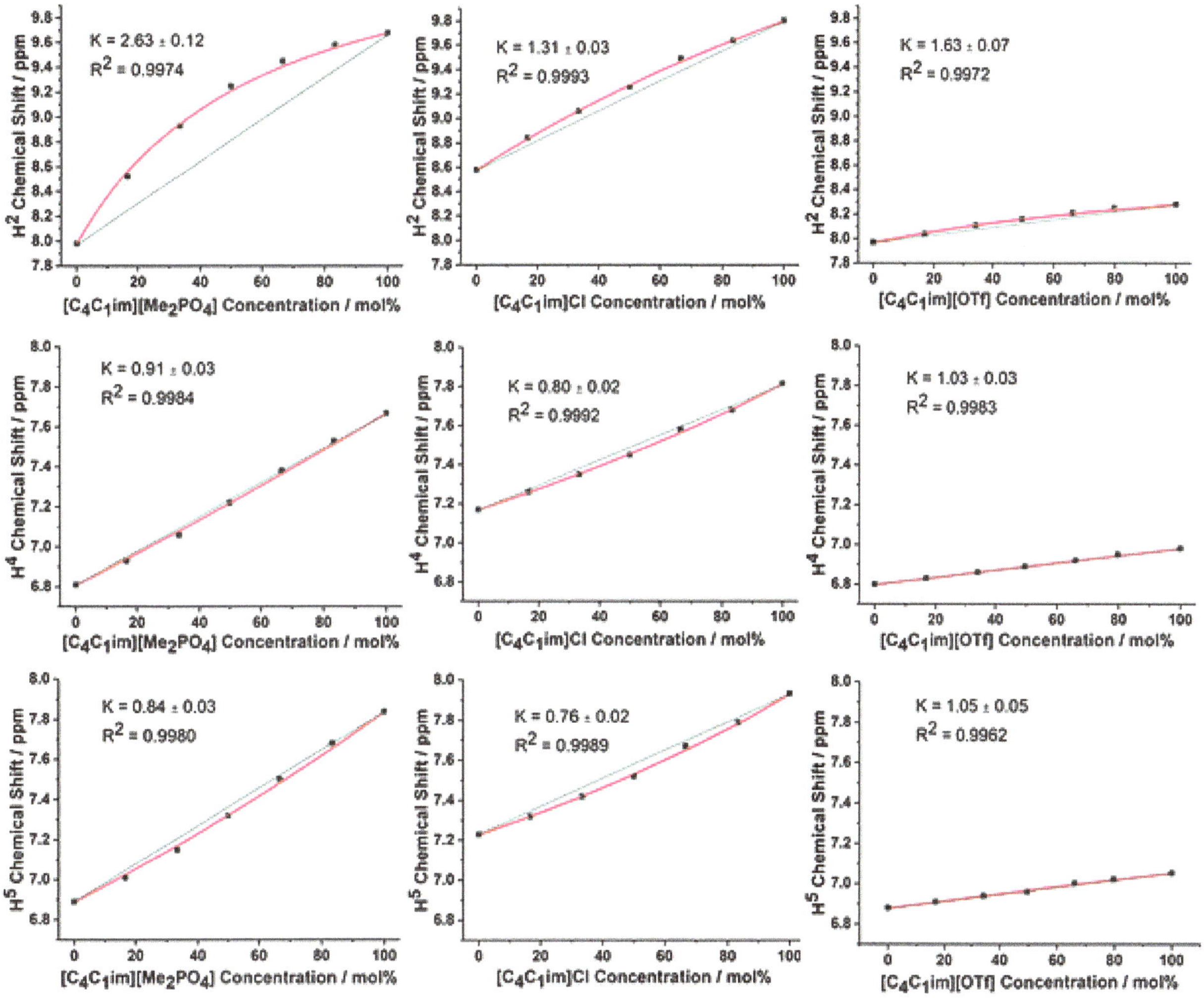

Figure. 2 Neat ^1H chemical shifts obtained for H^2 (top), H^4 (middle) and H^5 (bottom) of the $[C_4C_1im][Me_2PO_4]_x[NTf_2]_{1-x}$ mixtures (left), $[C_4C_1im]Cl_x[SCN]_{1-x}$ mixtures (middle) and $[C_4C_1im][OTf]_x[NTf_2]_{1-x}$ mixtures (right). Red lines represent the fitted model with given fitting

parameter (see below), straight black lines are drawn between the chemical shifts of the simple ILs as a guide for the eye.

From Fig. 2 and ESI, Fig. S24–S29 it is apparent that the chemical shifts for H^2, H^4 and H^5 follow a similar trend for all of the IL mixtures. All of the chemical shifts for these resonances lie within the range defined by the simple IL components of the mixture. Increasing the strength of H-bonding interactions has been associated with a downfield shift of the H-bond donating resonance in the ^1H NMR, i.e. an increased chemical shift. Hence, displacement of the chemical shift above the black line joining the chemical shift of both simple ILs is indicative of preferential H-bonding between the relevant hydrogen and the stronger H-bond acceptor, i.e. the ^1H environment interacts to a greater extent with the stronger H-bond accepting anion over the NMR timescale than would be expected from a random association of anions with that environment. These results indicate that H^2 preferentially interacts with the stronger H-bond accepting anion, while H^4 and H^5 either exhibit no preference for either anion or a slight preference for the more weakly H-bonding anion.

The impact of H-bonding on the ^{13}C NMR signals is more complex as there is an interplay between electron density changes due to the formation of the H-bond, and variations in the energy of the molecular orbitals. Despite experimental and theoretical results for ILs indicating that electron density increases on the carbon atom upon the formation of a stronger C–H···A⁻ H-bond, the ^{13}C signals of $C^{2/4/5}$ move downfield with increasing H-bond accepting ability of the anion.[47,48] This phenomenon has been attributed to the reduction in the natural bond order of the C–H bond dominating electron density effects. Correspondingly, the influence of stronger H-bonding on the chemical shift of ^{13}C NMR signals would be expected to follow a similar trend to that of the ^1H NMR, i.e. a downfield (higher ppm) shift. The ^{13}C NMR results for C^2 (Fig. 3 and Fig. S31–S36, Tables S35–S40, ESI†) are largely consistent with those obtained for H^2 for all of the mixtures.

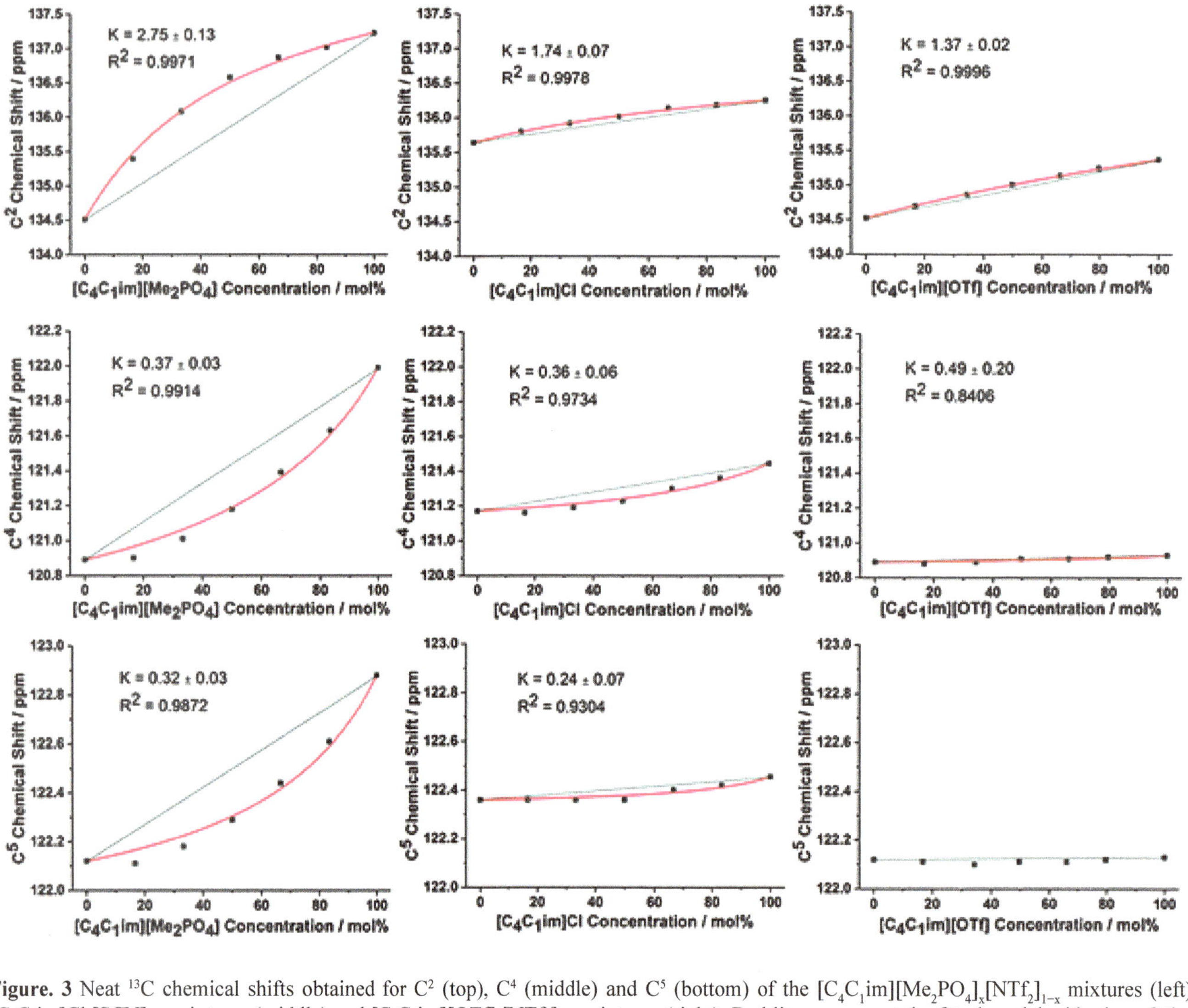

Figure. 3 Neat ^{13}C chemical shifts obtained for C^2 (top), C^4 (middle) and C^5 (bottom) of the $[C_4C_1im][Me_2PO_4]_x[NTf_2]_{1-x}$ mixtures (left), $[C_4C_1im]Cl_x[SCN]_{1-x}$ mixtures (middle) and $[C_4C_1im][OTf]_x[NTf_2]_{1-x}$ mixtures (right). Red lines represent the fitted model with given fitting

parameter (see below), straight black lines are drawn between the chemical shifts of the simple ILs as a guide for the eye.

The ^{13}C NMR results for C^4 and C^5, however, show a much greater affinity for the more weakly H-bond accepting anion than their H^4 and H^5 counterparts, i.e. a larger negative deviation from linearity, with the exception of $[C_4C_1im][OTf]_x[NTf_2]_{1-x}$ where the chemical shift change between the simple ILs was too small to determine a distinct trend. This suggests that the more weakly H-bond accepting anion has a greater preference for interactions with C^4/C^5 than H^4/H^5, consistent with the anion predominantly occupying a position above the imidazolium ring rather than an in-plane interaction with the back of the ring.

A quantum chemical study of ion-pair dimers (IP-dimers) which included $[C_1C_1im]_2[OTf]_2$ found that this IP-dimer only formed stable conformers with an alternating (diagonal anion) configuration, Fig. 4.[34] In these conformers the anion is positioned above the ring with one O-atom sitting over the C^4=C^5 bond at the rear of the ring. This conformation is consistent with the ^{13}C NMR data which suggests preferential interaction of weak H-bonding anions with the $C^{4/5}$ atoms. This can be rationalised by weakly H-bonding anions being typically more diffuse and better able to form dispersive interactions with the π-cloud of the imidazolium ring than strongly coordinating anions. In these calculated structures, there are no H-bonds with the rear of the ring, but there can be supporting H-bond interactions with the alkyl chain. The calculated structures do not feature a more strongly H-bond accepting anion which would interact preferentially with H^2 and could promote the interaction of the weakly coordinating anions with $H^{4/5}$ as is observed from the NMR data.

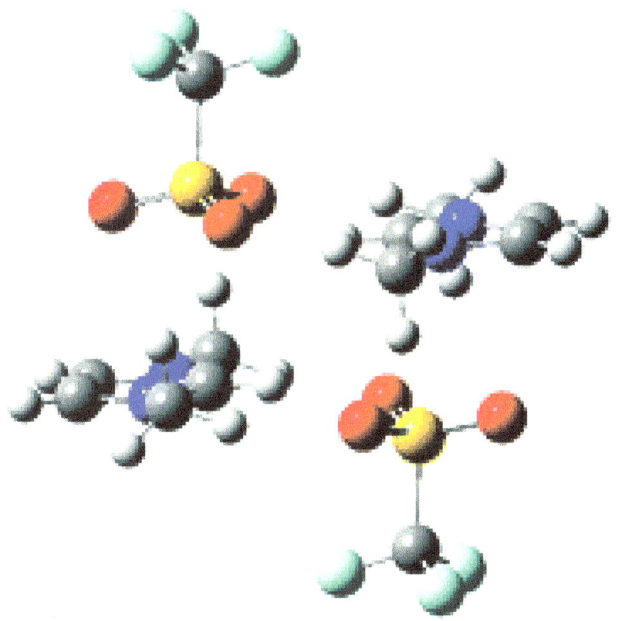

Figure. 4 Stable IP-dimer of $[C_1C_1im]_2[OTf]_2$ showing a strong anion–π^+ interaction, the effect on $C^{4/5}$ will be much stronger than the effect of the anion on $H^{4/5}$.

It is worth noting the contrast between the curves in NMR chemical shift obtained here (for H^2, H^4 and H^5) compared to the linear behaviour reported by Brüssel et al. for $[C_4C_1im]Cl_x[SCN]_{1-x}$ mixtures.[18] This discrepancy can be rationalised by consideration of the different temperatures used for the experiments which precluded, in their case, the examination of mixtures with mole fractions of Cl^- anions greater than 0.5. The very subtle variation from linearity observed in Fig. 2 would not be evident over the reduced composition range, leading to the initial finding of a linear relationship. Nonetheless, in both cases it is evident that within these mixtures any preferential interactions of either anion with a specific position on the cation ring are small in magnitude.

Another important comparison is between the non-linear NMR results obtained here and the linear dependence of the C^2–H^2 IR stretch obtained for similar mixtures containing the $[C_4C_1im]^+$ cation.[49] One of the major differences between these two techniques is the time scale, with IR providing measurements over a much shorter timescale than NMR. This suggests that perhaps the total number of H-bonds formed with each anion varies consistently with the anion composition of the mixture (leading to the linear dependence in IR) but the lifetime of H-bonds with the better H-bond accepting anion are longer (leading to a non-linear NMR dependence which represents a time weighted average of these interactions). Differences in H-bond dynamics have been investigated computationally for simple ILs previously,[50,51] but are beyond the scope of the current study.

To enable trends between mixtures to be more readily compared, a simple model is proposed, based on an equilibrium between the IL anions (A_1 and A_2) and an interacting atom on the cation (C_i, eqn (1) and (2)). This model is an adaptation of that introduced by Bosch and Rosés as well as Skwierczynski and Connors to analyse the preferential solvation of solvatochromic dyes in mixed solvent systems.[52,53] While this model has been used previously to analyse solvatochromic dyes and rate constants in binary IL mixtures,[54,55] to the best of our knowledge this is the first time that it has been applied to the modelling of NMR data.

$$C_iA_1 + A_2 \rightleftharpoons C_iA_2 + A_1 \qquad (1)$$

$$\delta(x) = \frac{Kx(\delta(A_1) - \delta(A_2))}{1 - x + Kx} + \delta(A_1) \qquad (2)$$

In eqn (2), $\delta(A_{1/2})$ are the chemical shifts of C_i in simple $[C_4C_1im]A_{1/2}$ respectively, x is the molar proportion of A_2 in the mixture and K is the fitted equilibrium constant for eqn (1).

From eqn (1) and (2), A_1 and A_2 were assigned such that A_2 was always the more strongly H-bond accepting anion, as determined by its Kamlet–Taft β value, so that more favourable interactions of this anion with a chemical environment on the cation would result in K > 1 and vice versa. It is important to note that this model implicitly assumes that the chemical shift can be described as an equilibrium between the local structures of the two simple ILs at a given cation environment, regardless of what those structures are, and does not necessarily imply the formation and exchange of well-defined ion pairs. Consequently the model is not able to account for conformations not present at the cation environment in the simple ILs (e.g. a bifurcated H-bond between different anions). However, despite these assumptions and the simplicity of this model, the majority of our data is well modelled by this approach. Full details of the underlying assumptions and limitations of the model are detailed in ESI, (Section D).

Fig. 2 and 3 detail the equilibrium constants obtained using the model for the $H^{2/4/5}$ atoms and $C^{2/4/5}$ atoms of $[C_4C_1im][Me_2PO_4]_x[NTf_2]_{1-x}$, $[C_4C_1im]Cl_x[SCN]_{1-x}$ and $[C_4C_1im][OTf]_x[NTf_2]_{1-x}$. The K values obtained represent a relative preference rather than absolute information. A large K value is not necessarily indicative of stronger H-bonding at that site, only that there is a greater preference for that cation environment to be dominated by interactions with the most strongly H-bond accepting anion. In general, it was found that the strong H-bond accepting anion interacts preferentially with H^2 and C^2 (K > 1.0) and that the more weakly coordinating H-bonding anion likely forms anion–π^+ interactions with $C^{4/5}$

(K ≪ 1.0) and weak preferential interactions with $H^{4/5}$ (K < 1.0).

To assess trends in preferential interactions across the range of IL mixtures examined, the free energy changes derived from the fitted values of K (using $\Delta G = -RT \ln K$) have been plotted against the difference in H-bond accepting strength of the anions, as determined by the difference in Kamlet–Taft β values (Fig. 5). The Kamlet–Taft $\Delta\beta$ values used for the mixtures are given in ESI, Table S42. The free energies derived from this model should be treated with some caution given the simplicity of the underlying model. Nonetheless, the excellent fits to the chemical shift data obtained from multiple independent experiments suggests that it should be a reasonable approximation of the true free energy difference.

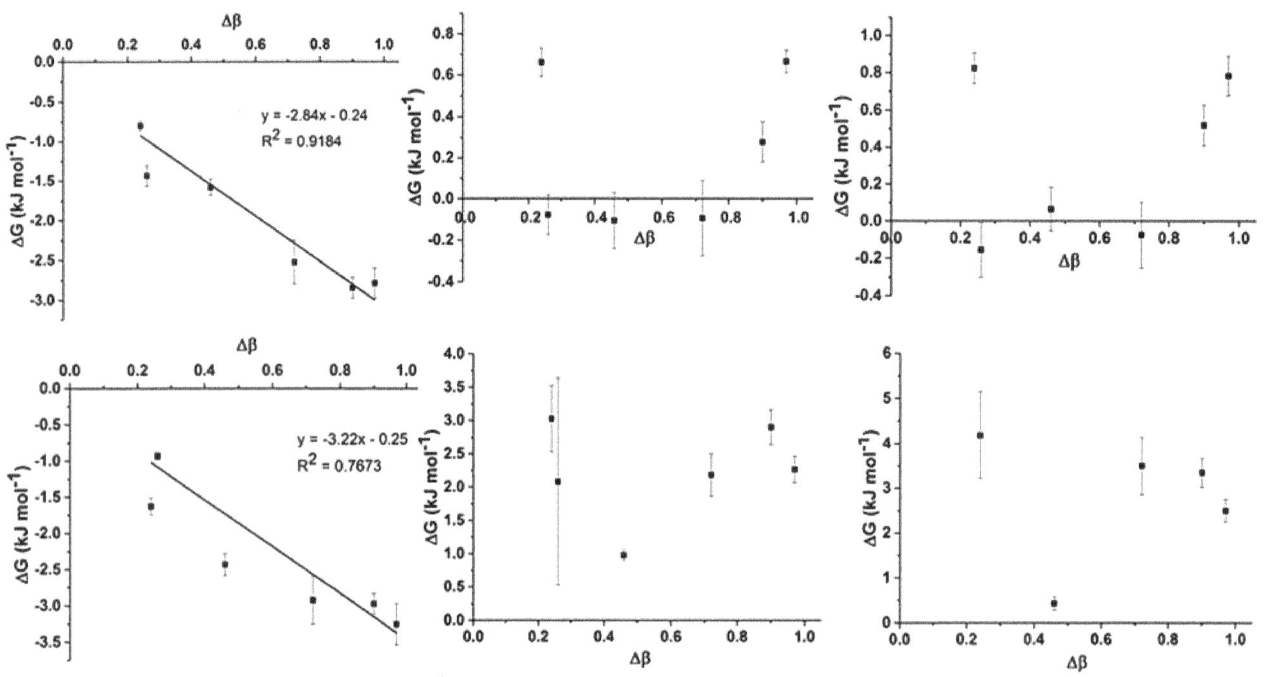

Figure. 5 Relationship between ΔG obtained from the preferential interaction model and $\Delta\beta$ of the IL mixture for H^2 (top left), H^4 (top centre), H^5 (top right), C^2 (bottom left), C^4 (bottom centre) and C^5 (bottom right). Where an approximately linear relationship was observed, a linear fit was made and the fitted parameters reported. Reported errors are fitting errors propagated through the free energy calculation.

From Fig. 5, some interesting trends emerge. The preferential interaction of H^2 and C^2 with the more strongly H-bonding anion increases as the difference in the H-bond acceptor capacity of the anion increases. This result is not surprising given the well-known ability of H^2 to be the most effective H-bond donor on the imidazolium cation.[33,48] However, it is more remarkable that this preference appears to be very small in magnitude (~3 kJ mol^{-1}) even for strongly interacting anions such as $[Me_2PO_4]^-$ and $[OAc]^-$. The computed interaction energies for ion pairs of $[C_1C_1im]Cl$ and $[C_1C_1im][Me_2PO_4]$ with H^4 (side conformer) and H^2 (front conformer) only differ by ≈10 kJ mol^{-1} in a CH_2Cl_2 CPCM and an IL SMD model which will be discussed in greater detail in the following section. These energy differences are close to that of thermal energy (RT) at 80 °C suggesting that $H^{2/4/5}$ will all be approximately equally populated over time, allowing the other anion within the mixture easy access. It is mutually supportive that both experimental and computational methods independently arrive at similar (very small) energy differences.

There is no generalised trend for preferential interactions with $H^{4/5}$ and $C^{4/5}$. There is a limited trend for the weaker anion to increasingly associate with $H^{4/5}$ for $\Delta\beta$ above an apparent threshold of $\Delta\beta > 0.72$. However, ΔG for these preferential interactions is less than 1 kJ mol^{-1}. A notable outlier is the $[SCN]^-$ anion in the $[C_4C_1im]Cl_x[SCN]_{1-x}$ mixtures ($\Delta\beta = 0.24$) which appears to preferentially interact with H^4 and H^5 despite both $[SCN]^-$ and Cl^- exhibiting similar H-bond accepting ability. The $[OTf]^-$ and $[NTf_2]^-$ anions have an analogously small $\Delta\beta$ (0.26) but $[NTf_2]^-$ does not exhibit a stronger preference over $[OTf]^-$ for $H^{4/5}$. The contrast between the pairs $Cl_x[SCN]_{1-x}$ and $[OTf]_x[NTf_2]_{1-x}$ may be due to the greater absolute H-bond accepting ability of $[SCN]^-$ and Cl^-, hence $[SCN]^-$ is more likely to favour the formation of H-bonding interactions than $[NTf_2]^-$.

C^4 and C^5 always favour interacting with the weaker H-bond accepting anion (all positive ΔG), although this preference is very small for $[C_4C_1im]Cl_x[OTf]_{1-x}$ (ΔG < 1 kJ mol^{-1}). $[C_4C_1im]Cl_x[OTf]_{1-x}$ is the only mixture in which $[OTf]^-$ is the weakest H-bond acceptor (the others feature $[SCN]^-$ or $[NTf_2]^-$) and therefore this result may indicate that the positioning of $[OTf]^-$ at C^4 and C^5 may be determined by factors other than simply H-bonding ability. This result could be rationalised by considering that $[OTf]^-$ is smaller and less diffuse than $[NTf_2]^-$, leading to a reduced propensity of this anion to form anion–π^+ interactions compared to $[NTf_2]^-$ in the $Cl_x[NTf_2]_{1-x}$ mixtures. Compared to the $Cl_x[SCN]_{1-x}$ mixtures, $[OTf]^-$ is a weaker H-bond acceptor and the polarisable sulfur atom is less accessible than that in $[SCN]^-$ leading to weaker H-bonding interactions with $H^{4/5}$ (incorporated in $C^{4/5}$ changes) as well as weaker dispersive interactions with the π system leading to the comparatively reduced $C^{4/5}$ preference for $[OTf]^-$ in $Cl_x[OTf]_{1-x}$ mixtures. MD simulations (discussed fully below) show that both the Cl^- and $[OTf]^-$ interact almost equally with C^4 and C^5 (ESI, Fig. S4), consistent with the small preference observed experimentally.

To isolate H-bonding effects from geometric considerations, the 1-butyl-2,3-dimethylimidazolium ($[C_4C_1C_1{}^2im]^+$) cation was employed in mixtures with Cl^- and $[NTf_2]^-$ anions. Methylating the C^2 position removes H^2, the most H-bond acidic site on the imidazolium cation, rendering H^4 and H^5 the most H-bond acidic positions, albeit with a lower propensity for H-bond donation than H^2 in the unsubstituted imidazolium ring. Results for the 2-methyl, ring protons and the corresponding ^{13}C signals are depicted in Fig. 6, with all NMR signals reported in ESI, Fig. S30, S37, Tables S34 and S41.

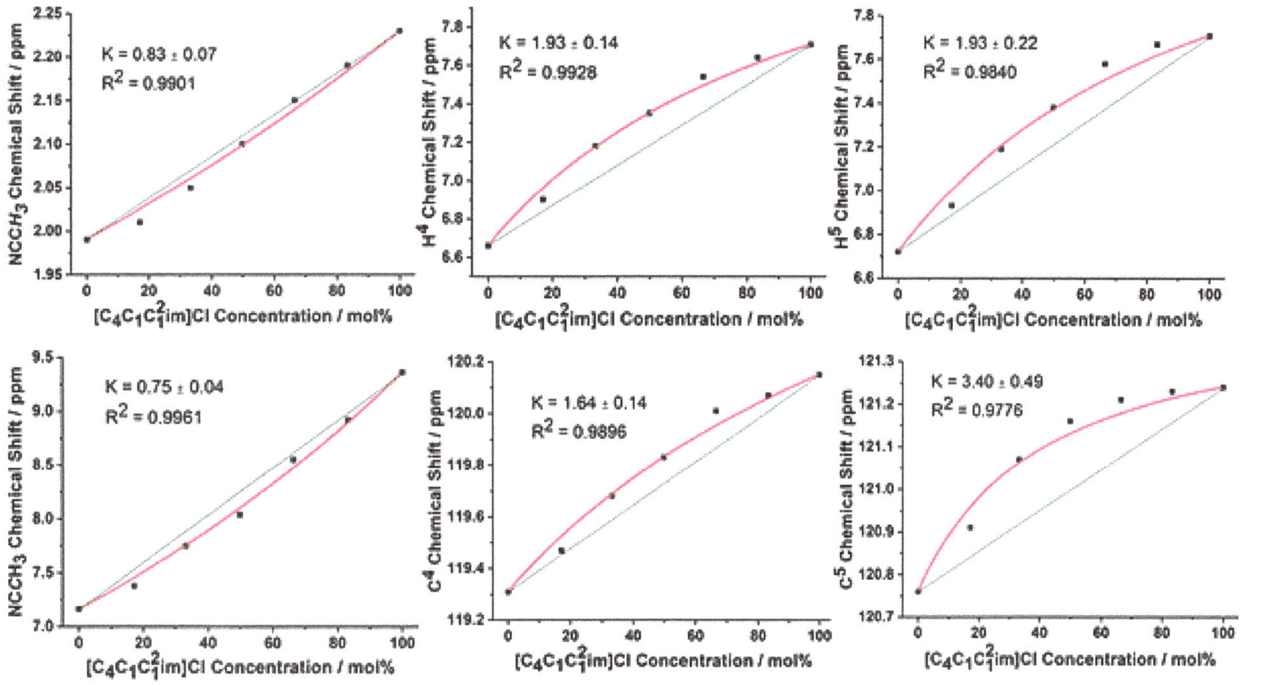

Figure. 6 Neat ^1H (top) and ^{13}C (bottom) NMR chemical shifts for the 2-methyl (denoted $NCCH_3$), C^4/H^4 and C^5/H^5 of the $[C_4C_1C_1{}^2im]Cl_x[NTf_2]_{1-x}$ mixtures. Red lines represent the fitted model, straight black lines are drawn between the chemical shifts of the simple ILs as a guide for the eye.

The behaviour for this set of mixtures is different to that of the mixtures based on the $[C_4C_1im]^+$ cation. Notably, the Cl^- anion now preferentially interacts with H^4 and H^5 at the back of the ring which are now the strongest available H-bond donor sites. Moreover, there is a slight preference for the $[NTf_2]^-$ anion to interact with the 2-methyl position at the front. The change in preferential interaction is directly in line with the H-bond strength of these positions and illustrates that the H-bond donor ability of the ring positions determines the extent of the preferential association with specific anions in the mixture.

Again, these preferential interactions are very small, with K values for the H^4/H^5 positions less than 2 ($\Delta G < 2$ kJ mol^{-1}). These values are also smaller than the analogous K and ΔGfound for the H^2 position of the $[C_4C_1im]Cl_x[NTf_2]_{1-x}$ mixtures, consistent with the higher H-bond donating ability of H^2. The ^{13}C NMR is subtly different from the ^1H NMR results with an increased preference of the Cl^- anion for C^5 compared to C^4 whereas identical results were obtained for $H^{4/5}$. As the C^5 position is likely to be more sterically hindered compared to C^4due to the adjacent butyl chain this may simply represent a decrease in the interaction of the much larger $[NTf_2]^-$ at this position rather than increased Cl^- association. The 2-methyl position exhibits a slight preference for $[NTf_2]^-$, in line with the ^1H NMR, however the C^2 resonance has the most favourable interaction with the $[NTf_2]^-$ anion (K = 0.34 compared to K = 0.75 for 2-methyl), indicative of anion–π^+ interactions with the opposite side of the imidazolium ring to the strongest H-bond donating groups.

Results so far have suggested that there is a slight preference for the association of anions with specific locations around the cation due to H-bonding. Stronger H-bond acceptors have a preference for interactions with H^2 in the front of the ring, and weaker H-bond acceptors have a preference for an anion–π^+ interaction (above the ring) and/or H-bonding with H^4 and H^5. However, we have not yet probed the mechanism for movement of ions to these locations. Two possibilities are that anions can diffuse into the preferred location or, alternatively, a cation surrounded by a selection of anions can rotate so as to preferentially orientate towards a specific anion.

To probe this more thoroughly, $[C_4C_1im]_{0.50}[C_4C_1pyrr]_{0.50}Cl_x[NTf_2]_{1-x}$ and $[C_4C_1im]_x[C_4C_1pyrr]_{1-x}Cl_x[NTf_2]_{1-x}$ mixtures have been examined to determine whether preferential anion clustering around the imidazolium or pyrrolidinium ion could be observed. The $[C_4C_1pyrr]^+$ cation was chosen because it is a weaker H-bond donor than $[C_4C_1im]^+$, thus it is anticipated that strongly H-bonding anions such as Cl^- would favour interacting with $[C_4C_1im]^+$ over $[C_4C_1pyrr]^+$ if ion translation was able to occur. Conversely, if preferential association is simply due to the rotation of the imidazolium cation within an established ionic matrix, the chemical shifts observed should be independent of the cation composition. The results for H^2,

H^4 and H^5 and the corresponding ^{13}C signals for the $[C_4C_1im]^+$ cation are depicted in Fig. 7. The other signals, including those for the $[C_4C_1pyrr]^+$ cation, can be found in ESI, Fig. S38, S39, S41, S42 and Tables S43–S50. No reliable data could be obtained for $x > 0.50$ for the $[C_4C_1im]_{0.50}[C_4C_1pyrr]_{0.50}Cl_x[NTf_2]_{1-x}$ mixtures due to their high melting points which prevented a neat NMR spectrum being obtained at 80 °C.

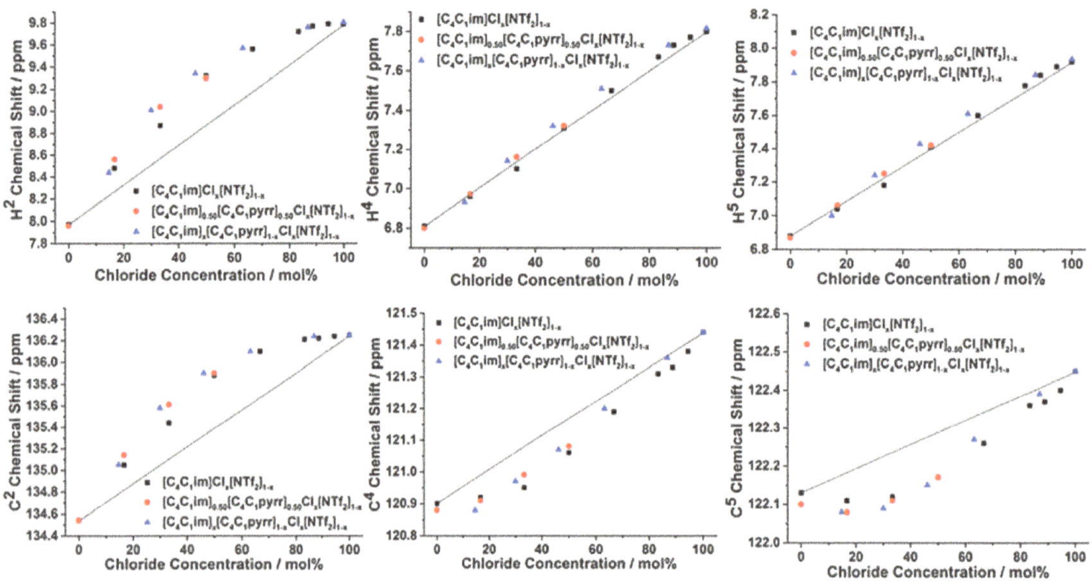

Figure. 7 Comparison of chemical shifts observed for $[C_4C_1im]Cl_x[NTf_2]_{1-x}$ (black squares), $[C_4C_1im]_{0.50}[C_4C_1pyrr]_{0.50}Cl_x[NTf_2]_{1-x}$ (red circles) and $[C_4C_1im]_x[C_4C_1pyrr]_{1-x}Cl_x[NTf_2]_{1-x}$ (blue triangles) mixtures as a function of the chloride ratio. The ^1H NMR for H^2, H^4 and H^5 are depicted on the top and ^{13}C NMR for C^2, C^4 and C^5 on the bottom. Black lines are drawn between the chemical shifts of the simple $[C_4C_1im]Cl$ and $[C_4C_1im][NTf_2]$ ILs as a guide to the eye.

From Fig. 7 it is apparent that the chemical shift observed is dependent only on the anion composition and not the relative proportion of cations present. The chemical shifts of $H^{2/4/5}$ or $C^{2/4/5}$ of $[C_4C_1im]^+$ do not change when $[C_4C_1pyrr]^+$ is present in the mixture, thus there is no local increase in the concentration of the strongly H-bonding Cl^- anions around the better H-bond donor $[C_4C_1im]^+$. Thus, there is no evidence for the preferential association or "clustering" of specific anions around the $[C_4C_1im]^+$ cation when the $[C_4C_1pyrr]^+$ cation is present. A similar outcome is observed from the NMR spectra of $[C_4C_1pyrr]^+$, with all ^1H and ^{13}C NMR signals being dependent only on the anion composition (ESI, Fig. S41 and S42). This result is important, it reinforces the hypothesis that there is an essentially random distribution of (mutually repulsive) anions and that preferential interactions arise from subtle perturbations to the orientation of the cations within the ionic framework, i.e. there is no net displacement of particular ions.[15]

However, preferential interactions of the imidazolium ring atoms with specific anions are identified, thus within a given solvation shell the cation is rotating so as to interact through H^2 with the best H-bond accepting anion, depicted schematically in Fig. 8. Blue cations represent the stronger H-bond donor and red anions the stronger H-bond acceptors. Within the random distribution of cations and anions, some "blue" cations find a "red" anion within their solvation shell and rotate so as to spend a larger proportion of time forming an H-bond with that anion. Simple rotation of the cation within a diffuse solvation shell is in accordance with the low energy (<3.5 kJ mol^{-1}) determined earlier using the equilibrium model. Fig. 8 illustrates a simplified snapshot and it is likely that the 'preference' observed by NMR relates more to H-bond dynamics than the absolute number of H-bonding interactions. Hence, sites with H-bonds between the best H-bond donating/accepting pair are longer lived than those between weaker pairs.

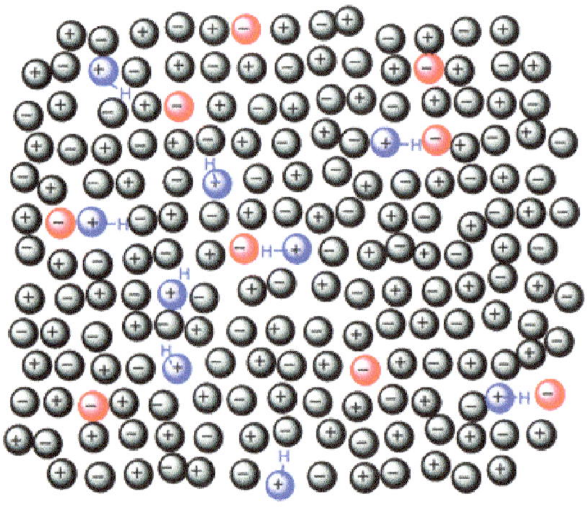

Figure. 8 Schematic 'snapshot' illustrating the preferential orientation of H-bonding within a randomly ordered ionic network for a reciprocal binary mixture $[C_1]_x[C_2]_{1-x}[A_1]_x[A_2]_{1-x}$ where blue cations represent those with good H-bond donating groups and red anions are the strongest H-bond acceptors.

NMR of the Simple Ionic Liquids in CD_2Cl_2

In the previous section, it was found that preferential interactions within the IL environment are relatively weak and relate to the rotation of cations within a randomly organised network of anions. As these effects are of relatively low energy and since ILs are well-known to possess a plethora of intermolecular forces, the influence of the formation of a mixture on other weak intermolecular interactions was studied. In particular, the presence of π^+–π^+ stacking and the role of the previously discussed anion–π^+ interactions are of interest.

In order to assess these weak interactions, it is necessary to eliminate contributions from stronger effects such as ion-pairing and H-bonding. Based on NMR and conductivity data, Avent et al. have proposed that imidazolium halide (Cl^-, Br^- or I^-) ILs in non-polar solvents, such as chloroform or dichloromethane, consist of associated ions, i.e. ion-pairs or other larger neutral ion clusters, rather than discrete solvated ions at all concentrations meaning that H-bonding and ion-pairing effects are not observed in NMR spectra with changing concentration.[33]However, as the concentration of the IL is increased the effects of clustering and aggregation of the ion-pairs can be observed.

In an NMR spectrum, where π^+–π^+ stacking occurs, the magnetic cone above and below the aromatic cation rings, induced by the delocalised π electrons, opposes the applied magnetic field, shielding nuclei within the cone and reducing the chemical shift. Nuclei outside of the cone are deshielded and have an increased chemical shift. Previously for IL solutions of [C_2C_1im]X (X = Cl, Br and I), an upfield shift of the H^2 signals with increasing IL concentration (in CD_2Cl_2) has been associated with the H^2 proton lying within the shielding cone of a neighbouring imidazolium ring. The H^4/H^5 signals typically shift downfield indicating they are outside the shielding cone. Taken together these results have been interpreted as the formation of rotated π stacked structures.[33]

It is worth emphasising that the NMR spectrum represents a weighted average over all conformations present within the liquid phase, so these results imply that rotated π stacked structures are preferentially formed or are longer lived than other conformations within the liquid.

Theoretical calculations on IL IP-dimers of imidazolium salts have identified two key motifs; a stacked cation conformer where the anions are located on the periphery but are positioned vertically (in the "middle") between the cations (Fig. 9a), and an alternating anion above or below the cation conformer where the anions are located on the "diagonal" and the cation rings are displaced relative to each other (Fig. 9b).[27,34] In the π^+–π^+ stacked IP-dimer structure, anions are positioned in front of H^2 and slightly out of plane of one imidazolium ring while the anions also lie slightly out of plane and interact via an H-bond with the rear $H^{4/5}$ of the other imidazolium ring. Calculations suggest that this arrangement stabilises an otherwise repulsive π^+–π^+ interaction and is facilitated by the strength of the H-bonds between the cation and anion.[34] Substitution of strong H-bonding anions (e.g. Cl^-, $[NO_3]^-$) with weakly coordinating anions (e.g. $[OTf]^-$ or $[BF_4]^-$) leads to a shift in the more favoured conformer from the π^+–π^+ stacked to alternating cation–anion conformer. A preference for the alternating structure has also been recently established for $[C_4C_1im]_2[NTf_2]_2$ IP-dimers as well as the $[C_1C_1im]_2[OTf]_2$ dimers introduced in the previous section.[56] In the alternating configuration anions form anion–π^+(dispersive) interactions, thus precluding stacking of the imidazolium π systems.[34] In contrast to the stacked conformer, the anions in the alternating conformer exhibit essentially no interaction with $H^{4/5}$ at the rear of the ring.

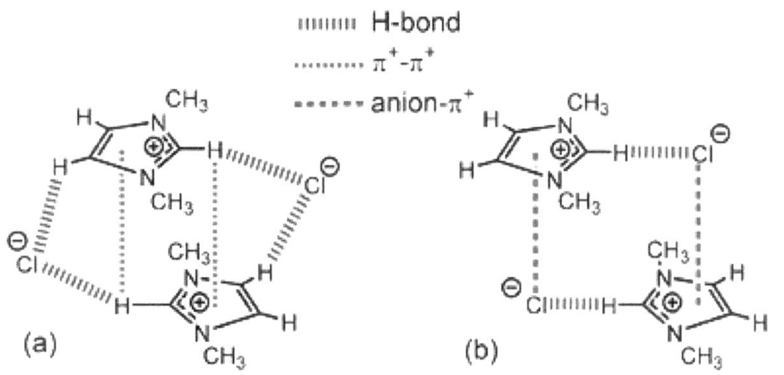

Figure. 9 (a) Stacked "middle" structure of [C_1C_1im]Cl showing C^2–H^2 positioned in the magnetic cone of the aromatic ring and (b) alternating "diagonal" structure of [C_1C_1im]Cl showing ion-pairs with a dominant front C^2–H^2 H-bond and weak anion–π^+ interaction.[34]Major H-bonding, anion–π^+ and π^+–π^+ interactions are depicted by dashed lines.

Hence, we have investigated the ^1H and ^{13}C NMR of ILs and IL mixtures in CD_2Cl_2 to give insight into the interplay between π^+–π^+ stacking and the anion–π^+ interactions identified as occurring in the undiluted IL mixtures. The NMR spectra for the simple ILs [C_4C_1im]Cl, [C_4C_1im][OTf], [C_4C_1im][SCN], [C_4C_1im][NTf_2] and [C_4C_1im][Me_2PO_4] are detailed in Fig. 10. Spectra range from those in highly dilute solutions to those with a concentration approaching 2.5 mol L^{-1}, which is close to that of a pure IL (cf. simple [C_4C_1im][NTf_2] has a concentration of 3.4 mol L^{-1} at 25 °C). The relative chemical shift between the ions depends primarily on the H-bonding ability of the anion, for example dilute [C_4C_1im][X] have chemical shifts at 11.13 ppm, 10.89 ppm, 9.55 ppm, 9.18 ppm and 8.68 ppm for X = Cl, Me_2PO_4, SCN, OTf and NTf_2 respectively. No significant change in chemical shift for the most dilute $H^{4/5}$ resonances are observed between ILs, consistent with an ion-pair H-bonding via H^2. The y-axis of Fig. 10 shows a uniform span of 1.4 ppm for H^2 and 1 ppm for $H^{4/5}$, allowing a comparison to be made between anions for the magnitude of the total change in chemical shift as the concentration of the IL is increased.

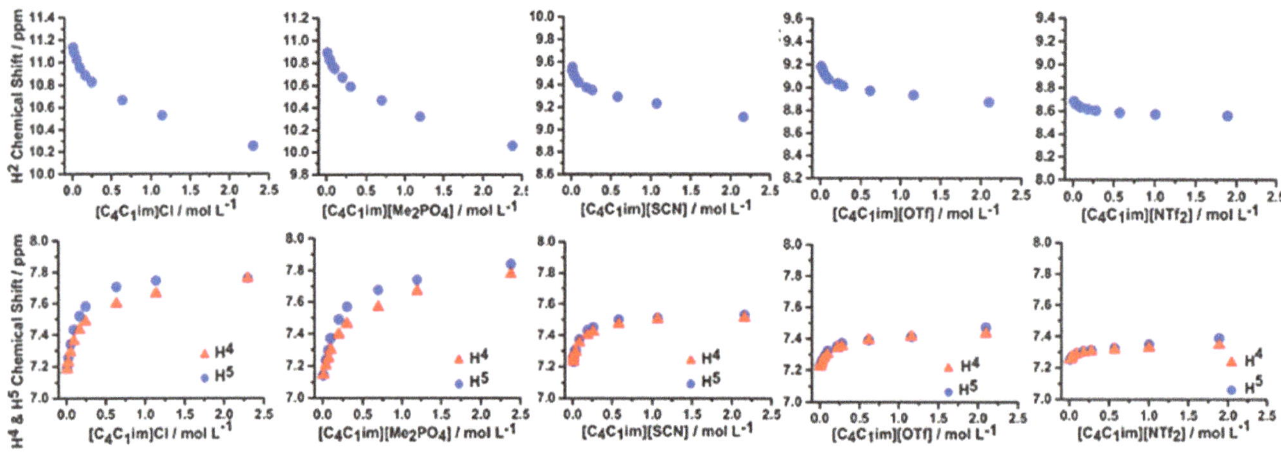

Figure. 10 ^{1}H chemical shifts of the H^2, H^4 and H^5 signals obtained in CD_2Cl_2 for various concentrations of the simple ILs.

The ILs studied here show a similar behaviour to that obtained for $[C_2C_1im]X$ (X = Cl, Br and I) in CD_2Cl_2.[33] The most pronounced changes in the ^{1}H chemical shifts of H^2 and $H^{4/5}$ are observed for $[C_4C_1im]Cl$ and $[C_4C_1im][Me_2PO_4]$, $[C_4C_1im][OTf]$ and $[C_4C_1im][SCN]$ show intermediate behaviour, and only a very weak effect is found for $[C_4C_1im][NTf_2]$. The observed upfield shift of H^2 with increasing concentration is indicative of π^+–π^+ stacking whereby H^2 lies above the shielding cone of an adjacent imidazolium ring with the corresponding downfield shift of $H^{4/5}$ due to positioning of these atoms outside of the shielding cone. However, if the conformers in Fig. 9 are considered it is evident that the transition from a "diagonal" to a "middle" structure is accompanied by an out of plane motion of the anion, which would reduce the H^2 H-bond strength (upfield chemical shift), coupled with an increase in $H^{4/5}$ H-bond strength (downfield chemical shift).

The magnitude of the H^2 and $H^{4/5}$ chemical shift change for $[C_4C_1im][Me_2PO_4]$ and $[C_4C_1im]Cl$ on increasing concentration is approximately 1 ppm which is of a similar order to the difference between the concentrated samples of the ILs containing a weakly ($[OTf]^-$) as opposed to strongly ($[Me_2PO_4]^-$) coordinating anion. The H-bond perturbation due to the changing anion is likely to be much greater than that due to a change in concentration (the effects of which should be minimal). Thus, the trend for changing chemical shift with concentration can be inferred to arise from the close approach of the imidazolium rings and not subtle H-bond effects.

The ^{13}C NMR results for C^2 (ESI, Fig. S19) follow a similar trend to those obtained for the ^{1}H chemical shifts of H^2, exhibiting a substantial upfield shift for $[C_4C_1im]Cl$ and $[C_4C_1im][Me_2PO_4]$ with increasing concentration, while $[C_4C_1im][SCN]$ and $[C_4C_1im][OTf]$ exhibit an intermediate shift (notably the shift for $[C_4C_1im][SCN]$ is greater than for $[C_4C_1im][OTf]$), and $[C_4C_1im][NTf_2]$ shows only a very small upfield shift. In contrast to H^2 and C^2, the ^{13}C NMR shifts of $C^{4/5}$ vary considerably compared to $H^{4/5}$ with both downfield and upfield shifts observed with increasing concentration for most ^{13}C signals.

The rationalisation of the $C^{4/5}$ results is that at low concentrations the imidazolium rings are offset enough for the $C^{4/5}$ regions to lie (on average) outside of the shielding cone, generating the initial downfield shifts with increasing concentration. Further aggregation appears to bring the C^4 and C^5 towards the shielding cone, particularly for $[C_4C_1im]Cl$ and $[C_4C_1im][SCN]$, as an upfield shift is observed at larger IL concentrations.

Despite significant movement first upfield and then downfield as the concentration is increased, the net change in chemical shift of $C^{4/5}$ between the most dilute and the most concentrated samples in $[C_4C_1im]Cl$ and $[C_4C_1im][SCN]$ is very small. The similarity of the initial and final chemical shifts indicate that C^4 and C^5 are in line with the shielding cone of the adjacent imidazolium ring. In contrast, $[C_4C_1im][Me_2PO_4]$, $[C_4C_1im][OTf]$ and $[C_4C_1im][NTf_2]$ undergo a net downfield shift indicating that the $C^{4/5}$ atoms lie outside the shielding cone of the adjacent imidazolium ring. These results suggest that larger anions interfere with the ability of the imidazolium rings to overlap, forcing the rings to lie slightly further offset. This is consistent with the MD findings which will be discussed in detail later.

The magnitude of the cumulative chemical shift changes for $[C_4C_1im][Me_2PO_4]$ is much greater than for the other two ILs with large anions indicating that the stronger H-bond accepting anion is more able to stabilise the π^+–π^+ interaction and enable closer approach of the imidazolium rings despite its size. Unlike the other simple ILs, $[C_4C_1im][Me_2PO_4]$ also shows a difference in the chemical shift trends between C^4 and C^5. C^4 shifts only downfield upon increasing the IL concentration while C^5 shifts downfield initially and then slightly upfield. Such an effect could arise from the formation of a rotated stacked structure where each parallel alternating imidazolium ring was rotated by less than 180°, positioning C^5 nearer the edge of the shielding cone than C^4.

As observed for the other signals, the Cl^- and $[Me_2PO_4]^-$ favour π^+–π^+ interactions, $[SCN]^-$ and $[OTf]^-$ exhibit intermediate behaviour and no significant evidence of π^+–π^+ interactions can be determined for $[NTf_2]^-$. The lack of π^+–π^+ stacking within $[C_4C_1im][NTf_2]$ could be due either to poor Coulombic stabilisation of the π^+–π^+ interaction by the diffuse $[NTf_2]^-$ anion or the formation of more favourable anion–π+ interactions caused by its weak H-bond accepting ability, both increasing the displacement between imidazolium rings (Fig. 11).

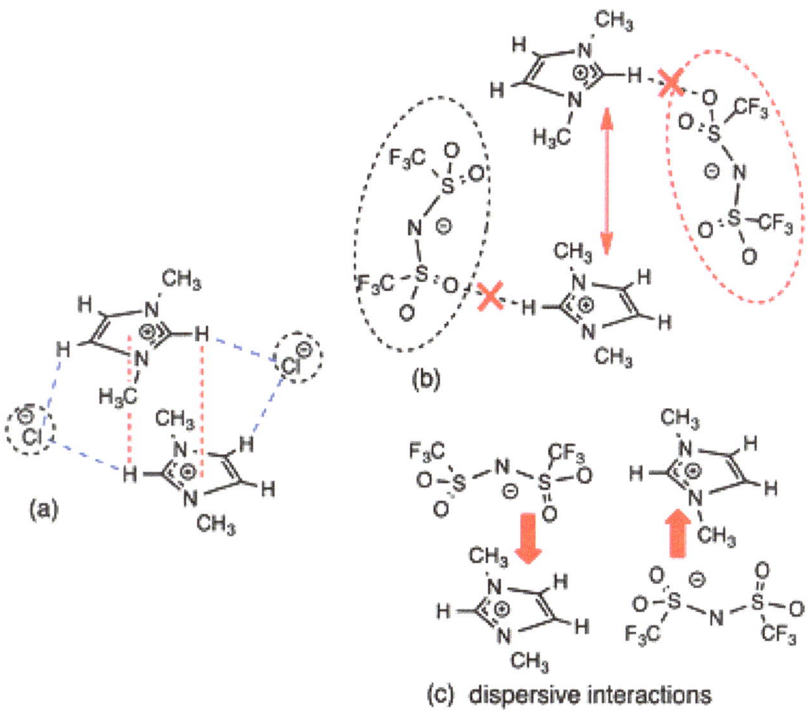

Figure. 11 (a) π^+–π^+ stacking with Cl^- (b) no stacking interactions due to larger more diffuse $[NTf_2]^-$ not stabilising the Coulombic repulsion of the imidazolium rings, forcing them further apart, and (c) the weaker H-bonding ability and stronger anion–π^+ interaction leading to $[NTf_2]^-$ preferring a top position and thus displacing the π^+–π^+ stacked cation.

In order to verify the nature of the π^+–π^+ stacking interactions for ILs based on the $[Me_2PO_4]^-$ and Cl^- anions, selected low energy π^+–π^+ stacked (anions in the middle M_FS_SF_A and M_FS_SF_R) and alternating (anions on the diagonal, D_FT_TF_A) structures of $[C_1C_1im]_2[Me_2PO_4]_2$ and $[C_1C_1im]_2Cl_2$ IP-dimers have been computed at the B3LYP-D3BJ/6-311+G(d,p) level, Fig. 12. The nomenclature for these structures has been described in a previous publication and is reproduced in the ESI,† (Section B).[27] The effect of a solvent environment has previously been shown to enhance π^+–π^+ stacking.[57] Thus, conformer energies and NMR parameters were recomputed employing (a) a CH_2Cl_2 CPCM continuum model to mimic the solvent environment of a dilute solution and (b) an IL SMD model to simulate the solvent environment of a concentrated solution, further computational details can be found in the ESI, (Section B), energies are reported in ESI, Table S1.

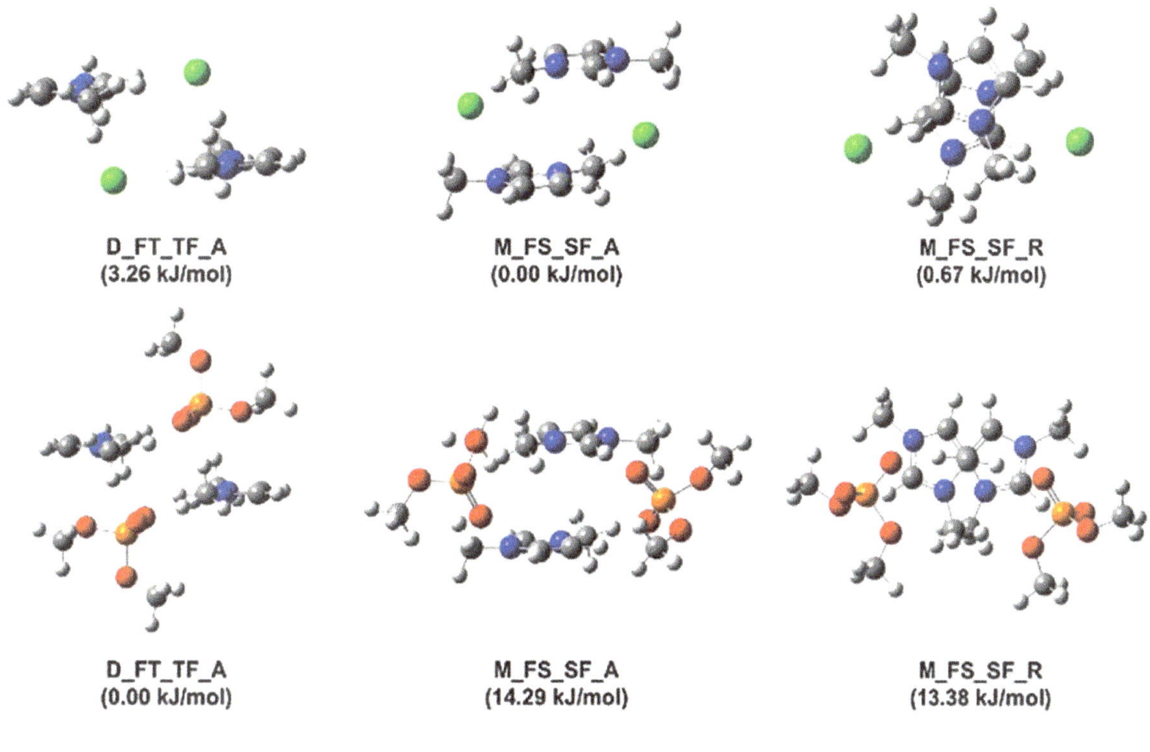

Figure. 12 Representative $[C_1C_1im]_2Cl_2$ and $[C_1C_1im]_2[Me_2PO_4]_2$ IP-dimer structures at the B3LYP-D3BJ/6-311+G(d,p) level. ZPE and BSSE corrected conformer energies are set relative to the respective lowest energy structure.

^{13}C NMR shifts are evaluated relative to the respective ion-pair chemical shifts of the front conformer. The computational results for the dilute (CH_2Cl_2) vs. concentrated (IL) should not be over-interpreted given the current level of theory employed and the limited number of structures examined. The experimental spectrum will be a weighted average of conformers, the stability of which changes as the solvent environment is altered. The NMR calculations presented here can be expected to provide qualitative rather than quantitative information.

Stable $\pi^+–\pi^+$ stacking structures have been obtained (in the gas phase) for $[C_1C_1im]_2[Me_2PO_4]_2$, but the energies are ≈14 kJ mol^{-1} higher than for the alternating structure (ESI,† Table S1). However, in CH_2Cl_2 the $\pi^+–\pi^+$ stacking IP-dimers are stabilised relative to the alternating IP-dimer, and under the greater dielectric screening of the concentrated IL the $\pi^+–\pi^+$ stacking and alternating structures are of comparable energy. For $[C_1C_1im]_2Cl_2$ the $\pi^+–\pi^+$ stacking and alternating structures are of similar energy in the gas-phase. However, the alternating conformer is increased in energy in CH_2Cl_2 and becomes unstable in the concentrated IL. Overall, we find that screening of the positive charge in the higher dielectric favours $\pi^+–\pi^+$stacking. Thus, as the concentration of the IL increases $\pi^+–\pi^+$ stacking will be favoured.

For $[C_1C_1im]_2Cl_2$ and $[C_1C_1im]_2[Me_2PO_4]_2$ the computed ^{13}C NMR chemical shift for C^2 and C^5 in the $\pi^+–\pi^+$ stacked conformers are upfield relative to the front ion-pair conformation, consistent with the upfield shift observed in the NMR spectra of the concentrated ILs, although somewhat larger than the small net effects observed experimentally. The computed ^{13}C NMR chemical shift for C^4 in the $\pi^+–\pi^+$ stacked conformers moves downfield, consistent with the downfield shift observed with increasing concentration for C^4 of $[C_4C_1im][Me_2PO_4]$ and the small net change for $[C_4C_1im]Cl$. The alternating structure shows essentially no change. These results support the hypothesis that $\pi^+–\pi^+$ stacking is causing the shifts in the ^{13}C spectra of C^2 and $C^{4/5}$.

Visual inspection of a plan view of the M_FS_SF_R conformer (Fig. 12), shows that for $[C_1C_1im]_2Cl_2$ the cation rings are essentially overlapped while for $[C_1C_1im]_2[Me_2PO_4]_2$ the rings are slightly displaced. This is consistent with the trends observed in the ^{13}C NMR $C^{4/5}$ spectra with increasing concentration (in CD_2Cl_2). $C^{4/5}$ of $[C_4C_1im]Cl$ exhibit similar behaviour because both C^4 and C^5 overlap the adjacent imidazolium ring. However, $C^{4/5}$ of $[C_4C_1im][Me_2PO_4]$ exhibit disparate behaviour because the rings are displaced slightly resulting in reduced overlap leading to C^5 aligning within the aromatic magnetic cone, but C^4 remaining outside. The computed ^{13}C NMR chemical shifts of C^4 move more substantially further downfield in $[C_1C_1im]_2[Me_2PO_4]_2$ compared to $[C_1C_1im]_2Cl_2$, in qualitative agreement with the experimental observations.

Two geometric parameters are of interest for the stacked structures; the overlap and the vertical distance between the cation rings. The shortest distance between any two atoms on the rings is closer for $[C_1C_1im]_2Cl_2$ by 0.016 Å and 0.074 Å for M_FS_SF_A and M_FS_SF_R respectively (details in ESI,† (Section B)). Examination of the optimised structures shows that the Cl^- anion must be positioned in the “middle” between the rings to enable H-bonding with both rings, this results in a significant loss of linearity in the H-bond ($C^{4/5}$–H–Cl angles ≈117–132°). In contrast, the oxygen atoms of the $[Me_2PO_4]^-$ anion easily form linear H-bonds due to the size of the anion ($C^{4/5}$–H–O angles ≈140–164°). However, the H-bonding is strengthened at the cost of forcing the rings further apart to accommodate the distance between the O atoms of $[Me_2PO_4]^-$, as can be seen in the structure M_FS_SF_A of Fig. 12. Thus both overlap and vertical distance facilitate $\pi^+–\pi^+$ stacking for $[C_1C_1im]_2Cl_2$ relative to $[C_1C_1im]_2[Me_2PO_4]_2$. The subtle differences between these ILs are not apparent from the NMR results where the magnitude of $\pi^+–\pi^+$ stacking was found to be similar for both ILs, nonetheless the stability of these $\pi^+–\pi^+$ stacked structures is in accordance with the experimental NMR data.

A QTAIM analysis has been performed on gas-phase dimers D_FT_TF_A and M_FS_SF_A for both $[C_1C_1im]_2Cl_2$ and $[C_1C_1im]_2[Me_2PO_4]_2$. $\rho_{BCP}(r)$ has been evaluated for the ring C–H H-bonding interactions (ESI,† Table S2 and Fig. S1). $\rho_{BCP}(r)$ reports the electron density at the bond critical point (BCP). The presence of a BCP is used to identify the existence of a bond and the magnitude of $\rho(r)$ at the BCP is a measure of the strength of a bond. The larger $\rho_{BCP}(r)$ for H^2 H-bonding within the $[C_1C_1im]_2[Me_2PO_4]_2$ IP-dimers (compared to $[C_1C_1im]_2Cl_2$) is indicative of stronger H-bonding. The existence of $\rho_{BCP}(r)$ for $H^{4/5}$ H-bonding within the IP-dimers indicates an increase in H-bonding relative to the front and top ion-pairs which exhibit no H-bonding at $H^{4/5}$.

Within $[C_1C_1im]_2Cl_2$ the aromatic rings overlap better, leading to both C^4 and C^5 being affected by the shielding cone of the adjacent imidazolium ring, whereas within $[C_1C_1im]_2[Me_2PO_4]_2$ the larger anion forces the aromatic rings to be forced slightly further apart and reduces the overlap, consistent with the different experimental observations for the $C^{4/5}$ resonances in the ^{13}C NMR. The combined experimental and computational results imply that the electrostatic stabilisation of the $\pi^+–\pi^+$ interaction provided by stronger H-bonding with the $[Me_2PO_4]^-$ anion is able to offset the steric effects induced by the larger anion which prevents the overlap of the imidazolium rings. This accounts for the similar magnitude of $\pi^+–\pi^+$ stacking behaviour observed experimentally between $[C_4C_1im]Cl$ and $[C_4C_1im][Me_2PO_4]$ despite the latter anion being significantly bulkier.

Overall, we find that the strong H-bonding ability and shape of an anion plays a distinct role in stabilising $\pi^+–\pi^+$ stacked interactions, consistent with the NMR results. There is experimental and computational evidence that strongly H-bond accepting anions such as $[Me_2PO_4]^-$ and Cl^- promote the closer approach of the imidazolium cations. Similar conclusions regarding the prevalence of $\pi^+–\pi^+$ stacking for Cl^- relative to $[SCN]^-$ based ILs have been obtained through MD simulations.[18] $[SCN]^-$ is smaller in size but a weaker H-bond acceptor than $[Me_2PO_4]^-$, hence the intermediate behaviour exhibited by $[SCN]^-$ indicates the extent to which H-bonding is required to stabilise $\pi^+–\pi^+$ interactions. The $[NTf_2]^-$ anion is both the most bulky and the weakest H-bond acceptor of the anions, hence it is not able to stabilise $\pi^+–\pi^+$ interactions electrostatically and if positioned between the imidazolium rings will increase their separation. Consequently, no evidence for $\pi^+–\pi^+$ stacking interactions is observed in this IL.

NMR of the Ionic Liquid Mixtures in CD_2Cl_2

Similar concentration dependent NMR studies were conducted for mixtures of ILs within CD_2Cl_2. These mixtures include $[C_4C_1im][Me_2PO_4]_x[NTf_2]_{1-x}$, $[C_4C_1im]Cl_x[NTf_2]_{1-x}$ and $[C_4C_1im][OTf]_x[NTf_2]_{1-x}$, as well as $[C_4C_1im]Cl_{0.50}[OTf]_{0.50}$. The ^1H and ^{13}C NMR data for $[C_4C_1im][Me_2PO_4]_x[NTf_2]_{1-x}$ and $[C_4C_1im][OTf]_x[NTf_2]_{1-x}$ are depicted in Fig. 13 and ESI,†Fig. S20 and S21 respectively, the ^1H NMR and ^{13}C NMR data for the other mixtures can be found in ESI, Tables S7–S27 and Fig. S17–S23.

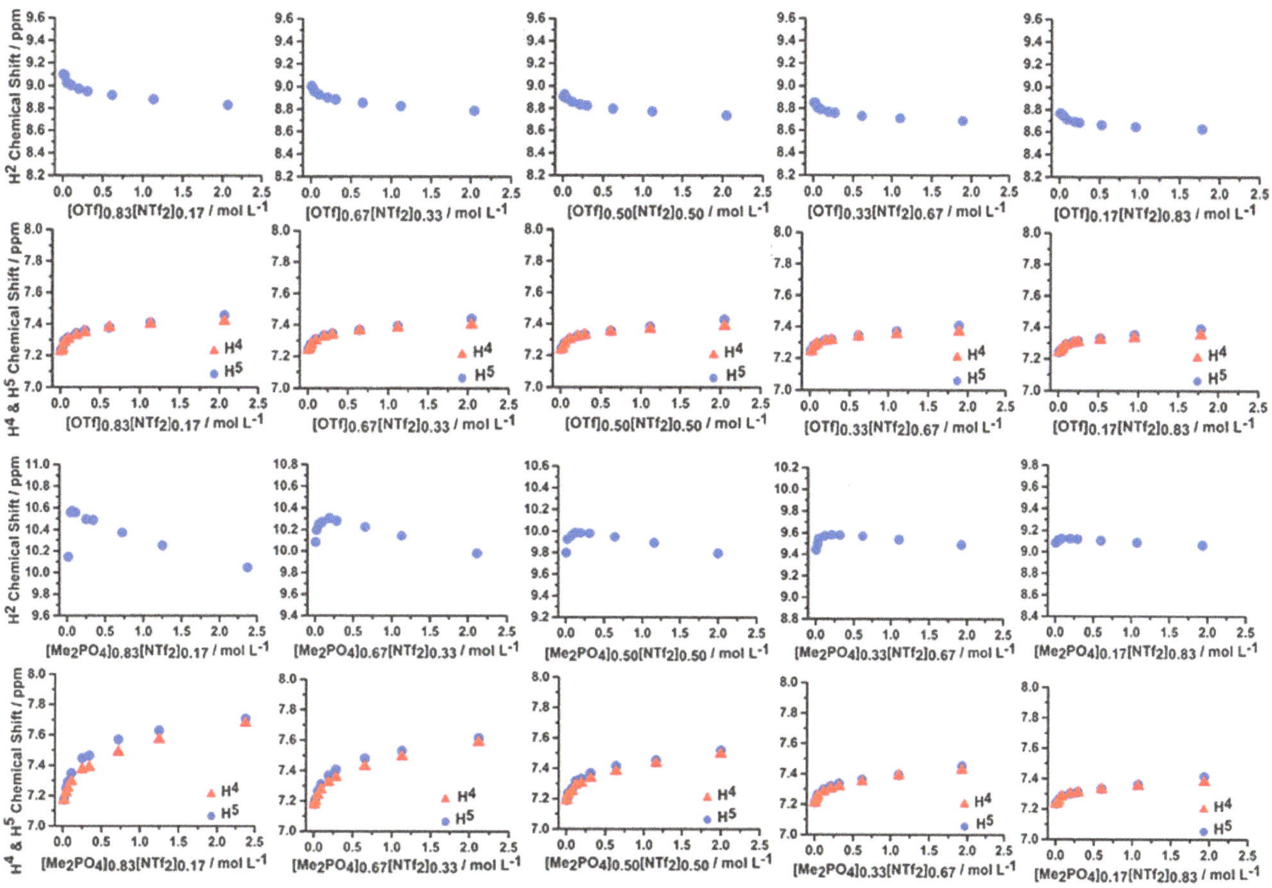

Figure. 13 ^1H NMR chemical shifts of H^2, H^4 and H^5 of the $[C_4C_1im][OTf]_x[NTf_2]_{1-x}$ mixtures (top) and $[C_4C_1im][Me_2PO_4]_x[NTf_2]_{1-x}$ mixtures (bottom) at various concentrations in CD_2Cl_2.

From Fig. 13 it is evident that the H^2, H^4 and H^5 signals in the $[C_4C_1im][OTf]_x[NTf_2]_{1-x}$ mixtures do not vary significantly with composition, and exhibit only a small change in curvature and chemical shift across the range of compositions examined. The magnitude of the chemical shifts and shapes of the curves are as expected based on an appropriately weighted average of the simple $[C_4C_1im][OTf]$ and $[C_4C_1im][NTf_2]$ ILs. The ^{13}C NMR results (ESI, Fig. S20 and S21) are similar with the only noticeable change being the slow disappearance of the upfield shift feature with increasing concentration for the C^4 and C^5 spectra as the proportion of $[NTf_2]^-$ increases. The magnitude of the chemical shift changes are small in all cases indicating that the π^+–π^+ interactions are not significant and that the distance between imidazolium cations does not change dramatically compared to $[C_4C_1im][OTf]$ and $[C_4C_1im][NTf_2]$.

In contrast, the $[C_4C_1im][Me_2PO_4]_x[NTf_2]_{1-x}$ mixtures are not simply additive and the H^2 signal exhibits substantially different behaviour compared to either of the simple ILs. In these mixtures, the H^2 signal initially shifts downfield then moves upfield at higher concentrations whereas it monotonically shifts upfield in the simple ILs. This is not observed in any of the other ^1H NMR spectra. The magnitude of the initial downfield shift decreases with a decreasing proportion of $[C_4C_1im][Me_2PO_4]$ within the mixture, while the concentration of the mixture in CD_2Cl_2 at which the maximum downfield shift is observed increases. The C^2 chemical shift of the mixture does not move downfield, instead only an upfield shift is observed. C^4 and C^5follow the pattern of an initial downfield shift, before either flattening out or moving slightly upfield with increasing IL concentration. The shape of the chemical shift change, particularly for the H^4/H^5 and C^4/C^5 positions, varies more rapidly with increasing $[NTf_2]^-$ up to $[C_4C_1im][Me_2PO_4]_{0.50}[NTf_2]_{0.50}$ where further increasing the $[NTf_2]^-$ anion concentration in the mixture leads to curve shapes consistent with simple $[C_4C_1im][NTf_2]$. The upfield shift of C^2 indicates there is π^+–π^+ stacking, however the initial downfield shift of H^2 indicates that either H^2 is outside of the magnetic cone (which is unlikely) or that a corresponding and more pronounced increase in H-bonding is occurring which dominates leading to an overall downfield shift. $[Me_2PO_4]^-$ containing ILs have much greater β-values than those based on $[NTf_2]^-$, indicating the improved H-bond accepting capacity of the $[Me_2PO_4]^-$.[58] A range of quantum chemical descriptors have also found H-bonding interactions with $[Me_2PO_4]^-$ to be much stronger than with $[NTf_2]^-$.[59] A possible rationalisation for the data is that as isolated ion pairs and ion clusters coalesce, the more strongly interacting $[Me_2PO_4]^-$ anion displaces $[NTf_2]^-$ anions from around H^2 leading to a downfield chemical shift at H^2 which is characteristic of strong H-bonding.[60,61] Once the H^2 sites have become saturated by $[Me_2PO_4]^-$ the aggregation of the IL proceeds in the conventional fashion with ring current effects leading to an upfield chemical shift.

The comparatively rapid changes in the shape of the curves with increasing $[NTf_2]^-$ proportion in the mixture indicate that the $[NTf_2]^-$ anion significantly disrupts the close interaction of the imidazolium rings. This disruption likely arises from favourable anion–π^+ interactions, such as those found for the $[NTf_2]^-$ anions in the undiluted NMR results, which then cause an increase in imidazolium ring displacement. The loss of a $[Me_2PO_4]^-$ anion may also reduce the stability of π^+–π^+ stacked structures (via the loss of Coulombic stabilisation and supportive H-bonding interactions), resulting in the translation of the imidazolium rings facilitated by a stabilising interaction with $[NTf_2]^-$ (ESI,† Fig. S2).

The $[C_4C_1im]Cl_x[NTf_2]_{1-x}$ mixtures (ESI,† Fig. S17 and S22) follow a similar, but less pronounced, trend to those observed for the $[C_4C_1im][Me_2PO_4]_x[NTf_2]_{1-x}$ mixtures. Mixtures with $[NTf_2]^-$ proportions above $[C_4C_1im]Cl_{0.67}[NTf_2]_{0.33}$ all exhibit changes in chemical shift on increasing the concentration similar to those found for $[C_4C_1im][NTf_2]$, reiterating the disruptive nature of the bulky $[NTf_2]^-$ anion on mixtures, and particularly those with more tightly coordinating anions. In a similar fashion, $[C_4C_1im]Cl_{0.50}[OTf]_{0.50}$ (ESI,† Fig. S18 and S23) displays behaviour reminiscent of $[C_4C_1im][OTf]$ in terms of the reduced magnitude of chemical shift. This signifies that the $[OTf]^-$ anion is capable of disrupting the interaction of imidazolium rings in a similar manner to $[NTf_2]^-$. The disruption of these interactions occurs despite: the closer approach of imidazolium cations in simple $[C_4C_1im][OTf]$, the smaller size and better H-bond accepting capacity of the $[OTf]^-$ anion, and the reduced propensity of the $[OTf]^-$ anion to engage in anion–π^+ interactions as determined by the undiluted NMR results. The behaviour of the $[C_4C_1im]Cl_{0.50}[OTf]_{0.50}$ mixture illustrates that π^+–π^+ stacking interactions in mixtures are readily disrupted by the presence of weakly coordinating anions, even those that are smaller and better H-bond acceptors than $[NTf_2]^-$.

In summary, the quantum chemical calculations of selected IP-dimers and the CD_2Cl_2 experiments indicate that the strongly coordinating anions (Cl^- and $[Me_2PO_4]^-$) lead to close interactions of the imidazolium cations, suggestive of π^+–π^+ stacking interactions. When present in mixtures with weakly coordinating anions, such as $[OTf]^-$ or $[NTf_2]^-$, the magnitude of ring current effects decrease dramatically even at relatively low (<50 mol%) proportions of the weakly coordinating anion. This illustrates that these anions disrupt the close stacking arrangement. Combined with the previous results for the undiluted ILs, the implication is that the interaction of the more strongly H-bond accepting anion with H^2 leads to the displacement of the weakly coordinating anion above and behind the ring, facilitating the greater displacement of imidazolium rings through anion–π^+ type interactions. When two weakly coordinating anions are present then additive behaviour is observed for the mixtures and no dramatic changes in imidazolium cation spacing are observed.

Molecular Dynamics Simulation of Ionic Liquid Mixtures

To elucidate the preferential interactions indicated by the NMR studies, MD simulations have been carried out for the mixtures exhibiting the most significant deviations from ideality. Thus the simple ILs and IL mixtures $[C_4C_1im][Me_2PO_4]_x[NTf_2]_{1-x}$ and $[C_4C_1im]Cl_x[OTf]_{1-x}$ have been simulated. Simulation details are provided in the ESI,† (Section B).

The spatial distribution functions (SDF) for $[C_4C_1im][Me_2PO_4]_x[NTf_2]_{1-x}$ (Fig. 14) provide a graphical representation of the relative position of anions around a cation. The shaded surfaces in Fig. 14 represent where the density of anions is at least 2.5 times greater than the background density. SDF for $[C_4C_1im]Cl_x[OTf]_{1-x}$ are provided in the ESI, Fig. S4.

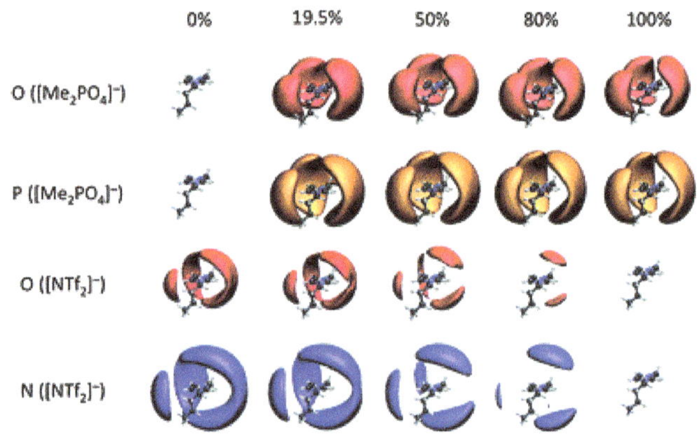

Figure. 14 Spatial density maps of selected atoms on the anions around the cation for several $[Me_2PO_4]^-$ concentrations in the $[C_4C_1im][Me_2PO_4]_x[NTf_2]_{1-x}$ mixtures. Isosurfaces: free $O([Me_2PO_4]^-)$ in red, $P([Me_2PO_4]^-)$ in orange, $O([NTf_2]^-)$ in matt red and $N([NTf_2]^-)$ in blue. Isosurfaces are shown at 2.5 times the normalised atomic density with the more strongly H-bonding anion to the front, and to have the weaker H-bonding anion positioned above or below the ring.

It is clear that in the mixture both anions interact with the ring H^2, H^4 and H^5 atoms. The strongly H-bond accepting anions can interact "in-plane" or in the "middle" between two rings, leading to the extension of the SDF surfaces above and below the plane around the periphery of an imidazolium cation. The more weakly H-bond accepting anions tend to occupy positions directly above and below the centre of the aromatic ring. This is more easily visualised in the related SDF of $[C_4C_1im]Cl_{0.50}[OTf]_{0.50}$ shown for increasing isovalue of an anion around the cation (ESI,† Fig. S5). The SDF for the N of $[NTf_2]^-$ (blue in Fig. 14) also shows a clear preference for $[NTf_2]^-$ to locate above or below the cation ring, particularly for low concentrations of $[NTf_2]^-$ in $[Me_2PO_4]^-$.

Fig. 14 shows that $[Me_2PO_4]^-$ is closely associated with the cation even in very low proportions. In contrast, the smaller surfaces for $[NTf_2]^-$ in Fig. 14 and $[OTf]^-$ in ESI, Fig. S5 demonstrate the much weaker interaction of these anions with the cation. These data are consistent with the NMR data observed for $[C_4C_1im][Me_2PO_4]_x[NTf_2]_{1-x}$ mixtures, where the addition of $[NTf_2]^-$ led to stronger apparent preferential interactions with $C^{4/5}$ rather than $H^{4/5}$, indicating an anion–π^+ interaction is dominating, rather than in-plane H-bonding at the rear of the ring. These data are also consistent with the NMR data observed for the $[C_4C_1im]_x[C_4C_1pyrr]_{1-x}Cl_x[NTf_2]_{1-x}$ mixtures. The first solvation shell of any cation is a random mixture of the anions present, this can be seen in snap shots from a single trajectory frame, as shown in ESI, Fig. S6 and S7. Within a solvation shell the cation will orient so as to interact given the similarity in shape and size of $[Me_2PO_4]^-$ and $[OTf]^-$ it is possible to question why $[C_4C_1im][Me_2PO_4]_x[NTf_2]_{1-x}$ shows larger deviations from ideality than $[C_4C_1im][OTf]_x[NTf_2]_{1-x}$. Some insight can be obtained from the charge parameters for the MD potentials; $[Me_2PO_4]^-$ is significantly more polarised than $[OTf]^-$. The charge on the H-bond accepting O atoms in $[Me_2PO_4]^-$ is −0.92 e, while those in $[OTf]^-$ are −0.63 e and those of $[NTf_2]^-$ carry even less charge (−0.53 e). Thus while the overall charge on the ion is −1 e, the polarisability of the central atom (P (+1.62) or S (+1.02)) can lead to a significantly larger charge present on the periphery of the anion.

Atom-atom pair radial distribution functions (RDFs) plot the distance between specific pairs of atoms over a simulation, and provide more detail than the SDFs. RDFs between the free O's (not OMe) of both $[Me_2PO_4]^-$ and $[NTf_2]^-$ with ring atoms (H^2, H^4 and H^5) for the various $[C_4C_1im][Me_2PO_4]_x[NTf_2]_{1-x}$ mixtures are presented in ESI,† Fig. S8 and those for C^2, $C^4$and C^5 are presented in ESI,† Fig. S9. Qualitatively similar results are found for the $[C_4C_1im]Cl_x[OTf]_{1-x}$ mixtures (ESI,† Fig. S10 and S11). The variation of cation–anion RDFs within IL mixtures as a function of the proportion of a constituent is not unique to these systems.[13,18,19]

In general, each RDF exhibits a first maximum at either 2–3 Å (H^n–X) or 3–4 Å (C^n–X) n = 2, 4, 5. The first peak corresponds to direct interaction of the anion with the respective hydrogen/carbon atom. The second peak mainly represents anions coordinated to the other ring atomic (H/C) sites. The position of the first maximum for each of the cation–anion RDFs is found to be independent of anion composition, indicating that the relative cation–anion distances of the simple ILs are retained on mixing. Thus the presence of a second anion does not weaken or strengthen the interactions to an extent large enough to influence the distance between the first anion and cation.

The higher g(r) maxima for $[Me_2PO_4]^-$ compared to $[NTf_2]^-$ indicate that the interaction of $[Me_2PO_4]^-$ with $[C_4C_1im]^+$ is much stronger than that of $[NTf_2]^-$. As the concentration changes, a smaller maximum in the RDF indicates a reduced probability of finding the anion in that specific location. For example, in the $[C_4C_1im][Me_2PO_4]_x[NTf_2]_{1-x}$ mixtures (ESI,† Fig. S8), the intensity of the first peak maximum of the distance between H^2 and the O of $[Me_2PO_4]^-$ systematically decreases from ≈9 to ≈5 with an increasing proportion of $[Me_2PO_4]^-$. Thus with increasing $[Me_2PO_4]^-$ concentration there is less interaction between the anion and all of the cation ring C–H groups. A decrease in H^2–O ($[NTf_2]^-$) from ≈2.5 to ≈1.5 is also observed with increasing $[Me_2PO_4]^-$ concentration. Thus, in contrast to $[Me_2PO_4]^-$, at high $[NTf_2]^-$ concentrations the $[NTf_2]^-$ anion is more likely to be interacting with the cation.

These data can be rationalised; for dilute $[Me_2PO_4]^-$ in $[C_4C_1im][NTf_2]$ there are strong directional H-bonds between the cation and $[Me_2PO_4]^-$, however as the concentration of $[Me_2PO_4]^-$ increases the cation becomes surrounded by $[Me_2PO_4]^-$ anions, reducing the preference for each site on $[C_4C_1im]^+$ to interact with a specific $[Me_2PO_4]^-$ anion. In contrast, for small amounts of $[NTf_2]^-$ in $[C_4C_1im][Me_2PO_4]$, the $[Me_2PO_4]^-$ dominates H-bonding interactions with the cation, and the weaker H-bonding $[NTf_2]^-$ is excluded from direct association, thus there is a reduced proportion of $[NTf_2]^-$ directly interacting with specific sites on the imidazolium ring. As the proportion of $[NTf_2]^-$ increases, the number of $[Me_2PO_4]^-$ anions H-bonding with, thereby restricting access to, the cation is reduced and the $[NTf_2]^-$ can undertake more structured H-bonding. Hence $[NTf_2]^-$ structuring around the imidazolium ring is more pronounced with a high proportion of $[C_4C_1im][NTf_2]$ while $[Me_2PO_4]^-$ structuring around the imidazolium ring is less pronounced as the ratio of $[C_4C_1im][Me_2PO_4]$ is increased.

Changes in the RDFs are less pronounced for H^5 than H^2, and are reduced even further for H^4. This result is in-line with the differences observed in the ^{13}C NMR of C^4 and C^5 for the mixtures in CD_2Cl_2. The differences between H^4 and H^5 are more pronounced for $[Me_2PO_4]^-$ than $[NTf_2]^-$. The stronger relative structuring around H^5 may be due to additional H-bonding or dispersive interactions of the anion with butyl chain groups facilitating anion association or the role of alkyl chain aggregation promoted by the more coordinating anion.

To gain further insight into the local cation–anion structure, angular distribution functions (ADFs) have been computed for each mixture relative to the cation ring protons. The data and insights obtained support the rationalisations presented here, and are discussed more fully in the ESI,† (Section B). The normalised first-shell coordination number of anions surrounding H^2has also been computed in a similar manner to Payal et al. for these mixtures (ESI,† Fig. S15 and S16).[13] The trends in these values accord with the previous discussion with slightly increased coordination of Cl^- and $[Me_2PO_4]^-$ with H^2 relative to $[OTf]^-$ and $[NTf_2]^-$ respectively. Further details and discussion are provided in the ESI,† (Section B).

The potential for π^+–π^+ stacking (or a change in the overall IL structure) has been investigated by computing RDFs between the centre-of-geometry (COG) of cation rings (i.e. excluding the alkyl chains) for both the $[C_4C_1im]Cl_x[OTf]_{1-x}$ and $[C_4C_1im][Me_2PO_4]_x[NTf_2]_{1-x}$ (Fig. 15). There are three distinct features, a pre-peak ≈4 Å, where π^+–π^+ stacking is observed, a primary maximum at ≈7 Å representing the first "solvation" shell of cations, and a second broad peak due to the second solvation shell of cations. The pre-peak has been assigned to π^+–π^+stacking as imidazolium rings positioned closer than in the purely charge ordered structures would require an attractive interaction (most likely π^+–π^+ stacking) for stabilisation. This assignment is consistent with that proposed by others.[13,18] As the proportion of Cl^- or $[Me_2PO_4]^-$ increases there is an increase in π^+–π^+ stacking, consistent with QM calculations,[56] and the NMR data reported here. The effect is significantly larger for $[Me_2PO_4]_x[NTf_2]_{1-x}$ compared to $Cl_x[OTf]_{1-x}$. As the proportion of $[NTf_2]^-$ increases there is a substantial increase in the cation–cation distances (not just those involved in π^+–π^+ stacking), for example, the first solvation shell peak shifts from ≈7 Å to broad flat topped peak ≈8–10 Å. This result is consistent with the NMR data which indicate that the $[NTf_2]^-$ rapidly disrupts π^+–π^+ stacking (<50 mol%). In addition, the increase in the first shell cation–cation distances, not observed within the $[C_4C_1im]Cl_x[OTf]_{1-x}$ mixtures, is consistent with a move from π^+–π^+ stacked structures to the alternating (diagonal anion) structural motifs found for the QC determined IP-dimers, Fig. 12and as motifs within the MD simulations (ESI,†

Fig. S14). In these structures, the positioning of the $[NTf_2]^-$ anion above and below the imidazolium ring would substantially increase the cation–cation distance.

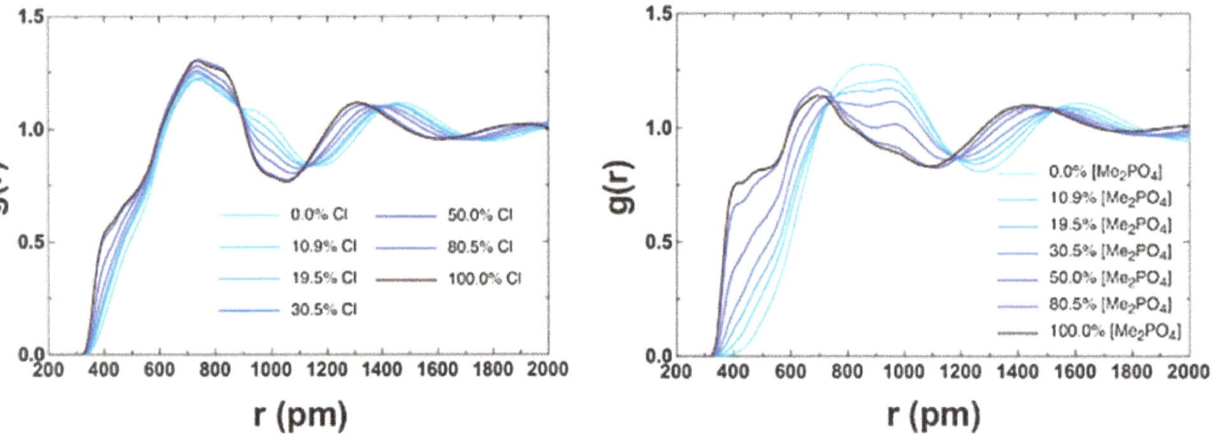

Figure. 15 Calculated RDFs between the centre of geometries of the cation rings for (left) $[C_4C_1im]Cl_x[OTf]_{1-x}$ and (right) $[C_4C_1im][Me_2PO_4]_x[NTf_2]_{1-x}$.

CONCLUSIONS

Preferential H-bonding between H^2 and the most strongly H-bond accepting anion has been identified in a range of IL mixtures through the analysis of NMR spectra and MD simulations of the undiluted ILs. This effect was more pronounced when the difference in H-bond accepting ability between the anions in the mixture was greater. These preferential H-bonds resulted in the more weakly H-bond accepting anion engaging in anion–π^+ interactions above the rear of the imidazolium ring and in some mixtures also participating in weak preferential H-bonding with $H^{4/5}$. The magnitude of these preferential interactions was small and it was found that the proportion of anions around a given cation was random with no evidence of anion clustering around favourable H-bonding sites. There is no evidence for the formation of new bonds with the imidazolium ring atoms within the mixture that cannot be found in the individual IL components of the mixture.

These results give rise to a visualisation of the structures of mixtures of the ILs studied in which there is a random distribution of anions and cations through the structure, as previously suggested.[15] Each cation is surrounded, on average, by a cage of anions (and vice versa) within which it is rotating. However, this rotation is not entirely random and the most strongly hydrogen bond donating protons will associate for a greater time with the most strongly hydrogen bond accepting anions than with the other anions in forming the cage.

The role of IL anions in π^+–π^+ stacking was also investigated by using NMR spectroscopy on IL solutions at various concentrations in CD_2Cl_2, quantum chemical calculations and MD simulations. It was established that strong H-bond accepting anions are required to provide the Coulombic stabilisation necessary for π^+–π^+ interactions to occur and that these interactions were enhanced in the presence of a higher dielectric. Larger more diffuse anions are both less able to provide the necessary level of Coulombic stabilisation and their propensity to form anion–π^+ interactions pushes the rings apart. The disruption of π^+–π^+ stacking by less coordinating anions occurs readily and at relatively low proportions (<50 mol%) of the weakly coordinating anion in mixtures. This was observed both for the bulky $[NTf_2]^-$ anion and the smaller $[OTf]^-$ and is consistent with previous computational findings for $Cl_x[SCN]_{1-x}$ mixtures. This disruption is promoted by the formation of anion–π^+ interactions by these anions.

As one of the motivating factors of this study, it is worth considering how these structures are linked with the physical properties of such mixtures. In particular, it was recently reported that $[C_4C_1im][Me_2PO_4]_x[NTf_2]_{1-x}$ mixtures are non-ideal whereas $[C_4C_1im][OTf]_x[NTf_2]_{1-x}$ mixtures gave ideal behaviour.[43] The results obtained here indicate that the non-ideality of the former mixtures may be due to the larger than anticipated ion separation at low proportions of $[NTf_2]^-$ due to the preferential positioning of $[NTf_2]^-$ above or below the imidazolium ring. This would account for the positive excess volumes observed. The cation–cation distances in $[C_4C_1im][OTf]_x[NTf_2]_{1-x}$ mixtures vary much less dramatically due to the greater initial distance between imidazolium rings for $[C_4C_1im][OTf]$ arising from the poorly stabilised π^+–π^+ interaction. Additionally, there is a weaker preferential association of each anion with specific positions on the cation which results in less reorganisation of the cation conformation on varying the composition of the mixture. This accords with the ideality of the measured physical properties.

The magnitude of non-ideality identified for most IL mixtures to date has generally been small despite the complexity of IL structure.[38–43] The fact that such effects are small can be accounted for by the structural considerations highlighted here. Mixtures of anions with quite substantial differences in H-bond accepting capacity only exhibit a slight preferential association with specific cation locations. These preferential associations are likely to be the main driving force for non-ideality in these solutions and the free energy changes estimated are approximately 2 orders of magnitude smaller than the average binding energies determined for similar ions. The main contributor to the structure of IL mixtures are the Coulombic interactions which are isotropic and are unlikely to change dramatically depending on anion choice (assuming the charges are the same). This was illustrated by the lack of ion clustering observed in the NMR experiments and MD simulations. Secondary effects such as H-bonding are much smaller in magnitude, particularly in the context of an IL mixture where it is not the absolute H-bond strength but the difference in H-bond strength between different ion combinations which governs the molecular orientation. Correspondingly, IL mixtures that consist of equivalently charged anions would be expected, compared to conventional organic molecules, to lead to mixtures that behave in a more ideal fashion. This also accounts for the IL mixtures that exhibit the greatest non-ideality arising from ions where ideal inter-ion distances are affected by steric factors, i.e. a disruption of the ionic network.

The dispersive interactions that form a component of π^+–π^+ stacking and anion–π^+ interactions are small but also competitive in magnitude to weak H-bonding interactions. This gives rise to a diversity of subtle local structural effects in these mixtures which could be of interest for various applications but do not greatly influence the bulk liquid properties outside of those that could be predicted by modelling the liquid as an ideal solution. There are likely to be some further exceptions to these considerations as discussed in the introduction, for example, ILs that possess a large degree of asymmetry, long alkyl chains leading to large dispersive interactions, fluorous phases or ILs that are functionalised resulting in the formation of very strong H-bonds. In these cases the liquids would behave less as mixtures of ions. Further investigation into these types of IL mixtures are currently underway within our laboratory.

ACKNOWLEDGEMENTS

The authors acknowledge an ERC Advanced Investigator Grant held by TW and PAH for funding.

NOTES AND REFERENCES

1. T. Welton, Chem. Rev., 1999, **99**, 2071–2083
2. J. P. Hallett and T. Welton, Chem. Rev., 2011, **111**, 3508–3576
3. V. I. Pârvulescu and C. Hardacre, Chem. Rev., 2007, **107**, 2615–2665
4. M. Armand, F. Endres, D. R. MacFarlane, H. Ohno and B. Scrosati, Nat. Mater., 2009, **8**, 621–629
5. C. C. Weber, A. F. Masters and T. Maschmeyer, Green Chem., 2013, **15**, 2655–2679.
6. T. Torimoto, T. Tsuda, K.-i. Okazaki and S. Kuwabata, Adv. Mater., 2010, **22**, 1196–1221
7. D. R. MacFarlane, N. Tachikawa, M. Forsyth, J. M. Pringle, P. C. Howlett, G. D. Elliot, J. H. Davis, Jr., M. Watanabe, P. Simon and C. A. Angell, Energy Environ. Sci., 2014, **7**, 232–250.
8. X. Han and D. W. Armstrong, Acc. Chem. Res., 2007, **40**, 1079–1086.
9. I. Minami, Molecules, 2009, **14**, 2286–2305.
10. M. Freemantle, Chem. Eng. News, 1998, **76**, 32–37
11. H. Niedermeyer, J. P. Hallett, I. J. Villar-Garcia, P. A. Hunt and T. Welton, Chem. Soc. Rev., 2012, **41**, 7780–7802.
12. R. Hayes, G. G. Warr and R. Atkin, Chem. Rev., 2015, **115**, 6357–6426
13. R. S. Payal and S. Balasubramanian, Phys. Chem. Chem. Phys., 2013, **15**, 21077–21083.
14. J.-M. Andanson, M. J. Beier and A. Baiker, J. Phys. Chem. Lett., 2011, **2**, 2959–2964.
15. M. Y. Lui, L. Crowhurst, J. P. Hallett, P. A. Hunt, H. Neidermeyer and T. Welton, Chem. Sci., 2011, **2**, 1491–1496.
16. A. P. Abbott, G. Frisch, H. Garrett and J. Hartley, Chem. Commun., 2011, **47**, 11876–11878.
17. O. Hollóczki, F. Malberg, T. Welton and B. Kirchner, Phys. Chem. Chem. Phys., 2014, **16**, 16880–16890.
18. M. Brüssel, M. Brehm, A. S. Pensado, F. Malberg, M. Ramzan, A. Stark and B. Kirchner, Phys. Chem. Chem. Phys., 2012, **14**, 13204–13215.
19. M. Brüssel, M. Brehm, T. Voigt and B. Kirchner, Phys. Chem. Chem. Phys., 2011, **13**, 13617–13620
20. A. B. Pereiro, J. M. M. Araújo, F. S. Oliveira, C. E. S. Bernardes, J. M. S. S. Esperança, J. N. Canongia Lopes, I. M. Marrucho and L. P. N. Rebelo, Chem. Commun., 2012, **48**, 3656–3658
21. F. S. Oliveira, A. B. Pereiro, J. M. M. Araújo, C. E. S. Bernardes, J. N. Canongia Lopes, S. Todorovic, G. Feio, P. L. Almeida, L. P. N. Rebelo and I. M. Marrucho, Phys. Chem. Chem. Phys., 2013, **15**, 18138–18147.
22. F. D'Anna, S. Marullo, P. Vitale and R. Noto, ChemPhysChem, 2012, **13**, 1877–1884.
23. H. Wang, S. P. Kelley, J. W. Brantley III, G. Chatel, J. Shamshina, J. F. B. Pereira, V. Debbeti, A. S. Myerson and R. D. Rogers, ChemPhysChem, 2015, **16**, 993–1002.
24. D. Xiao, J. R. Rajian, L. G. Hines, S. Li, R. A. Bartsch and E. L. Quitevis, J. Phys. Chem. B, 2008, **112**, 13316–13325.
25. D. Xiao, J. R. Rajian, S. Li, R. A. Bartsch and E. L. Quitevis, J. Phys. Chem. B, 2006, **110**, 16174–16178.
26. K. Fumino, A.-M. Bonsa, B. Golub, D. Paschek and R. Ludwig, ChemPhysChem, 2015, **16**, 299–304.
27. R. P. Matthews, T. Welton and P. A. Hunt, Phys. Chem. Chem. Phys., 2014, **16**, 3238–3253
28. J. D. Holbrey, W. M. Reichert and R. D. Rogers, Dalton Trans., 2004, 2267–2271.
29. J. S. Wilkes and M. J. Zaworotko, Supramol. Chem., 1993, **1**, 191–193
30. J. S. Wilkes and M. J. Zaworotko, J. Chem. Soc., Chem. Commun., 1992, 965–967
31. M. Deetlefs, C. Hardacre, M. Nieuwenhuyzen, A. A. H. Padua, O. Sheppard and A. K. Soper, J. Phys. Chem. B, 2006, **110**, 12055–12061.
32. A. Mele, G. Romanò, M. Giannone, E. Ragg, G. Fronza, G. Raos and V. Marcon, Angew. Chem., Int. Ed., 2006, **45**, 1123–1126
33. A. G. Avent, P. A. Chaloner, M. P. Day, K. R. Seddon and T. Welton, J. Chem. Soc., Dalton Trans., 1994, 3405–3413.
34. R. P. Matthews, C. Ashworth, T. Welton and P. A. Hunt, J. Phys.: Condens. Matter, 2014, **26**, 284112

35. R. P. Matthews, T. Welton and P. A. Hunt, Phys. Chem. Chem. Phys., 2015, **17**, 14437–14453.

36. B. Aoun, A. Goldbach, S. Kohara, J.-F. Wax, M. A. González and M.-L. Saboungi, J. Phys. Chem. B, 2010, **114**, 12623–12628

37. J. de Andrade, E. S. Böes and H. Stassen, J. Phys. Chem. B, 2008, **112**, 8966–8974

38. A. Podgoršek, A. S. Pensado, C. C. Santini, M. F. Costa Gomes and A. A. H. Pádua, J. Phys. Chem. C, 2013, **117**, 3537–3547

39. D. Song and J. Chen, J. Chem. Thermodyn., 2014, **77**, 137–143

40. J. N. Canongia Lopes, T. C. Cordeiro, J. M. S. S. Esperança, H. J. R. Guedes, S. Huq, L. P. N. Rebelo and K. R. Seddon, J. Phys. Chem. B, 2005, **109**, 3519–3525

41. A. Stoppa, R. Buchner and G. Hefter, J. Mol. Liq., 2010, **153**, 46–51.

42. P. Navia, J. Troncoso and L. Romani, J. Chem. Eng. Data, 2007, **52**, 1369–1374.

43. M. T. Clough, C. R. Crick, J. Gräsvik, P. A. Hunt, H. Niedermeyer, T. Welton and O. P. Whitaker, Chem. Sci., 2015, **6**, 1101–1114.

44. A. Arce, M. J. Earle, S. P. Katdare, H. Rodríguez and K. R. Seddon, Chem. Commun., 2006, 2548–2550

45. G. Annat, M. Forsyth and D. R. MacFarlane, J. Phys. Chem. B, 2012, **116**, 8251–8258.

46. S. Omar, J. Lemus, E. Ruiz, V. R. Ferro, J. Ortega and J. Palomar, J. Phys. Chem. B, 2014, **118**, 2442–2450.

47. T. Cremer, C. Kolbeck, K. R. J. Lovelock, N. Paape, R. Wölfel, P. S. Schulz, P. Wasserscheid, H. Weber, J. Thar, B. Kirchner, F. Maier and H.-P. Steinrück, Chem. - Eur. J., 2010, **16**, 9018–9033

48. P. A. Hunt, C. R. Ashworth and R. P. Matthews, Chem. Soc. Rev., 2015, **44**, 1257–1288.

49. S. Cha and D. Kim, Phys. Chem. Chem. Phys., 2015, **17**, 29786–29792.

50. W. Zhao, F. Leroy, B. Heggen, S. Zahn, B. Kirchner, S. Balasubramanian and F. Müller-Plathe, J. Am. Chem. Soc., 2009, **131**, 15825–15833.

51. I. Skarmoutsos, T. Welton and P. A. Hunt, Phys. Chem. Chem. Phys., 2014, **16**, 3675–3685.

52. R. D. Skwierczynski and K. A. Connors, J. Chem. Soc., Perkin Trans. 2, 1994, 467–472

53. E. Bosch and M. Rosés, J. Chem. Soc., Faraday Trans., 1992, **88**, 3541–3546.

54. C. C. Weber, A. F. Masters and T. Maschmeyer, Org. Biomol. Chem., 2013, **11**, 2534–2542.

55. C. C. Weber, A. F. Masters and T. Maschmeyer, J. Phys. Chem. B, 2012, **116**, 1858–1864

56. B. A. Marekha, O. N. Kalugin and A. Idrissi, Phys. Chem. Chem. Phys., 2015, **17**, 16846–16857.

57. E. Buisine, K. de Villiers, T. J. Egan and C. Biot, J. Am. Chem. Soc., 2006, **128**, 12122–12128

58. M. A. Ab Rani, A. Brant, L. Crowhurst, A. Dolan, M. Y. Lui, N. H. Hassan, J. P. Hallett, P. A. Hunt, H. Niedermeyer, J. M. Perez-Arlandis, M. Schrems, T. Welton and R. Wilding,Phys. Chem. Chem. Phys., 2011, **13**, 16831–16840.

59. H. Niedermeyer, C. Ashworth, A. Brandt, T. Welton and P. A. Hunt, Phys. Chem. Chem. Phys., 2013, **15**, 11566–11578.

60. R. Lungwitz and S. Spange, New J. Chem., 2008, **32**, 392–394.

61. P. Bonhôte, A.-P. Dias, N. Papageorgiou, K. Kalyanasundaram and M. Grätzel, Inorg. Chem., 1996, **35**, 1168–1178

CHAPTER 5

PREDICTIVE THERMODYNAMICS FOR IONIC SOLIDS AND LIQUIDS

Leslie Glasser and H. Donald Brooke Jenkins

INTRODUCTION

Consider what obliges one to attempt to estimate the properties of a known inorganic material or predict the properties of an as-yet unprepared material. Examine the complexities! There are about 100 chemical elements, which combine to form about 100 million already-known compounds, of which some 10% could be classified as inorganic/mineral.[1] There remains an almost limitless range of possible combinations as yet unexplored. By contrast, the most comprehensive current crystallographic databases report data on only about one million of these organic compounds and on only about one-quarter of a million inorganic compounds.[2] Ionic liquids are combinations of such a broad range of cations and anions so that the in-principle possible number of such liquids is of the order of 10^{18} (although the realistically possible number is orders of magnitude smaller) – already, some 1000 have been reported in the literature.[3]On the other hand, thermodynamic data is available for only some 30 000 compounds, of which about 60% (20 000) are inorganic (see ESI for a list of thermodynamic data compendia). Thus, the chance of finding the property data one seeks is miniscule; add to this, the need to obtain data on as-yet unprepared material, such as might be required for a proposed synthesis. As a consequence, a number of simple empirical rules have been developed for a variety of thermodynamic properties (see Table 1).

Table 1 Some empirical thermodynamic and volume rules

Rule	Thermodynamic property	Value	Ref.
Dulong and Petit and Neumann–Kopp summation: heat capacity	C_p, per mole for atoms in solids	≈3R ≈ 25 J K^{-1} mol^{-1} where R is gas constant	4
Trouton: vaporization	$\Delta_{vap}S°$ for organic molecules and non-polar liquids $\Delta_{vap}H°$ for organic molecules and non-polar liquids For hydrogen bonding in liquids	$\Delta_{vap}S°$/J K^{-1} mol^{-1} ≈ 85 $\Delta_{vap}H°$/kJ mol^{-1} ≈ 85T_b/K T_b = boiling point $\Delta_{vap}H°$/kJ mol^{-1} > 85T_b/K	
Walden: fusion	$\Delta_{fus}S°$ for rigid non-spherical molecules	≈20–60 J K^{-1} mol^{-1}	

Citation: Leslie Glasser and H. Donald Brooke Jenkins, Predictive thermodynamics for ionic solids and liquids, DOI: 10.1039/c6cp00235h.

Richards: fusion	$\Delta_{fus}S°$ for rigid spherical molecules	≈7–14 J K^{-1} mol^{-1}	5
Glasser & Jenkins: C_p, for silicate anions	Single ion heat capacity, C_p, for silicate anions	≈13.8n where n is the number of atoms contained within the silicate anion	6
Glasser & Jenkins: C_p for general anions	Single ion heat capacity, C_p for general anions	≈17n where n is the number of atoms contained within the anion	6
Westwell, Searle, Wales & Williams: sublimation enthalpy from melting point	$\Delta_{subl}H°$ for molecules that do not possess internal rotors, such as long chain organic molecules	$\Delta_{subl}H°$/kJ mol^{-1} = 0.188T_m/K + 0.522 T_m = melting point R^2 = 0.90	7
Westwell, Searle, Wales & Williams: sublimation enthalpy from boiling point	$\Delta_{subl}H°$ for molecules that do not possess internal rotors, such as long chain organic molecules	$\Delta_{subl}H°$/kJ mol^{-1} = 0.119T_b/K + 1.38 T_b = boiling point R^2 = 0.96	
Westwell, Searle, Wales & Williams: vaporization enthalpy of solids	$\Delta_{vap}H°$ for molecules that do not possess internal rotors, such as long chain organic molecules	$\Delta_{vap}H°$/kJ mol^{-1} = 0.166T_m/K − 3.99 T_m = melting point R^2 = 0.86	
Westwell, Searle, Wales & Williams: vaporization enthalpy of liquids	$\Delta_{vap}H°$ for molecules that do not possess internal rotors, such as long chain organic molecules	$\Delta_{vap}H°$/kJ mol^{-1} = 0.108T_b/K − 5.08 T_b = boiling point R^2 = 0.98	
Westwell, Searle, Wales & Williams: T_b from T_m	T_b from T_m for molecules that do not possess internal rotors, such as long chain organic molecules	T_b/K = 1.52T_m/K + 14.5 T_b = boiling point T_m = melting point R^2 = 0.86	
Westwell, Searle, Wales & Williams: vaporization and sublimation enthalpies	$\Delta_{vap}H°$ for molecules that do not possess internal rotors, such as long chain organic molecules	$\Delta_{vap}H°$/kJ mol^{-1} = 0.889$\Delta_{subl}H°$/kJ mol^{-1} − 4.75 R^2 = 0.98	
Kempster & Lipson; Glasser & Jenkins: formula unit volume	V_m mainly for organic solids	V_m ≈ 0.018 nm^3 per atom	8
	V_m for water	V_m ≈ 0.0245 nm^3	(seeTable 3)[9]

The most basic data that needs to be obtained for this essentially unlimited set of materials is thermodynamic because such data informs us of the stability of the materials, our ability to synthesise them, and to maintain their integrity. In the absence of published data, the question then arises as to how one should proceed in order to obtain that data.

The most fundamental approach would be through quantum mechanical (QM) calculation,[10] where one considers in detail how the fundamental particles of which a material consists, such as atoms and electrons, interact with one another through electrostatic forces, charge transfer, van der Waals (dispersion) interactions, electron correlation, and so forth. While such an approach has yielded important results, it is complex, uses expensive computation facilities, and requires considerable expertise in both application and interpretation. At a somewhat simpler level, density functional theory (DFT)[11] has reduced the complexity of QM methods and, hence, their cost by relating the energetics to the more readily computable electron density of the material and using functions of the electron density function (that is, functionals) to derive experimentally observable results. DFT has found increasing favour in recent years in providing useful results but difficulties remain in dealing with dispersion and electron correlation. Thus, in stark contrast to the VBT approach, these QM approaches require considerable expertise to execute and interpret reliably.

A rational response has been to collect data on related materials and use that data to extrapolate (or interpolate) in order to estimate the properties of the material under investigation. We illustrate this approach in some general terms first, and then focus on an approach which we have termed volume-based thermodynamics (VBT)[12,13] together with the thermodynamic difference rule (TDR),[9,14–16] both of which we and colleagues have developed and fostered over the last two decades. These empirical procedures have proven to have great generality and utility, and have been widely implemented for ionic solids and liquids,[17] as also illustrated in a list of applications in the final section of this paper. One further very successful method is the "Simple Sum Approximation" (SSA)[18] where the thermodynamic properties of a complex ionic, such as $MgSiO_3$, is treated as a sum of its components, being the oxides MgO and SiO_2 in this case.

Prediction basically relies on the combination, through the Gibbs relation, of enthalpy, H, and entropy, S, contributions:[19]

$$\Delta G = \Delta H - T\Delta S \qquad (1)$$

VBT and TDR provide estimates of values of standard enthalpy, $\Delta_f H^\circ$, standard entropy, S°_{298}, and hence, via eqn (1) lead to the prediction of $\Delta_f G^\circ$ for individual materials as well as $\Delta_r G^\circ$ for a reaction of interest. It is often of little concern that such estimates may not be highly precise, since the purpose of thermodynamic prediction may, in many instances, simply be one of assessing synthetic feasibility or otherwise, i.e., simply whether $\Delta_r G$ is negative (feasible) or positive (infeasible in principle, although a small positive value, say ~20 kJ mol^{-1}, does not preclude formation of useful proportions of product which can be extracted from the reaction system).[20] Furthermore, experimentally-derived thermodynamic values themselves can have considerable uncertainties.[21–23] The actual mathematics required is minimal yet quantitative interpretation results. This review summarises VBT, TDR, SSA, and single-ion additivity, and highlights many of their successes – indeed, the scope of their application has proved to be quite remarkable – and directs the reader to numerous applications where these procedures have played a significant role in yielding the thermodynamics.

In essence, VBT [together with its isomegethic ("equal size") rule,[24] which vastly extends its application to new and hypothesised materials] relates the formula unit volumes, V_m, of materials, however measured or estimated, to their thermodynamic quantities, thereby leading to practical prediction. TDR uses the differences between related materials to predict values for other similar materials, while the additive SSA and single-ion values (and also TDR, in a slightly more complex way) demonstrate that the properties of complex materials may be estimated by summing the corresponding properties of their component parts.

Group Additivity Methods

Group methods operate by assuming that a material (typically, but not exclusively, an organic molecule) contains independent entities (such as single bonds, double bonds, alcohol, amine, etc.) whose thermodynamic property values can be summed to produce a property value for the whole material, but supplemented by other terms to allow for the interactions between and among groups. (An extensive list of group methods is presented in our ESI.) The most developed of these are termed Benson group methods.[25] In order to permit broad application of the methods, it has been necessary to develop hundreds of group terms, with the further complication of the necessity for the user to identify suitable groups within the material under consideration. Many computer programs incorporate these methods, often as preliminary steps to a more complex analysis. The NIST WebBook[26] provides a free service which implements the Benson group additivity scheme for gas-phase organic molecules.

Group methods have also been developed for ionic systems, by identifying constituent cations and anions whose properties are summed to provide the overall property value sought. The results are most reliable when based on related materials. In general, these methods have not received wide acceptance.

Volume-based Thermodynamics

Early thermodynamic property-size relations were generally based upon ion radii since the alkali metal and halide monatomic ions of the most important alkali halides are spherical and radius, which could be quite readily established from X-ray data, was the most obvious measure of relative ion size. An important equation in this context was the Kapustinskii relation[27] for lattice potential energy, U_{POT}:

$$U_{POT} = \frac{A|\nu z_+ z_-|}{\langle r\rangle}\left(1 - \frac{\rho}{\langle r\rangle}\right) \qquad (2)$$

where z_+, z_-/electron units are the integer charges on the cations and anions, respectively, ν is the number of ions per formula unit, ρ is a compressibility constant (usually chosen as ρ = 0.345 nm), ⟨r⟩ is the sum of the cation and anion radii (which is often equated to the shortest cation–anion distance found in the lattice), and A (=121.4 kJ mol^{-1} nm) is an electrostatic constant. While the contact distance, ⟨r⟩, between cation and anion is a straightforward sum for simple ions, it becomes ill-defined when complex ions are present. In addition to this conceptual problem, the Kapustinskii equation cannot be applied beyond binary materials because (i) there is no provision for more than one type of cation–anion contact, nor (ii) can more than a pair of charge types be accommodated.

Mallouk, Bartlett and coworkers[28–31] presented some relationships between formula unit volume, V_m and thermodynamic properties (notably lattice energy, U_{POT}, as a function of $V_m^{-1/3}$) and standard entropy, S°_{298} (as a function of V_m) but only for a handful of MX (1 : 1) simple salts. Jenkins, Roobottom, Passmore and Glasser[32] explored these relationships further, by (i) replacing the distance sum with the equivalently-dimensioned cube-root of the formula unit volume, $V_m^{1/3}$, and (ii) using a generalisation of the charge product[33] into an ionic strength factor-type summation, I:

$$|\nu z_+ z_-| \Rightarrow \sum_i n_i z_i^2 = 2I \tag{3}$$

where n_i = number of ions of type i in the formula unit.

Scheme 1 summarises the processes and equations which use material volumes to produce thermodynamic values.

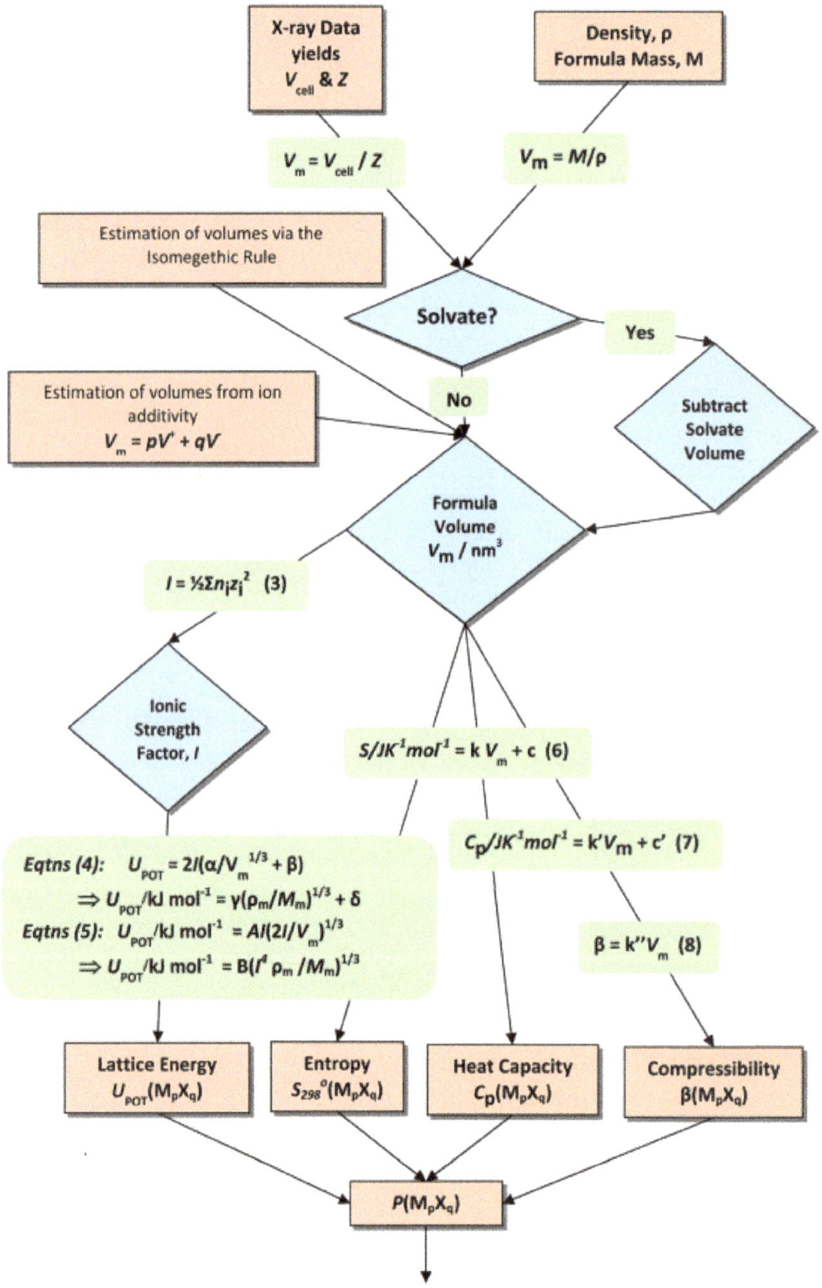

Scheme 1 Volume-based thermodynamics (VBT) flowchart. Data sources of various kinds are used to generate a formula unit volume, V_m, from which thermodynamic properties are estimated. $P(M_pX_q)$ represents any thermodynamic property, P, of the material M_pX_q. α, β, γ,δ, A, B, k, c , c′, k′ and k″ represent various constants obtained by fitting to experimental data. Eqn (4)–(8) appear in this scheme.

Lattice energies for a large database of simple ionic solids could be reliably correlated using this linear VBT function:[34]

$$U_{POT} = 2I\left(\frac{\alpha}{V_m^{1/3}} + \beta\right) \tag{4; Scheme 1}$$

where α and β are empirical constants which differ depending on stoichiometry, and have been determined by fitting to extensive experimental data.[32] It is noteworthy that the fitted constant, α, is found always to be close in value to the electrostatic factor, A, in eqn (2), above.

Equivalent equations may be couched in terms of density, ρ_m, and formula mass, M_m:

$$U_{POT}/\text{kJ mol}^{-1} = \gamma(\rho_m/M_m)^{1/3} + \delta \tag{4, Scheme 1}$$

$$U_{POT}/\text{kJ mol}^{-1} = B(I^4\rho_m/M_m)^{1/3} \tag{5, Scheme 1}$$

where B is a combined constant.

For lattice energies greater than 5000 kJ mol^{-1}, which includes most minerals, a limiting version[35] of this equation exists which contains no empirical constants whatsoever and yet satisfactorily predicts lattice energies up to 70 000 kJ mol^{-1} and probably beyond:

$$U_{POT} = AI(2I/V_m)^{1/3} \quad (5, \text{Scheme } 1)$$

Lattice energy is readily converted to lattice enthalpy[36] (as it needs to be if it is to be included in an enthalpy-based thermochemical cycle such as that in Fig. 1 below) using the equation:

$$\Delta_L H = U_{POT} + \sum_{i=1}^{n} s_i\left(\frac{c_i}{2} - 2\right)RT \quad (9)$$

where $\Delta_L H$ is the lattice enthalpy, n is the number of ion types in the formula unit, s_i is the number of ions of type i, and c_i is defined according to whether ion i is monatomic ($c_i = 3$), linear polyatomic ($c_i = 5$), or nonlinear polyatomic ($c_i = 6$).

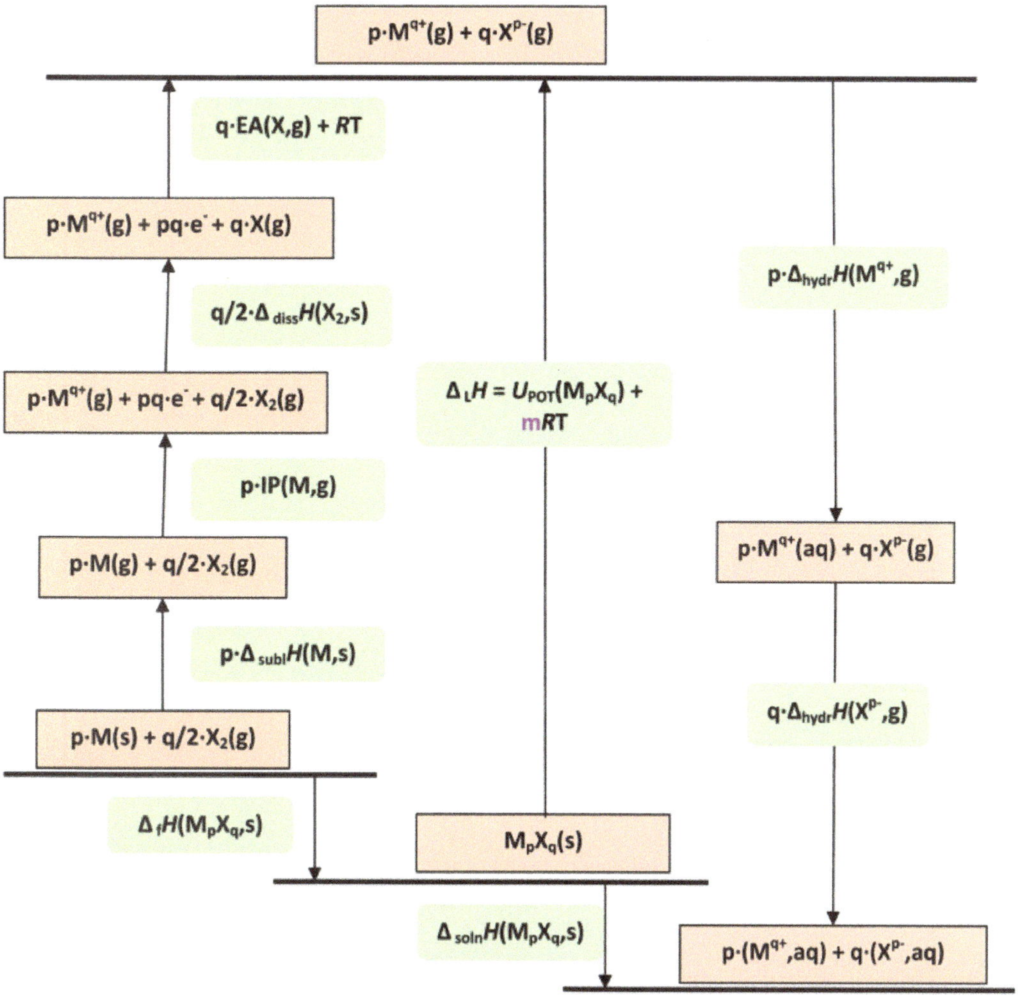

Figure 1 Born–Haber–Fajans cycle for solids and aqueous solutions of formula M_{pXq}. IP = ionisation potential; EA = electron affinity; lattice enthalpy (ΔLH) and enthalpies of formation (ΔfH), sublimation (ΔsublH), dissociation (ΔdissH), hydration (ΔhydrH), and solution (ΔsolnH) are involved. UPOT represents the lattice potential energy. In the formula for ΔLH, mdepends on p and q and the nature of the ions Mq+ and Xq− (see eqn (9) in text).36

Using these approaches, it becomes simple to evaluate the lattice energies, U_{POT} (and enthalpies of formation, $\Delta_f H°$, via the Born–Haber–Fajans relation) of ionic solids.

Thermodynamic Difference Rule (TDR)

The thermodynamic difference rule, TDR, is a complementary set of procedures which utilizes additive connections among related materials.[9,12–16]Scheme 2 shows the steps by which TDR is usually applied. The technique is extremely powerful as a result of its ability to enable estimates to be made of thermodynamic data not otherwise available.

$$P(M_pX_q{\cdot}nL,s) - P(M_pX_q,s) = n\theta_P(L, \text{s–s}) \quad (11a)$$

$$P(M_pX_q{\cdot}mL,s) - P(M_pX_q{\cdot}nL,s) = (m - n)\theta_P(L, \text{s–s}) \quad (11b)$$

Thus, various thermodynamic state properties may be estimated as, for example, the lattice energies of hydrates using the thermodynamic difference rule relation:

$$U_{POT}(M_pX_q{\cdot}nH_2O,c) - U_{POT}(M_pX_q,c) = n\theta_U(H_2O) \quad (12)$$

with $\theta_U(H_2O) = 54.3$ kJ mol^{-1}, as empirically determined. Table 2 lists values for the fitted constants for various groups of materials, while Table 3 lists values pertaining to the hydrate TDR rules. TDR constants for other solvates may be found in Table 1 in the literature referenced.[9] An important recent paper considers the thermodynamics of hydration in minerals.[37]

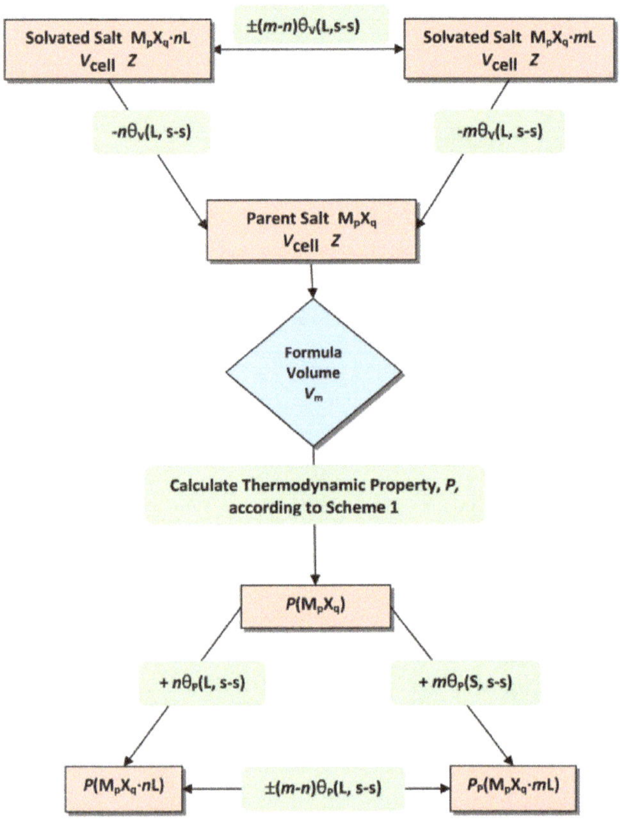

Scheme 2 Thermodynamic difference rule (TDR) flowchart. The volume of addend or solvate, L, is subtracted to generate the volume of the parent material:

$$V(M_pX_q \cdot nL) - nV(L) = V(M_pX_q) \qquad (10)$$

from which the thermodynamic property, P, is calculated. Finally, the thermodynamic property value, θ_P(L, s–s), of the addend or solvent (which may be water) is added (where "s–s" represents that the difference, θ_P, between materials each in the same phase, often solid).

Table 2 Constants for a selection of volume-based thermodynamic relations

Material	Ionic strength factor, I	α/kJ mol^{-1} (nm^3 formula unit^{-1})$^{-1/3}$	β/kJ mol^{-1}	Mean absolute error (%)
Lattice energy from volume data, U_{POT}/kJ mol^{-1} = $2I(\alpha/V_m^{1/3} + \beta)$ (4)				
MX (1:1)	1	117	52	4
MX_2 (2:1)	3	134	61	
M_2X (1:2)	3	165	−30	
MX (2:2)	4	119	60	
M_pX_q	$\frac{1}{2}(pq^2 + qp^2)$	139	28	
Material	**Ionic strength factor, I**	**γ/kJ mol^{-1} cm^{-1}**	**δ/kJ mol^{-1}**	**Mean absolute error (%)**
Lattice energy from density data, U_{POT}/kJ mol^{-1} = $\gamma(\rho/M)^{1/3} + \delta$ (5)				
MX (1:1)	1	1981.2	103.8	4
MX_2 (2:1)	3	8375.6	−178.8	
M_2X (1:2)	3	6764.3	365.4	
MX (2:2)	4	6864.0	732.0	
M_pX_q	$\frac{1}{2}(pq^2 + qp^2)$	2347.6 × I	55.2 × I	
		k/J K^{-1} mol^{-1} (nm^{-3} formula unit)	**c/J K^{-1} mol^{-1}**	**Mean absolute error (%)**
Entropy, S/J K^{-1} mol^{-1} = $kV_m + c$ (6)				
Anhydrous ionic salts		1360 ± 56	15 ± 6	12
Hydrated ionic salts		1579 ± 30	6 ± 6	7.4
Organic liquids		1133 ± 7	44 ± 2	5.7
Organic solids		774 ± 21	57 ± 6	10.4
		k'/J K^{-1} mol^{-1} (nm^{-3} formula unit)	**c'/J K^{-1} mol^{-1}**	**Mean absolute error (%)**
Heat capacity, C_p/J K^{-1} mol^{-1} = $k'V_m + c'$ (7)				
Non-framework silicates		1465	11	
General ionic solids[a]		1322	−0.8	24.5
Ionic liquids		1037	45	
		k''/GPa^{-1} (nm^{-3} formula unit)		**Mean absolute error (%)**
Isothermal compressibility, β/GPa^{-1} = $k''V_m$ (8)				
General ionic solids (no alkali halides)		0.634		12
Perovskites		0.472		7

[a] C_p values have been estimated for 799 materials, ranging from small ionics to large mineral structures. Poor outliers may be avoided by using the lesser of the calculated value from eqn (7) and the approximate limiting Dulong–Petit value of 25 × m J K^{-1} mol^{-1}, where m = number of atoms in the formula unit (*cf.* Table 1).

Table 3 Thermodynamic difference values, θ_P, for hydrates for property P: $P(M_pX_q \cdot nH_2O) - P(M_pX_q) = n\theta_P(H_2O,s\text{–}s)$[a]

Thermodynamic property, P	θ_P[9]
V_m/nm^3 (mol of $H_2O)^{-1}$	0.0245
U_{POT}/kJ (mol of $H_2O)^{-1}$	54.3
$\Delta_f H$/kJ (mol of $H_2O)^{-1}$	−298.6
$\Delta_f S$/J K^{-1} (mol of $H_2O)^{-1}$	−192.4
$\Delta_f G$/kJ (mol of $H_2O)^{-1}$	−242.4
S°_{298}/J K^{-1}(mol of $H_2O)^{-1}$	40.9
C°_{p298}/J K^{-1}(mol of $H_2O)^{-1}$	42.8[16]

[a] A more complete table, including values of θ_P for solvates other than H_2O, has been published.[16]

Room-temperature Ionic Liquids

Room-temperature ionic liquids (RTILs)[38–41] can replace organic solvents used for the dissolution of both polar and non-polar solutes and for processing or extraction of materials, while also having useful catalytic features.[42] ILs are low-polluting, with low combustibility, good thermal stability and low vapour pressures. They have high viscosities, and their liquid range can often cover several hundred degrees. The range of their applications has been extended by use of mixtures of IL's[43] and as supercritical fluids.[44]

As their name implies, they are usually liquid at ambient temperatures and consist solely of ionic species. In order to reduce the lattice energy of their crystalline state and hence their melting point, one or sometimes both of their cations and anions need to be large and their cations also often have low symmetry. The cations are generally organic with long-chain features and buried charges, such as the pyrollidinium, methylimidazolium and pyridinium cations (see Fig. 2) while the anions, such as BF_4^-, PF_6^-, or NTf_2^- [formula: $(CF_3SO_2)_2N^-$] have diffuse charges.

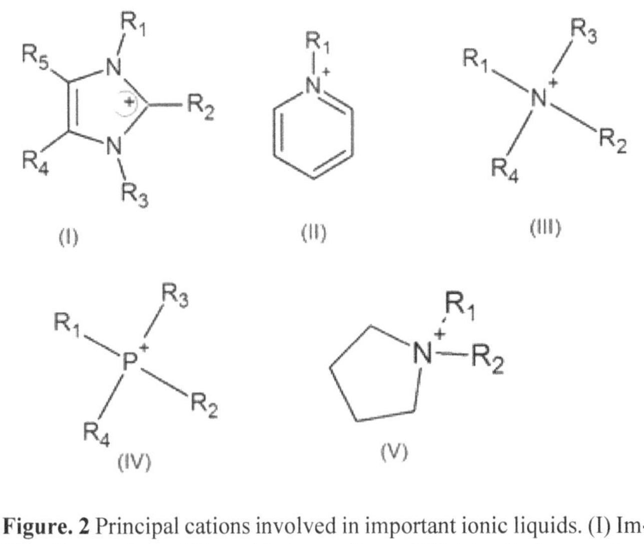

Figure. 2 Principal cations involved in important ionic liquids. (I) Imidazolium cations, (II) pyridinium cations, (III) tetraalkylammonium cations, (IV) tetraalkylphosphonium cations, and (V) pyrollidinium cations.

Since there are, as noted above, many possible combinations of cation and anion, it becomes possible to consider designing ionic liquids to purpose. Although early QSAR predictions[45–47] were not always regarded as satisfactory,[48] molecular volume[49] has emerged as an important observable. Thus, Glasser[38] has estimated a VBT-based entropy for ionic liquids, derived from correlations for both inorganic solids and organic liquids:

$$S\ (\mathrm{J\ K^{-1}\ mol^{-1}}) = 1246.5(V_m/\mathrm{nm}^3) + 29.5 \qquad (13)$$

Similarly, Gutowski, et al.,[50–52] have developed a lattice energy correlation for 1 : 1 ionic liquids, with amended constants (eqn (4), Scheme 1): $I = 1$, $\alpha = 8326$ kJ nm mol^{-1} and $\beta = 157$ kJ mol^{-1}.

VBT relations have now been used to establish other physical thermodynamic properties, independent of crystal structure, such as liquid entropy,[53] melting point,[54–56] heat capacity[57] and critical micelle concentration (c.m.c.).[58] A recent review by Beichel[59] cites relevant literature from the Krossing laboratory; in order to emphasize lack of reliance on any experimental input at all, this group has introduced the term "augmented volume-based thermodynamics".

An Explanation for the Simple Volume Relations

The striking simple relations to volume involved in VBT, almost independent of structure, upon which we have reported invite some explanation. We suggest that the bulk thermodynamic properties derive from the interactions between the particles involved (complex ions or even molecules) rather than within those particles.[60] For ionic materials, the interactions are largely coulombic (electrostatic or Madelung energies) with lesser contributions from specific repulsion and van der Waals-type interactions so that (to the approximation inherent in our correlations) the interactions are similar, independent of the specific species involved and also independent of structure. At this level of approximation we have found that the thermodynamic values that emerge prove adequate, in the majority of cases, for deciding questions of alternative synthetic routes for the preparation of inorganic materials.

As we have noted above, an important consequence of the independence of structure is that these relations apply equally to pure liquids as to solids, so that they can be applied to the increasingly important class of ionic liquids.[38]

Madelung Energies – for Known Structures

The coulombic (or Madelung) energy, E_M, of a material of known structure is readily calculated by means of standard computer programs, such as GULP[61] and EUGEN.[62] This energy corresponds to separating the constituent ions into independent gas phase ions against coulombic forces only. We have observed[63] that the resulting Madelung energy is closely related to the corresponding lattice energy, in the form

$$U_{POT}/\text{kJ mol}^{-1} = 0.8518E_M + 293.9 \quad (14)$$

Eqn (14) thus provides a further simple direct means for obtaining lattice energies, apart from VBT. However, the Madelung calculation comes into its own when applied to an ionic system with structures containing covalently-bonded complexes, such as K_2SnCl_6. If the complex ion, $SnCl_6^{2-}$ in this case, is treated as a "condensed ion", with all the ion charge placed on the central atom (thus Sn^{2-}) and the ligands given zero charge (Cl^0), then we are effectively dealing with a system which decomposes to $2K^+(g) + Sn^{2-}(g)$ when the Madelung energy, nowE_M', is supplied. For this system

$$U_{POT}/\text{kJ mol}^{-1} = 0.963E_M' \quad (15)$$

From these values, we can determine a formation energy for the "condensed ion" complex (see example below).

Isomegethic Rule

Our isomegethic ("equal size") rule[24] states that "ionic salts of the same empirical chemical formula having identical charge states (i.e., lattice ionic strength factors, I) will have approximately equal formula unit volumes, V_m." Since I, V_m and stoichiometry are then approximately identical, isomegethic compounds will have almost identical lattice potential energies too.

As an illustration, consider the relation:

$$V_m(NO^+ClO_4^-) \approx V_m(NO_2^+ClO_3^-) \approx V_m(ClO_2^+NO_3^-) \quad (14a)$$

and hence:

$$U_{POT}(NO^+ClO_4^-) \approx U_{POT}(NO_2^+ClO_3^-) \approx U_{POT}(ClO_2^+NO_3^-) \quad (14b)$$

Good examples of the scope of the isomegethic rule in providing multiple estimates for the volumes of ions are given in detail in ref. 24 and 64. Relations of this kind provide enormous scope for estimation of formula unit volumes and lattice energies, which is especially useful for hypothesised materials. If the enthalpies of formation of the individual gaseous component ions are known[65,66] then the enthalpy of formation of the isomegethic compound may usually be estimated, thus taking us into the full compass of the thermodynamics of the material concerned.

Single-ion Values

Since it is seldom that all the desired thermodynamic values are available to generate the desired data, we have prepared sets of internally consistent single-ion values which may be used additively to generate otherwise absent data, collected in Table 4. An early example of this procedure is provided by the work of Latimer[67,68] in developing single ion entropy estimates.

Table 4 Single-ion additive volumes, entropies and heat capacities for a range of inorganic cations and anions. Note that the values listed for H_2O were used as reference values in the optimizations to generate these single-ion values

Cation	Volume/ nm^3	S°_{ion}/J K^{-1} mol^{-1}	$C_p(298)_{ion}$/ J K^{-1} mol^{-1}	$-\Delta_fH(298)$/ kJ mol^{-1}	$-\Delta_fG(298)$/kJ mol^{-1}	Anion	Volume/ nm^3	S°_{ion}/J K^{-1} mol^{-1}	$C_p(298)_{ion}$/ J K^{-1} mol^{-1}	$-\Delta_fH(298)$/ kJ mol^{-1}	$-\Delta_fG(298)$/ kJ mol^{-1}
NH_4^+	0.0356	67.0				F^-	0.014	20.6		5	5.0
Li^+	0.0067	19.4	21.1	308.7	259.8	Cl^-	0.0298	36.1	23.5	110.8	113.7
Na^+	0.0158	37.2	29.7	278.3	232.2	Br^-	0.0363	48.6	26.6	66.3	70.6
K^+	0.0277	50.5	31.4	316	258.7	I^-	0.0488	56.8	28.3	−8.2	0.0
Rb^+	0.0341	63.1	31.3	311.7	257.4	N_3^-	0.0416	54.8			
Cs^+	0.042	69.2	31.7	315.4	260.4	O^{2-}	0.0134	7.4			
Ag^+			29.9			OH^-	0.0184	20.7	21.2	228.6	247.9
Mg^{2+}	0.0049	20.5	26.1	491.4	374	S^{2-}	0.032	22.1			
Ca^{2+}	0.0201	32.5	28.9	579.8	468.2	ClO_3^-			70.1	68.5	
Sr^{2+}	0.0213	37.9	30.3	627.2	558.7	BrO_3^-			76.1	39.7	
Ba^{2+}	0.027	55.1	30.2	655	570.1	CO_3^{2-}	0.0426	52.6	57.5	594.3	582.7
Mn^{2+}			30.4	262.9	161.3	NO_3^-	0.0492	78.5			
Fe^{2+}	0.0067	38.6	31.8	138.5	32.3	PO_4^{3-}	0.057	71.8			
Zn^{2+}	0.0125	37.3				SO_4^{2-}	0.0611	74.3	71.5	812.2	802.3
Cu^{2+}	0.0053	33.5				ClO_4^-	0.0619	105.8	134.9		
Ni^{2+}	0.0004	29.6		126.7		MnO_4^-	0.0665				
Co^{2+}	0.0019	39.5	32.0	161.8		AsO_4^{3-}	0.0658				
Fe^{3+}	0.0061	28.2		208.2	60.0	VO_4^{3-}	0.0663				
Al^{3+}		17.3	20.9	630.1	462.5	$Fe(CN)_6^{4-}$			219.8		
La^{3+}			28.8	785.9	595.3						
Cr^{3+}			31.0	282.1	85.6						
Fe^{3+}			30.7			SiO_3^{2-}			57.58	1056.9	1090.1
Sc^{3+}	0.0035					SiO_4^{4-}			68.93	1198.5	1300.8
Lu^{3+}	0.0102					SiO_5^{6-}			82.18	1344.4	1520.9
Yb^{3+}	0.0111					$Si_2O_5^{2-}$			101.86	1962.7	1920.0
Tm^{3+}	0.011					$Si_2O_7^{6-}$			127.4	2246.4	2361.2
Er^{3+}	0.0126					$Si_2O_{10}^{12-}$			162.2		
Y^{3+}	0.0131					$Si_3O_{10}^{8-}$			182.77	3315.1	3476.7
Dy^{3+}	0.0137					$Si_5O_{18}^{16-}$			317.21	5660.3	6050.0
Tb^{3+}	0.0148					$Si_4O_{11}^{6-}$			213.09	4098.6	4108.7
Gd^{3+}	0.0133										
Eu^{3+}	0.0146										
Sm^{3+}	0.0164										
Si^{4+}		24.5									
Ti^{4+}		35.2				**H_2O**	**0.0245**	**40.9**	**41.3**	**285.8**	**237.1**

An Example Set of Predictive Thermodynamic Calculations: K_2SnCl_6

We here provide a set of results for the material dipotassium hexachlorostannate, K_2SnCl_6, where we have deliberately selected the difficult case of a partially covalent material for which experimental thermodynamic values are available for comparison. This demonstrates some of the weaknesses of these predictive methods against some of their strengths in that they may provide a wide range of otherwise unavailable thermodynamic values.

Simple Salt Approximation (SSA)

Table 5 demonstrates the features of the "Simple Salt Approximation" in generating results by combining reaction components.

Table 5 Additive thermodynamic values for K_2SnCl_6, following the "Simple Salt Approximation".[18] Data (columns 2, 3, 4 and 7) from HSC Chemistry 8[69]

	KCl	$SnCl_4$	K_2SnCl_6	2KCl + $SnCl_4$	% Diff	2KCl + $SnCl_4$ → K_2SnCl_6
$\Delta_fH°$(25 °C,s)/kJ mol^{-1}	−436.7	−517.0	−1482.0	−1390.4	−6.2	−91.6
S°(25 °C,s)/J K^{-1} mol^{-1}	82.7	265.0	371.0	430.4	16.0	−59.4
$\Delta_fG°$(25 °C,s)/kJ mol^{-1}	−461.3	−596.0	−1592.6	−1518.7	−4.6	−73.9
C_p(25 °C,s)/J K^{-1} mol^{-1}	51.4	157.2	221.1	260.1	17.6	−39.0

For the SSA to be accurate, it is necessary that the reaction to form product should yield zero (or small) thermodynamic differences. As may be seen from the final column in Table 5, the reaction 2KCl + $SnCl_4$ → K_2SnCl_6 produces non-zero differences, so that the SSA results (3rd last column) are not accurate, but may be useful as a general guide when the thermodynamic values are unknown.

Volume-based Thermodynamics (VBT)

VBT is a specifically ionic-based set of empirical procedures, correlated against strongly ionic materials such as simple halides and more complex oxides. We examine its application to K_2SnCl_6 with its covalent central ion, $SnCl_6^{2-}$.

The constants used in the following calculations are selected from Table 2.

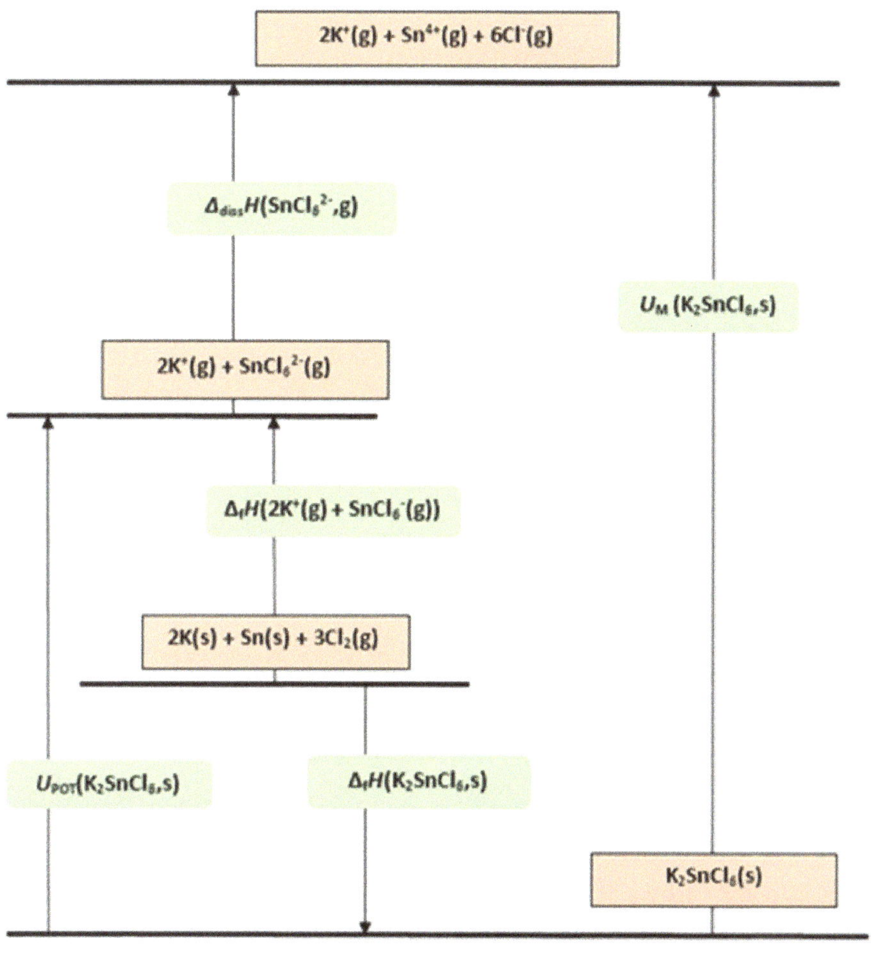

Figure. 3 Born–Haber–Fajans cycle for K_2SnCl_6.

- Formula unit volume,[70] $V_m/nm^3 = V_{cell}/Z = 1.0057/4 = 0.2514$
- Ionic strength factor, $I = 1/2\sum n_i z_i^2 = 3$ (eqn (3))
- Lattice (potential) energy, U_{POT}/kJ mol^{-1} = $2I(\alpha/V_m^{1/3} + \beta)$ (eqn (4)) = 2 × 3 (165/0.2514$^{1/3}$ − 30) = 1389 (cf. Born–Haber–Fajans cycle value[71] = 1390, Fig. 3)
- Standard entropy, S/J K^{-1} mol^{-1} = $kV_m + c$ (eqn (6)) = 1360 × 0.2514 + 15 = 356.9 (cf. 371.0, diff. 3.9%, Table 5)
- Heat capacity, C_p/J K^{-1} mol^{-1} = $k'V_m + c'$ (eqn (7)) = 1322 × 0.2514 − 0.8 = 331.6 (cf. 221.1, diff. −33%, Table 5) (Neumann–Kopp atom additive value[6] for 9 atoms = 25 × 9 = 225: diff. 1.7%)
- Comment: the VBT value calculated for C_p considerably exceeds the 9-atom limiting Neumann–Kopp value, which we propose is a preferred value (see Table 2). This suggests that the rigid covalent $SnCl_6^{2-}$ structure corresponds to too-large a volume compared with a close-packed strictly ionic system. Correspondingly, the predicted entropy is also too large. By contrast, in calculating the lattice energy, any volume error is minimised by the use of a cube-root volume.
- $\Delta_fH(SnCl_6^{2-},g)$/kJ mol^{-1} = $U_{POT}(K_2SnCl_6,s)$[71] + $\Delta_fH(K_2SnCl_6,s)$[69] − $2\Delta_fH(K^+,g)$[71] = (1390) + (−1482) − 2(514) = −1120 (cf.[65] −1156, diff. 3.1%)
- Madelung energy[72] for $K_2SnCl_6(s)$, E_M/kJ mol^{-1} = 9322 converted to U_M/kJ mol^{-1} = 0.8518 × 9322 + 293.9 = 8234
- Comment: see Born–Haber–Fajans cycle, Fig. 3.
- Madelung energy, assuming the "condensed ion"[72] $SnCl_6^{2-}$, E_M'/kJ mol^{-1} = 1613.7 converted to U_{POT}/kJ mol^{-1} = 0.963 × 1614 = 1554
- Comment: this result may be compared with the value noted in (c) above of 1390 kJ mol^{-1} (+11%).
- Comment: see Born–Haber–Fajans cycle, Fig. 3.
- By difference, dissociation energy to independent ions, $\Delta_{diss}H(SnCl_6^{2-},g)$ = 5199 kJ mol^{-1}
- Comment: this is a "new" value, not previously reported.
- Lattice energy/kJ mol^{-1} to form independent gaseous ions 2 × K^+ (Δ_fH = 2 × 541.0),[69] 6 × Cl^- (Δ_fH = 6 × −227.6),[69] and Sn^{4+} (Δ_fH = 9320.7)[65] = 7501, diff. 10% from (g).

CONCLUSION

Neither volume-based thermodynamics (VBT) nor the thermodynamic difference rule (TDR), together with their supporting quantities, require a high level of computational expertise nor expensive high-performance computation tools – a spreadsheet will usually suffice – yet the techniques are extremely powerful and accessible to non-experts. Table 2 summarises correlation equations between formula unit volume, V_m, and various thermodynamic properties, together with measures of anticipated errors. These correlation equations provide ready access to otherwise unavailable thermodynamic data, as also rapid checks of published data. The results should always be treated with appropriate caution by checking against known values for related materials.

Applications of Volume-Based Thermodynamics, VBT, and the Thermodynamic Difference Rule, TDR

In rough chronological order, we present a selection of applications of VBT and TDR which have enabled input, usually in the form of thermodynamics, to be brought to bear on a range of topical problems, often "state of the art". These present a diverse range of applications for our techniques. The message for the reader is that:

- VBT and TDR can be applied in numerous situations;
- Their application can lead to surprising new results as well as confirmatory ones;
- The basic application is usually very straightforward (Table 6).

Table 6 Applications of volume-based thermodynamics, VBT, and the thermodynamic difference rule, TDR

Volume Based Thermodynamics (VBT) or Difference Rule (TDR) specific application	Topic treated	Ref.
Thermodynamics of new homopolyatomic cation salts of sulfur and selenium.	A study designed to investigate hitherto unknown homopolyatomic cation species of sulfur and selenium (e.g., S_8^{2+} and Se_8^{2+}). The crystalline salt, $S_8(AsF_6)_2$ and its selenium analogue were predicted, prior to synthesis, to be lattice-stabilised in the solid state relative to the corresponding AsF_6^- salts of their stoichiometrically possible dissociation.	73
The fluoride ion affinity of AsF_5 and standard enthalpy of formation, $\Delta_fH(AsF_6^-,g)$.	Two different values for $\Delta_fH°(AsF_6^-,g)$ had been published in the literature at the time of our investigation. The first estimate was made by Bartlett, et al.Jenkins, Krossing, Passmore and co-workers estimated $\Delta_fH(AsF_6^-,g)$ using the thermochemical data for the salts $[NF_4^+][BF_4^-]$, $[NF_4^+][SbF_6^-]$, and $[NF_4^+][AsF_6^-]$, obtaining a value which was lower than Bartlett et al.'s value. This lower standard enthalpy of formation implied a lower fluoride ion affinity of AsF_5. In order to decide which value (433 or 467 kJ mol^{-1}) is likely to be the more accurate, we modelled the fluoride ion affinity using ab initio methods and concluded a most likely value to be: 430.5 ± 5.5 kJ mol^{-1}. In a further, later, study further FIA values were assigned.	74–76

Thermodynamics of stabilisation of S^{2+}, $Se^{2+}$and Te^{2+} ions in the solid state.	In a study on the solvated salt $S_4(AsF_6)_2{\cdot}AsF_3$ we found, using the TDR, that the solvated salt: $S_4(AsF_6)_2{\cdot}AsF_3$ was more stable than the unsolvated salt, $S_4(AsF_6)_2$ or more stable than the decomposition products of $S_4(AsF_6)_2{\cdot}AsF_3$ (which are $2S_2AsF_6$ (c) and AsF_3(l)). Such systems could not have been examined thermodynamically in this way other than by using VBT and TDR because of the complexity of the lattices involved.	77
Study of fluoride ion donor ability.	Here was studied the ability of fluoride ion donors, C^+F^-(s), to donate a fluoride ion to an acceptor, A(g). The trend among experimentalists attempting to synthesise increasingly more effective fluoride ion donors: $C^+F^-(s) + A(g) \rightarrow C^+AF^-(g)$ had been to increase the cation size further and further but this study, guided by VBT, showed that there was very little gained by size increases beyond a critical level. Here VBT played a defining role in that it led to an abortion of certain synthetic work which was taking a particular (unfruitful) avenue of development.	78 and 79
Application of VBT and TDR to assessment of the relative stabilities of bisulfite (HSO_3^-) and metabisulfite ($S_2O_5^{2-}$) salts.	Although the bisulfites, $MHSO_3$ (M = Li, Na, and K), are widely believed to be stable compounds, they have never been obtained as solids. VBT suggests that the cation volumes, $V_m(M^+)$ are insufficiently large to stabilize the HSO_3^- ion relative to the metabisulfite, $S_2O_5^{2-}$ ion. Crystalline compounds, originally thought to be $NaHSO_3$ and $KHSO_3$, were eventually identified to be metabisulfites, $Na_2S_2O_5$ and $K_2S_2O_5$. The fact that $RbHSO_3$ and $CsHSO_3$can be isolated, whilst attempts to crystallize $MHSO_3$ (M = Li, Na, K) lead to the formation of $M_2S_2O_5$: $2MHSO_3(s) \rightarrow M_2S_2O_5(s) + H_2O(l)$ has been explained using volume-based thermodynamics, VBT.	80
Use of VBT to identify thermodynamics of complex reactions and to assist in discerning the probable synthetic route of formation of products.	This application illustrates the level of sophistication of the thermochemistry that can be dealt with using VBT. Here the synthesis of the salts $NO[Al(OR)_4]$, R = $C(CF_3)_2Ph$ and $C(CF_3)_3$ by metathesis reaction of $NO[SbF_6]$ and the corresponding $Li[Al(OR)_4]$ salts in liquid sulfur dioxide solution were studied and the complete thermodynamics involved in the reaction (in the liquid sulfur dioxide medium): $3NO(g) + NO_2(g) + Li[Al(OR)_4](s) \rightarrow NO[Al(OR)_4](s) + LiNO_3(s) + N_2O(g)$ was analysed using VBT, whereupon it was seen that an unfavourable disproportionation reaction: $2NO_2(g) \rightarrow NO^+(g) + NO_3^-(g)$ is more than compensated for by the disproportionation enthalpy of the reaction: $3NO(g) \rightarrow N_2O(g) + NO_2(g)$ and the lattice energy of $LiNO_3$(s). Evidence is presented that the reaction proceeds via a complex of Li^+ with NO, NO_2 (or their dimers) and N_2O.	81
Assessment of the stability of N_5^+ salts.	The nitrogen-rich energetic salts, $N_5^+N_3^-$ and $N_5^+N_5^-$ were initially seen as having enormous potential as environmentally clean (friendly) rocket propellants and aircraft fuels offering the possibility of cheap travel. Their potential captured the excitement of the media. However, in a theoretical study, VBT coupled with ab initio calculations for the species N_3, N_3^-, N_5^+ and N_5^- showed that neither $N_5^+N_3^-$ or $N_5^+N_5^-$ could be stabilised thermodynamically and that they would decompose instantaneously into N_3 radicals and nitrogen gas.	82 and 83

VBT and Prussian blue – a contribution to evolving technology.	This study examined the solid-state thermodynamics of the cation exchanges that occur in electrochromic reactions of Prussian Blue. These electrochemically induced changes of valence that result in striking color changes constitute the process of electrochromism, a modern evolving technology in which final "best formulations" are yet being sought. Our study provided an incisive thermodynamic clarification of an ill-understood ion exchange that accompanies the early electrochromic cycles. The results showed by how much the exchange of interstitial Fe^{3+} ions by alkali metal ions, as exemplified by K^+, is thermodynamically favoured.	84
Use of VBT to investigate the thermodynamics of a new binary fluoro metal dianion $[Ti_4F_{19}]^{2-}$ prepared by autoionisation of TiF_4 by cation complexation with crown ethers.	This application of VBT shows how the methodology can be applied to investigate new and complex problems in thermodynamics. Following the successful preparation of the new $Ti_4F_{19}^{2-}$ anion, this application was used to guide synthetic endeavour by studying whether the $Ti_4F_{19}^{2-}$ anion is more able to form salts with mono or dications. To investigate this we considered two reactions: $2MTi_2F_9(s) \rightarrow M_2Ti_4F_{18}(s)$ (the target reaction, for which we needed to estimate $\Delta H \approx \Delta G$) and the gas phase dimerisation reaction: $2Ti_2F_9^-(g) \rightarrow Ti_4F_{18}^{2-}(g)$.	85
Hypothetical compounds – nitryl chlorate and its possible preparation.	Volume-based thermodynamics (VBT) has examined the attempted preparation of the salt, nitryl chlorate: $[NO_2][ClO_3]$. Nitrosyl perchlorate, $[NO][ClO_4]$, is a known and stable salt which provoked attempts to synthesize the isomeric nitryl chlorate, $[NO_2][ClO_3]$. Our isomegethic rule assigns these two materials to have similar lattice potential energies. Silver chlorate, $Ag[ClO_3]$, was reacted with nitryl hexafluoroantimonate, $[NO_2][SbF_6]$ (in nitromethane) at room temperature. Silver hexafluoroantimonate, $Ag[SbF_6]$ was identified as one of the reaction products but no $[NO_2][ClO_3]$ was detected. Nitric acid, HNO_3, was found as a side product and it was concluded that this was possibly caused by the presence of traces of water within the nitromethane solvent employed. After three attempts, we failed to prepare the target material, $[NO_2][ClO_3]$. However VBT calculations tend to confirm that $[NO_2][ClO_3]$ is formed but then decomposes in neat $Ag[ClO_3]$ and $[NO_2][SbF_6]$.	86
Novel, state-of-the-art, selenium chemistry	This study represents state of the art chemistry and an extremely complex problem. Here ^{77}Se NMR spectroscopy, DFT and MO computations, coupled with VBT and TDR estimations, were utilised to investigate the reversible dissociation of solid $Se_6I_2(AsF_6)_2{\cdot}2SO_2$ in liquid SO_2 into solutions containing 1,4-$Se_6I_2^{2+}$ in equilibrium with Se_4^{2+}, Se_8^{2+} and Se_{10}^{2+} and seven binary selenium–iodine cations. The study also provided preliminary evidence for the existence and stability of the 1,1,4,4-$Se_4Br_4^{2+}$ and cyclo-Se_7Br^+ cations.	87
Predicted stabilities for noble-gas fluorocation salts using VBT.	In the few situations in noble gas (Ng) chemistry where thermochemical facts are known, VBT can be used to validate and confirm these. This provides further evidence that, in predictive mode, results from VBT can provide a reliable guide to thermodynamic possibilities. The lattice energies, standard enthalpies, and Gibbs energies of formation for salts containing the NgF^+, $Ng_2F_3^+$, XeF_3^+, XeF_5^+, $Xe_2F_{11}^+$, and $XeOF_3^+$ (Ng = Ar, Kr, Xe) cations were estimated using crystallographic and estimated ion volumes. VBT was able, in this way, to provide estimates and predict stabilities – albeit sometimes with quite large uncertainties in the estimated data. While, for example, the stabilities of $[XeF_n][Sb_2F_{11}]$ (n = 2, 3) and $[XeF][AsF_6]$ were confirmed with respect to dissociation to the xenon fluoride and the pnictogen pentafluoride, the stabilities of $[XeF_3][AsF_6]$ and $[XeF_3][As_2F_{11}]$ were shown to be marginal under standard conditions.	88
The thermodynamic hydrate difference rule (HDR) applied to materials at the inorganic–organic interface.	The thermodynamic hydrate difference rule (HDR) is shown here also to apply equally well to formate, carbonate, acetate, glycolate and oxalate salts and their hydrates.	89

The thermodynamic difference rule and the thermodynamics of hydration (and solvation) of inorganic solids and the existence/absence of certain hydrates.	The thermodynamics of the formation of solid and liquid inorganic hydrates and ammoniates was examined in this application, suggesting that hydration is always marginally thermodynamically favourable. More detailed consideration further demonstrated that the mean value of $\Delta_r G$ per mole of water addition, from anhydrous parent to hydrate within a sequence, increases consistently toward zero, becoming progressively less favorable as the degree of hydration, n, increases, and is also broadly independent of any structural features of the materials.	90
VBT determination of ionicity or covalency within structures.	The question as to why $[P(C_6H_5)_4]^+N_3^-$ and $[As(C_6H_5)_4]^+N_3^-$ exist as ionic salts whilst $[Sb(C_6H_5)_4]^+N_3^-$ and $[Bi(C_6H_5)_4]^+N_3^-$ are covalent solids was examined. The estimations provided an unexpected answer!	91
Interpolation of thermodynamic data for sulfur compounds and use of VBT to examine possible synthetic routes and subsequent stability.	Using VBT the values of $\Delta_f G$ led us to a tentative proposal for the synthesis of Na_2SO_2. The stability of Na_2SO_2 however raises doubts and Na_2SeO_2 emerges as a more attractive target material for synthesis. We predict its synthesis will be easier and that it is stable to disproportionation into Na_2Se and Na_2SeO_4. (Subsequently to the publication of this paper it was preliminarily reported to Vegas verbally that Na_2SO_2 has indeed been possibly identified in aqueous solution).	92
Tetraanionic nitrogen-rich tetrazole-based salts	Study of energetics using VBT	93
(TDAE)(O_2SSSSO_2)(s) examined using VBT and containing the very first polythionite anion, $OSSSSO_2^{2-}$.	Synthesis of (TDAE)(O_2SSSSO_2).	94
Examination of VBT in respect of estimation of lattice energies of salts containing the 5,5′-(tetrazole-1N-oxide)$^{2-}$ anion.	Investigation of energetics of defence materials.	95
Use of TDR to examine thermochemical data for arsenic and phosphorus compounds.	Arsenic oxides/hydrates conform exactly to the TDR whilst analogous phosphorus oxides/hydrates do not. Comments on enthalpies of formation of phosphates.	96 and 97

ACKNOWLEDGEMENTS

LG and HDBJ thank their respective institutions for providing facilities for the execution of this work. Witali Beichel (Freiburg) is thanked for helpful discussion on ILs. The Curtin University Library is thanked for providing the means by which this article is available in Open Access.

REFERENCES

1. Chemical Abstracts Service, CAS Content, https://www.cas.org/content/cas-databases, accessed May, 2016.
2. L. Glasser, Crystallographic Information Resources, J. Chem. Educ., 2016, **93**, 542–549.
3. E. I. Izgorodina, Towards large-scale, fully ab initio calculations of ionic liquids, Phys. Chem. Chem. Phys., 2011, **13**, 4189–4207.
4. G. W. Castellan, Physical Chemistry, Addison-Wesley, Reading, MA, 2nd edn, 1971
5. B. Wunderlich, Pure Appl. Chem., 1989, **61**, 1347–1351.
6. L. Glasser and H. D. B. Jenkins, Ambient Isobaric Heat Capacities, $C_{p,m}$, for ionic solids and liquids: an Application of Volume-Based Thermodynamics (VBT), Inorg. Chem., 2011,**50**, 8565–8569.
7. M. S. Westwell, M. S. Searle, D. J. Wales and D. H. Williams, J. Am. Chem. Soc., 1995, **117**, 5013–5015.
8. C. J. E. Kempster and H. Lipson, Acta Crystallogr., Sect. B: Struct. Crystallogr. Cryst. Chem., 1972, **28**, 3674
9. H. D. B. Jenkins and L. Glasser, Difference rule - A new thermodynamic principle: Prediction of standard thermodynamic data for inorganic solvates, J. Am. Chem. Soc., 2004, **126**, 15809–15817.
10. L. Piela, Ideas of quantum chemistry, Elsevier, Amsterdam, 2013
11. W. Koch and M. C. Holthausen, A chemist's guide to density functional theory, Wiley-VCH, Weinheim, New York, 2001.
12. L. Glasser and H. D. B. Jenkins, Predictive thermodynamics for condensed phases, Chem. Soc. Rev., 2005, **34**, 866–874
13. L. Glasser and H. D. B. Jenkins, Volume-Based Thermodynamics: A Prescription for Its Application and Usage in Approximation and Prediction of Thermodynamic Data, J. Chem. Eng. Data, 2011, **56**, 874–880.
14. H. D. B. Jenkins and L. Glasser, Thermodynamic Difference Rules: A Prescription for their Application and Usage to Approximate Thermodynamic Data, J. Chem. Eng. Data, 2010,**55**, 4231–4238.
15. H. D. B. Jenkins, L. Glasser and J. F. Liebman, The Thermodynamic Hydrate Difference Rule (HDR)

Applied to Salts of Carbon-Containing Oxyacid Salts and Their Hydrates: Materials at the Inorganic–Organic Interface, J. Chem. Eng. Data, 2010, **55**, 4369–4371.

16. L. Glasser and H. D. B. Jenkins, The thermodynamic solvate difference rule: Solvation parameters and their use in interpretation of the role of bound solvent in condensed-phase solvates, Inorg. Chem., 2007, **46**, 9768–9778

17. L. Glasser, Thermodynamic estimation: Ionic materials, J. Solid State Chem., 2013, **206**, 139–144.

18. C. H. Yoder and N. J. Flora, Geochemical applications of the simple salt approximation to the lattice energies of complex materials, Am. Mineral., 2005, **90**, 488–496.

19. H. D. B. Jenkins, Chemical Thermodynamics - at a glance, Blackwell Publishing, Oxford, 2008.

20. L. Glasser, Correct Use of Helmholtz and Gibbs Function Differences, ΔA and ΔG: The van't Hoff Reaction Box, J. Chem. Educ., 2016, **93**, 978–980.

21. H. S. Elliott, J. F. Lehmann, H. P. A. Mercier, H. D. B. Jenkins and G. J. Schrobilgen, X-ray Crystal Structures of $XeF \cdot MF_6$ (M = As, Sb, Bi), $XeF \cdot M_2F_{11}$ (M = Sb, Bi) and Estimated Thermochemical Data and Predicted Stabilities for Noble-Gas Fluorocation Salts using Volume-Based Thermodynamics, Inorg. Chem., 2010, **49**, 8504–8523.

22. L. V. Gurvich, Reference books and data banks on the thermodynamic properties of individual substances, Pure Appl. Chem., 1989, **61**, 1027–1031.

23. H. Nyman, T. Talonen, A. Roine, M. Hupa and J. Corander, Statistical Approach to Quality Control of Large Thermodynamic Databases, Metall. Mater. Trans. B, 2012, **43**, 1113–1118.

24. H. D. B. Jenkins, L. Glasser, T. M. Klapötke, M. J. Crawford, K. K. Bhasin, J. Lee, G. J. Schrobilgen, L. S. Sunderlin and J. F. Liebman, The ionic isomegethic rule and additivity relationships: Estimation of ion volumes. a route to the energetics and entropics of new, traditional, hypothetical, and counterintuitive ionic materials, Inorg. Chem., 2004, **43**, 6238–6248.

25. N. Cohen and S. W. Benson, Estimation of heats of formation of organic compounds by additivity methods, Chem. Rev., 1993, **93**, 2419–2438.

26. P. J. Linstrom and W. G. Mallard, NIST Chemistry WebBook, http://webbook.nist.gov/chemistry/, accessed April, 2016.

27. A. F. Kapustinskii, Lattice energy of ionic crystals, Q. Rev., Chem. Soc., 1956, **10**, 283–294.

28. T. E. Mallouk, G. L. Rosenthal, R. Muller, R. Brusasco and N. Bartlett, Fluoride ion affinities of germanium tetrafluoride and boron trifluoride from thermodynamic and structural data for $(SF_3)_2GeF_6$, ClO_2GeF_5, and ClO_2BF_4, Inorg. Chem., 1984, **23**, 3167–3173.

29. T. E. Mallouk, PhD thesis, University of Berkeley, CA, 1983.

30. C. Shen, R. Hagiwara, T. E. Mallouk and N. Bartlett, in ACS Symp. Ser. No. 555, ed. J. Thrasher and S. Strauss, ACS, Washington, DC, 1994, pp. 26–39.

31. D. Gibler, PhD thesis, Princeton University, NJ, 1974.

32. H. D. B. Jenkins, H. K. Roobottom, J. Passmore and L. Glasser, Relationships among ionic lattice energies, molecular (formula unit) volumes, and thermochemical radii, Inorg. Chem., 1999, **38**, 3609–3620.

33. L. Glasser, Lattice Energies of Crystals with Multiple Ions - a Generalised Kapustinskii Equation, Inorg. Chem., 1995, **34**, 4935–4936.

34. H. D. B. Jenkins and L. Glasser, Volume-based thermodynamics: Estimations for 2 : 2 salts, Inorg. Chem., 2006, **45**, 1754–1756.

35. L. Glasser and H. D. B. Jenkins, Lattice energies and unit cell volumes of complex ionic solids, J. Am. Chem. Soc., 2000, **122**, 632–638.

36. H. D. B. Jenkins, Thermodynamics of the Relationship between Lattice Energy and Lattice Enthalpy, J. Chem. Educ., 2005, **82**, 950–952.

37. P. Vieillard, Thermodynamics of Hydration in Minerals: How to Predict These Entities, in Thermodynamics - Fundamentals and Its Application in Science, ed. R. Morales-Rodriguez, InTech, 2012.

38. L. Glasser, Lattice and phase transition thermodynamics of ionic liquids, Thermochim. Acta, 2004, **421**, 87–93

39. H. D. B. Jenkins, Ionic liquids – an overview, Sci. Prog., 2011, **94**, 265–297

40. H. D. B. Jenkins, Recent initiatives in experimental thermodynamic studies on ionic liquids [IL] – the emergence of a standard thermochemical database for ILs, Sci. Prog., 2011, **94**, 184–210.

41. C. Robelin, Models for the thermodynamic properties of molten salt systems: Perspectives for ionic liquids, Fluid Phase Equilib., 2016, **409**, 482–494.

42. J. P. Hallett and T. Welton, Room-Temperature Ionic Liquids: Solvents for Synthesis and Catalysis. 2, Chem. Rev., 2011, **111**, 3508–3576.

43. H. Niedermeyer, J. P. Hallett, I. J. Villar-Garcia, P. A. Hunt and T. Welton, Mixtures of ionic liquids, Chem. Soc. Rev., 2012, **41**, 7780–7802.

44. S. Keskin, D. Kayrak-Talay, U. Akman and Ö. Hortaçsu, A review of ionic liquids towards supercritical fluid applications, J. Supercrit. Fluids, 2007, **43**, 150–180.

45. A. R. Katritzky, R. Jain, A. Lomaka, R. Petrukhin, M. Karelson, A. E. Visser and R. D. Rogers, Correlation of the Melting Points of Potential Ionic Liquids (Imidazolium Bromides and Benzimidazolium Bromides) Using the CODESSA Program [J. Chem. Inf. Comput. Sci., 2002, **42**, 225–231], J. Chem. Inf. Model., 2005, **45**, 533–534.

46. A. R. Katritzky, A. Lomaka, R. Petrukhin, R. Jain, M. Karelson, A. E. Visser and R. D. Rogers, QSPR Correlation of the Melting Point for Pyridinium Bromides, Potential Ionic Liquids, J. Chem. Inf. Comput. Sci., 2002, **42**, 71–74.

47. M. Deetlefs, K. R. Seddon and M. Shara, Predicting physical properties of ionic liquids, Phys. Chem. Chem. Phys., 2006, **8**, 642–649.

48. D. Wileńska, I. Anusiewicz, S. Freza, M. Bobrowski, E. Laux, S. Uhl, H. Keppner and P. Skurski, Predicting the viscosity and electrical conductivity of ionic liquids on the basis of theoretically calculated ionic volumes, Mol.

Phys., 2015, **113**, 630–639

49. J. M. Slattery, C. Daguenet, P. J. Dyson, T. J. S. Schubert and I. Krossing, How to predict the physical properties of ionic liquids: A volume-based approach, Angew. Chem., Int. Ed., 2007, **46**, 5384–5388.
50. K. E. Gutowski, J. D. Holbrey, R. D. Rogers and D. A. Dixon, Prediction of the formation and stabilities of energetic salts and ionic liquids based on ab initio electronic structure calculations, J. Phys. Chem. B, 2005, **109**, 23196–23208
51. K. E. Gutowski, R. D. Rogers and D. A. Dixon, Accurate thermochemical properties for energetic materials applications. I. Heats of formation of nitrogen-containing heterocycles and energetic precursor molecules from electronic structure theory, J. Phys. Chem. A, 2006, **110**, 11890–11897.
52. K. E. Gutowski, R. D. Rogers and D. A. Dixon, Accurate thermochemical properties for energetic materials applications. II. Heats of formation of imidazolium-, 1,2,4-triazolium-, and tetrazolium-based energetic salts from isodesmic and lattice energy calculations, J. Phys. Chem. B, 2007, **111**, 4788–4800
53. H. D. B. Jenkins and L. Glasser, Standard absolute entropy, S°_{298}, values from volume or density. 1. Inorganic materials, Inorg. Chem., 2003, **42**, 8702–8708.
54. I. Krossing, J. M. Slattery, C. Daguenet, P. J. Dyson, A. Oleinikova and H. Weingartner, Why are ionic liquids liquid? A simple explanation based on lattice and solvation energies, J. Am. Chem. Soc., 2006, **128**, 13427–13434.
55. U. Preiss, S. Bulut and I. Krossing, In Silico Prediction of the Melting Points of Ionic Liquids from Thermodynamic Considerations: A Case Study on 67 Salts with a Melting Point Range of 337 °C, J. Phys. Chem. B, 2010, **114**, 11133–11140.
56. U. P. Preiss, W. Beichel, A. M. T. Erle, Y. U. Paulechka and I. Krossing, Is Universal, Simple Melting Point Prediction Possible?, ChemPhysChem, 2011, **12**, 2959–2972
57. U. Preiss, J. M. Slattery and I. Krossing, In Silico Prediction of Molecular Volumes, Heat Capacities, and Temperature-Dependent Densities of Ionic Liquids, Ind. Eng. Chem. Res., 2009, **48**, 2290–2296.
58. U. Preiss, C. Jungnickel, J. Thoming, I. Krossing, J. Luczak, M. Diedenhofen and A. Klamt, Predicting the Critical Micelle Concentrations of Aqueous Solutions of Ionic Liquids and Other Ionic Surfactants, Chem. - Eur. J., 2009, **15**, 8880–8885.
59. W. Beichel, U. P. Preiss, S. P. Verevkin, T. Koslowski and I. Krossing, Empirical description and prediction of ionic liquids' properties with augmented volume-based thermodynamics, J. Mol. Liq., 2014, **192**, 3–8.
60. L. Glasser and H. A. Scheraga, Calculations on Crystal Packing of a Flexible Molecule: Leu-Enkephalin, J. Mol. Biol., 1988, **199**, 513–524.
61. J. D. Gale, GULP: A computer program for the symmetry-adapted simulation of solids, J. Chem. Soc., Faraday Trans., 1997, **93**, 629–637.
62. E. I. Izgorodina, U. L. Bernard, P. M. Dean, J. M. Pringle and D. R. MacFarlane, The Madelung Constant of Organic Salts, Cryst. Growth Des., 2009, **9**, 4834–4839 CAS.
63. L. Glasser, Solid-State Energetics and Electrostatics: Madelung Constants and Madelung Energies, Inorg. Chem., 2012, **51**, 2420–2424.
64. H. D. B. Jenkins and J. F. Liebman, Volumes of solid state ions and their estimation, Inorg. Chem., 2005, **44**, 6359–6372.
65. Y. Marcus, Ion Properties, Marcel Dekker, New York, 1997.
66. Y. Marcus, Ions in Solution and their Solvation, Wiley, 2015
67. W. M. Latimer, Methods of Estimating the Entropies of Solid Compounds, J. Am. Chem. Soc., 1951, **73**, 1480–1482.
68. P. J. Spencer, Estimation of thermodynamic data for metallurgical applications, Thermochim. Acta, 1998, **314**, 1–21.
69. Outotec, HSC Chemistry 8, http://www.outotec.com/en/Products--services/HSC-Chemistry/, accessed April, 2016.
70. National Institute for Materials Science, AtomWork Inorganic Material Database, http://crystdb.nims.go.jp/index_en.html, accessed June, 2016.
71. in CRC Handbook of Chemistry and Physics, ed. W. M. Haynes, CRC Press, Boca Raton, 2015.
72. L. Glasser, Simple Route to Lattice Energies in the Presence of Complex Ions, Inorg. Chem., 2012, **51**, 10306–10310.
73. T. S. Cameron, R. J. Deeth, I. Dionne, H. B. Du, H. D. B. Jenkins, I. Krossing, J. Passmore and H. K. Roobottom, Bonding, structure, and energetics of gaseous E_8^{2+} and of solid $E_8(AsF_6)_2$ (E = S, Se), Inorg. Chem., 2000, **39**(25), 5614–5631.
74. S. Brownridge, I. Krossing, J. Passmore, H. D. B. Jenkins and H. K. Roobottom, Recent advances in the understanding of the syntheses, structures, bonding and energetics of the homopolyatomic cations of Groups 16 an 17, Coord. Chem. Rev., 2000, **197**, 397–481.
75. H. D. B. Jenkins, H. K. Roobottom and J. Passmore, Estimation of enthalpy data for reactions involving gas phase ions utilizing lattice potential energies: Fluoride ion affinities (FIA) and pF(-) values of $mSbF_5$(l) and $mSbF_5$(g) (m = 1, 2, 3), AsF_5(g), $AsF_5{\cdot}SO_2$(c). Standard enthalpies of formation: $\Delta H_f°(Sb_mF_{5m+1}^-$, g) (m = 1, 2, 3), $\Delta H_f°(AsF_6^-$, g), and $\Delta H_f°(NF_4^+$, g), Inorg. Chem., 2003, **42**(9), 2886–2893.
76. H. D. B. Jenkins, I. Krossing, J. Passmore and I. Raabe, A computational study of Sb_nF_{5n} (n = 1–4) - Implications for the fluoride ion affinity of $nSbF_5$, J. Fluorine Chem., 2004,**125**(11), 1585–1592.
77. T. S. Cameron, I. Dionne, H. D. B. Jenkins, S. Parsons, J. Passmore and H. K. Roobottom, Preparation, X-ray Crystal Structure Determination, Lattice Potential Energy, and Energetics of Formation of the Salt $S_4(AsF_6)_2{\cdot}AsF_3$

Containing the Lattice-Stabilized Tetrasulfur [2+] Cation. Implications for the Understanding of the Stability of M_4^{2+} and M^{2+} (M = S, Se, and Te) Crystalline Salts, Inorg. Chem., 2000, **39**(10), 2042–2052.

78. K. O. Christe and H. D. B. Jenkins, Quantitative measure for the "nakedness" of fluoride ion sources, J. Am. Chem. Soc., 2003, **125**(31), 9457–9461

79. K. O. Christe and H. D. B. Jenkins, Quantitative measure for the "nakedness" of fluoride ion sources, J. Am. Chem. Soc., 2003, **125**, 14210.

80. H. D. B. Jenkins and D. Tudela, New methods to estimate lattice energies – Application to the relative stabilities of bisulfite (HSO_3^-) and metabisulfite ($S_2O_5^{2-}$) salts, J. Chem. Educ., 2003, **80**(12), 1482–1487.

81. A. Decken, H. D. B. Jenkins, G. B. Nikiforov and J. Passmore, The reaction of $LiAl(OR)_4$ R = $OC(CF_3)_2Ph$, $OC(CF_3)_3$ with NO/NO_2 giving $NO{\cdot}Al(OR)_4$, $LiNO_3$ and N_2O. The synthesis of $NO{\cdot}Al(OR)_4$ from LiAl(OR) and $NO.SbF_6$ in sulfur dioxide solution, Dalton Trans., 2004,(16), 2496–2504.

82. D. A. Dixon, D. Feller, K. O. Christe, W. W. Wilson, A. Vij, V. Vij, H. D. B. Jenkins, R. M. Olson and M. S. Gordon, Enthalpies of Formation of Gas-Phase N_3, N_3^-, N_5^+, and $N_5^-$from Ab Initio Molecular Orbital Theory, Stability Predictions for $N_5^+N_3^-$ and $N_5^+N_5^-$, and Experimental Evidence for the Instability of $N_5^+N_3^-$, J. Am. Chem. Soc., 2004, **126**(3), 834–843.

83. K. O. Christe, A. Vij, W. W. Wilson, V. Vij, D. A. Dixon, D. Feller and H. D. B. Jenkins, $N_5^+N_5^-$ allotrope is California dreaming, Chem. Br., 2003, **39**(9), 17.

84. D. R. Rosseinsky, L. Glasser and H. D. B. Jenkins, Thermodynamic clarification of the curious ferric/potassium ion exchange accompanying the electrochromic redox reactions of Prussian blue, iron(III) hexacyanoferrate(II), J. Am. Chem. Soc., 2004, **126**(33), 10472–10477.

85. A. Decken, H. D. B. Jenkins, C. Knapp, G. B. Nikiforov, J. Passmore and J. M. Rautiainen, The autoionization of $[TiF_4]$ by cation complexation with [15]crown-5 to give $[TiF_2([15]crown\text{-}5)][Ti_4F_{18}]$ containing the tetrahedral $[Ti_4F_{18}]^{2-}$ ion, Angew. Chem., Int. Ed., 2005, **44**, 7958–7961

86. K. K. Bhasin, M. J. Crawford, H. D. B. Jenkins and T. M. Klapötke, The volume-based thermodynamics (VBT) and attempted preparation of an isomeric salt, nitryl chlorate: $NO_2{\cdot}ClO_3$, Z. Anorg. Allg. Chem., 2006, **632**(5), 897–900.

87. S. Brownridge, L. Calhoun, H. D. B. Jenkins, R. S. Laitinen, M. P. Murchie, J. Passmore, J. Pietikainen, J. M. Rautiainen, J. C. P. Sanders, G. J. Schrobilgen, R. J. Suontamo, H. M. Tuononen, J. U. Valkonen and C. M. Wong, ^{77}Se NMR Spectroscopic, DFT MO, and VBT Investigations of the Reversible Dissociation of Solid $(Se_6I_2)[AsF_6]_2{\cdot}2SO_2$ in Liquid $SO_2$to Solutions Containing 1,4-$Se_6I_2^{2+}$ in Equilibrium with Se_n^{2+} (n = 4, 8, 10) and Seven Binary Selenium Iodine Cations: Preliminary Evidence for 1,1,4,4-$Se_4Br_4^{2+}$ and cyclo-Se_7Br^+,Inorg. Chem., 2009, **48**(5), 1938–1959.

88. H. S. Elliott, J. F. Lehmann, H. P. A. Mercier, H. D. B. Jenkins and G. J. Schrobilgen, X-ray Crystal Structures of $XeF.MF_6$ (M = As, Sb, Bi), $XeF{\cdot}M_2F_{11}$ (M = Sb, Bi) and Estimated Thermochemical Data and Predicted Stabilities for Noble-Gas Fluorocation Salts using Volume-Based Thermodynamics, Inorg. Chem., 2010, **49**(18), 8504–8523.

89. H. D. B. Jenkins, L. Glasser and J. F. Liebman, The Thermodynamic Hydrate Difference Rule (HDR) Applied to Salts of Carbon-Containing Oxyacid Salts and Their Hydrates: Materials at the Inorganic–Organic Interface, J. Chem. Eng. Data, 2010, **55**(10), 4369–4371.

90. L. Glasser and F. Jones, Systematic Thermodynamics of Hydration (and of Solvation) of Inorganic Solids, Inorg. Chem., 2009, **48**(4), 1661–1665.

91. K. O. Christe, R. Haiges, J. A. Boatz, H. D. B. Jenkins, E. B. Garner and D. A. Dixon, Why Are $[P(C_6H_5)_4]^+N_3^-$ and $[As(C_6H_5)_4]^+N_3^-$ Ionic Salts and $Sb(C_6H_5)_4N_3$ and $Bi(C_6H_5)_4N_3$Covalent Solids? A Theoretical Study Provides an Unexpected Answer, Inorg. Chem., 2011, **50**, 3752–3756.

92. A. Vegas, J. F. Liebman and H. D. B. Jenkins, Unique thermodynamic relationships for $\Delta_fH°$ and $\Delta_fG°$ for crystalline inorganic salts. I. Predicting the possible existence and synthesis of Na_2SO_2 and Na_2SeO_2, Acta Crystallogr., Sect. B: Struct. Sci., 2012, **68**, 511–527.

93. D. Srinivas, V. Ghule, K. Muralidharan and H. D. B. Jenkins, Tetraanionic Nitrogen-Rich Tetrazole-Based salts, Chem. – Asian J., 2013, **8**, 1023–1028.

94. P. Bruna, A. Decken, S. Greer, F. Grein, H. D. B. Jenkins, B. Mueller, J. Passmore, T. A. P. Paulose, J. M. Rautiainen and S. Richardson, Synthesis of (TDAE) (O_2SSSSO_2) and Discovery of (TDAE)(O_2SSSSO_2)(s) containing the first polythionite anion, Inorg. Chem., 2013, **52**, 13651–13662.

95. L. Glasser, H. D. B. Jenkins and T. M. Klapötke, Is the VBT Approach valid for estimation of the lattice enthalpy of salts containing the 5,5′-(Tetrazole-1N-oxide)$^{2-}$?, Z. Anorg. Allg. Chem., 2014, **640**, 1297–1299.

96. H. D. B. Jenkins, D. Holland and A. Vegas, Thermodynamic data for crystalline arsenic and phosphorus compounds, $M_2O_5{\cdot}nH_2O$ reexamined using the Thermodynamic Difference Rules, Thermochim. Acta, 2016, **633**, 24–30.

97. H. D. B. Jenkins, D. Holland and A. Vegas, An assessment of the TDR for mixed inorganic oxides and comments on the enthalpies of formation of phosphates, Thermochim. Acta, 2015, **601**, 63–67.

CHAPTER 6 HYDROGEN BONDING IN A MIXTURE OF PROTIC IONIC LIQUIDS

Dietmar Paschek, Benjamin Golub and Ralf Ludwig

INTRODUCTION

During the past 15 years ionic liquids (ILs) have received great attention from the scientific community.[1–4] More recently, also the ability to design targeted liquid properties by forming binary and ternary mixtures has become a focus of interest.[5,6] In particular, it has been recognized that the balance between different competing interactions, such as Coulomb interactions, van der Waals interactions, and hydrogen bonds is of great importance. In a situation, where strong (Coulombic) forces compete with each other and partly cancel each other out, also weaker interactions, such as hydrogen-bonds, might be able to significantly affect the properties of the fluid.[7,8] This is particularly important for mixtures of ionic liquids, where different constituents (ions) might compete with each other for specific interactions. The mixing behavior of several ionic liquids has been discussed recently with a focus on their thermodynamic and transport properties.[5,6,9–13] However, mostly the behavior of aprotic ionic liquids has been considered so far. We would like to point out that physical properties reported in these studies, such as the density and the viscosity, were mostly exhibiting ideal-like mixing behavior in the sense that the properties vary almost linearly with the mixture composition. In addition, also the molar mixing-enthalpies and -entropies indicated only a small deviation from the ideal mixture-behavior.[10,11] Recently, Welton and co-workers have commented on the tendency for ideal-mixture behavior in mixtures of ionic liquids and reviewed IL mixtures with a focus on property design an opportunities for applications.[5,9]

The situation seems to be different for protic ionic liquids, where strongly directional hydrogen-bonds (can) exist.[14,15] We could show very recently, that for a mixture of triethylammonium-methylsulfonate (TEAMS) and triethylammonium-triflate (TEATF), which are both sharing the same cation, the two different anions compete with each other for a hydrogen bond to the triethylammonium-cation (TEA).[16] Using far-infrared spectroscopy, we could assign signatures distinguishing between hydrogen-bonded aggregates of TEA to each of the two anions within the mixture.[16] A population analysis as a function of mixture-composition is suggesting a favorable interaction of TEA with the methylsulfonate-ion. In addition, also the experimentally determined viscosity is showing a strongly non-linear mixture dependence, pointing at the changing size of aggregates in the liquid as a function of the mixture-composition.[16] With the present study we would like to complement our earlier endeavour with an in-depth analysis of a detailed molecular model of the same binary TEATMS–TEATF mixture, focusing on the local solvation-structure of the ions, and, in particular, on the hydrogen-bonding. In addition, we will show that the size of the hydrogen-bonded aggregates can be predicted and explained by a simple combinatorial lattice model, which is based on an approximate coordination number of the ions, and is using statistical weights that mostly account for the fact that each anion can only accept up to three hydrogen bonds.

Finally, we would like to point out that Rogers and co-workers have argued recently that ions in a mixture of ILs do not seem to retain their individual nature and proposed to consider them as double salt ionic liquids.[6] Double salts are salts containing more than one cation or anion, obtained by combination of two different salts which were crystallized in the same regular ionic lattice. Although this definition is not strictly applicable to the case of an amorphous state such as a liquid, we can provide evidence for double-salt-like behavior of the TEAMS–TEATF mixture.

Citation: Dietmar Paschek, Benjamin Golub and Ralf Ludwig, Hydrogen bonding in a mixture of protic ionic liquids, DOI: 10.1039/c4cp05432f

METHODS

Molecular Dynamics Simulations

We employ molecular dynamics (MD) simulations in the isobaric isothermal (NPT) ensemble using the Nosé–Hoover thermostat[17,18] and the Rahman–Parrinello barostat[19,20] with coupling-times $\tau_T = 0.5$ ps and $\tau_p = 2$ ps (assuming the isothermal compressibility to be $\chi_T = 5 \times 10^{-5}$ bar^{-1}), respectively. The electrostatic interactions are treated in the "full potential" approach by the smooth particle mesh Ewald summation[21] with a real space cutoff of 1.0 nm and a mesh spacing of approximately 0.12 nm and 4th order interpolation. The Ewald convergence factor $\alpha$was set to 3.38 nm^{-1} (corresponding to a relative accuracy of the Ewald sum of 10^{-5}). A 2.0 fs timestep was used for all simulations and the constraints were solved using the SETTLE procedure.[22] During the simulations all bond-length were kept fixed. All simulations reported here were carried out using the GROMACS 4.6 program.[23,24] Statistical errors in the analysis were computed using the method of Flyvbjerg and Petersen.[25] For all reported mixtures and temperatures initial equilibration runs of 1 ns length were performed using the Berendsen weak coupling scheme for pressure and temperature control $\tau_T = \tau_p = 0.5$ ps.[26] A total of 16 different compositions were investigated with compositions of $x = x_{CH3SO3} = \{0.0, 0.125, 0.25, 0.3125, 0.40625, 0.5, 0.5625, 0.625, 0.6875, 0.75, 0.8125, 0.84375, 0.875, 0.90625, 0.9375, 1.0\}$ at T = 300 K, 300 K, and 400 K. In each case the simulated system consisted of 256 ion pairs in a cubic box with periodic boundary conditions. As shown in Fig. 1, the density of the model-system follows rather well the experimental density, both as a function of temperature and composition. Production runs of 120 ns (300 K), 48 ns (350 K), and 24 ns (400 K) simulation-length were finally recorded and analyzed.

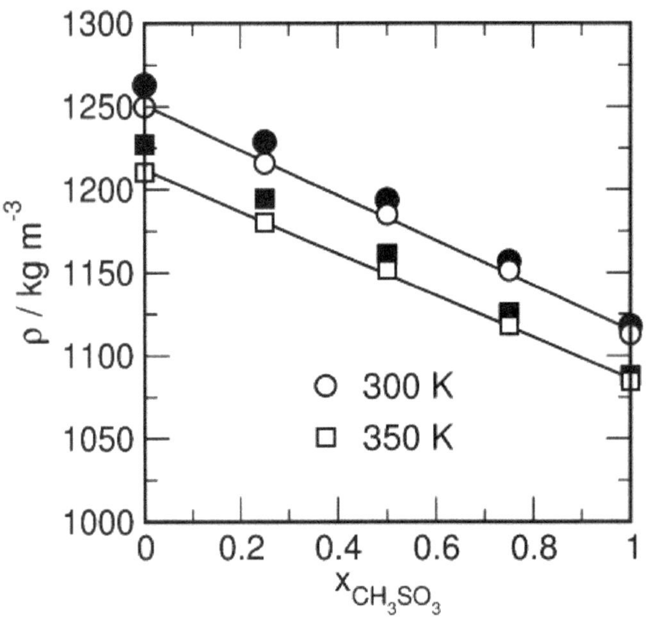

Figure. 1 Densities of selected TEATF–TEAMS mixtures as a function of their composition as obtained from the molecular dynamics simulations at 300 K and 350 K, respectively. The closed symbols refer to experimental densities obtained at 298 K and 343 K.[16]

The Molecular Model

For the description of the potential model, a classical forcefield approach has been used, similar to the OPLS-model of Jorgensen et al.[27] All employed partial charges are atom-centered and were determined from ab initio calculations of the individual ions by applying many body perturbation theory (MP2) and using the 6-311++G** basis set. All reported ab initio calculations were performed with the Gaussian 09 program.[28] The charges were fitted to the electrostatic potential surrounded by the atoms using the ESP[29] and RESP[30] methods. Due to the rather small size of the ions no significant changes were found between the ESP and RESP charges. CH_3- and CH_2-groups treated as united atoms and represented by a single interaction site. Nonbonded Lennard-Jones parameters for the triflate-ions were taken from the [NTF_2] forcefield-model of Köddermann et al.,[31] while the starting point for the nonbonded parameters for the triethylammonium were parameters reported by Krienke et al.[32] for the ammonium group, in addition to the TraPPE-parameters of Martin and Siepmann[33] for the united-atom CH_3- and CH_2-groups. The σ-parameter for the nitrogen atom needed to be resized to 3.25 Å in order to match the H···O distance of the hydrogen bonded ion pair in the gas-phase with the distance obtained for the energy minimized structure using ab initio calculations. Constant pressure simulations of the model describes the density of triethylammonium-triflate quite satisfactorily. For the methylsulfonate-ion, the Lennard-Jones parameters of Köddermann were also used for the SO_3-group. The starting point for the Lennard-Jones parameters for the methyl-group was again the forcefield of Martin and Siepmann.[33] In this case, the Lennard-Jones σ of the methyl-group needed to be resized to match the density of the liquid phase of neat triethylammonium-methylsulfonate. Due to a significantly stronger bond-polarization of the S–O bond within the methylsulfonate-ion, compared to the triflate-ion, larger partial charges on the oxygens were obtained for the methylsulfonate, leading to stronger hydrogen-bonding with the triethylammonium ion. In order to represent molecular configurations of the ions properly, the equilibrium bond distances and angles were adjusted in such a way that they lead to minimum energy configurations close to the structures obtained from the ab initiocalculations. All nonbonded Lennard-Jones parameters and partial charges are summarized in Table 1. The bond-length and bond-bending parameters can be found in Tables 2 and 3. Dihedral potentials for intramolecular torsions around the N–C2-bond in the triethylammonium ion and the S–C-bond in the triflate ions were fitted to ab initio calculations to represent the molecular conformations accurately. The values are summarized in Table 4. The full torsion potential also include nonbonded 1–4-interactions. Here both, the nonbonded Lennard-Jones and Coulomb 1–4-interactions are scaled by a factor of 0.5.

Table 1 Non-bonding Lennard-Jones interaction parameters with $V_{ij} = 4\varepsilon_{ij}[(\sigma_{ij}/r_{ij})^{12} - (\sigma_{ij}/r_{ij})^{6}]$ and partial charges used in the MD Simulations of the PIL-mixtures. Lorentz–Berthelot mixing-rules with $\sigma_{ij} = (\sigma_{ii} + \sigma_{jj})/2$ and $\varepsilon_{ij} = (\varepsilon_{ii}\cdot\varepsilon_{jj})^{1/2}$ were applied for determining the cross-parameters

Atom *i*	σ_{ii}/Å	$\varepsilon_{ii}\cdot k_B^{-1}$/K	q_i/|e|
Triflate			
S	4.08	37.3	1.40
O	3.46	31.7	−0.73
C	3.15	10.0	0.48
F	2.66	8.0	−0.23
Methylsulfonate			
S	4.08	37.3	1.620
O	3.46	31.7	−0.825
C3	2.90	98.0	−0.145
Triethylammonium			
H	0.0	0.0	0.37
N	3.25	75.0	−0.39
C2	3.95	46.0	0.28
C3	3.75	98.0	0.06

Table 2 Harmonic bond-stretching parameters employed in the forcefield model using $V^b_{ij} = (k^b_{ij}/2)\cdot(r_{ij} - r^0_{ij})^2$

Bond *i–j*	r^0_{ij}/Å	k^b_{ij}/kJ mol^{-1} Å^{-2}
Triflate		
C–F	1.347	3700.0
C–S	1.860	1850.0
O–S	1.469	5850.0
Methylsulfonate		
C3–S	1.8040	1850.0
O–S	1.4816	5850.0
Triethylammonium		
H–N	1.020	2500.0
C2–N	1.500	2500.0
C2–C3	1.540	2500.0

Table 3 Harmonic bond-bending parameters employed in the forcefield model using $V^a_{ijk} = (k^a_{ijk}/2)\cdot(\theta_{ijk} - \theta^0_{ijk})^2$

Angle *i–j–k*	**θ^0_{ijk}/degrees**	**k^a_{ijk}/kJ mol^{-1} rad^{-2}**
Triflate		
F–C–F	108.4	650.0
F–C–S	110.5	420.0
C–S–O	102.6	620.0
O–S–O	119.5	850.0
Methylsulfonate		
C3–S–O	104.26	620.0
O–S–O	114.14	850.0
Triethylammonium		
H–N–C2	108.4	460.0
C2–N–C2	108.4	460.0
C3–C2–N	108.4	460.0

Table 4 Dihedral-potential for the rotation around the N–C2 bond in the triethylammonium-ion and for the C–S bond in the triflate-ion according to $V_{ijkl} = \sum k^d_m[1 + \cos(n_m \cdot \phi)]$

m(i–j–k–l)	***n_m***	**k^d_m/kJ mol^{-1}**
Triflate F–C–S–O		
1	3.0	0.8619
Triethylammonium H–N–C2–C3		
1	1.0	8.734
2	2.0	3.224
3	3.0	5.514
4	4.0	−0.465
5	5.0	−0.481
6	6.0	0.756

DISCUSSION

Structure of the Ionic Liquid Mixture

To provide an impression of the structure of the discussed ionic liquid mixture, a snapshot of the 50 : 50 mixture of TEATF–TEMS at T = 300 K is given in Fig. 2. Indicated are the positions of the centers of mass of the ions, which can be identified via their respective colors. From visual inspection it is evident that the charge of the ions impose a long-range ion-alternating order, which is leading to a significant structural homogeneity of the ionic liquid. This behavior is typical for ionic liquids and has been pointed out earlier by Maroncelli and co-workers.[34] The ion-alternating order is a consequence of the fact that the ions are predominantly surrounded by a solvation shell of ions with opposite charge. This defining structural feature is well represented by the location of the first peak of the center-of-mass pair distribution functions of the ions: in Fig. 3a and b the center-of-mass pair distribution functions for all ion-pair combinations are shown for a 50 : 50 mixture obtained at 300 K. From Fig. 3a we can conclude that the first peak of the anion–cation pair correlation function is shifted to significantly lower values, with the dominant first peak of the TEA–MS-pair found at 0.41 nm, and the peak of the TEA–TF-pair at 0.54 nm. The first TEA–TEA-peak is found at a much larger distance of 0.70 nm. The anion–anion pair correlation functions are shown in Fig. 3b: here, we would like to point out that, besides the dominant main first peak around 0.8 nm, there also exists pre-peak at a distance of about 0.55 nm, which is most strongly pronounced for the correlation between the two methylsulfonate ions, and is only weakly pronounced in the pair-correlation of the triflate-ions. In addition, Fig. 3b also shows that this structural feature of the first peak is completely absent for the TEA–TEA correlations. The pre-peak in the anion–anion pair correlation function is indicating the existence of direct anion–anion contacts, whereas direct cation–cation contacts are apparently avoided. Moreover, also the snapshot in Fig. 2reveals the presence of anion–anion-pairs in close proximity on many occasions.

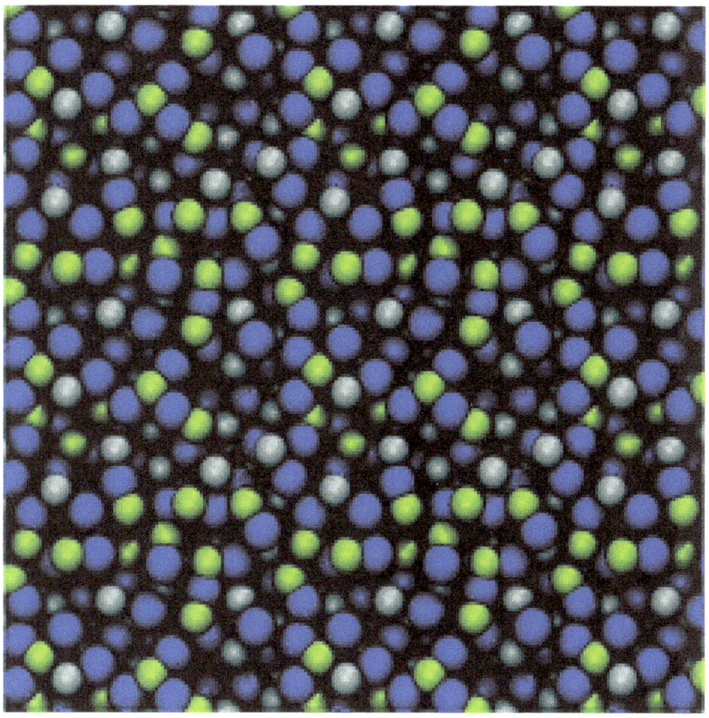

Figure. 2 Snapshot of a 50 : 50 mixture of TEATF–TEAMS at T = 300 K. The spheres represent the centers of gravity of the ions. The TEA-positions are indicated by blue spheres. The triflate- and methylsulfonate-positons are color-coded in white and yellow, respectively.

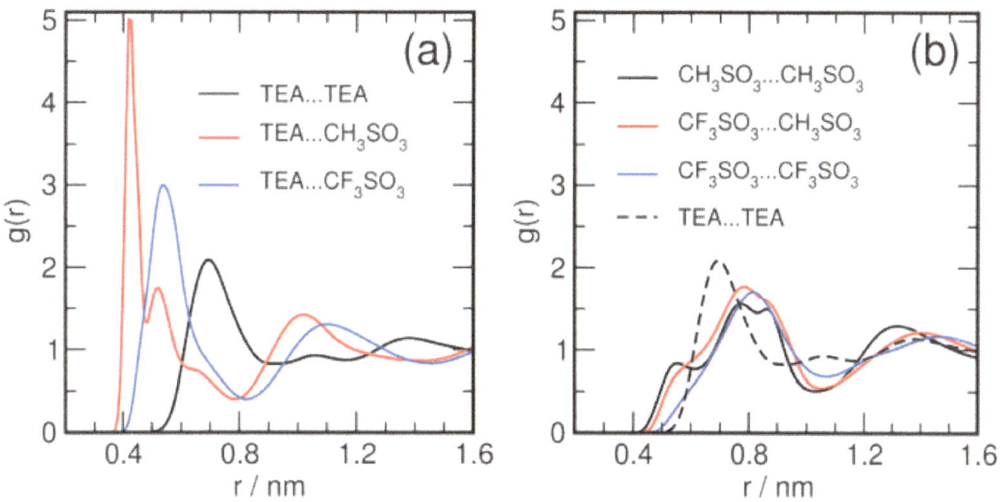

Figure. 3 Complete set of ion–ion center of mass pair distribution functions g(r) obtained for the 50 : 50 mixture of TEATF–TEMS at T = 300 K.

The coordination number of the ion-solvation is deduced from the anion–cation pair correlation functions, shown in Fig. 4a and b for both pure ionic liquids. First, we would like to point at the characteristic three-sub-peak nature of peak of the first solvation-shell, which is present in both ionic liquids, although significantly more "structured" in the case of TEAMS. The integral over the pair correlation function reveals the number of neighbors as function of their distance

$$N(r) = 4\pi\rho\int_0^r g(r')r'^2\mathrm{d}r'. \qquad (1)$$

Here ρ represents the number density of the selected ion-species. The data for N(r) are also given in Fig. 4. Integrating over the entire first peak leads to a consistent number of nearest neighbors of about 6.8 to 6.9 counter-ions for both ionic liquids. Here, the first sub-peak corresponds in both cases to about one neighbor, strongly suggesting that this is indicative of a hydrogen-bonded ion-pair configuration. The second and third sub-peaks are contributing about three additional neighbors each. Both are apparently related to configurations with different orientational order. As shown in Fig. 5, the average first-shell counter-ion coordination number of about 6.8 to 6.9 remains constant over the entire mixture range.

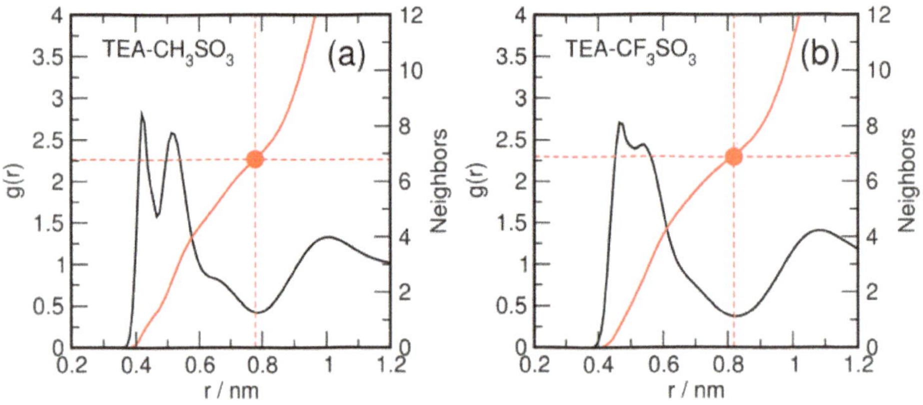

Figure. 4 Anion–cation center of mass pair distribution functions g(r) obtained for (a) pure TEAMS and (b) pure TEATF at T = 300 K. The position of the first minimum is used as a cut-off distance for the nearest neighbor computation (TEA–MS: 0.78 nm, TEA–TF: 0.81 nm).

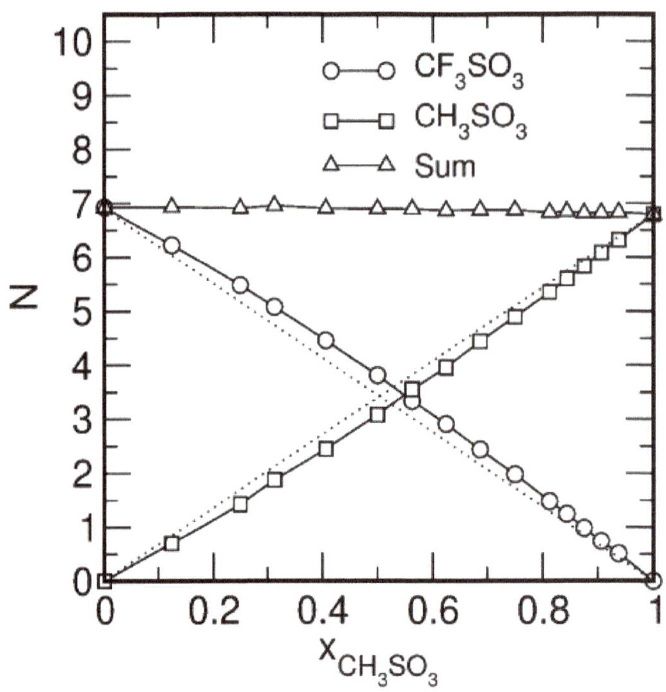

Figure. 5 Average number of both anion species N, as well as their sum, found in the first solvation shell of a TEA-cation as a function of mixture composition at T = 300 K.

At this point we would like to discuss apparent deviations from an ideal mixture-behavior in these ionic liquids: a significant structural difference between the pure liquids and the mixture is observed for first sub-peak of the anion–cation pair distribution functions for the 50 : 50 TEATF–TEAMS mixture, shown in Fig. 3a, compared to the pure liquids shown in Fig. 4. From Fig. 4 it is evident that the first sub-peak has about the same height for both pure ionic liquids. In the 50 : 50 mixture (Fig. 3a), however, the first peak is very strongly pronounced for the TEA–MS pair, and is almost completely absent for the TEA–TF pair, which is suggesting a preference for the formation of TEA–MS hydrogen bonded conformations. This preference is also evident from Fig. 2, where many of the supposedly hydrogen-bonded TEA–MS-pairs can be identified as dumbbell-like configurations, formed by a pair of blue and yellow spheres. This imbalance of the hydrogen-bond strengths has also a surprising consequence for the structure of solvation-shells of the ions in the mixture. In Fig. 5 the average coordination-number of the anions surrounding a central cation is depicted as a function of the mixture-composition. While the total number of anions varies only slightly between 6.8 and 6.9 as a function of mixture composition, the build-up of the individual solvation shells changes more strongly: as shown in Fig. 5, in a 50 : 50 mixture the average number of TF-anions (N_{TF} = 3.82) in the first solvation shell of a TEA-cation is significantly larger than the number of MS-anions (N_{MS} = 3.08). This imbalance is also represented in the solvation shell of the anions: Fig. 6 shows the distribution of the number of the TEA-cations in the first solvation shell around each anion. For the pure ionic liquids (given in Fig. 6a), the distribution functions are almost identical with a maximum at 7 neighbors. In the 50–50 mixture, however, the distribution for methylsulfonate is shifted to lower values, with a maximum at 6, whereas the distribution for triflate is shifted to larger values. Apparently, the triflate-ion is able to "steal away" cations from the methylsulfonate in the mixture. This effect might be just due to the larger size of triflate-ion compared to the methylsulfonate. However, the reduced number of TEA-cations in the solvation shell of methylsulfonate leaves the methylsulfonate-ion prone to direct anion–anion contacts in the mixture as observed in Fig. 3. The number of TEA-cations attracted by the triflate- and methylsulfonate-ions is related with an average free energy of association, which has, of course, both entropic and an enthalpic contributions. The temperature dependence of the distribution functions shown in Fig. 7 indicates that the "binding" of TEA to the triflate-ion is significantly more temperature-sensitive than that of TEA to the methylsulfonate-ion and, hence, can be considered more weakly.

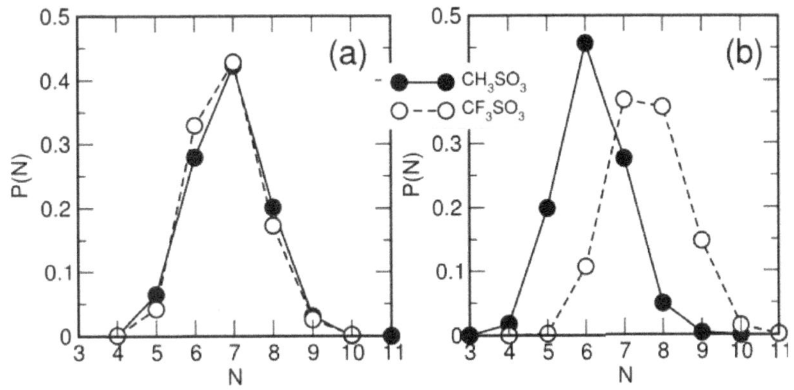

Figure. 6 Distribution of the number N of TEA-cations in the first solvation shell of the triflate- and methylsulfonate-anions at T = 300 K. (a) Pure ionic liquids (b) 50 : 50 mixture of TEATF–TEAMS.

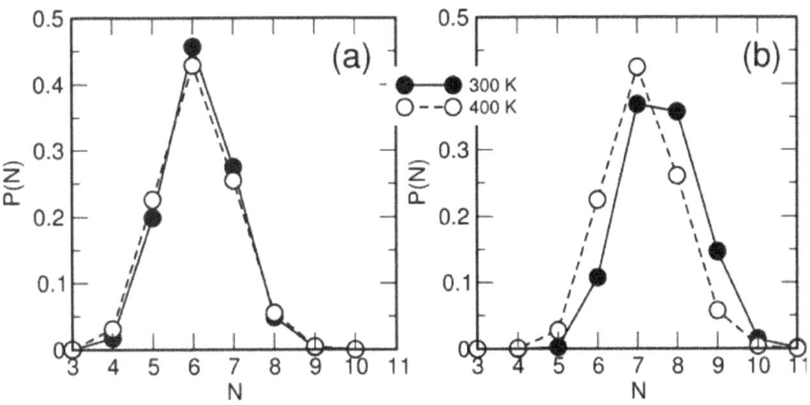

Figure. 7 Distribution of the number N of TEA-cations in the first solvation shell of the triflate- and methylsulfonate-anions in a 50 : 50 mixture of TEATF–TEAMS obtained for T = 300 K and T = 400 K. (a) Methylsulfonate–TEA (b) triflate–TEA.

Hydrogen Bonding

Numerous geometrical procedures for the determination of hydrogen bonds have been proposed[35] in the past. Mostly, they rely on distance criteria in addition to some angular constraints.[35,36] We will see that for the case of protic ionic liquids discussed here, the use of angular constraints can even be omitted, since the distance distribution between the ammonium–hydrogen and the sulfonate–oxygen itself does very well discriminate between hydrogen bonded and non-hydrogen bonded states. As shown in Fig. 8, the pair correlation function between an ammonium–hydrogen and the sulfonate–oxygen exhibits a very sharp narrow first peak which goes almost down to zero for both, the pure ionic liquids, as well as for the mixtures. The number of nearest oxygen-neighbors around a hydrogen atom as a function of distance has a very well defined step-like form with a plateau at 1 for the pure liquids, which indicates that an N–H-group in both ionic liquids stays practically always involved in a hydrogen-bond. Here we use a distance criterion of $r_{H\text{–}O} \leq 0.24$ nm to define a hydrogen bond for both, the triflate- and the methylsulfonate-ions. When applying this cutoff-distance of 0.24 nm, the sum of the computed number of oxygen-neighbors N for both anion-species add up to one with $\delta N < |\pm 0.01|$ for all investigated mixtures and temperatures. To further substantiate that the intermolecular H–O distance is a proper criterion for a hydrogen-bonded configuration, we have computed a probability density map of the distance $r_{H\text{–}O}$ and the cosine of the angle $\alpha_{N\text{–}H/S\text{–}O}$ between the N–H bond on the TEA and the respective S–O-bond on the methylsulfonate anion, depicted in Fig. 9. The density-plot in Fig. 9 shows that distances $r_{H\text{–}O}$ < 0.24 nm correspond to an anti-parallel alignment, essentially confined to angles between 120 and 180 degrees. A second basin located at larger distances is well separated from the hydrogen-bonded basin by a free energy barrier at about $r_{H\text{-}O} \approx 0.24$ nm. The minimum of the second basin is located at a 60 degree angle, which is certainly incompatible with hydrogen bonded configurations. In fact, these configurations correspond to correlations with oxygen-sites adjacent to the oxygen involved in a hydrogen bond. An alternative representation is given in Fig. 10a, showing those angle-distributions for selected $r_{H\text{-}O}$-distance intervals. Distances with$r_{H\text{-}O} > 0.44$ nm show almost no orientational bias with respect to the N–H and S–O bonds involved. For comparison, Fig. 10b shows the distribution of the cosine of the angle between the intermolecular N–O-vector and the N–H-bond $\alpha_{N\text{-}H/N\text{-}O}$, which has been used previously to identify linearly hydrogen-bonded configurations.[36] An angular cutoff of 30 degrees is suitable for hydrogen-bonds in liquid water.[35,36] In the ionic liquid systems discussed here, the computed $\alpha_{N\text{-}H/N\text{-}O}$ fall well below this threshold. In fact, the computed $\alpha_{N\text{-}H/N\text{-}O}$ complies with a much narrower threshold-value of about 18 degrees, suggesting an even better defined linear hydrogen-bond configuration as in liquid water. We conclude that the O–H distance is a suitable measure to identify hydrogen bonds for the discussed systems. This has to be attributed to the peculiarities of the investigated IL, where we only have a single hydrogen-bond donor located on the cation, which is partially shielded by the ethyl-groups, in combination with an abundance of rather strong hydrogen-bond acceptor sites on the anions.

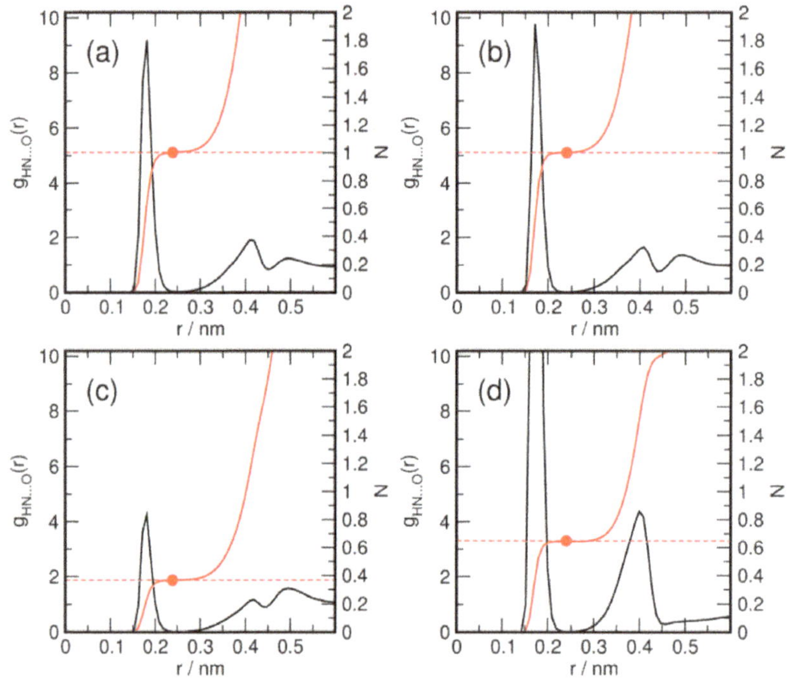

Figure. 8 H–O intermolecular site–site radial pair distribution functions g(r) obtained for pure TEATF and TEAMS ionic liquids and their 50 : 50 mixture at 300 K. Also shown are the number of nearest neighbors as a function of distance N(r). (a) pure TEATF (b) pure TEAMS (c) H–O correlation between TEA and TF in the $x_{MS} = 0.25$ mixture (d) H–O correlation between TEA and MS in the $x_{MS} = 0.25$ mixture.

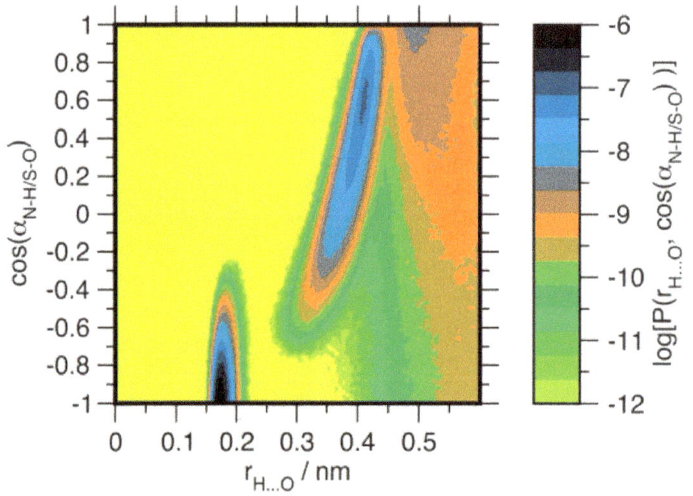

Figure. 9 Logarithmic representation of the probability-density of finding an intermolecular H–O distance $r_{H\text{-}O}$ and the cosine angle between the respective N–H and S–O vectors $\cos(\alpha_{N\text{-}H/S\text{-}O})$ obtained for the TEA–methylsulfonate pairs in a 50 : 50 TEAMS–TEATF-mixture at 300 K. The pairs were preselected with respect to their intermolecular N–O distance using$r_{N\text{-}O} < 0.8$ nm.

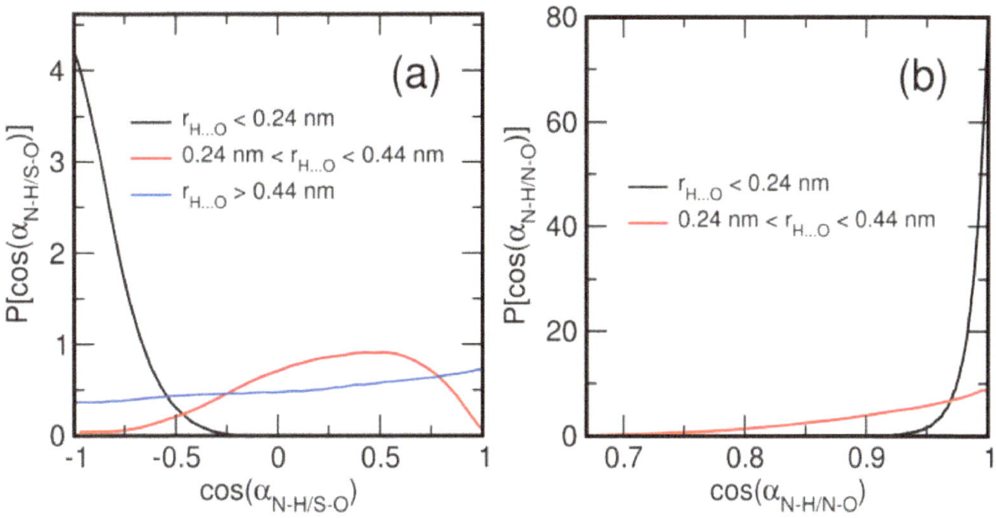

Figure. 10 Distribution of the cosine angles between two vectors for selected $r_{H\text{-}O}$-distance ranges. (a) Cosine angle between the N–H-vector in TEA and S–O vector in methylsulfonate $\cos(\alpha_{N\text{-}H/S\text{-}O})$ (b) cosine angle between N–H-vector in TEA and intermolecular N–O vectors $\cos(\alpha_{N\text{-}H/N\text{-}O})$. Data is shown for TEA–methylsulfonate pairs in a 50 : 50 TEAMS–TEATF-mixture at 300 K. The pairs were preselected with respect to their intermolecular N–O distance using $r_{N\text{-}O} < 0.8$ nm.

In the previous section we have seen that hydrogen bonding also affects the center-of-mass pair correlation functions and is assumed to be responsible for the splitted first peak of the anion–cation pair correlation function (shown in Fig. 4). As shown in Fig. 8c and d, the hydrogen-bond donated by the cation in the mixture is not shared by both anions according to their abundance in the mixture ($x_{MS} = 0.25$). In fact, the hydrogen bond to the methylsulfonate-ion appears to be formed in the majority of cases, shifting the equilibrium towards the formation of hydrogen bonds with the methylsulfonate-ion. This effect is not unexpected, given the larger negative partial charges on the methylsulfonate-oxygen sites. To quantify the shift, we have computed the percentage of TEA-cations involved in a hydrogen bond with respect to either anion (see Fig. 11). The data shown in Fig. 11 demonstrates the shift in equilibrium, and also indicates only a weak temperature dependence. Moreover, from the equimolar mixture we can directly estimate the difference in strength of the hydrogen-bond of a TEA-ion to either anion as a free-energy difference at 300 K according to $\Delta G^{\circ}_{HB} = -RT \ln K_{HB} \approx 5$ kJ mol^{-1} with $K_{HB} = y_{HB,TF}/y_{HB,MS} \approx 0.13$.

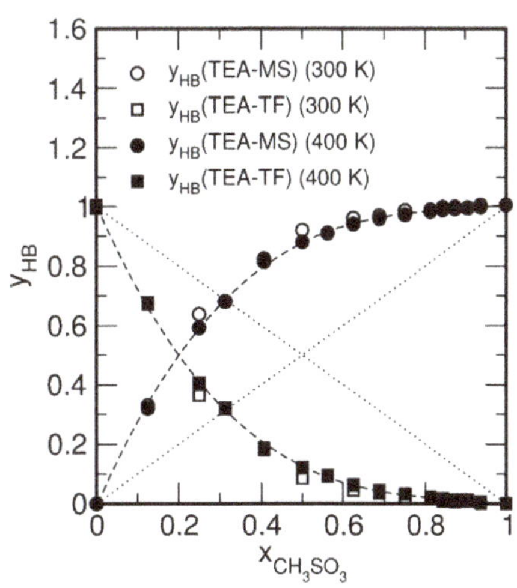

Figure. 11 Percentage of TEA-cations connected via a hydrogen bond to either a triflate- or a methylsulfonate-ion as a function of the mixture-composition at 300 K and 400 K, respectively.

The fact that both anions can accept up to three hydrogen bonds and that these hydrogen bonds are differently strong leads to a formation of hydrogen-bonded aggregates of varying size as a function of the composition of the mixture. First, we discuss the total number of hydrogen bonds accepted by each of the anions as a function of the mixture-composition shown in Fig. 12. In the pure liquids, each anion accepts on average one single hydrogen bond. Since each sulfonate group can accept up to three hydrogen-bonds, the strong TEA–MS hydrogen bonds causes the methylsulfonate-ion to accepting about up to ≈2.8 hydrogen bonds in the diluted solution ($x_{MS} \rightarrow 0$). By increasing the methylsulfonate-concentration, this number drops monotonously, since the methylsulfonate-ions start to compete with each other for TEA hydrogen-bonds.

The triflate-ion in the dilute triflate-solution ($x_{MS} \rightarrow 1$) only receives about ≈0.1 hydrogen-bonds on average, and is hence almost completely stripped of hydrogen bonds.

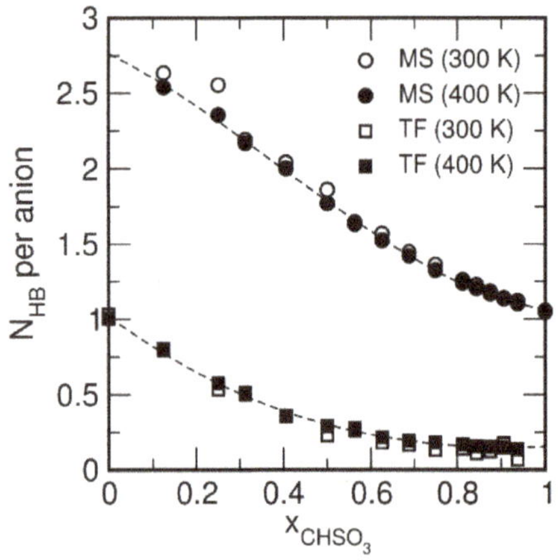

Figure. 12 Average number of hydrogen bonds accepted by the triflate- and methylsulfonate-ions at 300 K and 400 K, respectively.

The distribution of the hydrogen-bonded aggregates are finally shown in Fig. 13. Indicated are the percentages of each of the anions connected to either no, one, two or three TEA-cations via hydrogen-bonding. Note that for both pure ionic liquids, the distribution of aggregate sizes is almost identical: single hydrogen-bonded (n = 1) aggregates contribute about ≈50%, whereas the configurations with no (n = 0) and two hydrogen bonds (n = 1) are both near at about ≈24%. Configurations with three hydrogen bonds (n = 3) are rare and contribute only to ≈2–3%. With increasing methylsulfonate-concentration, the population of triflate-ions with no hydrogen-bonds ("free" triflate-ions) is increasing strongly to a plateau value of about 85% to 90% for $x_{MS} \rightarrow 1$. At the same time the percentage of configurations with one hydrogen bond drops down to 10% to 15%. Configurations with more than one hydrogen bond are practically non-existent. With increasing triflate-concentration, on the other hand, the methylsulfate-ions attract more and more hydrogen-bonds, leading to an increasing percentage of MS-ions attached to two (n = 2) and three (n = 3) TEA-cations. Here the configurations with two hydrogen-bonds first increase, but are getting increasingly replaced by configurations with three hydrogen-bonds. For $x_{MS} \rightarrow 0$, finally, the n = 3-fraction represents the majority of configurations with ≈75%, and n = 2 representing the rest with ≈25%. Configurations with no or one hydrogen-bond have practically vanished.

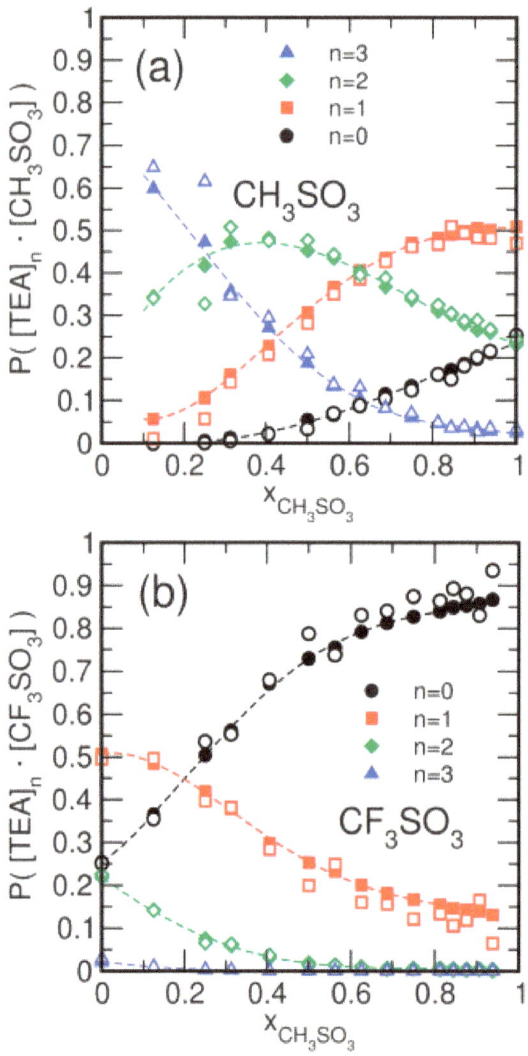

Figure. 13 Percentage of triflate- and methylsulfonate-ions hydrogen-bonded to zero, one, two, and three TEA-cations as a function of the mixture-composition. Open symbols: 300 K. Closed symbols: 400 K (a) methylsulfonate (b) triflate.

We summarize that the exact size and distribution of the hydrogen-bonded aggregates is apparently determined by the strengths of the hydrogen-bonds, the number of hydrogen-bonded donors and acceptors on the ions, as well as their respective coordination-numbers. In the following section, we will introduce a "cartoon-like" representation of the ionic liquid, which is based on a minimal set of assumptions extracted from our MD-simulation, and can be used compute the observed cluster-sizes as a function of the mixture-composition in a semi-quantitative fashion.

A "Cartoon-like" Model of Hydrogen Bonding in the IL Mixture

For a better understanding of the hydrogen-bonded cluster-sizes in the IL mixtures, we would like to introduce a simplified description of the hydrogen bond equilibrium. For this purpose we construct a simplified model of the ionic liquid as outlined in Fig. 14. Here, the liquid is represented by a cubic lattice, where the anions and cation occupy sites in an alternating fashion, leading to a octahedral coordination of each ion with a coordination number of c = 6. In addition, we assume that each TEA-cation stays always involved in a hydrogen-bond, and is pointing with its N–H-bond towards one of the c surrounding anions. In a pure ionic liquid, the cation–anion hydrogen-bond is, of course, supposed to connect the TEA-cation randomly with one of its c anion neighbors. If each anion would be able to accept c hydrogen-bonds from the surrounding TEA-cations equally well, the probability of finding an anion accepting nhydrogen-bonds simultaneously $P_{c,n}$ would exactly follow a binomial distribution

$$P_{c,n} = \frac{c!}{n!(c-n)!} \cdot p^n \cdot (1-p)^{c-n}, \tag{2}$$

where p represents the probability of a TEA-cation pointing with its N–H bond towards the central anion. Assuming a purely random orientation of the TEA–N–H-bond on the lattice, pcorresponds to the inverse coordination number of the TEA-ion ($p = 1/c$). However, to account for the fact that the triflate and methylsulfonate-anions can each accept only up to three hydrogen-bonds, we have to modify the distribution function by introducing configuration-specific weights-factors $w_{c,n}$ with $w_{c,n} = 0$ for $n > 3$.

$$P_{c,n} = \frac{c!}{n!(c-n)!} \cdot Q^{-1} \cdot w_{c,n} \cdot p^n \cdot (1-p)^{c-n}. \tag{3}$$

To ensure proper normalisation, the modified weights are counter-balanced by a constant Q, which has to be computed numerically according to

$$Q = \sum_{n=0}^{c} \frac{c!}{n!(c-n)!} \cdot w_{c,n} \cdot p^n \cdot (1-p)^{c-n}. \tag{4}$$

In a first step, we will discuss a situation, where the non-zero statistical weights are all equal ($w_{c,n} = 1$ for $n \leq 3$). In addition, we will assume that the chosen weights do not change as function of the mixture-composition. In the mixture, however, the hydrogen-bond of TEA to each of the two anion-species varies in strength. This will, hence, also modify the hydrogen-bond acceptance probability for each of the anions. In all cases, the TEA cation points to one of the c surrounding neighbors with a site-specific probability p_i. Of course, those probabilities do add all up to unity:

$$1 = \sum_{i=1}^{c} p_i. \tag{5}$$

Since each of the sites i is assumed to be statistically occupied by one of the two anion-species according to their mixture composition $x = x_{MS}$, the average probability p_i of accepting a hydrogen-bond is hence a composition-weighted sum of the individual probabilities of the anions according to

$$p_i = x \cdot p_{MS} + (1-x) \cdot p_{TF} \tag{6}$$

From eqn (5) and (6), the universal anion-specific hydrogen-bond-acceptance probabilities follow as

$$p_{MS} = \frac{1}{cx + c(1-x)f}$$
$$p_{TF} = \frac{1}{cxf^{-1} + c(1-x)}, \tag{7}$$

with $f = p_{TF}/p_{MS}$. From eqn (7) we can directly determine the fraction of times, a N–H bond of a TEA-cation points to either a methylsulfonate- or triflate-ion as

$$y_{HB,MS} = cx \cdot p_{MS} = \frac{1}{1 + f(1-x)x^{-1}} \tag{8}$$

and

$$y_{HB,TF} = c(1-x) \cdot p_{TF} = \frac{1}{1 + f^{-1}(1-x)^{-1}x}. \tag{9}$$

The ratio f can be determined from MD simulation data by taking the values for the fraction of hydrogen-bonds of one TEA-molecule to each of the two anion-species from the equimolar mixture. Using the data shown in Fig. 11, we determine $f = y_{HB,TF}/y_{HB,MS} \approx 0.13$. Assuming that f does not change significantly as a function of the composition x, we can now compute the hydrogen-bonded cluster sizes $P_{c,n}$ for both, the methylsulfonate and the triflate anions, as well as the average number of hydrogen bonds accepted by each anion according to

$$N_{HB} = \sum_{n=0}^{c} n \cdot P_{c,n} \tag{10}$$

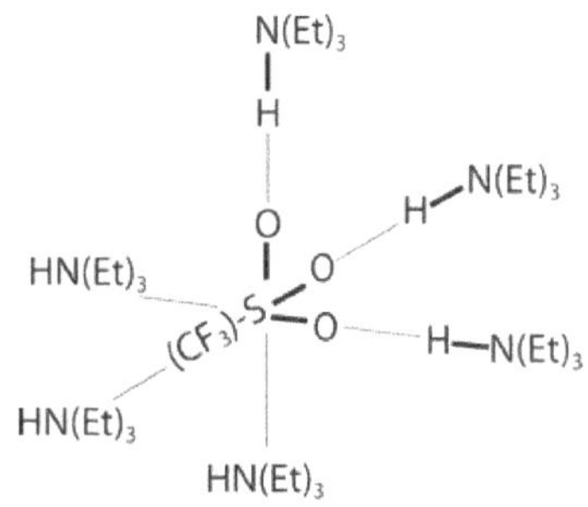

Figure. 14 "Cartoon-like" representation of the ionic liquid: the liquid is represented by a cubic lattice, where the anions and cations occupy sites in an alternating fashion. Each ion is octahedrally coordinated and surrounded by c = 6 counter-ions. Each anion (either triflate or methylsulfonate) can accept up to three hydrogen bonds. The N–H groups of each of the TEA-molecules are involved in a hydrogen-bond to one of their c surrounding neighbors.

Fig. 15 shows the prediction of the sizes of the hydrogen-bonded aggregates according the model using equal weights with $w_{6,n} = 1$ for $n \leq 3$ and $w_{6,n} = 0$ for $n > 3$. Essential features of the distributions obtained from the MD-simulation, such as the existence of a maximum of MS–TEA aggregates with two hydrogen bonds, are well recovered by just using an approximate coordination number, and taking into account that both anions can accept no more than three hydrogen bonds. Also, the percentage of hydrogen-bonds of TEA to either anion-species, as well as the total number of accepted hydrogen bonds (Fig. 16) are nicely captured. However, the distribution of aggregate-sizes in the pure TEAMS and TEATF liquids are not correctly represented. In particular, the model underestimates the single-bonded configurations and overestimates the configurations with no hydrogen-bond. To better represent the distribution within the pure liquids and to improve the model, we adjust the non-zero weight-factors $w_{c,n}$ for $n \leq 3$ to match the MD simulation results (as shown in Fig. 13) for the respective pure ionic liquids with $P_{6,0} = 0.24$, $P_{6,1} = 0.51$, $P_{6,2} = 0.22$, and $P_{6,3} = 0.03$. For simplicity, we use the same values for both, TEATF and TAMS. Those probabilities are recovered by using statistical weights of $w_{6,0} = 0.75$, $w_{6,1} = 1.3$, $w_{6,2} = 1.12$, and $w_{6,3} = 0.5$ for both anions. The weight-factors indicate that the anions in the pure IL feel most comfortable with just one hydrogen bond pointing towards them. Situations with three hydrogen bonds have a significantly lower statistical weight, most likely due to orientational restraints, introduced by a configuration involving three hydrogen bonds to adjacent ions. Particularly, the situation without any hydrogen-bonds exhibits a significantly reduced statistical weight compared to the random distribution of hydrogen-bonds. The distribution of clusters shown in Fig. 17 demonstrate a significant improvement, leading to a semi-quantitative description of the MD-data. Moreover, the good overall description of the simulation-data by our model is suggesting that the two anion-species are statistically distributed in the liquid. However, also certain weaknesses of the model are evident, particularly for the percentage of hydrogen-bonds of TEA to either anion-species in the range $x_{MS} < 0.3$. By introducing a composition-dependent factor $f(x_{MS})$ a better representation of the data shown in Fig. 16a and 18a could be achieved (not shown). Interestingly, this does, in fact, worsen the description of the distribution of cluster-sizes, indicating the need to also account for the changing coordination-numbers, and thus pointing at the limits of the model.

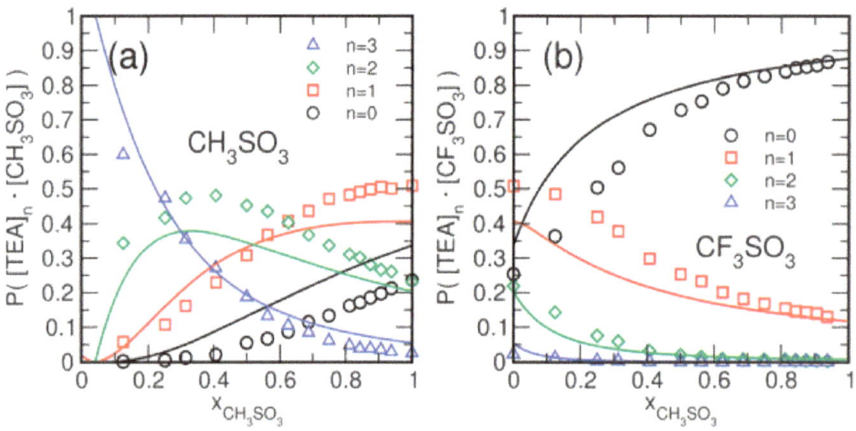

Figure. 15 Percentage of methylsulfonate- (a) and triflate-ions (b) hydrogen-bonded to zero, one, two, and three TEA-cations, as a function of the mixture-composition at 400 K. Symbols: MD-simulation. Solid lines: model using equal statistical weights with $w_{6,n} = 1$ for $n \leq 3$ and $w_{6,n} = 0$ for $n > 3$.

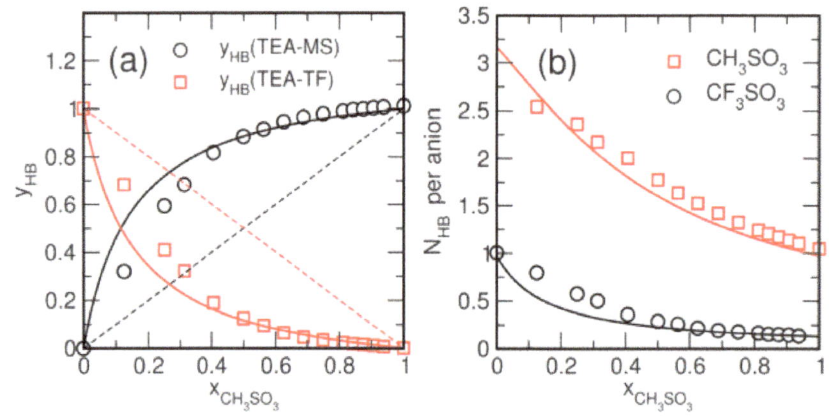

Figure. 16 (a) Percentage of TEA-cations connected via a hydrogen bond to either a triflate-ion or a methylsulfonate-ion as a function of the mixture-composition. (b) Average number of hydrogen bonds accepted by the triflate- and methylsulfonate-ions. Open symbols: MD-data obtained at 400 K. Solid lines: model using equal statistical weights with $w_{6,n} = 1$ for $n \leq 3$ and $w_{6,n} = 0$ for $n > 3$.

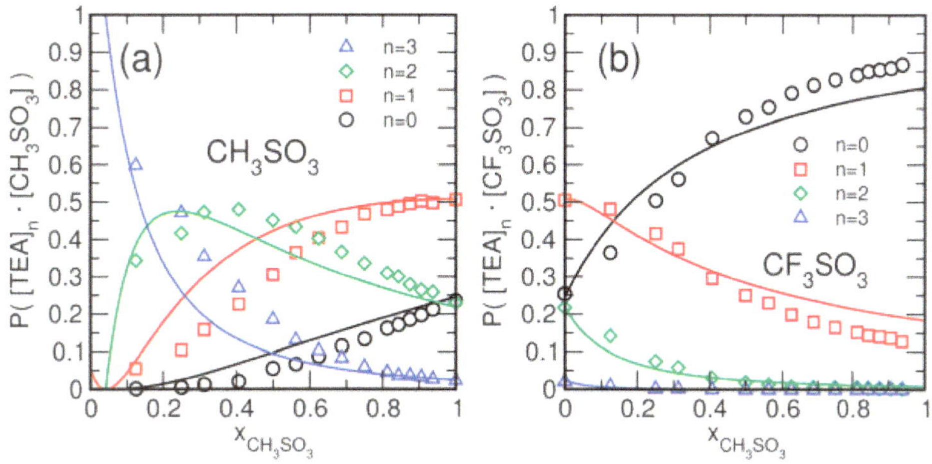

Figure. 17 Percentage of methylsulfonate- (a) and triflate-ions (b) hydrogen-bonded to zero, one, two, and three TEA-cations, as a function of the mixture-composition at 400 K. Symbols: MD-simulation. Solid lines: model using statistical weights adjusted to the distribution in the pure liquids with $w_{6,0} = 0.75$, $w_{6,1} = 1.3$, $w_{6,2} = 1.12$, $w_{6,3} = 0.5$, and $w_{6,n} = 0$ for $n > 3$.

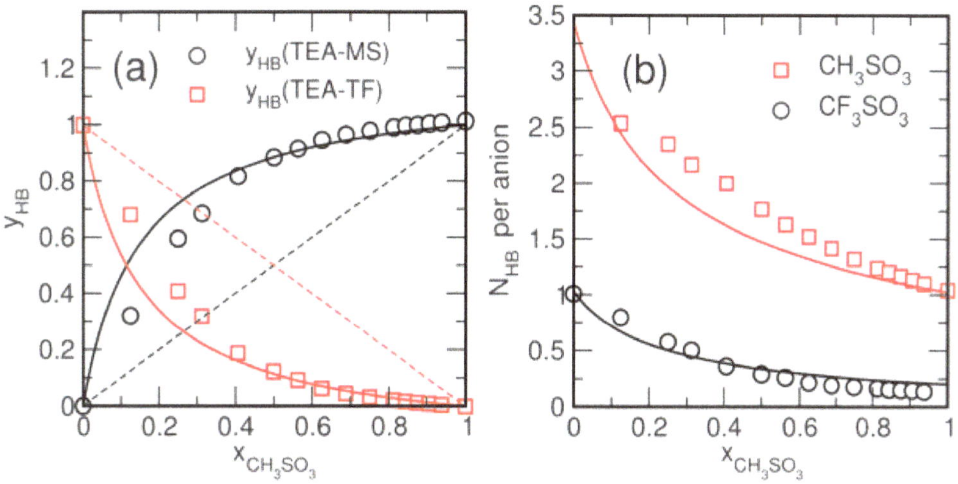

Figure. 18 (a) Percentage of TEA-cations connected via a hydrogen bond to either a triflate-ion or a methylsulfonate-ion as a function of the mixture-composition. (b) Average number of hydrogen bonds accepted by the triflate- and methylsulfonate-ions. Open symbols: MD-data obtained at 400 K. Solid lines: model using statistical weights adjusted to the distribution in the pure liquids with $w_{6,0} = 0.75$, $w_{6,1} = 1.3$, $w_{6,2} = 1.12$, $w_{6,3} = 0.5$, and $w_{6,n} = 0$ for $n > 3$.

CONCLUSIONS

We have reported results of molecular dynamics simulations characterising the hydrogen bonding in mixtures of two protic ionic liquids sharing the same cation: triethylammonium-methylsulfonate and triethylammonium-triflate. The triethylammonium-cation acts as a hydrogen-bond donor, being able to donate a single hydrogen bond. Both, the methylsulfonate- and the triflate-anions can act as hydrogen-bond acceptors, which can accept multiple hydrogen bonds via their SO_3-groups. Replacing a methyl-group in the methylsulfonate by a trifluoromethyl-group in the triflate significantly weakens the strength of a hydrogen bond from an adjacent triethylammonium-cation to the oxygen-site in the SO_3-group of the respective anion. Our MD simulations show that these rather subtle differences in hydrogen bond strength significantly affect the formation of differently-sized hydrogen-bonded aggregates in the mixtures as a function of the mixture-composition. A defining structural feature of ionic liquids in general and of the studied protic ionic liquid in particular is the charge-alternating order of the ions, leading to counter-ion coordination numbers between six and eight. This fact is represented in a simple "cartoon-like" lattice model of the liquid, placing ions on a cubic lattice in an alternating fashion with an approximate counter-ion coordination number of six. By introducing statistical weights that mostly account for the fact that each anion can only accept up to three hydrogen bonds, and by adjusting the balance between of the hydrogen bond of TEA to either of the two anions, the simple model is able to qualitatively predict the distribution of differently-sized hydrogen-bonded aggregates as well as the total number of hydrogen bonds accepted by the anions as a function of the mixture-composition. By adjusting the statistical weights for each cluster in such a way that the cluster sizes match the distribution observed in the pure liquids, an even semiquantitative description of the data obtained from MD simulation emerges. Finally, we would like to comment of the issue of the "doublesaltiness" of the studied ionic liquid mixture as pointed out by Rogers et al.[6] Their line of reasoning is that in a mixture of ILs, the constituents (ions) do not seem to retain their individual nature. Hence, they proposed to consider them as double-salt ionic liquids. We think that our model-system is a particularly nice demonstration of the concept: although the pure ionic liquids and their mixtures exhibit great similarity with respect to the overall ion-alternating structure, each of the statistically distributed anions experiences a completely different local environment.

ACKNOWLEDGEMENTS

This work was supported by the Deutsche Forschungsgemeinschaft (DFG) Priority Programme SPP 1191 "Ionic Liquids" and the DFG Collaborative Research Center SFB 652 "Strong correlations and collective effects in radiation fields: Coulomb systems, clusters and particles"

REFERENCES

1. P. Wasserscheid and T. Welton, Ionic Liquids in Synthesis, Wiley-VCH, 2nd edn, 2008.
2. F. Endres and S. Z. El Abedin, Phys. Chem. Chem. Phys., 2006, **8**, 2101–2116.
3. H. Weingärtner, Angew. Chem., Int. Ed., 2008, **47**, 654–670.
4. N. V. Plechkova and K. R. Seddon, Chem. Soc. Rev., 2008, **37**, 123–150.
5. H. Niedermeyer, J. P. Hallett, I. J. Villar-Garcia, P. A. Hunt and T. Welton, Chem. Soc. Rev., 2012, **41**, 7780–7802.
6. G. Chatel, J. F. B. Pereira, V. Debbeti, H. Wang and R. D. Rogers, Green Chem., 2014, **16**, 2051–2083.
7. K. Fumino, A. Wulf and R. Ludwig, Angew. Chem., Int. Ed., 2008, **47**, 8731–8734.
8. K. Fumino and R. Ludwig, J. Mol. Liq., 2014, **192**, 94–102
9. M. Y. Lui, L. Crowhurst, J. P. Hallett, P. A. Hunt, H. Niedermeyer and T. Welton, Chem. Sci., 2011, **2**, 1491–1496.
10. P. Navia, J. Troncoso and L. Romani, J. Chem. Eng. Data, 2007, **52**, 1369–1374
11. A. Arce, M. J. Earle, S. P. Katdare, H. Rodriguez and K. R. Seddon, Chem. Commun., 2006, 2548–2550.
12. J. N. Canongia Lopes, T. C. Cordeiro, J. M. S. S. Esperanca, H. J. R. Guedes, S. Huq, L. P. N. Rebelo and K. R. Seddon, J. Phys. Chem. B, 2005, **109**, 3519–3525.
13. D. Xiao, J. R. Rajian, L. G. Hines, S. Li, R. A. Bartsch and E. L. Quitevis, J. Phys. Chem. B, 2008, **112**, 13316–13325.
14. K. Fumino, V. Fossog, K. Wittler, R. Hempelmann and R. Ludwig, Angew. Chem., Int. Ed., 2013, **52**, 2368–2372
15. P. Stange, K. Fumino and R. Ludwig, Angew. Chem., Int. Ed., 2013, **52**, 2990–2994.
16. K. Fumino, A. M. Bonsa, B. Golub, D. Paschek and R. Ludwig, ChemPhysChem, 2015, **16**, 299–304.
17. S. Nosé, Mol. Phys., 1984, **52**, 255–268.
18. W. G. Hoover, Phys. Rev. A: At., Mol., Opt. Phys., 1985, **31**, 1695–1697.
19. M. Parrinello and A. Rahman, J. Appl. Phys, 1981, **52**, 7182–7190.
20. S. Nosé and M. L. Klein, Mol. Phys., 1983, **50**, 1055–1076.
21. U. Essmann, L. Perera, M. L. Berkowitz, T. A. Darden, H. Lee and L. G. Pedersen, J. Chem. Phys., 1995, **103**, 8577–8593.
22. S. Miyamoto and P. A. Kollman, J. Comput. Chem., 1992, **13**, 952–962.
23. E. Lindahl, B. Hess and D. van der Spoel, J. Mol. Model., 2001, **7**, 306–317.
24. M. Abraham, E. Apol, R. Apostolov, H. J. Berendsen, A. van Buuren, P. Bjelkmar, R. van Drunen, A. Feenstra, S. Fritsch, G. Groenhof, C. Junghans, J. Hub, P. Kasson, C. Kutzner, B. Lambeth, P. Larsson, J. A. Lemkul, E. Marklund, P. Meulenhoff, T. Murtola, S. Pall, S. Pronk, R. Schulz, M. Shirts, A. Sijbers, P. Tieleman, M. Wolf, B. Hess, D. van der Spoel and E. Lindahl, GROMACS - Groningen Machine for Chemical Simulations User Manual Version 4.6.1, 2013.
25. H. Flyvbjerg and H. G. Petersen, J. Chem. Phys., 1989, **91**, 461–466.
26. H. J. C. Berendsen, J. P. M. Postma, W. F. van Gunsteren, A. DiNola and J. R. Haak, J. Chem. Phys., 1984, **81**, 3684–3690.
27. W. L. Jorgensen and M. Ibrahim, J. Am. Chem. Soc., 1981, **103**, 3976–3985.
28. M. J. Frisch, G. W. Trucks, H. B. Schlegel, G. E. Scuseria, M. A. Robb, J. R. Cheeseman, G. Scalmani, V. Barone, B. Mennucci, G. A. Petersson, H. Nakatsuji, M. Caricato, X. Li, H. P. Hratchian, A. F. Izmaylov, J. Bloino, G. Zheng, J. L. Sonnenberg, M. Hada, M. Ehara, K. Toyota, R. Fukuda, J. Hasegawa, M. Ishida, T. Nakajima, Y. Honda, O. Kitao, H. Nakai, T. Vreven, J. A. Montgomery Jr, J. E. Peralta, F. Ogliaro, M. Bearpark, J. J. Heyd, E. Brothers, K. N. Kudin, V. N. Staroverov, R. Kobayashi, J. Normand, K. Raghavachari, A. Rendell, J. C. Burant, S. S. Iyengar, J. Tomasi, M. Cossi, N. Rega, J. M. Millam, M. Klene, J. E. Knox, J. B. Cross, V. Bakken, C. Adamo, J. Jaramillo, R. Gomperts, R. E. Stratmann, O. Yazyev, A. J. Austin, R. Cammi, C. Pomelli, J. W. Ochterski, R. L. Martin, K. Morokuma, V. G. Zakrzewski, G. A. Voth, P. Salvador, J. J. Dannenberg, S. Dapprich, A. D. Daniels, Ö. Farkas, J. B. Foresman, J. V. Ortiz, J. Cioslowski and D. J. Fox, Gaussian 09, Revision D.01, Gaussian Inc, Wallingford, CT, 2009.
29. U. C. Singh and P. Kollman, J. Comput. Chem., 1984, **5**, 129–145.
30. C. I. Bayly, P. Cieplak, W. D. Cornell and P. A. Kollman, J. Phys. Chem., 1993, **97**, 10269–10280.
31. T. Köddermann, D. Paschek and R. Ludwig, ChemPhysChem, 2007, **8**, 2464–2470.
32. H. Krienke, V. Vlachy, G. Ahn-Ercan and I. Bako, J. Phys. Chem. B, 2009, **113**, 4360–4371.
33. M. G. Martin and J. I. Siepmann, J. Phys. Chem. B, 1998, **102**, 2569–2577.

34. D. Roy, N. Patel, S. Conte and M. Maroncelli, J. Phys. Chem. B, 2010, **114**, 8410–8424.

35. R. Kumar, J. R. Schmidt and J. L. Skinner, J. Chem. Phys., 2007, **126**, 204107.

36. A. Luzar and D. Chandler, Nature, 1996, **379**, 55–57.

CHAPTER 7

ELECTRIC DIPOLE MOMENTS AND CHEMICAL BONDING OF DIATOMIC ALKALI–ALKALINE EARTH MOLECULES

Johann V. Pototschnig, Andreas W. Hauser and Wolfgang E. Ernst

INTRODUCTION

In the last years ultracold molecules have attracted a lot of attention due to their outstanding properties.[1–6] The low temperature reduces the decoherence, and ensembles of ultracold molecules can be used to simulate quantum systems[7–12] or perform quantum computations.[13] Most degrees of freedom are frozen out at these temperatures, which allows precise control of chemical reactions.[14–16] Another advantage is the reduction of the linewidth and noise in measurements of fundamental physical properties and constants[17] such as the proton to electron mass ratio,[18–20] the electron electric dipole moment[21,22] or the fine structure constant.[23,24]

The first molecules produced in the ultracold temperature regime were homonuclear alkali-dimers,[25–28] which do not have a permanent electric dipole moment. A few years later, ultracold heteronuclear alkali-dimers were produced.[29–34] Their permanent electric dipole moment gives experimentalists an additional handle for manipulation via electric fields. In this sense, the next logical step is the production of ultracold molecules with a dipole moment plus an additional magnetic moment, originating from an unpaired electron spin. This magnetic moment causes additional interactions and enables control of these molecules via magnetic fields. Currently, the most promising candidates are alkali (AK)–alkaline earth (AKE) and AK–Yb diatomic molecules because the involved atoms are well under control in ultracold atomic physics. Bose–Einstein condensation, for example, has been achieved for most AK atoms,[35]Ca,[36] Sr[37] and Yb.[38] In a series of recent studies combining experimental and computational efforts[39–43] we found permanent electric dipole moments (PEDM) with opposite signs in the lowest Σ^+ states of the doublet and the quartet multiplicity for AK–AKE molecules. In this work we analyze this behavior and discuss the trends in greater detail. The magnitude of the permanent electric dipole moment (PEDM) is of special importance for future applications. A significant value is necessary to align molecules in an optical lattice,[44,45] to control interactions with an external microwave field,[4,11,46] and it might even be useful to reach low vibrational levels.[47]

Early on, there has been interest in the prediction of dipole moments for alkali halides and alkaline earth halides. These two groups of molecules with predominantly ionic bonding could successfully be described by electrostatic interaction models.[48–50] The Rittner model[48] for alkali halides assumes a complete charge transfer of one electron and considers the polarizabilities of the ions. Törring et al.[51,52] applied a similar model to AKE halides. More sophisticated computations using ligand field theory were performed for AKE halides[53] and lanthanum containing diatomics.[54,55] A model taking orbital mixing into account was applied to SrBr in ref. 56. However, these techniques assume a transfer of at least one electron and can therefore not be applied to AK–AKE molecules. Recently, an empirical model for the relation between dipole moment and vibrational frequencies was presented.[57]

There are a few theoretical and experimental studies concerned with the properties of AK–AKE molecules. Bauschlicher et al.[58] performed one of the first calculations. Two theoretical studies examined the ground state of Li–AKE molecules by coupled cluster calculations.[59,60] The ground states of neutral AK–Sr molecules and several states of the molecular ions have been investigated by a configuration interaction method.[61,62] In ref. 63, the ground states of AK–Mg diatomic molecules were calculated at the coupled cluster level of theory. Using the same approach, Gopakumar et al.[45] recently

Citation: Johann V. Pototschnig, Andreas W. Hauser and Wolfgang E. Ernst, Electric dipole moments and chemical bonding of diatomic alkali–alkaline earth molecules, DOI: 10.1039/c5cp06598d.

investigated the AK–AKE molecules (Na, K, Rb)–(Ca, Sr). The combination of barium with an alkali atom has also been considered in the ground state.[64] Some older studies analyzed the ground and several excited states of LiBa[65,66] and NaBa.[67] The most thoroughly studied representative of these diatomic molecules is LiCa, with experimental,[68–70] theoretical[71,72] and combined investigations.[39,73] The first realization of a combined quantum degenerate mixture of Rb and Sr[74] was followed by a series theoretical and experimental publications.[40–42,75–77] The excited states of LiSr,[72] RbBa,[78] LiMg,[79] and RbCa[43] were calculated recently.

In this study, we investigate the energetically lowest electronic states of neutral AK (Li, Na, K, Rb)–AKE (Be, Mg, Ca, Sr) molecules in the doublet and quartet manifold. Throughout the article, the coordinate system is defined such that the AKE atom is on the negative side of the axis and the AK atom on the positive side. Therefore, a positive value of the PEDM corresponds to a dipole moment pointing in axis direction from the AKE to the AK atom (i.e., the center of the negative charge lies closer to the AKE atom).

This manuscript is structured as follows: in the next section details of the computational method are given, then an overview of the results and a comparison to previous results. After discussing trends for AK–AKE molecules we give a qualitative description of the situation for the lowest Σ^+ states in AK–AKE molecules. Finally, we present an empirical model in Section 3.4.

COMPUTATIONAL DETAILS

All ab initio calculations were carried out with the MOLPRO[80] software package. A multiconfigurational self consistent field calculation (MCSCF[81,82]) was performed, followed by multireference configuration interaction including Davidson correction (MRCI[83,84]). The quintuple-zeta basis set family of Peterson et al. covers most atoms of our study and was used where possible. All electrons basis sets[85] were applied for Li, Na, Be and Mg. For the heavier atoms the corresponding effective core potential basis sets[86] of Peterson were combined with the recommended effective core potentials[87,88] of Lim et al. For K and Rb, however, no Peterson basis sets were available. In these cases, we fell back on the basis sets provided by Lim,[87,88] decontracted them, and added several diffuse functions to obtain a description of similar accuracy.‡ The same strategy has recently been applied to RbCa,[43] yielding excitation energies in good agreement with experimental measurements.

We further applied core polarization potentials (CPPs) as introduced by Müller and Meyer[89] due to their importance for correct magnitudes of the permanent electric dipole moments[90,91] (see Fig. 1). The parameters of the core polarization potential are given (see Table 1) and were optimized for our effective core potentials and basis sets. This approach only allows basis functions with an angular quantum number below 5. Therefore, basis functions with higher quantum numbers were neglected. No core polarization was added for Li and Be due to their small cores and large energy difference between the core orbitals and valence orbitals compared to the other elements. With this approach the energy difference between the first excited state ($^2P_{AK}/^3P_{AKE}$) and the ground state agrees within 150 cm^{-1} (1%) with the atomic values[92] in the asymptotic limit.

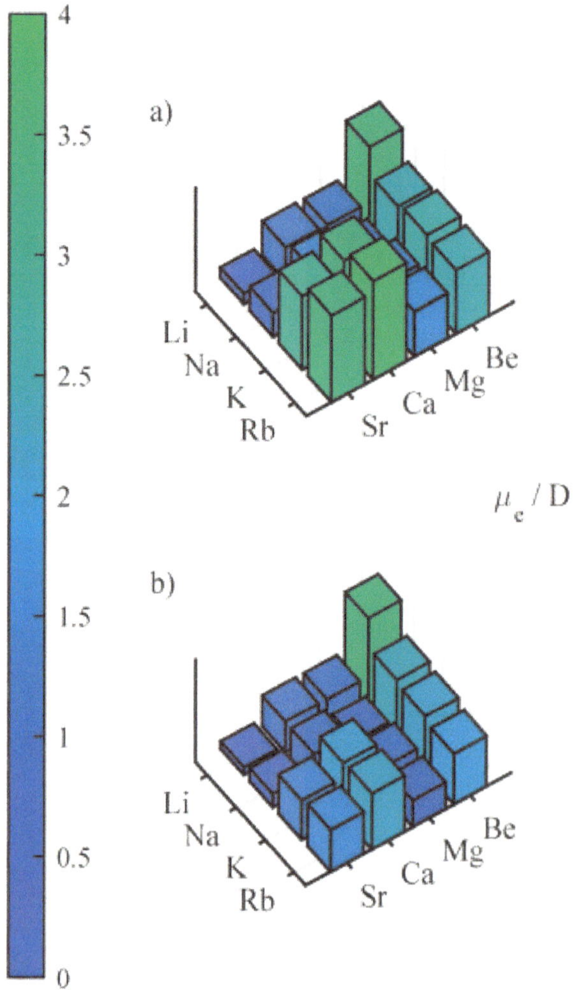

Figure. 1 The permanent electric dipole moment of the doublet ground state at its equilibrium distance, obtained at the MCSCF + MRCI level of theory, without (a) and with (b) the inclusion of core polarization potentials.

Table 1 Parameters of the CPPs used in this study; α_{core} is the dipole polarizability and ρ is the cutoff parameter. The CPPs account for the polarizability of the innermost shells. These parameters have been adjusted for the best reproduction of atomic excitation energies

AK	α_{core}/a.u.	ρ/Å^{-2}	AKE	α_{core}/a.u.	ρ/Å^{-2}
Na	1.00	1.27	Mg	1.20	0.60
K	5.61	0.46	Ca	7.16	0.25
Rb	9.74	0.35	Sr	9.28	0.28

It is known that the correlation or polarization of the core electrons has a significant influence on the dipole moment.[90,91] The importance of this contribution can be estimated by looking at Fig. 1, where the results without (a) and with (b) core polarization potentials are compared. Typically, the PEDM is overestimated if the polarization of the core is neglected, which agrees well with the observations of ref. 90. This error increases with atomic mass. The largest deviation was found for RbSr, where the dipole moment is reduced by 1.65 D to about half its value after inclusion of core polarization. The PEDMs for the computations including core polarization potentials (see plot b of Fig. 1) were obtained by finite field calculations,§ which we carried out for different field strengths varying from 10^{-5} to 5×10^{-3} a.u. for all 16 molecules. For each field strength the PEDMs and polarizabilities were computed by three point and five point finite field approximations and compared to each other to estimate the size of numerical error. The values were then taken where the smallest deviation was observed. (The values and the field strengths for their calculation are listed in the ESI.)

The PEDMs for the lowest Σ^+ state of the quartet multiplicity were also changed significantly by the core polarization potential, especially for heavy molecules, but without a clear trend. An improvement of the potential depth of LiCa was achieved by a new set of recently adjusted core polarization parameters. The potential depth for LiCa obtained in this work (2613 cm^{-1}) now agrees better with the experimental value (2605 cm^{-1})[69] than our previous calculation.[39] However, part of the improvement might originate also from the slightly different basis set. On average, the equilibrium distance was reduced by the core polarization potential, although some minor deviations occurred, especially for the quartet states.

In the MCSCF and MRCI calculation the same active space was applied, but it was necessary to adapt it to the different basis sets and effective core potentials. Details on the size and definition of the active space for the various diatomics can be found in the ESI. Note that the lowest Σ^+ state in the quartet multiplicity is an electronically excited state accompanied by several excited states of the doublet manifold. Therefore, several excited states were calculated in both spin configurations to obtain a meaningful description of this state. In a combined MCSCF calculation 16 to 32 states of doublet multiplicity and 3 or 8 states of quartet multiplicity were optimized. The number of states was kept the same in the MRCI calculation for the quartet multiplicity, but the number of doublet states was reduced to 12–17 states. With this setup for the state-averaged calculations we obtained smooth potential energy curves up to an internuclear separation of 10 Å. At this distance, the deviation for the ground state from the asymptotic value is only 10 cm^{-1} for light atoms and about 25 cm^{-1} for heavy atoms (ESI).

RESULTS

Overview

The properties of 16 AK–AKE molecules have been calculated by MCSCF + MRCI and the values for the ground state are listed in Table 2. Results for the lowest Σ^+ state in the quartet multiplicity are given in Table 3. Besides the ab initio calculations performed in this work, several other theoretical predictions as well as three experimental studies are listed in Table 2. Overall agreement between the various studies is good except for significant differences in dissociation energies. LiCa, the best studied molecule in the literature, with data available from 8 different sources, shows a large variation of the dissociation energy. Our value and the value of ref. 60 agree best with recent experimental findings.[69] Two older experimental results[93,94] are also listed in Table 2. A recent experiment of our group[40] yielded a vibrational constant (ω_e) of (42 ± 5) cm^{-1} for RbSr, which is in good agreement with the theoretical results in Table 2. The calculated PEDMs are all positive and differ on average by 0.16 D (13%) from previous computations, with the largest difference observed for the PEDM of KCa in ref. 45. Note that the PEDM values listed in Table 2 were evaluated for the respective equilibrium distance. When zero-point motion is considered, the PEDM is on average reduced by 0.03 D for the doublet states and by 0.06 D for the quartet states (details are provided in the ESI†). The difference to other equilibrium distances is on average 0.05 Å (1%) with the largest deviations occurring for RbSr and RbCa when compared to ref. 45. This deviation in the latter cases also explains the slight difference in the PEDM as the listed values are taken at the equilibrium separation. A small disagreement can also be found between the parallel polarizabilities (along the internuclear axis) listed in Table 2. Note that only the polarizabilities ($\alpha_\parallel$) at the equilibrium distances along the internuclear axis are given here. A more detailed treatment of polarizabilities and their experimental significance is provided in ref. 45 for AK–AKE molecules or in ref. 44 for AK–AK molecules.

Table 2 Results for the $^2\Sigma^+$ ground state, obtained at the MRCI level of theory, including core polarization potentials. Where available, values from the literature are given for comparison

Mol.	Ref.	μ_e/D	r_e/Å	D_e/cm^{-1}	ω_e/cm^{-1}	$\alpha_\parallel$/a.u.
LiBe	This work	3.47	2.60	2427	313	365
	60	3.58	2.58	2406	300	
	58		2.61	2339	300	
	93		2.59		295	
LiMg	This work	1.18	3.10	1538	181	470
	79	1.22	3.10	1395	174	
	63	1.02	3.10	1445	180	
	59	0.86[c]	3.12	1332	187	482
	60	1.12	3.11	1432	206	
	58		3.11	1613	183	
	94			1331	190	

LiCa	This work	1.19	3.39	2613	200	588
	39		3.34	2883	203	
	72		3.40	2131	206	
	59	1.10[c]	3.40	2260	210	599
	60	1.12	3.36	2607	207	
	73		3.40	2178	204	
	71		3.30	2355	196	
	69		3.36	2605	202	
LiSr	This work	0.28	3.57	2471	181	653
	72		3.58	2074	181	
	59	0.31[c]	3.53	2223	196	640
	60	0.24	3.55	2401	184	
	61	0.34	3.48	2587	185	
NaBe	This work	2.33	2.97	1291	173	397
	58		3.05	1129	164	
NaMg	This work	0.86	3.47	946	90	432
	63	0.72	3.53	825	83	
	58		3.56	887	85	
NaCa	This work	1.18	3.67	1792	103	577
	45	1.01	3.72	1453	97	594
NaSr	This work	0.51	3.84	1728	87	636
	45	0.49	3.89	1441	82	633
	61	0.62	3.82	1597	85	
KBe	This work	2.22	3.51	920	122	628
KMg	This work	1.08	3.99	779	64	656
	63	0.83	4.09	647	58	
KCa	This work	2.12	4.20	1455	71	869
	45	1.67	4.32	974	61	892
KSr	This work	1.52	4.39	1358	57	925
	45	1.27	4.53	964	48	942
	61	1.52	4.41	1166	52	
RbBe	This work	1.97	3.70	816	104	631
RbMg	This work	1.04	4.17	744	54	664
	63	0.79	4.29	586	47	

RbCa	This work	2.19	4.37	1337	57	922
	43	2.54	4.37	1406	58	
	45	1.75	4.53	921	49	961
RbSr	This work	1.64	4.56	1279	43	972
	42	1.80	4.59	1274	42	
	77		4.67	1041	38	
	76		4.67	1018	36	
	45	1.41	4.72	916	36	1009
	61	1.54	4.60	1073	32	

[a] Polarizabilities along the internuclear axis for the diatomic molecules.

[b] Experimental results.

[c] d_0 values are given.

Table 3 Results for the lowest $^4\Sigma^+$ state, obtained at the MRCI level of theory, including core polarization potentials. Where available, values from the literature are given for comparison

Mol.	**Ref.**	μ_e/D	r_e/Å	D_e/cm^{-1}	ω_e/cm^{-1}	$\alpha_{\parallel}$/a.u.
LiBe	This work	−0.61	5.51	59	25	223
LiMg	This work	−2.52	4.72	222	32	423
LiCa	This work	−4.58	4.59	522	55	1031
	39		4.33	608	67	
	72		5.01	265	30	
LiSr	This work	−6.61	4.59	692	61	1340
	72		4.60	401	47	
NaBe	This work	−0.45	5.84	50	18	229
NaMg	This work	−0.99	5.71	131	21	326
NaCa	This work	−1.62	5.74	237	24	686
NaSr	This work	−2.00	5.81	287	23	1200
KBe	This work	−0.72	5.99	48	16	300
KMg	This work	−1.58	5.73	157	20	338
KCa	This work	−2.50	5.73	343	26	950
KSr	This work	−2.99	5.95	376	22	1523
RbBe	This work	−0.79	6.03	47	15	363
RbMg	This work	−1.66	5.84	170	18	645
RbCa	This work	−2.13	6.05	316	19	922
	43		5.98	336	19	
RbSr	This work	−2.66	6.15	343	16	1491
	42		6.17	390	16	
	77		6.25	329	15	
	77		6.15	336	15	

[a] Polarizabilities along the internuclear axis for the diatomic molecules.

A similar compilation of data for the lowest $^4\Sigma^+$ state is given in Table 3, although fewer theoretical and experimental values are available.¶ In general, the equilibrium radius is larger and the potential depth and vibrational constant are smaller than corresponding quantities for the ground state. For the polarizability no clear trend can be observed.

Trends for AK–AKE Molecules

A diatomic molecule formed by a single alkali atom and a single alkaline earth atom comprises three valence electrons. If excitations of core electrons and ionizations are neglected, these three electrons can only form a doublet or a quartet state. The AK atom, providing one unpaired electron, is always in a doublet state in the asymptotic limit of separated atoms. In contrast, the AKE atom can be either in a singlet or in a triplet state. A doublet state of the molecule is therefore either realized by the combination of a singlet AKE atom with the doublet AK atom, or by the combination of a triplet AKE atom with an anti-aligned electron from the AK atom. Quartet multiplicity can only arise if a triplet state of the AKE atom is combined with an aligned electron on the AK atom. The ground state of the molecule must be a doublet state, since the lowest state of the AKE atom is of singlet character. Only an excited triplet state of the AKE can result in a quartet state. For the AKE atoms considered in this study (Be–Sr), this is a ^{3}P state with excitation energies between 22 000 cm^{-1} and 14 500 cm^{-1}. For Ba and Ra the lowest excitation has ^{3}D character. The lowest excitation of the AK atom is a ^{2}P state with energies ranging from 12 500 cm^{-1} to 17 000 cm^{-1}. The energy difference between the lowest excitation of the AKE (^{3}P) and the AK atom (^{2}P) is plotted in graphics (a) of Fig. 2. Note that there are negative values for combinations of light alkali atoms and heavy alkaline earth atoms. Rather than the ^{3}D-excitations, which are the lowest ones, the ^{3}P-excitations were used in Fig. 2 for Ba and Ra. In plot (b) of Fig. 2 the difference in electronegativities (ENs) is shown with a behavior similar to the differences in excitation energies. A positive value of the electronegativity difference indicates that the electron density is pulled towards the AKE atom. Only combinations of light AK atoms with heavy AKE atoms have negative differences. Within our test set only LiSr has a negative EN difference. The difference in excitation energy or electronegativities characterizes the direction and magnitude of the pull at the electron density. Additionally, the polarizability is important as it measures the flexibility of the electron density. The atomic values for the polarizability range from 37 to 248 a.u. for AKE atoms and from 164 to 313 a.u. for AK atoms,[95] with an increasing value for heavier atoms. The change in the orbital of the AK atom can be expected to be larger than the change for the AKE atom if atoms of similar mass are combined. Such a behavior was observed and is discussed in Section 3.3.

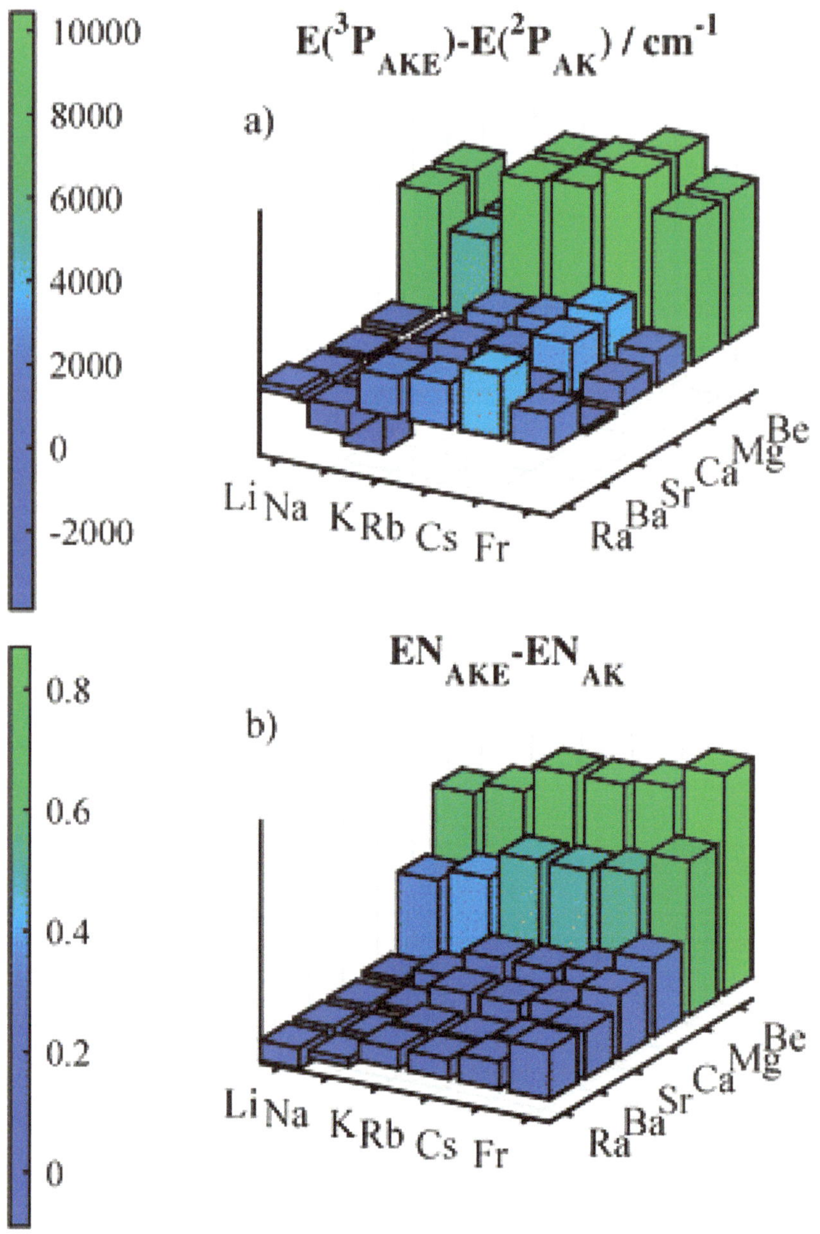

Figure. 2 Differences between AK and AKE atoms for two atomic properties are shown in the figure, in the upper plot energy differences of the first P-excitations[92] and in the lower plot the differences in the electronegativities according to Pauling's scale.[96]

Fig. 3 shows the PEDMs (a) and dissociation energies (b) for the ground state of the 16 AK–AKE molecules within our test set. The smallest PEDM was obtained for LiSr, but unlike the difference in ENs, it still is positive. The largest EN difference for the calculated molecules is found for the RbBe molecule, but LiBe shows the largest dipole moment. The increase by going from light to heavy AK atoms and AKE atoms as seen in the difference of ENs in plot (b) of Fig. 2 is not seen for the PEDMs. For KMg and RbMg small PEDMs are observable inFig. 3. The small PEDMs for Mg containing molecules are probably related to the significantly smaller electronegativity difference for Mg than for Be, although they have similar differences in excitation energies (Fig. 2). Also, the potential depths are different. In plot (b) of Fig. 3 the potential depths are shown and the values for molecules containing Mg are noticeably smaller than for the other AKE atoms, with no other trend being obvious for AKE atoms. However, there is a trend for AK atoms. The potential depth decreases seemingly exponentially in going from light to heavy AK atoms. The largest potential depth was determined for LiCa (2613 cm^{-1}), the smallest one for RbMg (744 cm^{-1}). In contrast to the PEDM and the equilibrium radii, the core polarization potential changes the potential depth in both directions. In cases where the AK atom is significantly lighter than the AKE atom the potential depth is reduced. However, if the AKE atom is lighter, the potential depth is increased. The equilibrium radii increase with inclusion of core polarization potentials. This effect is larger for heavier AK atoms, but shows no recognizable dependence on the AKE atom (ESI†).

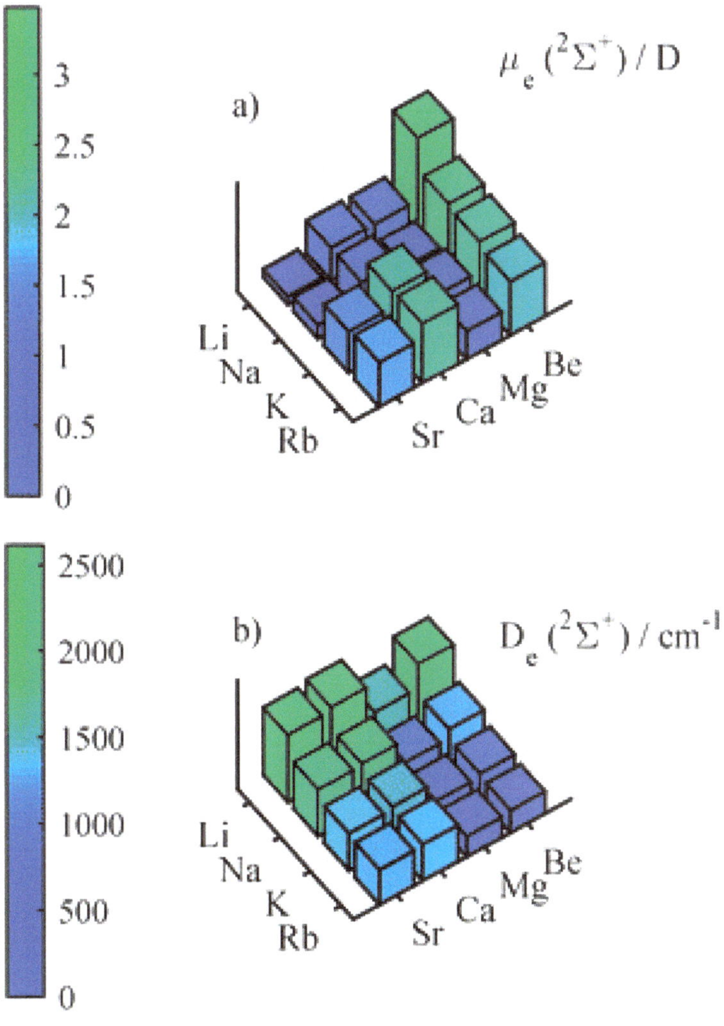

Figure. 3 The permanent electric dipole moment (a) and the potential depth (b) for several AK–AKE molecules in their ground state ($^2\Sigma^+$) are shown.

In contrast to the strictly positive PEDM for the doublet ground states the PEDMs for the lowest Σ^+ state in the quartet manifold are negative, as can be seen in Fig. 4. The PEDMs for the quartet states show a correlation with the difference in ENs (b) in Fig. 2. The most negative PEDM is found for the only molecule with a negative EN difference (LiSr), while small PEDMs are obtained for strongly positive EN differences. Small PEDMs of the $^4\Sigma^+$ state are observable in plot (a) in Fig. 4 for Na containing diatomic molecules including NaBe with the smallest value. This seems to be related to the first excitation energy of Na, which is the largest of all AK atoms. The large excitation energy leads to the small – and for some combinations with Na even negative – differences in the excitation energy, as can be seen in graphics (a) of Fig. 2. The potential depth turns out to be a reliable descriptor for the magnitude of the PEDM, as is evident by comparing plot (a) and (b) of Fig. 4, but is less useful for the estimation of PEDMs in the doublet manifold. For the $^4\Sigma^+$ state the dissociation energy and the PEDM have a correlation coefficient of $R = -0.97$.

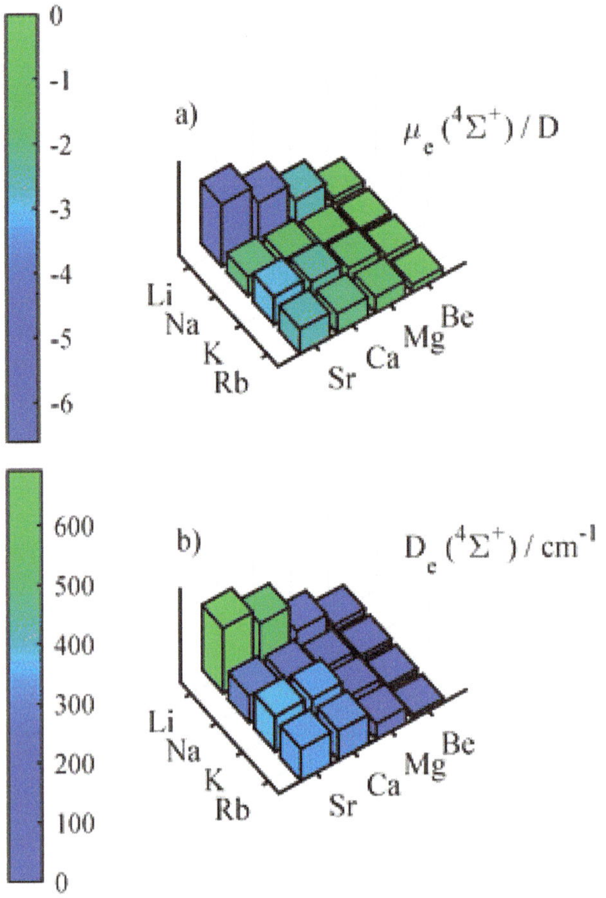

Figure. 4 The upper plot shows the PEDMs for several AK–AKE molecules in the lowest $^4\Sigma^+$state. Note that the axis is reversed so that the highest bar corresponds to the most negative value. The graphics below shows the corresponding dissociation energy for a direct comparison.

The equilibrium internuclear distances of the lowest Σ^+ states in the two multiplicities are given in the Tables 2 and 3. The r_e values for the doublet state are all below 4.57 Å, those for the quartet states lie above this value. The equilibrium distances can be compared to the tabulated covalent[97] and van der Waals[98] radii (ESI). The internuclear distances for the ground state minimum lie in between the covalent and van der Waals values. On average, they are off by 0.4 Å from the covalent radii and by 1.4 Å from the van der Waals radii. Them being closer to the covalent radii indicates a covalent or ionic bond in the lowest state of the doublet manifold. The lowest $^4\Sigma^+$ states have equilibrium internuclear separations that are on average 0.5 Å larger than predicted by the tabulated van der Waals values, which indicates a weak bond.

The fraction of ionic character[48] can be estimated by dividing the determined dipole moment by the dipole moment obtained by shifting one electron by the equilibrium distance: $f_{\text{Ion}} = \frac{\mu}{e \cdot r_e}$. For the ground state, this fraction lies below 21%, except for molecules containing Be, where the values are always above 21%. As is obvious from the method of calculation, this quantity is correlated with the determined dipole moment and should be interpreted with caution. For the quartet states, particularly high values are observed for Li containing molecules.

With finite field calculations, which we performed in order to determine the dipole moment, we could also determine the polarizabilities of the molecules along the internuclear axis. Graphics (a) of Fig. 5 shows the polarizabilities for the ground state, which increase with atomic mass. These values are about 3.7 (with an variation from 3.4 to 4.0) times larger than the mean of the polarizability of the atoms involved. The polarizabilities for the quartet states show stronger variations (from 1.8 to 7.7 of the mean of the atomic pol.) than for the doublet states, as can be seen in graphics (b) of Fig. 5. Furthermore, there is a pronounced increase of polarizability for heavy AKE atoms, but no clear trend for the AK atoms. This dependence is probably related to the lowering of the first excitation energy when going from light to heavy AKE atoms. An increased mixing of the s-orbital of the AK atom and the p-orbital of the AKE atom facilitates a change in the electron density distribution.

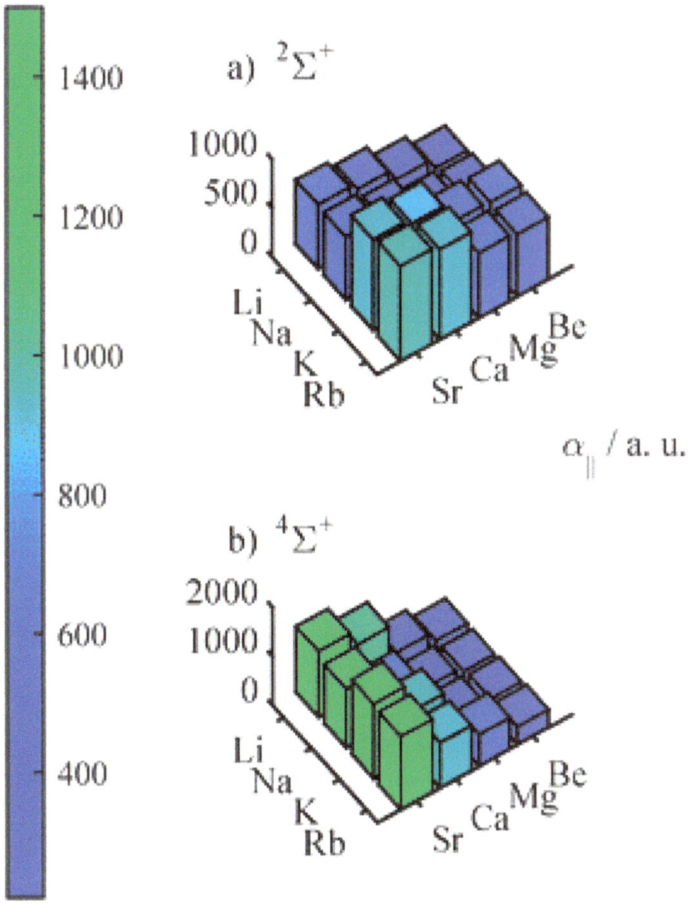

Figure. 5 Finite-field polarizabilities in the doublet ground state (upper graphics) and the lowest $^4\Sigma^+$ state (lower graphics), calculated at the corresponding equilibrium geometry.

Qualitative Picture

Typical potential energy curves and PEDM curves for the lowest Σ^+ states in the two multiplicities are shown in Fig. 6. The lowest $^2\Sigma^+$ is strongly bound with a positive PEDM. The PEDM approaches zero for large internuclear distances as expected for neutral molecules. In contrast, the lowest $^4\Sigma^+$ state is weakly bound with a negative PEDM. In general, the maximum ($^2\Sigma^+$) or minimum ($^4\Sigma^+$) of the PEDM curve is found for internuclear distances smaller than the equilibrium distance. The results displayed in Fig. 6 refer to LiBe, but the same behavior was previously observed for LiCa,[39] RbCa,[42] and RbSr.[43] In the current study, all 16 AK–AKE molecules confirm this trend. However, this behavior is not very pronounced for LiSr, and a different sign of the dipole moment can be expected in the ground state if Ba or Ra atoms are combined with Li. In fact, this has been observed by Gou et al.[64]

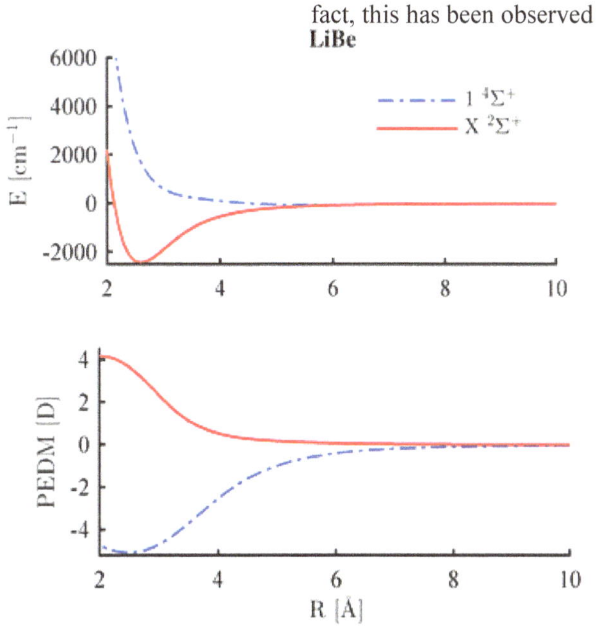

Figure. 6 Potentials and PEDMs of the lowest Σ^+ state in the doublet and quartet multiplicity for the LiBe molecule.

Our findings are summarized in Fig. 7. The ground state, displayed in graphics (a), shows a smaller bond distance. The dipole moment is positive, the vector of the dipole moment points from the AKE atom (++) to the AK atom (+). This is the result of the center of charge of the electrons being closer to the AKE atom than the center of positive charge. The quartet state has a larger bond distance and a negative dipole moment. The center of electron charge is now closer to the AK atom than the center of positive charge. We explain this behavior by the orbital occupation indicated on the right side of Fig. 7. In the ground state (a) the 1st orbital, mainly an s-orbital around the AKE atom,|| is doubly occupied. The 2nd orbital, mainly a mixture of s-orbitals from the AK atom and p-orbitals from the AKE atom, is singly occupied. Because the AKE orbital mixes into it, it is not centered at the AK atom. Therefore, the center of negative charge is closer to the AKE atom, and a positive PEDM is obtained in our coordinate system. This agrees with the situation indicated by the atomic polarizabilities. In general, the AK atoms have larger polarizabilities than AKE atoms. In the ground state, the highest occupied AK orbital should deform more easily and mix with AKE orbitals. The situation is different for the lowest Σ^+ state in the quartet multiplicity (b) in Fig. 7. Since all spins need to be aligned, the Pauli principle does not allow for a double occupation of the 1st orbital. Therefore, one electron is transferred from this orbital into the 3rd orbital, which is a mixture of s-orbitals from the AK atom and p-orbitals from the AKE atom. For this reason, it has a probability density close to the AK atom, which changes the direction of the PEDM compared to the ground state. The 3rd orbital also mixes s- and p-orbital contributions in a similar way as the 2nd orbital does. However, unlike the 2nd orbital, it has an additional node between the atoms, not only at the center of the AKE core.|| It corresponds, at least approximately, to the antibonding counterpart of the 2nd orbital, which explains the small dissociation energy of the lowest $^4\Sigma^+$ state.

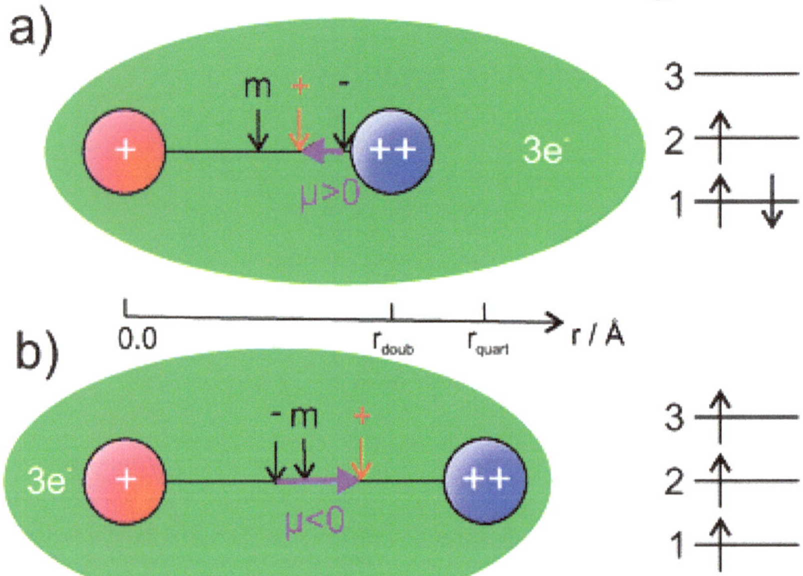

Figure. 7 A visualization of the charge distribution in our set of 16 AK–AKE molecules (no absolute scale). The AK (+) atom is shown in red, the AKE (++) atom in blue. Center of mass, positive and negative charge are indicated by arrows. On the right side the orbital occupation discussed in the text is depicted. Calculated isosurfaces of the corresponding orbitals can be found in the ESI.

Dipole Moment Predictions

In this section we attempt to describe and predict the behavior of the dipole moment in the ground state due to its experimental importance for several applications mentioned in the introduction. The PEDMs of the lowest Σ^+ state of the quartet multiplicity are well correlated with the dissociation energy, which is not the case for the ground state. However, we find the following empirical equation to reproduce the dipole moments surprisingly well for all species under investigation:

$$\mu_e = a_1 \cdot \mathrm{sign}(\Delta \mathrm{EN}) \cdot \sqrt{|\Delta \mathrm{EN}|} + a_2 \cdot \bar{P}_{\mathrm{at}} + \frac{a_3}{D_e} \qquad (1)$$

This model requires the difference of electronegativities ΔEN, the mean of the atomic polarizabilities $\bar{P}_{at}$, and the dissociation energy D_e as input for the evaluation of the dipole moment. We note that it does not have yet a clear theoretical justification, as it was derived from an analysis of correlations between the dipole moment and atomic properties by a simple trial and error approach. However, it agrees very well with the available ab initio data, as can be seen from Fig. 8, although it only requires three fitting parameters a_1, a_2 and a_3. At least the occurrence of the atomic descriptors in the formula can be readily justified. The difference in ENs accounts for the pull on the electron density, and the corresponding contribution to the dipole moment must therefore be related to a charge transfer. We do not have an explanation for the square root dependence, which made it necessary to also introduce the sign function in order to have the sign of the EN difference correlate with the direction of dipole moment. Because the value of the EN difference covers only a region from −0.1 to 0.9 other functions might also be suitable to capture the behavior. Slightly worse results are obtained for functions like $a_1 \cdot \left(\Delta \mathrm{EN} - \frac{\Delta \mathrm{EN}^2}{2}\right)$ or the bounded $a_3 \cdot \tanh(\Delta \mathrm{EN})$. The increase of the PEDM with increasing mean polarizability is easily understood by the greater spatial flexibility of the electron distribution. However, an explanation for the third term in eqn (1) is less straight-forward. The corresponding fitting parameter a_3 takes a negative value in the model, therefore reducing the dipole moment proportional to the inverse of the corresponding dissociation energy. In other words, the weaker the bond, the more effectively this term reduces the dipole moment. This last contribution is not fully understood, but both properties are related to the spatial distribution of the orbitals.

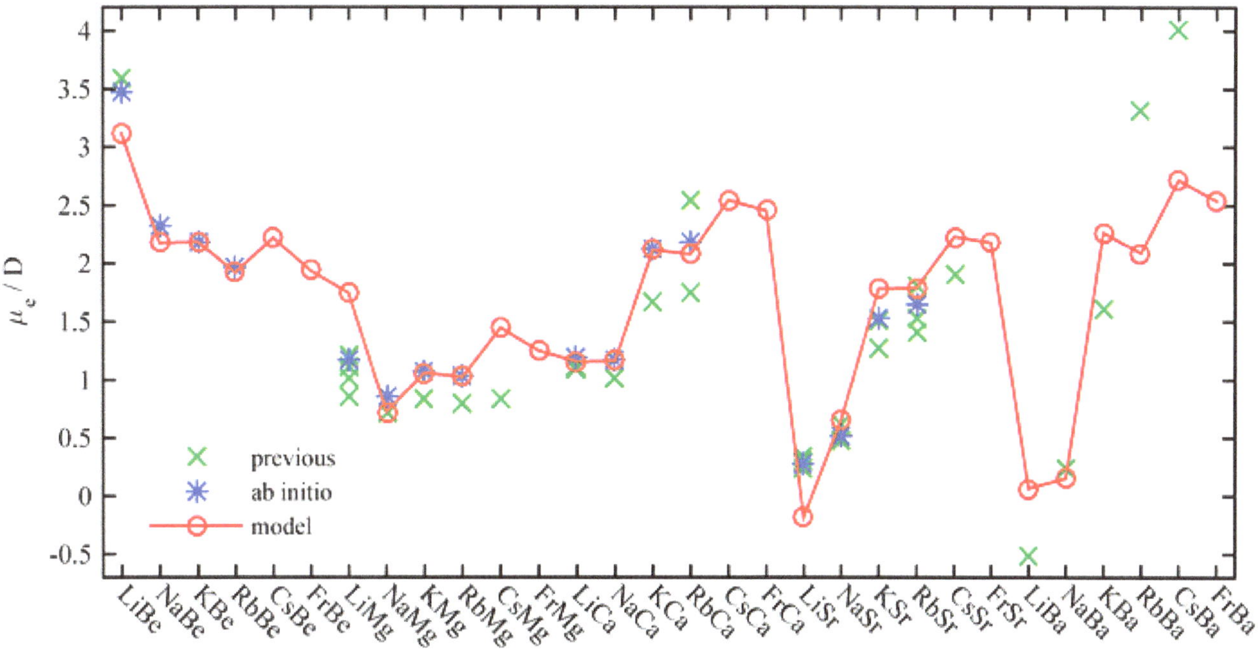

Figure. 8 Dipole moment predictions as described in the text (○). The three parameters of the model were determined by fitting to the ab initio results of this work (*). The resulting model shows good agreement with results from previous theoretical treatments (×) listed inTable 2. Additional values have been added for the CsMg (ref. 61), CsSr (ref. 63) and AK–Ba molecules (ref. 64).

This empirical model allows us to predict the dipole moments for other AK–AKE molecules. However, while the required ENs and the atomic polarizabilities can be found in ref. 95 and 96, the dissociation energies need to be estimated. For Ba, the values from ref. 64 have been used. The potential depth of molecules including Cs or Fr were extrapolated with an exponential function from values for smaller AK atoms, since an exponential dependence was observed for the dissociation energy in Fig. 3. These results, together with previous theoretical results for AK–AKE molecules beyond our test set,[61,63,64] are shown in Fig. 8. We note that Gou et al.[64] neglected the polarization of the cores, which might explain their larger amplitudes, similar to what we observed without core polarization potentials (Fig. 1). A detailed comparison between predicted and previous values is given in the ESI.

CONCLUSION

The lowest Σ^+ states in the doublet and quartet manifold for a selection of 16 AK–AKE molecules were analyzed at the MCSCF + MRCI level of theory. Trends for the PEDMs, equilibrium internuclear separations, dissociation energies, and polarizabilities along the internuclear axis were discussed. We find that the PEDMs of the lowest Σ^+ state in the doublet and quartet multiplicity are pointing in opposite directions for all molecules under investigation, which is in line with previous calculations. This behavior was explained in a molecular orbital picture. We further presented an empirical formula which describes the PEDM in the $^2\Sigma^+$ state as a function of two atomic descriptors (electronegativity, polarizability) and the dissociation energy. After fitting three parameters of the formula to our test set, we used this model to successfully predict the PEDM of several AK–AKE molecules beyond our initial test set.

The PEDM is important for applications of ultracold molecules, since it is the essential parameter for the interaction with microwaves, and, in combination with the polarizability, for the alignment in electric fields and optical lattices. A high PEDM is favorable because lower intensities and field strengths are sufficient to manipulate the molecules. In our study we find the highest PEDM for combinations with Be. However, handling these molecules might be a problem in the experiment due to the toxicity of beryllium. Good alternatives with high PEDM are RbCa, KCa, and RbSr, with the latter being the most promising candidate at present.[74] Our extrapolation formula indicates that CsCa might also be a promising choice.

ACKNOWLEDGEMENTS

The authors gratefully acknowledge support from NAWI Graz. This research was supported by the Austrian Science Fund (FWF) under Grant No. 22962-N20.

REFERENCES

1. J. Ulmanis, J. Deiglmayr, M. Repp, R. Wester and M. Weidemuller, Chem. Rev., 2012, **112**, 4890–4927.
2. G. Quéméner and P. S. Julienne, Chem. Rev., 2012, **112**, 4949–5011
3. I. Bloch and P. Zoller, in Ultracold Bosonic and Fermionic Gases, ed. A. L. F. Kathryn Levin and D. M. Stamper-Kurn, Elsevier, 2012, vol. 5, pp. 121–156.
4. L. D. Carr, D. DeMille, R. V. Krems and J. Ye, New J. Phys., 2009, **11**, 055049.
5. J. Doyle, B. Friedrich, R. Krems and F. Masnou-Seeuws, Eur. Phys. J. D, 2004, **31**, 149–164 .
6. M. Baranov, L. Dobrek, K. Goral, L. Santos and M.

Lewenstein, Phys. Scr., 2002, **T102**, 74–81 .

7. L. Santos, G. V. Shlyapnikov, P. Zoller and M. Lewenstein, Phys. Rev. Lett., 2000, **85**, 1791–1794 .

8. K. Goral, L. Santos and M. Lewenstein, Phys. Rev. Lett., 2002, **88**, 170406 .

9. L. D. Carr, G. V. Shlyapnikov and Y. Castin, Phys. Rev. Lett., 2004, **92**, 150404 .

10. M. A. Baranov, L. Dobrek and M. Lewenstein, Phys. Rev. Lett., 2004, **92**, 250403 .

11. H. P. Buchler, E. Demler, M. Lukin, A. Micheli, N. Prokof'ev, G. Pupillo and P. Zoller, Phys. Rev. Lett., 2007, **98**, 060404 .

12. M. A. Baranov, M. Dalmonte, G. Pupillo and P. Zoller, Chem. Rev., 2012, **112**, 5012–5061 .

13. D. DeMille, Phys. Rev. Lett., 2002, **88**, 067901 .

14. B. K. Stuhl, M. T. Hummon and J. Ye, Annu. Rev. Phys. Chem., 2014, **65**, 501–518 .

15. R. V. Krems, Phys. Chem. Chem. Phys., 2008, **10**, 4079–4092 .

16. E. R. Hudson, C. Ticknor, B. C. Sawyer, C. A. Taatjes, H. J. Lewandowski, J. R. Bochinski, J. L. Bohn and J. Ye, Phys. Rev. A: At., Mol., Opt. Phys., 2006, **73**, 063404 .

17. C. Chin, V. V. Flambaum and M. G. Kozlov, New J. Phys., 2009, **11**, 055048 .

18. M. Kajita, G. Gopakumar, M. Abe and M. Hada, J. Phys. B: At., Mol. Opt. Phys., 2013, **46**, 025001 .

19. M. Kajita, G. Gopakumar, M. Abe and M. Hada, J. Mol. Spectrosc., 2014, **300**, 99–107 .

20. V. V. Flambaum and M. G. Kozlov, Phys. Rev. Lett., 2007, **99**, 150801 .

21. E. R. Meyer and J. L. Bohn, Phys. Rev. A: At., Mol., Opt. Phys., 2009, **80**, 042508 .

22. E. R. Meyer, J. L. Bohn and M. P. Deskevich, Phys. Rev. A: At., Mol., Opt. Phys., 2006, **73**, 062108 .

23. K. Beloy, A. W. Hauser, A. Bohevsky, V. V. Flambaum and P. Schwerdtfeger, Phys. Rev. A: At., Mol., Opt. Phys., 2011, **84**, 062114 .

24. E. R. Hudson, H. J. Lewandowski, B. C. Sawyer and J. Ye, Phys. Rev. Lett., 2006, **96**, 143004 .

25. A. Fioretti, D. Comparat, A. Crubellier, O. Dulieu, F. Masnou-Seeuws and P. Pillet, Phys. Rev. Lett., 1998, **80**, 4402–4405 .

26. T. Takekoshi, B. M. Patterson and R. J. Knize, Phys. Rev. Lett., 1998, **81**, 5105–5108 .

27. T. Takekoshi, B. M. Patterson and R. J. Knize, Phys. Rev. A: At., Mol., Opt. Phys., 1999, **59**, R5–R7 .

28. A. N. Nikolov, E. E. Eyler, X. T. Wang, J. Li, H. Wang, W. C. Stwalley and P. L. Gould, Phys. Rev. Lett., 1999, **82**, 703–706 .

29. A. J. Kerman, J. M. Sage, S. Sainis, T. Bergeman and D. DeMille, Phys. Rev. Lett., 2004, **92**, 153001 .

30. A. J. Kerman, J. M. Sage, S. Sainis, T. Bergeman and D. DeMille, Phys. Rev. Lett., 2004, **92**, 033004 .

31. J. M. Sage, S. Sainis, T. Bergeman and D. DeMille, Phys. Rev. Lett., 2005, **94**, 203001 .

32. M. W. Mancini, G. D. Telles, A. R. L. Caires, V. S. Bagnato and L. G. Marcassa, Phys. Rev. Lett., 2004, **92**, 133203 .

33. D. Wang, J. Qi, M. F. Stone, O. Nikolayeva, H. Wang, B. Hattaway, S. D. Gensemer, P. L. Gould, E. E. Eyler and W. C. Stwalley, Phys. Rev. Lett., 2004, **93**, 243005 .

34. D. Wang, J. Qi, M. F. Stone, O. Nikolayeva, B. Hattaway, S. D. Gensemer, H. Wang, W. T. Zemke, P. L. Gould, E. E. Eyler and W. C. Stwalley, Eur. Phys. J. D, 2004, **31**, 165–177 .

35. A. J. Leggett, Rev. Mod. Phys., 2001, **73**, 307–356 .

36. S. Kraft, F. Vogt, O. Appel, F. Riehle and U. Sterr, Phys. Rev. Lett., 2009, **103**, 130401 .

37. S. Stellmer, M. K. Tey, B. Huang, R. Grimm and F. Schreck, Phys. Rev. Lett., 2009, **103**, 200401 .

38. Y. Takasu, K. Maki, K. Komori, T. Takano, K. Honda, M. Kumakura, T. Yabuzaki and Y. Takahashi, Phys. Rev. Lett., 2003, **91**, 040404 .

39. G. Krois, J. V. Pototschnig, F. Lackner and W. E. Ernst, J. Phys. Chem. A, 2013, **117**, 13719–13731 .

40. F. Lackner, G. Krois, T. Buchsteiner, J. V. Pototschnig and W. E. Ernst, Phys. Rev. Lett., 2014, **113**, 153001 .

41. G. Krois, F. Lackner, J. V. Pototschnig, T. Buchsteiner and W. Ernst, Phys. Chem. Chem. Phys., 2014, **16**, 22373–22381 .

42. J. V. Pototschnig, G. Krois, F. Lackner and W. E. Ernst, J. Chem. Phys., 2014, **141**, 234309 .

43. J. V. Pototschnig, G. Krois, F. Lackner and W. E. Ernst, J. Mol. Spectrosc., 2015, **310**, 126–134 .

44. J. Deiglmayr, M. Aymar, R. Wester, M. Weidemueller and O. Dulieu, J. Chem. Phys., 2008, **129**, 064309 .

45. G. Gopakumar, M. Abe, M. Hada and M. Kajita, J. Chem. Phys., 2014, **140**, 224303 .

46. A. Micheli, G. K. Brennen and P. Zoller, Nat. Phys., 2006, **2**, 341–347 .

47. S. Kotochigova, Phys. Rev. Lett., 2007, **99**, 073003 .

48. E. S. Rittner, J. Chem. Phys., 1951, **19**, 1030–1035 .

49. P. Brumer and M. Karplus, J. Chem. Phys., 1973, **58**, 3903–3918 .

50. R. L. Matcha and S. C. King, J. Am. Chem. Soc., 1976, **98**, 3420–3432 .

51. T. Törring, W. E. Ernst and J. Kändler, J. Chem. Phys., 1989, **90**, 4927–4932 .

52. T. Törring, W. E. Ernst and S. Kindt, J. Chem. Phys., 1984, **81**, 4614–4619 .

53. S. F. Rice, H. Martin and R. W. Field, J. Chem. Phys., 1985, **82**, 5023–5034 .

54. J. Schamps, M. Bencheikh, J. C. Barthelat and R. W. Field, J. Chem. Phys., 1995, **103**, 8004–8013 .

55. H. Schall, M. Dulick and R. W. Field, J. Chem. Phys., 1987, **87**, 2898–2912 .

56. W. E. Ernst and J. O. Schröder, Z. Phys. D: At., Mol. Clusters, 1986, **1**, 103–112 .

57. S. Hou and P. F. Bernath, J. Phys. Chem. A, 2015, **119**, 1435–1438 .

58. C. W. Bauschlicher, S. R. Langhoff and H. Partridge, J. Chem. Phys., 1992, **96**, 1240–1247 .

59. G. Gopakumar, M. Abe, M. Kajita and M. Hada, Phys. Rev. A: At., Mol., Opt. Phys., 2011, **84**, 062514 .

60. S. Kotochigova, A. Petrov, M. Linnik, J. Klos and P. S. Julienne, J. Chem. Phys., 2011, **135**, 164108 .

61. R. Guerout, M. Aymar and O. Dulieu, Phys. Rev. A: At., Mol., Opt. Phys., 2010, **82**, 042508 .

62. M. Aymar, R. Guerout and O. Dulieu, J. Chem. Phys., 2011, **135**, 064305 .

63. L. Augustovičová and P. Soldán, J. Chem. Phys., 2012, **136**, 084311 .

64. D. Z. Gou, X. Y. Kuang, Y. F. Gao and D. M. Huo, J. Chem. Phys., 2015, **142**, 034308 .

65. A. R. Allouche and M. Aubert-Frécon, J. Chem. Phys., 1994, **100**, 938–944 .

66. J. D'Incan, C. Effantin, A. Bernard, G. Fabre, R. Stringat, A. Boulezhar and J. Vérges, J. Chem. Phys., 1994, **100**, 945–949 .

67. N. Boutassetta, A. R. Allouche and M. Aubert-Frécon, Chem. Phys., 1994, **189**, 33–39 .

68. M. Ivanova, A. Stein, A. Pashov, A. V. Stolyarov, H. Knockel and E. Tiemann, J. Chem. Phys., 2011, **135**, 174303 .

69. A. Stein, M. Ivanova, A. Pashov, H. Knöckel and E. Tiemann, J. Chem. Phys., 2013, **138**, 114306 .

70. C. H. Wu, H. R. Ihle and K. A. Gingerich, Int. J. Mass Spectrom., 1983, **47**, 235–238 .

71. A. R. Allouche and M. Aubert-Frécon, Chem. Phys. Lett., 1994, **222**, 524–528 .

72. G. Gopakumar, M. Abe, M. Hada and M. Kajita, J. Chem. Phys., 2013, **138**, 194307 .

73. L. M. Russon, G. K. Rothschopf, M. D. Morse, A. I. Boldyrev and J. Simons, J. Chem. Phys., 1998, **109**, 6655–6665 .

74. B. Pasquiou, A. Bayerle, S. M. Tzanova, S. Stellmer, J. Szczepkowski, M. Parigger, R. Grimm and F. Schreck, Phys. Rev. A: At., Mol., Opt. Phys., 2013, **88**, 023601 .

75. P. S. Zuchowski, J. Aldegunde and J. M. Hutson, Phys. Rev. Lett., 2010, **105**, 153201 .

76. T. Chen, S. B. Zhu, X. L. Li, J. Qian and Y. Z. Wang, Phys. Rev. A: At., Mol., Opt. Phys., 2014, **89**, 063402 .

77. P. S. Zuchowski, R. Guerout and O. Dulieu, Phys. Rev. A: At., Mol., Opt. Phys., 2014, **90**, 012507 .

78. S. Knecht, L. K. Sorensen, H. J. A. Jensen, T. Fleig and C. M. Marian, J. Phys. B: At., Mol. Opt. Phys., 2010, **43**, 055101 .

79. Y. F. Gao and T. Gao, Mol. Phys., 2014, **112**, 3015–3023 .

80. H.-J. Werner, P. J. Knowles, G. Knizia, F. R. Manby, M. Schütz, P. Celani, T. Korona, R. Lindh, A. Mitrushenkov, G. Rauhut, K. R. Shamasundar, T. B. Adler, R. D. Amos, A. Bernhardsson, A. Berning, D. L. Cooper, M. J. O. Deegan, A. J. Dobbyn, F. Eckert, E. Goll, C. Hampel, A. Hesselmann, G. Hetzer, T. Hrenar, G. Jansen, C. Köppl, Y. Liu, A. W. Lloyd, R. A. Mata, A. J. May, S. J. McNicholas, W. Meyer, M. E. Mura, A. Nicklass, D. P. O'Neill, P. Palmieri, D. Peng, K. Pflüger, R. Pitzer, M. Reiher, T. Shiozaki, H. Stoll, A. J. Stone, R. Tarroni, T. Thorsteinsson and M. Wang, MOLPRO, version 2012.1, a package of ab initio programs, 2012, see http://www.molpro.net Search .

81. P. J. Knowles and H.-J. Werner, Chem. Phys. Lett., 1985, **115**, 259–267 .

82. H.-J. Werner and P. J. Knowles, J. Chem. Phys., 1985, **82**, 5053 .

83. H.-J. Werner and P. J. Knowles, J. Chem. Phys., 1988, **89**, 5803–5814 .

84. P. J. Knowles and H.-J. Werner, Theor. Chem. Acc., 1992, **84**, 95–103 .

85. B. P. Prascher, D. E. Woon, K. A. Peterson, T. H. Dunning and A. K. Wilson, Theor. Chem. Acc., 2011, **128**, 69–82 .

86. H. D. Li, H. Feng, W. G. Sun, Y. Zhang, Q. C. Fan, K. A. Peterson, Y. M. Xie and H. F. Schaefer, Mol. Phys., 2013, **111**, 2292–2298 .

87. I. S. Lim, P. Schwerdtfeger, B. Metz and H. Stoll, J. Chem. Phys., 2005, **122**, 104103 .

88. I. S. Lim, H. Stoll and P. Schwerdtfeger, J. Chem. Phys., 2006, **124**, 034107 .

89. W. Müller, J. Flesh and W. Meyer, J. Chem. Phys., 1984, **80**, 3297–3310 .

90. W. Müller and W. Meyer, J. Chem. Phys., 1984, **80**, 3311–3320 .

91. D. A. Fedorov, A. Derevianko and S. A. Varganov, J. Chem. Phys., 2014, **140**, 184315 .

92. Y. Ralchenko, A. Kramida, J. Reader and NIST ASD Team, National Institute of Standards and Technology, Gaithersburg, MD, 2015, available: http://physics.nist.gov/asd.

93. R. Schlachta, I. Fischer, P. Rosmus and V. E. Bondybey, Chem. Phys. Lett., 1990, **170**, 485–491 .

94. K. R. Berry and M. A. Duncan, Chem. Phys. Lett., 1997, **279**, 44–49 .

95. J. Mitroy, M. S. Safronova and C. W. Clark, J. Phys. B: At., Mol. Opt. Phys., 2010, **43**, 202001 .

96. CRC Handbook of Chemistry and Physics, ed. D. R. Lide, CRC Press, 85th edn, 2004 Search .

97. B. Cordero, V. Gomez, A. E. Platero-Prats, M. Reves, J. Echeverria, E. Cremades, F. Barragan and S. Alvarez, Dalton Trans., 2008, 2832–2838 .

98. S. Alvarez, Dalton Trans., 2013, **42**, 8617–8636 .

CHAPTER 8

FUNCTIONAL SYSTEMS WITH ORTHOGONAL DYNAMIC COVALENT BONDS

Adam Wilson , Giulio Gasparini and Stefan Matile

INTRODUCTION

The creation of functional systems from scratch is a central objective in organic chemistry. Covalent, non-covalent and dynamic covalent bonds[1–3] are available as tools for the construction of such "active" molecular or supramolecular architectures (Fig. 1). The importance of both covalent and non-covalent bonds is well appreciated. Among non-covalent interactions, hydrogen bonds, ion pairs, hydrophobic interactions, π–π interactions and cation–π interactions[4] can be named as the basic set. The expansion of this surprisingly limited toolkit attracts much current scientific attention. Leading examples include anion–π interactions[5–7] and halogen bonds,[8,9] the underrecognized counterparts of cation–π interactions[4] and hydrogen bonds, respectively.

non-covalent	dynamic covalent	covalent
hydrogen bonds	disulfides	C-C
ion pairs	hydrazones	C-H
hydrophobic	boronic esters	C-O
π-π	imines	C-N
cation-π	thioesters	C-S
-------------------	-------------------	C-P
		C-B
anion-π	hemi-/acetals	O-H
halogen bonds	hemiaminals	O-N
dipole-dipole	hemithioacetals	O-S
ion-dipole	dithianes	O-P
...	α-aminonitriles	O-B
	nitrones	...
	alkenes	
	enones	
	nitroolefins	
	...	

Figure 1. In supramolecular functional systems, dynamic covalent bonds can combine the advantages of non-covalent and covalent bonds. The most popular members of each class are listed first, whereas promising and in part just emerging alternatives appear below the dashed line.

Citation: Adam Wilson, Giulio Gasparini and Stefan Matile, Functional systems with orthogonal dynamic covalent bonds, DOI: 10.1039/C3CS60342C.

The interest in dynamic covalent bonds may be seen as having originated from the same desire to discover new tools for the creation of function. Dynamic covalent bonds are unique in the sense that they combine the characteristics of covalent and non-covalent bonds.[1–3] Under certain conditions, they can reversibly form and break like non-covalent bonds. Under different conditions, they can be as strong and permanent as covalent bonds. The disulfide bond, for example, is stable under neutral and acidic conditions, but under reductive or basic conditions in the presence of thiolates, disulfides exchange rapidly. Hydrazones are stable under neutral and basic conditions but readily hydrolyze under acidic conditions and exchange in the presence of hydrazides, aldehydes or ketones. Boronic esters are less stable. They hydrolyze quite easily and exchange in the presence of vicinal diols and catechols. Imines, the less stable homologs of hydrazones, are obtained from amines and aldehydes or ketones. Like disulfides, thioesters exchange rapidly with thiols.

Beyond these top five, there exists a rich collection of less frequently used dynamic covalent bonds, each with its own specific characteristics. Efforts to apply dynamic covalent bonds such as hemiacetals,[10] thiohemiacetals[11] or dithianes[12] in functional systems are ongoing. This review will also cover examples using aminonitriles,[13] alkenes that equilibrate with Cope rearrangements[14] and chirality sensors employing hemiaminals.[15] It is also important to emphasize that a certain grey area exists in the definition of the dynamic covalent bonds. The term may be extended to include many bonds that one would normally call covalent, but that become labile under particular, highly specific conditions.

The dual nature of dynamic covalent bonds offers unique advantages for the creation of function. The recent, rapid growth of dynamic covalent chemistry has provided many wonderful examples, the most visible of which is their use in dynamic covalent libraries.[1–3] These libraries are composed of mixtures of cyclic and linear oligomers with rapidly exchanging subcomponents. The presence of templates can shift their composition toward a desired distribution. In contrast to most supramolecular systems assembled with non-covalent bonds, active structures arising from templated dynamic libraries can usually be purified and isolated under conditions that do not permit further equilibration.

The recent progress with dynamic covalent libraries confirms that dynamic covalent bonds are ideal tools for generating diversity and exploring the possibilities of template and substrate-induced amplification. Dynamic covalent bonds further excel when used in challenging topics such as self-repair, self-healing, adaptive self-sorting, replication and transcription, often operating with proximity effects and sometimes leading to highly sophisticated multicomponent architectures.[1] Many elegant applications of these advantages to create functional systems with single dynamic covalent bonds exist, from catenanes that bind acetylcholine[16] to enzyme inhibitors or sensors of the heterogeneity of lipid bilayer membranes.[17]

Lessons from nature provide superb examples of how dynamic covalent bonds may be used in functional systems. Disulfide exchange during protein folding is arguably the most impressive illustration of dynamic covalent bonds at work in biology, beautifully exploited in daily life by the hairdresser.[18] The same chemistry accounts for many of the properties of rubber.[18] In addition, thioester exchange is employed to perfection in polyketide biosynthesis,[19] dynamic imines are essential in the chemistry of vision, dynamic enone chemistry ensures that hot spices feel really hot,[20] hemiacetal chemistry is not limited to sugars but also plays an important role in membrane-active natural products such as amphotericin, nystatin or monensin, and so on.

In functional systems, non-covalent bonds are routinely used together. For example, the self-assembly of DNA duplexes involves a combination of hydrogen bonding, π–π interactions, hydrophobic interactions and charge repulsion. The orthogonality of hydrophobic interactions and hydrogen bonding, increasing and decreasing with increasing polarity of the environment, respectively, is of particular importance for the function of double-stranded DNA. All principal non-covalent interactions contribute to protein secondary and tertiary structures, and the design of synthetic functional systems is inconceivable without the insightful combination of orthogonal non-covalent bonds. One might thus assume that the use of orthogonal dynamic covalent bonds would be commonplace in synthetic functional systems. Surprisingly, quite the opposite is true. Only a few examples presently exist in the literature. In sharp contrast to this rare combination of two or more dynamic covalent bonds, dynamic covalent bonds are routinely used together with non-covalent bonds. Sequence-adaptive thioester peptide nucleic acids (PNAs) discussed in the following can serve as a telling example for this approach to generate function.[21] Coordination to metals is also very frequently used in combination with dynamic covalent bonds.

The concept of orthogonality has already appeared in a number of reviews with emphasis on the combination of dynamic covalent bonds with coordination chemistry or in the general context of reactivity and supramolecular interactions.[22–24] However, compelling experimental support of the concept of organic orthogonal dynamic covalent bonds is very rare and recent. The following will comprise a brief summary of structural studies that demonstrate the scope and applicability of orthogonal dynamic covalent chemistry, followed by a review of functional systems with orthogonal dynamic covalent bonds. Hopefully, these illustrations will stimulate the more frequent use of orthogonal dynamic covalent bonds to create function.

ORTHOGONAL DYNAMIC COVALENT BONDS

The concept of orthogonality of dynamic covalent bonds first appeared in 2001 in combination with coordination to metals.[25] In 2005, two true, *i.e.*, organic, orthogonal dynamic covalent bonds were placed for the first time in dynamic libraries.[26] In this study, disulfide exchange produced acyclic or cyclic oligomers, whereas thioester exchange capped the acyclic oligomers and prevented cyclization. The longest oligomer in a mixture of at least eight molecules was found to be compound **1** (Fig. 2). Three years later, a similar study was published, focusing on disulfides and hydrazones.[27] A simple molecule equipped with a disulfide bond and two hydrazides was used. By exposing this molecule to an aqueous solution containing both aldehydes and thiols, hydrazone or disulfide exchange could be turned on and off simply by changing the pH (2.5 and 8.5, respectively). Moreover, the authors were able, using intermediate pH (4.5) and in the presence of a catalyst (aniline), to force both moieties to undergo dynamic exchange simultaneously, albeit at a lower rate. They also pointed out that the system reached different product distributions depending on the order of activation of the two exchange reactions.

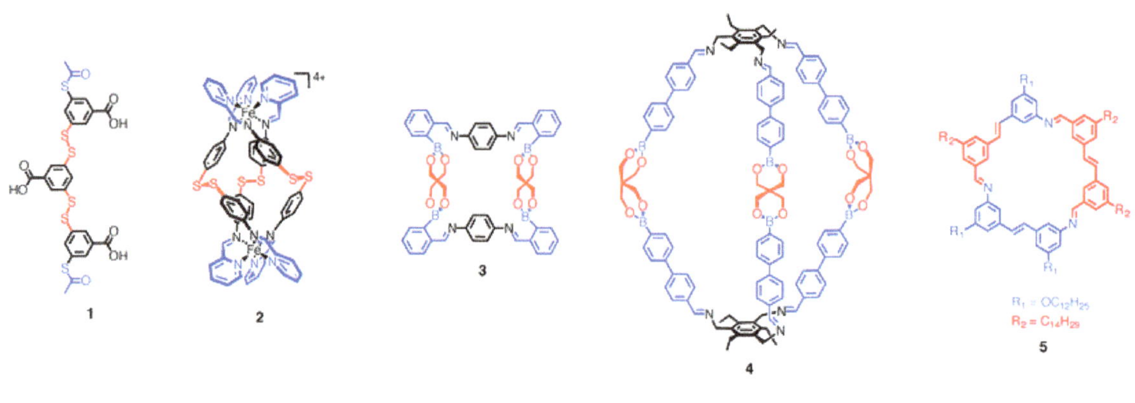

Figure 2 Early and advanced architectures with orthogonal dynamic covalent bonds.

A similar approach was used in a study published back-to-back with this report. Here, a building block carrying both a protected aldehyde and a disulfide bond was introduced to study the exchange of thiols and hydrazides at different TFA/TEA ratios.[28] The results confirmed that the two chemistries could be merged, opening the field of dynamic libraries in which the two exchange processes can be carried out sequentially and repetitively. In a follow-up study, a dynamic combinatorial library containing three different exchangeable motifs – hydrazones, disulfides and thioesters – was introduced.[29] While the hydrazones were dynamic in the presence of a catalytic amount of acid, the disulfides and thioesters only underwent exchange simultaneously under basic catalysis. The dynamic covalent library was generated by exposing a building block, containing both a disulfide bond and an aldehyde, to an excess of two different hydrazides, a thioester and a free thiol. Two different exchange protocols were used. In the first, hydrazone exchange in acidic conditions was followed by thiol–thioester exchange on the addition of an excess of base. In the second, the library was first exposed to base and then an excess of acid was added. The final compositions of the dynamic combinatorial library obtained using the two different chemical pathways were again different, due to the fact that product distributions can be shifted away from the one corresponding to the overall energy minimum.

The first application of orthogonal dynamic covalent chemistry in a functional system, an NMR shift reagent based on imines and boronic esters, was presented in a memorable departmental lecture by Tony James in Geneva on 18 January 2007.[30,31] This breakthrough inspired the construction of multicomponent architectures with orthogonal dynamic covalent bonds in the group of Jonathan Nitschke, then in Geneva;[32,33] it also led to the use of orthogonal dynamic covalent bonds in porous biosensors in the Matile group (see below).[34] The same year, Otto and Nitschke reported the triple helix **2**, combining orthogonal dynamic disulfide and imine bonds with coordination to iron.[32] Soon after, the Nitschke group also reported metal-free architectures constructed from imines and boronic esters, including macrocycle **3**.[33] From there, the chemistry of multicomponent architectures with orthogonal dynamic covalent bonds was advanced to a very high standard by the group of Kay Severin in Lausanne.[35–37] The giant macrobicycle **4** is just one of many marvelous structures reported over the years.

Most recently, macrocycles of different shapes and sizes have been synthesized using two different orthogonal dynamic covalent reactions, *i.e.*, imine and olefin metathesis.[38] For this study, a small collection of vinyl anilines and vinyl benzaldehydes was used. The synthesis of macrocycles such as **5** was tried by simply mixing these compounds in the presence of TFA (trifluoroacetic acid; to improve imine exchange) and Grubbs–Hoveyda second-generation catalyst. Unfortunately, even if the conversion was higher than 95%, polymeric species predominated. Good results were obtained by running the two different reactions in sequence, *i.e.*, imine metathesis followed by olefin metathesis. The amines and aldehydes were allowed to react until the imine formation was complete; then Grubbs–Hoveyda catalyst was added to the system, giving the desired macrocycles in good yield. Following this procedure several different macrocycles were synthesized, their size and shape tuned simply by selecting the appropriate building blocks.

These are just a few examples of structural studies with orthogonal covalent dynamic bonds. However, there are not many more in the literature. In the following, the use of orthogonal dynamic covalent bonds to build systems with interesting functions will be described in more detail. This summary is almost comprehensive because the literature is not very rich on this topic: the field is in its early stages and hopefully will grow in the future.

FUNCTIONAL SYSTEMS WITH ORTHOGONAL DYNAMIC BONDS

Molecular Motion

The molecular "walkers" developed by the Leigh group are one of the most impressive examples of orthogonal dynamic covalent bonds in action (Fig. 3).[39] The objective is to move a molecular component along a molecular track in one direction. Several molecular walkers had already been realized based on DNA nanotechnology.[40,41] To perform the same task with small molecules, the introduction of orthogonal dynamic covalent bonds was the key. Starting with oligomer **6**, the "walker" subunit is on the left side, bound to the track with one disulfide and one hydrazone bridge (Fig. 3, red). Disulfide exchange from the first arene to the third arene from the left is unfavorable as long as the double bond between the second and the third arene is in *trans* configuration. Light-induced isomerization from *trans* to *cis* enables directional disulfide exchange under basic conditions from arene one in **7** to arene three in **8**. According to the analysis by nuclear magnetic resonance (NMR) spectroscopy, this process occurs with an efficiency of greater than 60%. Light-induced isomerization back from *cis* to *trans* adds strain to the central macrocycle of **9**. This strain can be released by hydrazone exchange under acidic conditions. In this directional process, the walker moves from arene two in **9** to arene four in **10**. The yield of this translocation is still above 50%. Control experiments confirm that the correct sequence of operations is needed for directional molecular motion. The walker can be made to walk in the opposite direction simply by reversing the sequence of stimuli. The perspective to see the same process taking place on a longer polymer is fascinating.

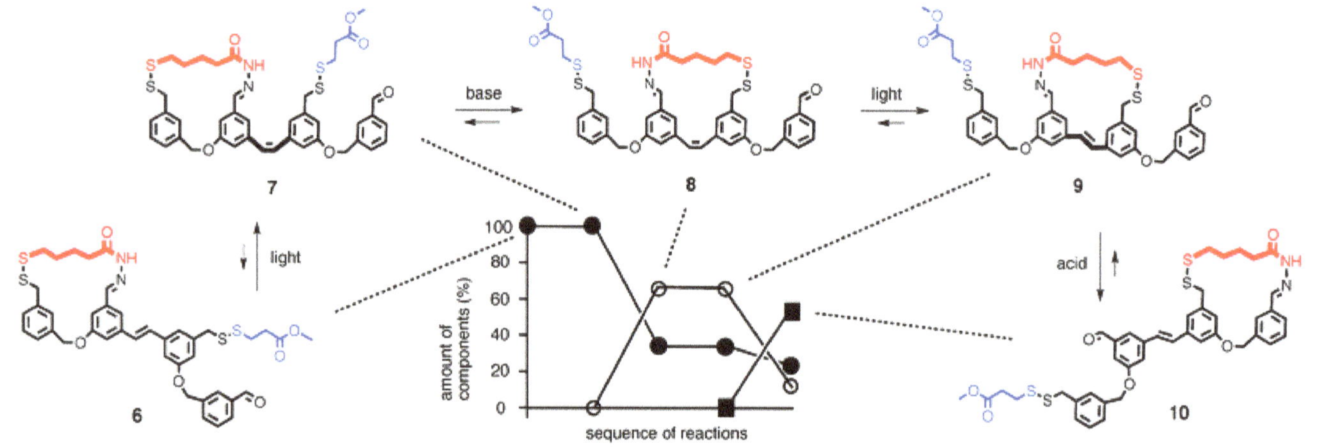

Figure 3 Directional molecular motion achieved with orthogonal dynamic disulfide and hydrazone bridges and photoinduced olefin *cis–trans* isomerization.

Directional walking down the molecular track in **6** is the highlight of a series of studies on molecular motion from the group of David Leigh. Earlier reports focused on the orthogonality of disulfide and hydrazone exchange[42] and random walking without directionality.[43,44] In a complementary system, random walking was explored by the spontaneous intramolecular migration of α-methylene-4-nitrostyrene along oligoamine tracks.[45,46] This system operates with an elegant reversible intramolecular Michael–retro-Michael addition mechanism. The presence of orthogonal dynamic covalent bonds would presumably be useful to achieve directional walking in this system, although attraction of the walker fragment toward a thermodynamic sink has been demonstrated.

"Living" Organic Materials, Multicomponent Photosystems

With regard to functional materials, expectations from dynamic covalent bonds are high because their "living" nature makes them ideally suited for easily introducing structural complexity and, at the same time, promises access to challenging topics such as self-repair, self-healing, adaptive self-sorting, replication and transcription. Self-healing is very nicely documented in a recent system with orthogonal dynamic covalent bonds.[47] Disulfide-containing bishydrazones and trialdehydes were used to generate polymer network **11** where two dynamic bonds are present, *i.e.*, disulfide and hydrazone bridges (Fig. 4a). The resulting dynamic hydrogels undergo hydrazone and/or disulfide exchange to repair damages, in acidic and basic environments respectively. The polymer retains its self-healing character even in neutral conditions (where both hydrazone and disulfide exchange are kinetically hampered) if a catalytic amount of base (aniline) is added during the preparation. Moreover, this new material displays sol–gel transition in response to pH and/or redox external stimuli.

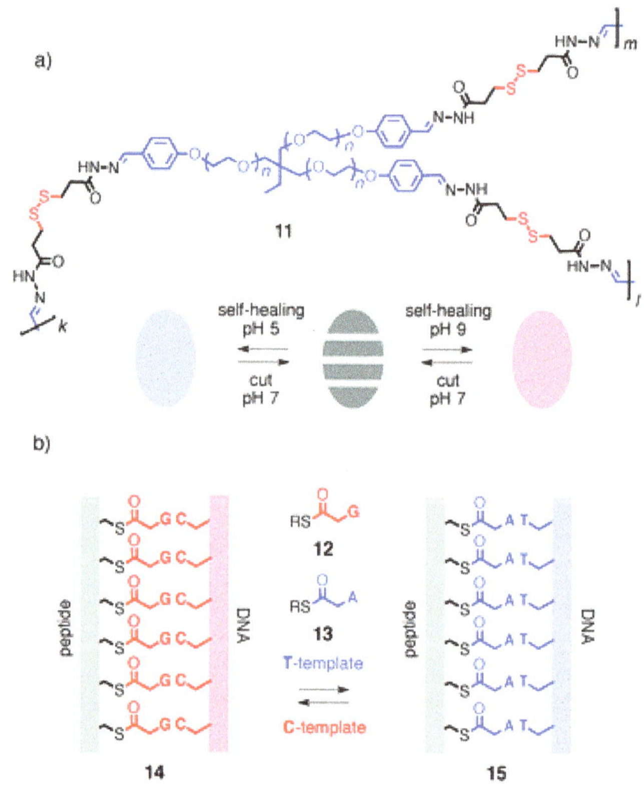

Figure 4 (a) Supramolecular systems with orthogonal dynamic bonds that form gels that, after being cut at pH 7 (grey, striped symbol), can self-heal under basic conditions (disulfide exchange, red symbol) or acidic conditions (hydrazone exchange, blue symbol). (b) Self-repair has also been nicely demonstrated with single dynamic thioesters in sequence adaptive peptide nucleic acids (G = guanine, C = cytosine, A = adenine, T = thymine).

Sequence adaptive thioester PNAs must be mentioned at this point as examples of dynamic covalent chemistry applied to function. Although they operate with only one dynamic bond in combination with the weak interactions involved in DNA technology (Fig. 4b),[21] they nevertheless provide the perfect example to illustrate self-repair with dynamic covalent bonds at the molecular level. Thioester PNAs were constructed from peptides with a cysteine in every second position. Thioester exchange with 12 and 13 randomly attached the two nucleobases (adenine A or guanine G) along the peptide scaffold in an overall ratio of 1 : 1. Thioester exchange with 12 and 13 in the presence of a single-stranded DNA dC_{20} generated duplex 14 with a G-rich PNA (G : A ratios up to 87 : 13). Templated thioester exchange with the complementary dT_{20} gave the duplex 15 with A-rich PNAs (G : A = 11 : 89). In the transcription plot for increasing cytosine C in the DNA template against increasing A in the resulting PNA, a linear dependence was found. The key experiment for self-repair was designed as follows. An A-rich PNA was prepared by incubation with a dT_{20} template. The dT_{20} template was then replaced by a dC_{20} template, and thioester exchange with 12 and 13 was reinitiated. A spectacular inversion of the G : A ratio from 12 : 88 to 70 : 30 demonstrated the existence of self-repair in the presence of dynamic covalent thioester bonds. Similar fidelity was observed for the sequence specific formation of oligoimines along complementary DNA templates.[48]

The advantages of using dynamic covalent bonds to build surface architectures have begun to be recognized only recently.[49,50] Orthogonal dynamic covalent bonds are particularly attractive for growing functional multicomponent architectures directly on solid surfaces. For this purpose, self-organizing surface-initiated polymerization (SOSIP)[51,52] and templated stack exchange (TSE) [53] have been introduced as general synthetic methods that operate with disulfide exchange and hydrazone exchange, respectively (Fig. 5 and 6).

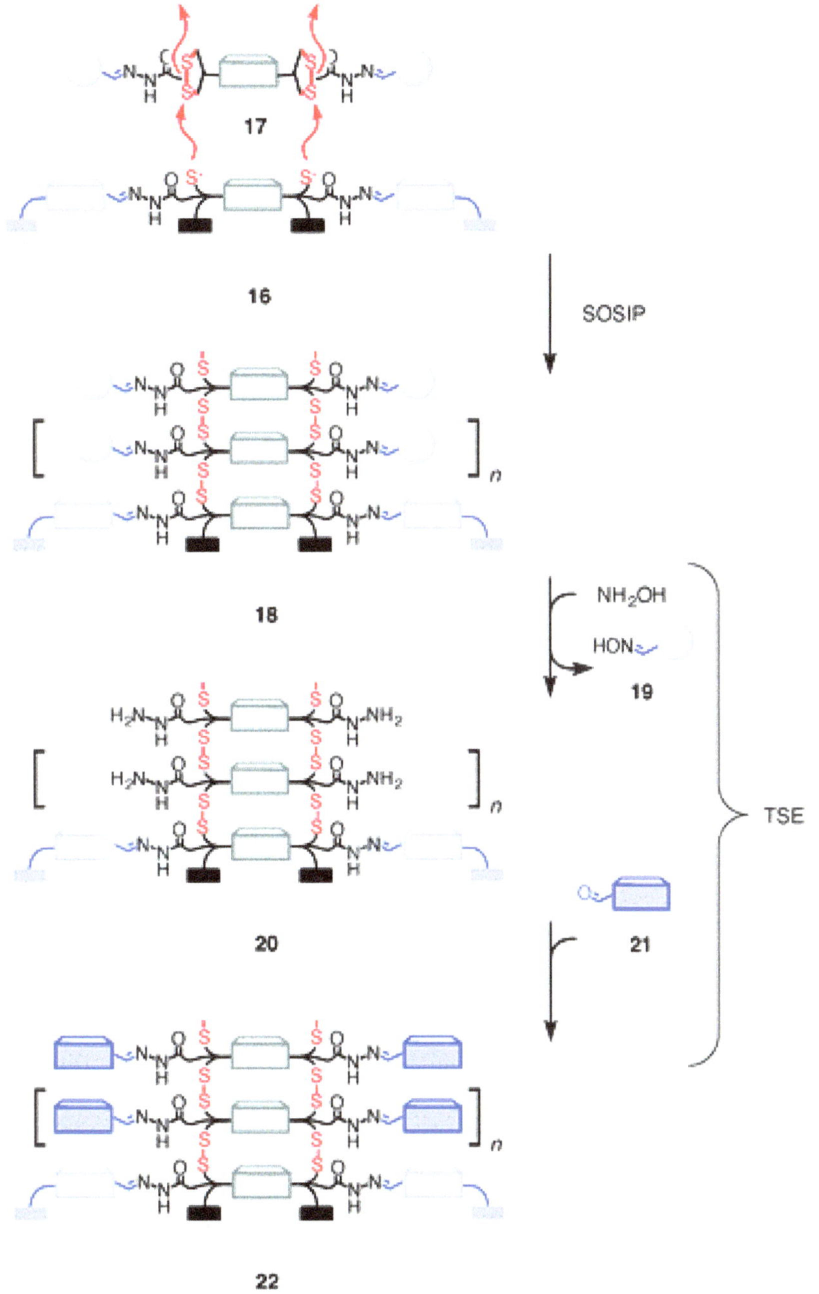

Figure 5. General synthetic access to functional multicomponent surface architectures is secured by self-organizing surface-initiated polymerization (SOSIP) using disulfide chemistry followed by templated stack exchange (TSE) with orthogonal hydrazone chemistry.

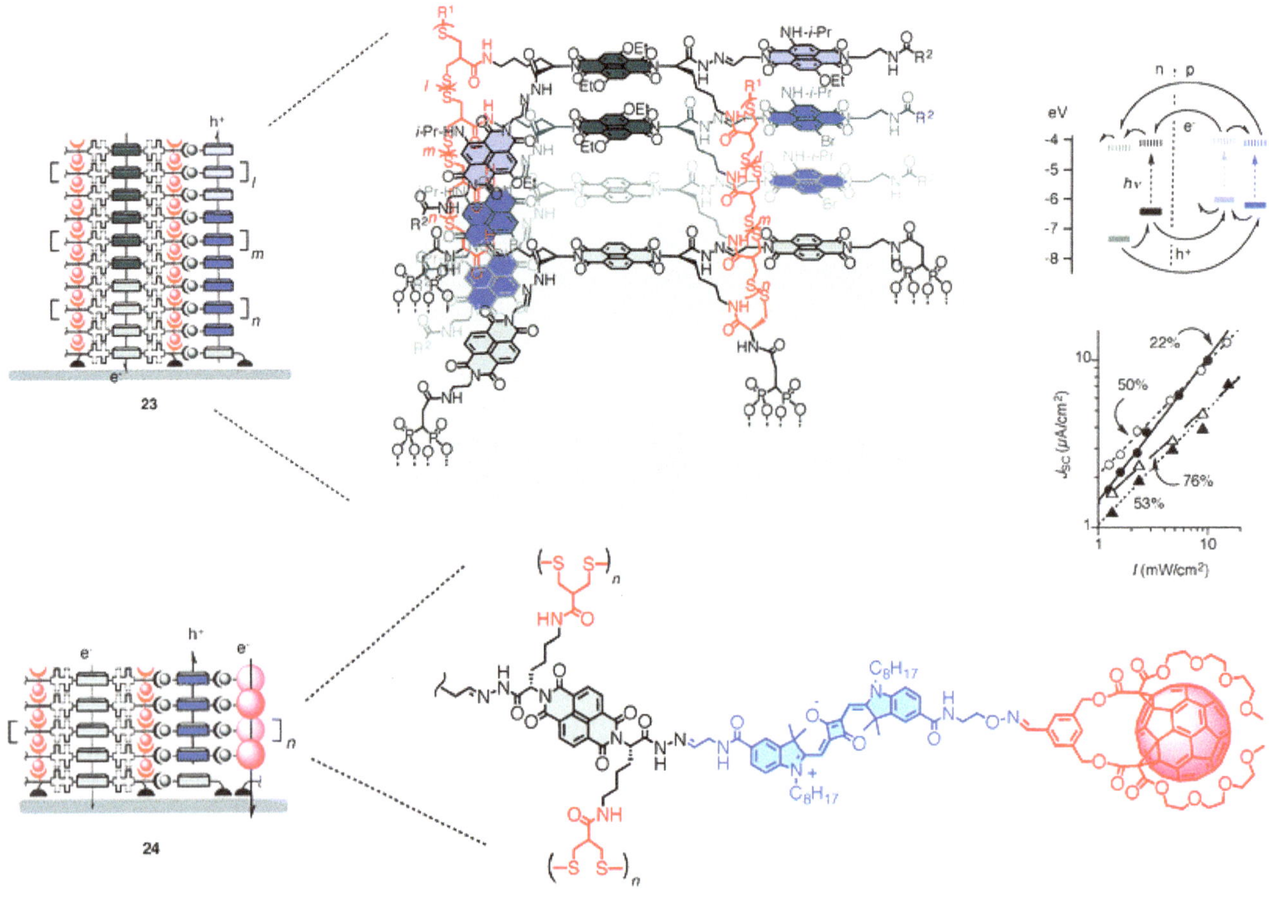

Figure 6. Representative double-channel photosystems with antiparallel redox gradients (top) and triple-channel photosystems (bottom) obtained by SOSIP–TSE on ITO surfaces (grey). Frontier molecular orbital levels and recombination efficiencies are given for **23** (filled circles) in comparison with controls without (empty circles, filled triangles) and misoriented gradients (empty triangles). Adapted from ref. 53 with permission, © 2011 American Chemical Society.

In SOSIP, initiator **16** and propagator **17** are required to contain a functional component in the middle (naphthalenediimide core) able to generate π-stacks, so they can transport electrons efficiently. This core is embedded in self-organizing (lysine-derived diamides) and polymerizing components (cysteine in initiator **16** and a strained cyclic disulfide in propagator **17**). The initiator incorporates two diphosphonate "feet" that allow it to bind to an oxide surface (usually indium tin oxide, ITO), after which its thiols are deprotected and deprotonated to initiate SOSIP. Self-organization is expected to place the two strained cyclic disulfides in propagator **17** directly on top of the initiator's thiolates. Ring-opening disulfide exchange binds the propagator to the surface and regenerates two thiolates for continuing SOSIP. The resulting ladderphane[54] surface architectures **18** – smooth, self-organized over long distances down to the molecular level without significant defects – have been characterized in many variations by many different methods.

Quite extensive screening of other existing methods suggests that the newly invented surface-initiated disulfide-exchange polymerization is an excellent means for constructing complex surface architectures in a facile and general way. Consistent with expectations from dynamic covalent chemistry, the occurrence of self-repair during SOSIP has been demonstrated.[52] The dynamic nature of the disulfide bond also accounts for intrinsic templation efficiencies up to 97% achieved for self-sorting when SOSIP is carried out with a mixture of different propagators (co-SOSIP).[55] Similar to templated polymerization in solution (*e.g.*, sequence-adaptive PNAs **14** and **15**,[21,48] or the central dogma in biology), the development of synthetic methods that operate on the transcription of information will likely be unavoidable for the creation of highly sophisticated multicomponent surface architectures.

Orthogonal dynamic covalent bonds have been the key to the introduction of TSE, another synthetic method conceived to build increasingly complex multicomponent surface architectures. For TSE, SOSIP is performed using initiator and propagator monomers bearing hydrazone-based cleavable templates that become incorporated into the π-stacks formed (blue spheres in **18**, Fig. 5). The templates along the SOSIP stack in **18** are then removed with excess hydroxylamine, releasing **19**. The holes left behind in **20** are filled with aldehydes **21** of free choice. The hydrazone exchange used for TSE is orthogonal to the disulfide exchange used for SOSIP. Dynamic covalent chemistry is thought to be as essential for TSE as for SOSIP. Multiple equilibria will be expected when drilling the deep pores in the surface architectures **20**. The obtained architectures **22** contain co-axial stacks of variable nature, a motif that is of high interest in materials science, particularly optoelectronics, including organic solar cells.

The power of SOSIP–TSE has been exemplified with many variations.[53,55–58] Highlights include double-channel photosystem **23** with antiparallel redox gradients (so-called

OMARG-SHJs, supramolecular *n*/*p*-heterojunctions with oriented multicomponent antiparallel redox gradients, Fig. 6).[53] Thanks to orthogonal dynamic covalent bonds, the directional synthesis of surface architectures as complex as **23** is fairly straightforward. It begins with SOSIP using a propagator based on a colorless, unsubstituted naphthalenediimide (NDI) core. To create a redox gradient in the SOSIP stack, the ITO electrode is simply dipped into a new solution of a propagator based on a yellow, bis-ethoxy-substituted NDI core. Both NDIs transport electrons along their stacks. The NDI cores› respective LUMO energies direct the flow of electrons toward the solid surface (LUMO: lowest unoccupied molecular orbital). With TSE, a two-component stack of red NDIs is installed with an oriented antiparallel gradient that directs the flow of holes in the HOMOs (highest occupied molecular orbitals) away from the solid surface. Once light-generated excitons charge-separate in photosystem **23**, the resulting electrons and holes are transported in opposite directions in a manner reminiscent of biological photosystems. The functional importance of antiparallel gradients was confirmed by the dependence of the short-circuit photocurrent density J_{SC} on the power of irradiation. The recombination efficiency, calculated from the slope, is as low as 22% in photosystem **23**. With only one or no gradient, recombination efficiencies rise to 50–53%; with misoriented gradients, recombination becomes dominant.

Another advance in the evolution of SOSIP–TSE is the very recent construction of triple-channel architectures such as **24** (Fig. 6).[58] In this architecture, central SOSIP stacks are built with NDIs using disulfide exchange chemistry as described. A central squaraine channel is attached by TSE using hydrazone exchange. The third channel composed of fullerene stacks is attached *in situ* before TSE *via* an oxime bond. Oximes are much more stable than hydrazones or imines and often not considered as dynamic covalent bonds. Compared to the poorly active photosystem with one NDI and one squaraine channel only, the addition of the fullerene channel increased photocurrent generation more than 10 times.

Very recently, orthogonal dynamic covalent bonds have been employed in the construction of resin-bound dynamic combinatorial libraries.[59,60] These libraries on solid support have been introduced to simplify the analytical problems associated with solution-phase libraries. In a typical screening strategy, one half of a dimer is immobilized on a synthesis resin; its complementary half is selected from a mixture in solution. The binding of fluorescent target molecules to the solid library can then be easily used to identify biologically active inhibitors.[59] To expand this approach to libraries of three-component inhibitors, orthogonal dynamic covalent disulfide and hydrazone bridges have been tested by the Miller group in a proof-of-principle study.[60] Good conditions to form, isolate and disassemble trimers such as **25** in solution and on solid support have been determined (Fig. 7).

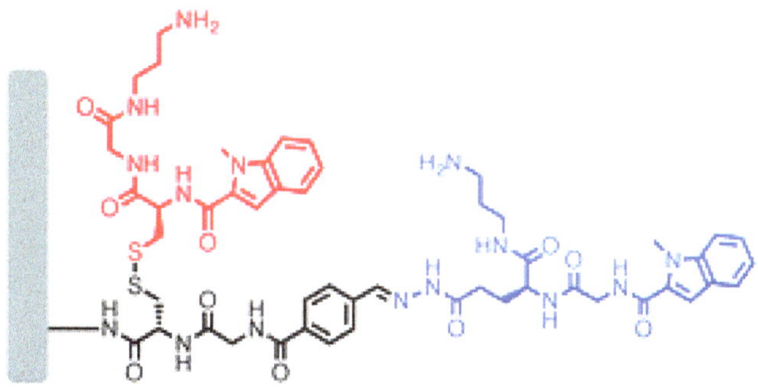

25

Figure 7. A model product for a resin-bound dynamic covalent three-component library with orthogonal disulfide and hydrazone bridges.

Molecular Recognition and Transformation

Freely exchanging dynamic combinatorial libraries can explore large regions of chemical space, allowing them to find the global thermodynamic minimum. When this thermodynamic minimum involves a supramolecular binding event to a particular guest, the system can rearrange so that it contains the best possible host for that guest. The work of optimization is effectively done by the library, often returning surprising structures as a result.

The scope of this process may be appreciated by considering the range of substrates for which receptors have been found. The Furlan group used a two-step evolutionary process to develop the macrocyclic lithium receptor **26** (Fig. 8a).[61] A hydrazone- and disulfide-bearing monomer was allowed to equilibrate under hydrazone-exchange conditions in the presence of lithium bromide. The resulting macrocycle was then subjected to a second round of optimization in which disulfide exchange was used to modify the macrocycle›s periphery. The surprising degree of lithium incorporation of macrocycle **26**, as determined by mass spectrometry, was attributed to rigidification of the macrocycle by the bridging disulfide bond. This bridge is thought to constrain the macrocycle into a conformation ideal for lithium binding.

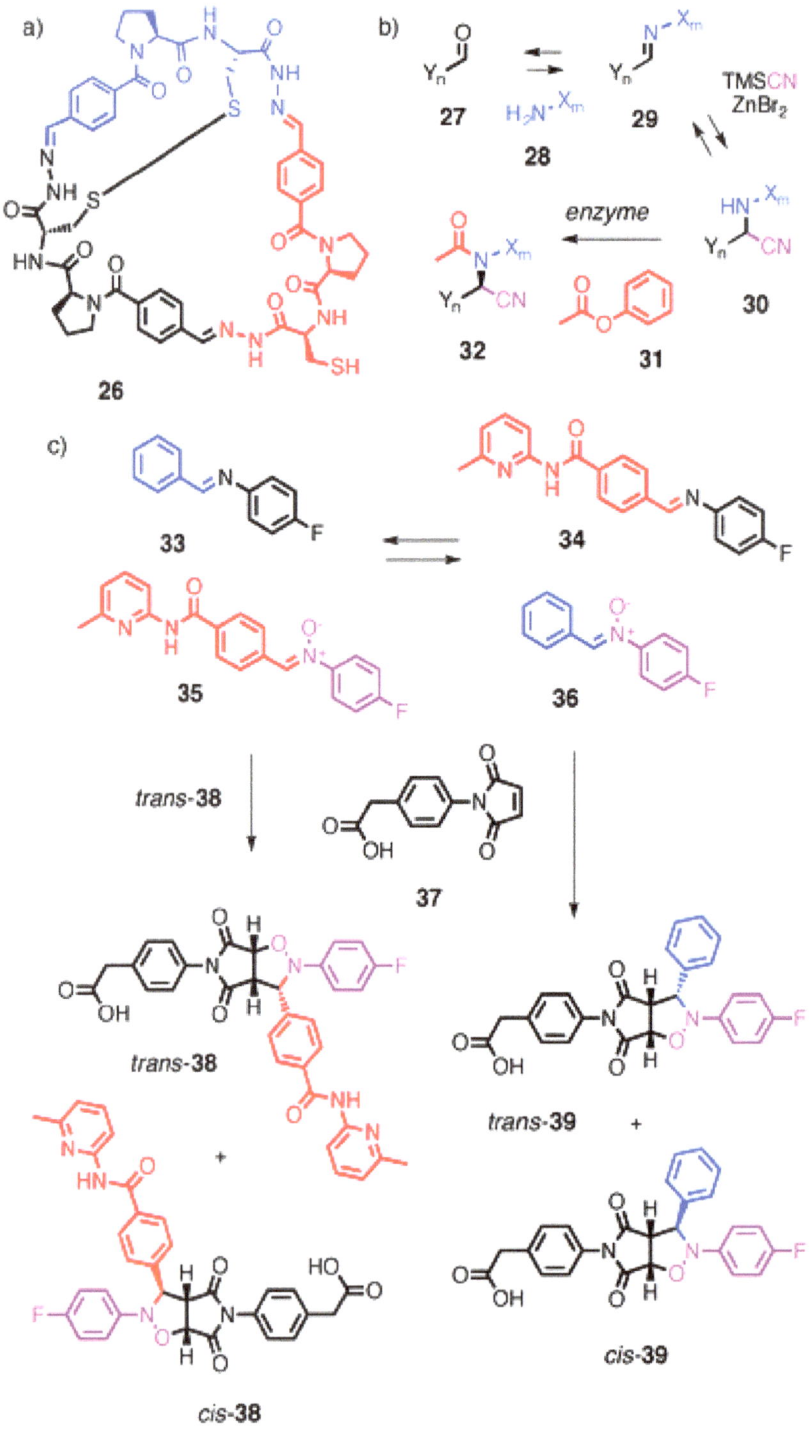

Figure 8. Molecular recognition and transformations with formal orthogonal dynamic covalent bonds.

Combinatorial chemistry and dynamic kinetic resolution have been elegantly employed to form carbon–carbon bonds around a stereogenic center with high enantiopurity (Fig. 8b).[62] A mixture of aldehydes 27 and amines 28 under dynamic exchange conditions resulted in a combinatorial library of imines 29. Addition of a cyanide source, acetic acid and $ZnBr_2$ effectively coupled the reversible transimination to a reversible Strecker reaction, resulting in a combined library of imines 29 and Strecker adducts 30, all freely exchanging at equilibrium. While the library initially generated through reversible transimination was made up exclusively of achiral imine molecules, the Strecker adducts 30 are chiral. The library formed was a racemic mixture, and as such was amenable to dynamic kinetic resolution. A chiral lipase derived from *Pseudomonas cepacia* was used to catalyze an irreversible reaction with phenyl acetate 31, collapsing the library into just three products 32. The library generated chemical diversity and populated the chemical space, allowing the enzyme to select from this library the members which react with lowest activation energy barriers. This is a nice example of chemical transformations with formal orthogonal dynamic covalent bonds because they are coupled in series rather than in parallel as usual.

In a system reported by Sadownik and Philp, the catalyst and substrate were identical, and a library with formally orthogonal dynamic covalent bonds was used to find the ideal autocatalyst (Fig. 8c).[63] A dynamic combinatorial library of two imines 33 and 34 and two nitrones 35 and 36 was allowed to react with maleimide 37 to form a variety of possible products 38 and 39. Of these products, *trans*-38 was shown to be an excellent catalyst for its own formation, accelerating the cycloaddition reaction between 35 and 37 by a factor of more than 100. Because the supply of 35 is continually being renewed through dynamic exchange, autocatalyst *trans*-38 is able to self-replicate at the expense of other species until the cycloaddition product mixture consists almost entirely of *trans*-38. This minimalist dynamic, autocatalytic system is not only of interest to the development of functional materials but also has a bearing on theories of the origin of life.

Formal orthogonal dynamic covalent bonds have also been used in chemical protein synthesis where joining two or more large peptide sequences cleanly and efficiently is not a trivial task. Recently, the commonly used strategy of native chemical ligation has been expanded by incorporating a cyclic disulfide (Fig. 9).[64] Native chemical ligation is a two-step process beginning with the reversible attack of a thioester 40 by a thiol 41; the cyclic disulfide in 41 is stable under these conditions. Once the reversible thioester exchange has been kinetically captured by a vicinal amine in 42, the cyclic disulfide in product 43 can be orthogonally reduced to the pair of thiols in 44. Intramolecular amide–thioester exchange converts 44 into 45 for a subsequent native chemical ligation with thiol-terminated peptide 46. Intermolecular thioester exchange yields 47, which is intramolecularly transformed into the final peptide 48. This strategy, relying fundamentally on reversible reactions that are independently addressable, has been employed to construct large, biologically active oligopeptide sequences in a one-pot sequence of reactions. Although the dynamic nature of disulfides is not really used, it is nevertheless an interesting example, both for the promise of this kind of chemistry in synthetic transformations and also because of the loose and somewhat incomplete definition of orthogonal dynamic covalent bonds.

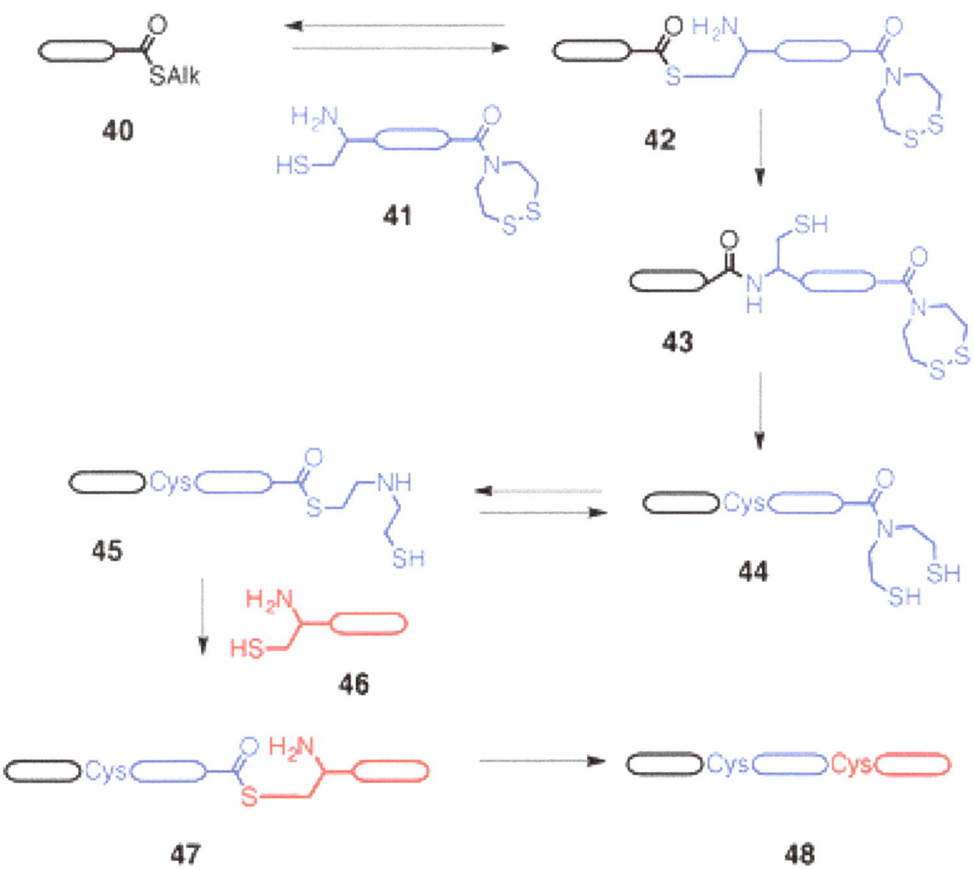

Figure 9 Formal orthogonal dynamic covalent bonds in advanced native chemical ligation.

Chemo- and Biosensors

The use of dynamic covalent bonds in sensing systems is quite common. A wonderful early example is a colorimetric chemosensor to determine the age of whisky (Fig. 10).[65]

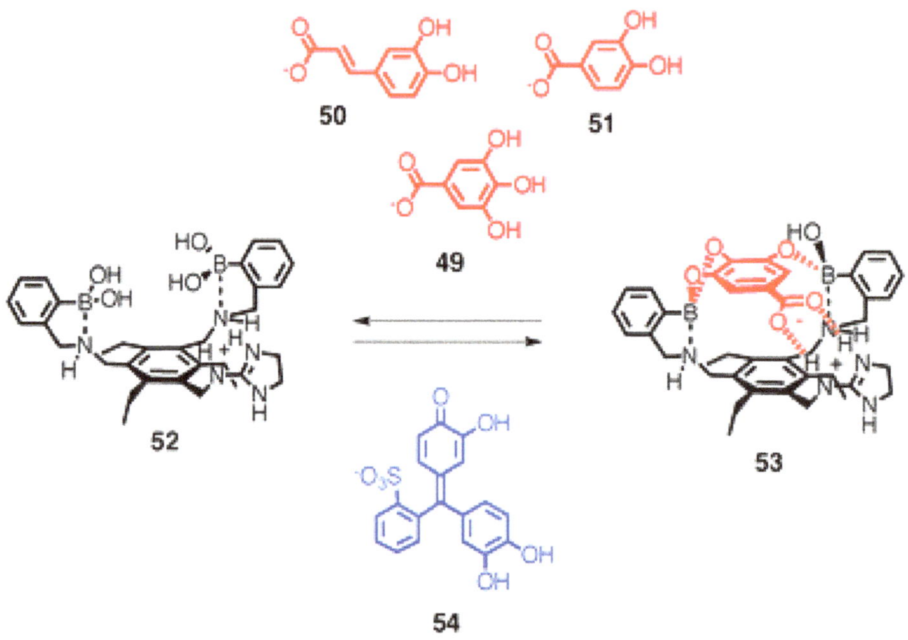

Scotch	age (years)	IDA	HPLC
Vintage Islay	5	4.0	4.0
Caol Lia	7	5.2	2.7
Laphroaig	10	10.7	5.3
Macallan	12	14.5	12.0
Lagavulin	16	18.7	8.0

Figure 10 A sensor for the age of Scotch that operates with indicator displacement assay (IDA) based on dynamic covalent boronic ester exchange. Results are compared to HPLC data for gallate only. Data are normalized to 4 for the lowest value.

The age of Scotch is best determined from the concentration of natural products extracted over the years from the oak barrels. The most abundant gallate 49 would be an obvious analyte, but high-pressure liquid chromatography (HPLC) analysis does not correspond well with the reported age of the whiskies tested. The problem is that although the total concentration of all members of the gallate family appears to remain constant, the concentrations of individual members tend to fluctuate. Other family members include coffeic acid 50 or 3,4-dihydroxybenzoic acid 51.

The bis-boronic acid 52 was designed to sense the entire gallate family of natural products, exemplified here with 49–51. Dynamic covalent bonds with the catechols contribute the most to form product 53. π–π Interactions and ion pairing between carboxylate and guanidinium complete the binding. For colorimetric detection, the indicator displacement assay (IDA) was used. In this assay, pioneered by the Anslyn group, the receptor is loaded with a chromophore of related structure. Pyrocatechol violet 54 was selected because it reproduces the catechol motif and the negative charge of the gallate family. Binding to receptor 52 changes the color of 54. Displacement of 54 by the gallate analyte restores the original color of the free chromophore. With the use of IDA, the determination of the age of Scotch gives correct trends and is possible with the naked eye.

A similar catechol sensing system that makes use of orthogonal dynamic covalent bonds was recently reported by the Bode group (Fig. 11).[14] In chemosensor 55, two boronic acids have been attached to a bullvalene core. These bicycles are wonderful examples of dynamic covalent chemistry because spontaneous Cope rearrangements lead to the coexistence of a large number of configurationally distinct isomers. Analysis of these dynamic mixtures is very demanding, although the isolation of a single pure bullvalene isomer has been reported last year.[66]Previously, bullvalenes have been equipped with two porphyrins for the adaptive sensing of fullerenes.[67–69] To sense polyphenols, the ^{13}C labeled bullvalene 55 was prepared.[14] Equilibration of the 860 possible isomers produced a signature in the ^{13}C NMR spectrum that could be treated as a fingerprint of the free sensor. In the presence of epigallocatechin gallate (EGCG) 56 as a model polyphenol, this ^{13}C fingerprint changes because the formation of two boronic esters as in 57 changes the equilibria among the bullvalene isomers. Tests with a representative collection of polyphenols suggested that the obtained ^{13}C signatures are unique for each analyte.

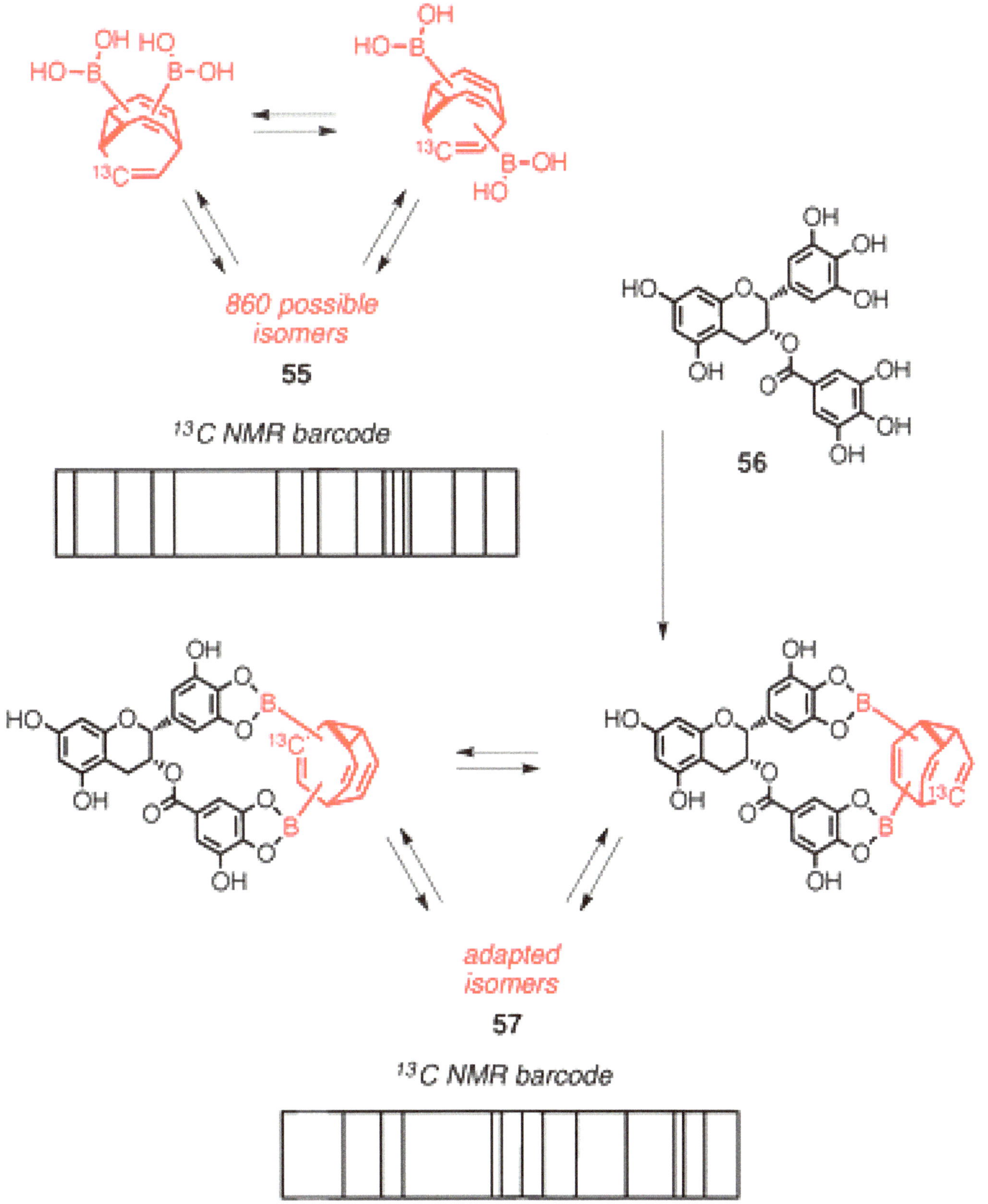

Figure 11 A polyphenol sensor that operates with orthogonal dynamic boronic esters and shapeshifting bullvalenes. A ^{13}C label generates analyte-specific signatures in the ^{13}C NMR spectra.

Similar polyphenol sensors that operate with orthogonal dynamic covalent bonds have been reported previously.[34] They are based on synthetic multifunctional pores that work in lipid bilayer membranes. The pores used were artificial β-barrels with π-acidic clamps or tweezers at their inner surface to catch π-basic analytes passing through the pore. The formation of donor–acceptor complexes within the pore is then reported as a decrease in fluorescence activity due to the hindered efflux of self-quenched fluorophore through the pore. To sense non-aromatic analytes, signal amplifiers such as **58** were introduced (Fig. 12). These bifunctional amplifiers contain a π-basic anthracene core for pore inactivation, a negative charge for solubility, and a hydrazide to covalently capture aldehyde- and ketone-containing analytes. This dynamic covalent approach was then combined with biosensing, a term that stands for the use of enzymes for signal generation. For lactate or citrate sensing in actual samples, pyruvate **59**, the product of enzymatic signal generation, was covalently captured by hydrazide **58**.[70]Pore blockage by hydrazone **60** was then reported as a decrease in fluorescence. The introduction of dynamic covalent signal amplifiers was the key breakthrough needed to create a molecular "artificial tongue" that operates in lipid bilayer membranes (and uses dynamic covalent chemistry)[20] like the receptors of our own tongue.[70]

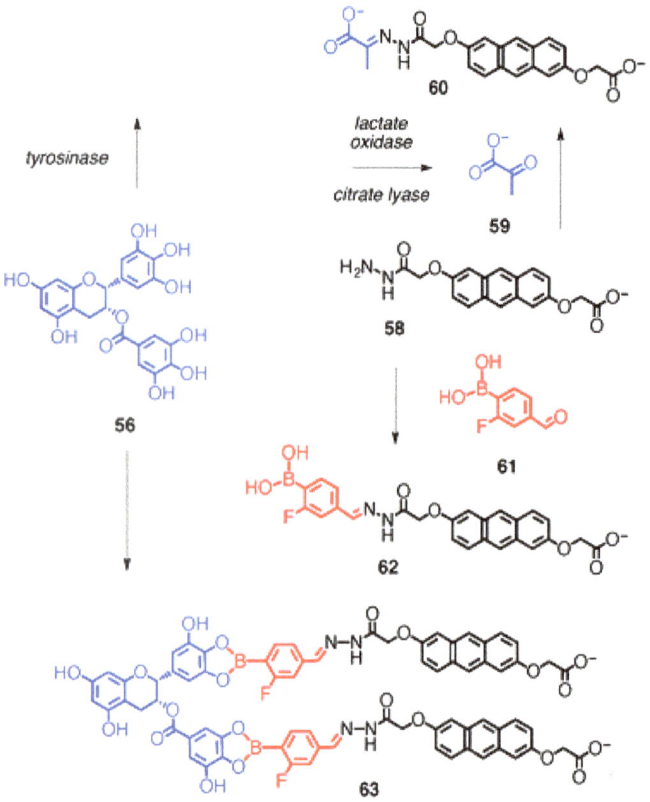

sample	polyphenol expected (mg/g)	found
green tea bags	72	≥ 49
Shincha leaves		≥ 111

Figure 12. A polyphenol sensor that operates with synthetic multifunctional pores in fluorogenic vesicles that respond to boronic ester and hydrazone exchange.

An additional, dynamic orthogonal boronic ester was needed for the sensing of polyphenols such as EGCG 56 with synthetic multifunctional pores. Reaction of signal amplifier 58 with benzaldehyde 61 generated boronic acid 62*in situ*, which was then used to covalently capture polyphenols such as 56. This process was monitored as pore activation because, unlike the effective blocker 62, the resulting bis-boronic ester 63 was too large to enter the synthetic multifunctional pore. To confirm that the polyphenols were responsible for the observed pore activation, tyrosinase was used to selectively remove them and thus reactivate the pore. The sensor was then calibrated with polyphenon, a commercially available polyphenol extract, and used to show that high quality Shincha leaves contain about twice as much polyphenol than green tea bags from Geneva supermarkets. The surprisingly good sensitivity in the low micromolar range prompted the study of the formation of bis-boronic ester **63** in more detail.[71] Because of the powerful exciton coupling between the two anthracenes, circular dichroism (CD) spectroscopy could be used to selectively detect dimers. The new biosensing strategy was further expanded to include catechol-free polyphenols such as resveratrol,[72] the key polyphenol in red wine that has been evoked to account for the "French paradox".

Dynamic covalent chemistry was also crucial for the construction of differential sensors that operate in lipid bilayer membranes.[73] This was surprisingly difficult to realize considering that many chemosensors and also biological olfactory systems operate this way. Pattern generation in the nose combined with pattern recognition in the brain is the only way to discriminate more than 10 000 odorants with ~350 receptors. The difficulty faced when attempting to realize differential sensing with synthetic transport systems in lipid bilayer membranes was pattern generation. The solution to this problem was dynamic covalent chemistry. For pattern generation with muscone **64** as a representative "Geneva" odorant, cationic peptide dendrons such as **65** with one to six hydrazides were prepared (Fig. 13a). Hydrazone formation then yielded cationic amphiphiles such as **66** that were able to activate DNA as cation transporters in fluorogenic vesicles.

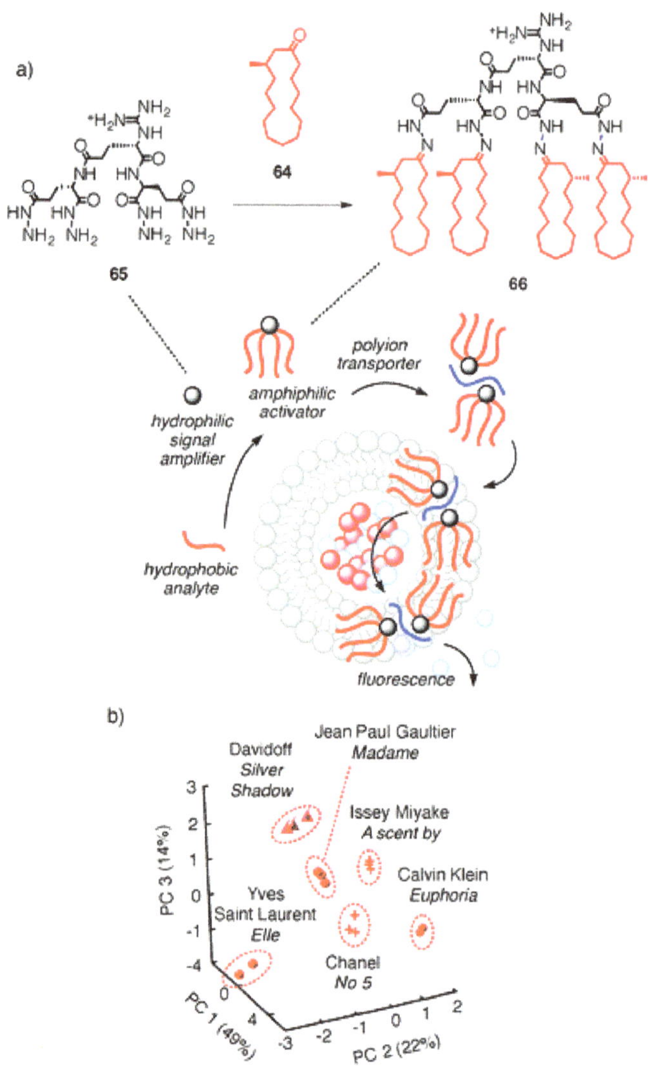

Figure 13. An artificial nose that operates with dynamic hydrazone bridges. (a) Odorants such as muscone **64** are covalently captured by charged hydrazides such as **65**. The obtained amphiphiles **66** activate DNA as a transporter in fluorogenic vesicles.

(b) Different activities obtained with different hydrazides are used to generate patterns, principal component (PC) analysis for pattern recognition of odorants or perfumes in the virtual PC space. Adapted from ref. 73 with permission, © 2011 Royal Society of Chemistry.

The differences in activity obtained with different peptide dendrons were then used to generate patterns. Routine pattern recognition in the virtual principal component space revealed overlap-free discrimination for all tested odorants, including enantiomers and *cis–trans* isomers (cucumber aldehyde). The obvious choice to probe compatibility of this "artificial nose" with samples from supermarket or hospital was perfumes. The results were excellent (Fig. 13b). Early examples for the extension of this approach to orthogonal dynamic bonds will be mentioned in the context of applications in the field of cellular uptake.[74,75]

One of the first examples of functional orthogonal dynamic bonds uses formylphenyl boronic acid as in polyphenol sensor 62 (Fig. 12).[30,31] It focuses on a simple method to assess the enantiomeric excess of different chiral primary amines by NMR spectroscopy. The approach relies on the rapid formation of a three-component adduct between the achiral bifunctional connector 67, (*P*)-binol 68 as an enantiopure aromatic diol, and a chiral primary amine 69 (Fig. 14a). The resulting diastereomeric iminoboronates 70 were rigid enough to generate an anisotropic effect on the amine moiety, leading to different chemical shifts in the ^{1}H NMR spectra. This approach is not limited to α chiral amines, but can also be applied to primary amines containing remote stereogenic centers.

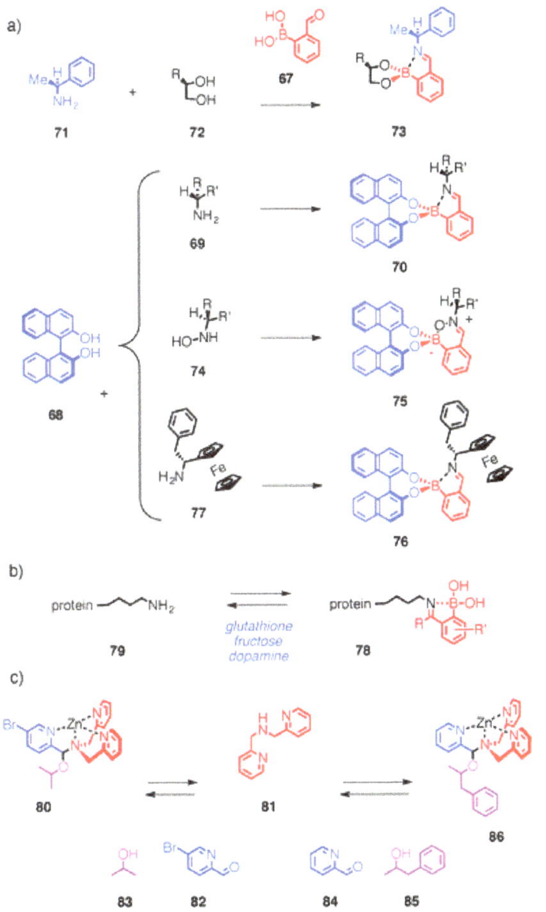

Figure 14. Bifunctional chirality sensors that work with orthogonal dynamic covalent bonds. Twofold covalent capture by benzaldehyde 67 is used to (a) determine the enantiopurity of chiral diols 72, amines 69 or 77, and hydroxyamines 74 by NMR spectroscopy or cyclic voltammetry, and to (b) reversibly label accessible lysine residues of proteins. (c) Dynamic covalent multicomponent systems such as 86 sense the chirality of secondary alcohols.

In a complementary approach, an enantiopure primary amine, (*S*)-(α)-methylbenzylamine 71, was reacted with 2-formylphenylboronic acid 67 and different chiral diols 72, yielding diastereomeric iminoboronate esters 73. The quantification of the enantiomeric purity of the tested diols (1,2-, 1,3- and 1,4-diols were tested) was achieved *via* the direct integration of the signals due to each diastereomer, with ee values up to 98% accurately reported. The same approach was successful in determining the enantiopurity of chiral hydroxylamines 74.[76] Simple mixing with 2-formylphenylboronic acid 67 and binol 68 gave the three-component diastereomeric nitrono-boronate esters 75. The means of detection was not limited to NMR spectroscopy. Established alternatives include CD spectroscopy and, remarkably, even cyclic voltammetry.[77] For this purpose, the chiral ferrocene derivative 76 was prepared from the enantiopure ferrocene amine 77 and 2-formylphenylboronic acid 67. The resulting Schiff base was coupled with (*P*/*M*)-binol 68 to give diastereomeric ferrocene boronate esters 76. These diastereomeric esters showed different behavior in square wave voltammetry and cyclic voltammetry. It was then possible to discriminate between mixtures with different degrees of enantiomeric excess of binol, simply by monitoring the potential values. The described method was powerful enough for an accurate determination of ee in the range of 60–98%, with an estimated error lower than 3%.

The same bifunctional 2-formylphenylboronic acid **67** has recently been used to achieve reversible protein modification (Fig. 4b).[78] This compound was able to form iminoboronates **78**between free amino groups present in the protein (*N*-terminal and ε-amino group of lysine, represented in **79**). Iminoboronate formation is possible in water at room temperature with high efficiency only when the carbonyl group is placed in proximity (*ortho* position) to the boronic acid moiety, due to the stabilization of the dative nitrogen–boron bond (confirmed by computational simulations). The approach was successfully applied both to a small natural neuropeptide (somatostatin) and to model proteins (lysozyme, cytochrome c, ribonuclease A and myoglobin). Moreover, the reversibility of the approach was demonstrated: The presence of endogenous molecules such as glutathione, fructose or dopamine promotes imine hydrolysis, releasing the protein in the original unmodified form.

A recent, elegant approach to determine the enantiopurity of secondary alcohols makes innovative use of dynamic covalent chemistry (Fig. 14c).[15] The multicomponent system **80** is composed of the secondary amine **81** that is able to form an unstable imminium with aldehyde **82**, which in turn reacts with isopropanol **83**. This multicomponent system is able to incorporate secondary alcohols with high sensitivity because the hemiaminal ether is stabilized by the coordination to a zinc cation placed directly above by the three N-heterocyclic ligands. Subcomponent exchange experiments covered different pyridine ligands, including **84**, and the incorporation of chiral alcohols such as **85**, 1-phenylethanol, 3-methyl-2-butanol or 2-butanol induces a twist in the ligand. This twist then results in strong exciton coupling in the CD spectra of the resulting multicomponent systems such as **86**, thus revealing the presence and enantiopurity of the chiral secondary alcohol.

Uptake and Release

The use of dynamic covalent bonds has been extensively explored in the field of cellular uptake. The general idea is to destroy the active transport system under specific (acidic or reductive) conditions, either during or after uptake, to minimize cytotoxicity and to release the substrate. Cell-penetrating poly(disulfide)s are currently regarded as the transporters of the future because they are depolymerized by glutathione in the cytosol.[18] Last year, it was reported that cell-penetrating poly(disulfide)s could be grown directly on substrates of free choice by ring-opening disulfide-exchange polymerization.[79] Given the promise of cell-penetrating poly(disulfide)s, it is surprising to note that disulfide-exchange chemistry has not yet been successfully complemented by orthogonal dynamic covalent bonds. The only reported example that really matches the topic of this review is cell-penetrating poly (disulfide) **87** (Fig. 15).[80] The polymer contains the biodegradable poly(disulfide) backbone and positive charges in the backbone and side chains to ensure transport activity in lipid bilayers. Finally, a large fraction of the lateral amines were converted into amides to install phenyl boronic acids. They were expected to form dynamic boronic esters with glycosaminoglycans (GAGs) present on the cell surface. The main consequence of the introduction of boronic acids in **87** was an increase in toxicity. However, this observed toxicity is not intrinsic for boronic acids. Marvelous examples exist of cellular uptake mediated by boronic esters at the cell surface without other orthogonal dynamic covalent bonds being involved.[81] Disulfide exchange chemistry at the cell surface is currently suspected to be responsible for mediating cellular uptake in a similar manner.[82]

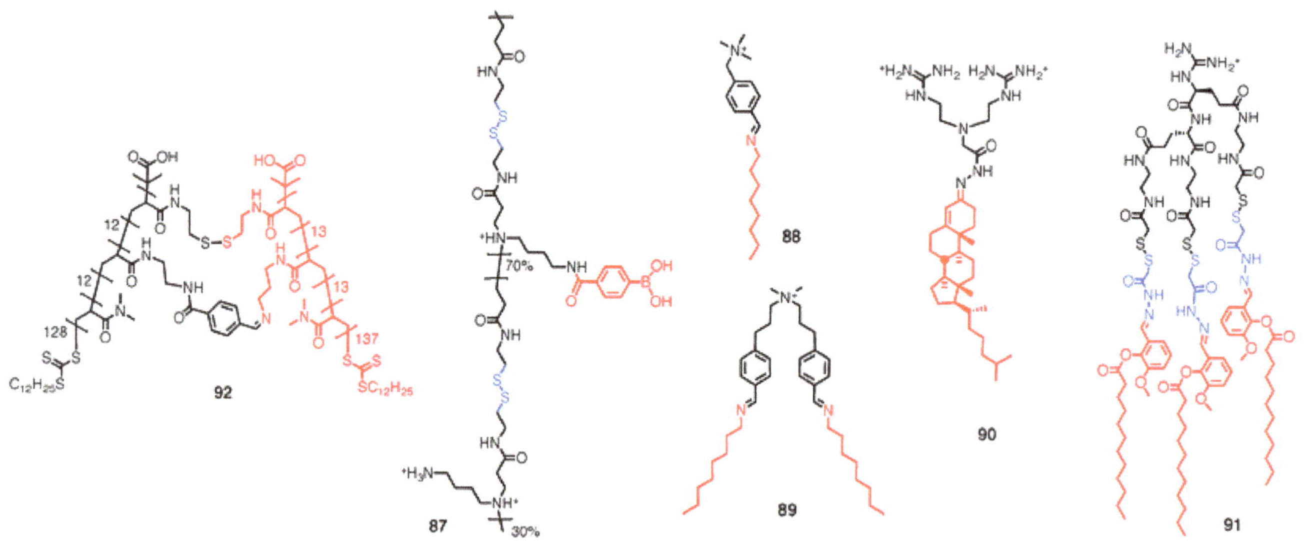

Figure 15. Functional architectures with orthogonal dynamic covalent bonds for cellular uptake (**87**, **91**) and release (**92**), together with cationic amphiphiles with only one dynamic bond (**88–90**).

The use of hydrazone bridges in cationic amphiphiles has already been mentioned in the context of differential sensing (see above).[73–75] Dynamic imine amphiphiles such as 88 and 89have also been explored in the context of stimuli-responsive micelles and vesicles.[83,84] For cellular uptake, amphiphiles with acid-labile hydrazone bridges including the doubly charged steroid conjugate 90 have been explored to deliver genes and drugs.[85,86] Their dynamic nature has provided facile access to large amphiphile libraries.[75] GFP-knockdown assays have been developed for the automated screening of these libraries for small interfering RNA (siRNA) uptake into HeLa cells (GFP: green fluorescent protein). The best hits found were almost twice as active as the best commercially available control (lipofectamine). The few examples with orthogonal disulfide and hydrazone bridges were moderately active. Amphiphile 91 with one positive charge, three tails and an additional biodegradable ester bond was among the best.[75]

The usefulness of dynamic covalent bonds to achieve stimuli-responsive substrate release is not limited to cellular uptake. One of the most spectacular applications of hydrazones focuses on the slow release of aldehyde- and ketone-containing odorants.[87] Chemoorthogonal release with dynamic imine and disulfide bridges in nanoparticles generated by **92** has been reported recently.[88] Here, orthogonal reversible reactions were used in tandem to allow precise control of the release of an encapsulated dye molecule. The nanoparticles consist of two polymers that contain disulfides and aldehydes or amines that are blended together in the presence of the hydrophobic dye Nile Red. Crosslinking between the two polymers may be achieved either by disulfide exchange in the presence of the reducing agent dithiothreitol (DTT) or by imine formation at pH < 5.5. These orthogonal processes may be carried out independently of each other, creating nanoparticles that are held together either by only imine bonds, by only disulfide bridges, or by both. In order to release the encapsulated dye molecules, all that is required is to cleave the crosslinks and destroy the nanoparticles. The orthogonality of the crosslink-forming reactions is mirrored exactly by the conditions required to cleave them. It was found that the imine-crosslinked particles could be disassembled to release the encapsulated dye by treatment with acid, and disulfide-crosslinked particles could be disassembled by treatment with a reducing agent.

The nanoparticles formed by **92**, containing both disulfide and imine bonds, would disassemble only in the presence of both a reducing agent and a pH < 5.5. The system **92**is therefore a chemical AND gate, releasing its cargo when the two decoupling stimuli are applied simultaneously but remaining intact when either stimulus is applied in isolation.

EPILOGUE

In the field of supramolecular chemistry, dynamic covalent chemistry has attracted more and more interest in recent years. This is due to the fact that it combines the robustness of "classical" covalent bonds with the flexibility typical for non-covalent interactions, permitting the development of complex nanostructures. Despite the large number of functional systems in the literature that operate with one type of dynamic covalent bond, functional systems based on more than one type of dynamic covalent bond are rare. This is surprising considering that non-covalent bonds are used best together to achieve function. With this review we have tried to underline the possibility for chemists to generate complex functional systems with orthogonal dynamic covalent bonds. A remarkably broad variety of activities have been covered in a short time. However, what really stands out is that most studies ask unusual, complex and highly original questions that are hard to address with other approaches. Emphasis is on molecular systems that can not only sense, transport and transform like ordinary functional systems often do but also self-sort, self-heal, adapt, template, amplify, exchange, replicate, transcribe, or even walk and "think" (logic gates).

These pioneering off-road adventures nicely outline perspectives for the future. Namely, the impact of orthogonal dynamic covalent bonds will not be limited to a specific topic but will concern diverse topics without direct relation except that they all address unusual questions that are otherwise difficult to address. With the perspective to couple self-healing with self-sorting and logic gates, applications towards the materials sciences appear most obvious and most exciting. The results available today further imply, perhaps more surprisingly, much potential for sensing applications. The general promise is to couple signal generation with signal transduction and detection. Similar cascade processes are presumably going to benefit most from orthogonal dynamic covalent bonds in catalysis. Applications of orthogonal dynamic covalent bonds toward the life sciences, finally, will focus on the coupled activation and inactivation of chemical interferants, probably identified by *in situ* selection. Sometimes, orthogonal dynamic covalent bonds appear a bit like molecular versions of the brave little tailor, catching two flies in one blow.[89] Although "seven with one blow" seems quite far away and maybe not really needed, the extension from bifunctionality toward multifunctionality with multiple dynamic covalent bonds at work together is certainly one of the most appealing perspectives for the future. The brave little tailor, after all, ended up as a king.

ACKNOWLEDGEMENTS

We warmly thank all past and present coworkers and collaborators who contributed in our research on the topic of this review. We also thank Jonathan Nitschke (Cambridge), Sijbren Otto (Groningen), Eric Anslyn (Austin) and Kay Severin (Lausanne) for advice, and the University of Geneva, the Swiss National Science Foundation (NSF), the National Centre of Competence in Research (NCCR) in Chemical Biology, and the European Research Council (ERC, Advanced Investigator) for financial support.

REFERENCES

1. J. Li, P. Nowak and S. Otto, *J. Am. Chem. Soc.*, 2013, 135, 9222–9239 .
2. F. B. L. Cougnon and J. K. M. Sanders, *Acc. Chem. Res.*, 2012, 45, 2211–2221 .
3. J.-M. Lehn, *Top. Curr. Chem.*, 2012, 322, 1–32.
4. D. A. Dougherty, *Acc. Chem. Res.*, 2013, 46, 885–893 .
5. A. Frontera, P. Gamez, M. Mascal, T. J. Mooibroek and J. Reedijk, *Angew. Chem., Int. Ed.*, 2011, 50, 9564–9583 .
6. P. Ballester, *Acc. Chem. Res.*, 2013, 46, 874–884 .
7. Y. Zhao, Y. Domoto, E. Orentas, C. Beuchat, D. Emery, J. Mareda, N. Sakai and S. Matile, *Angew. Chem., Int. Ed.*, 2013, 52, 9940–9943 .
8. P. Metrangolo, F. Meyer, T. Pilati, G. Resnati and G. Terraneo, *Angew. Chem., Int. Ed.*, 2008, 47, 6114–6127 .
9. A. Vargas Jentzsch and S. Matile, *J. Am. Chem. Soc.*, 2013, 135, 5302–5303 .
10. D. Drahonovsky and J.-M. Lehn, *J. Org. Chem.*, 2009, 74, 8428–8432 .
11. R. Caraballo, H. Dong, J. P. Ribeiro, J. Jimenez-Barbero and O. Ramström, *Angew. Chem., Int. Ed.*, 2010, 49, 589–593 .
12. G. Joshi and E. V. Anslyn, *Org. Lett.*, 2012, 14, 4714–4717 .
13. P. Vongvilai and O. Ramström, *J. Am. Chem. Soc.*, 2009, 131, 14419–14425 .
14. J. F. Teichert, D. Mazunin and J. W. Bode, *J. Am. Chem. Soc.*, 2013, 135, 11314–11321 .
15. L. You, J. S. Berman and E. V. Anslyn, *Nat. Chem.*, 2011, 3, 943–948 .
16. T. S. R. Lam, A. Belenguer, S. L. Roberts, C. Naumann, T. Jarrosson, S. Otto and J. K. M. Sanders, *Science*, 2005, 308, 667–669 .
17. S. Turkyilmaz, P. F. Almeida and S. L. Regen, *Langmuir*, 2011, 27, 14380–14385 .
18. E.-K. Bang, M. Lista, G. Sforazzini, N. Sakai and S. Matile, *Chem. Sci.*, 2012, 3, 1752–1763.
19. M. C. Walker, B. W. Thuronyi, L. K. Charkoudian, B. Lowry, C. Khosla and M. C. Chang, *Science*, 2013, 341, 1089–1094 .
20. L. J. Macpherson, A. E. Dubin, M. J. Evans, F. Marr, P. G. Schultz, B. F. Cravatt and A. Patapoutian, *Nature*, 2007, 445, 541–545 .
21. Y. Ura, J. M. Beierle, L. J. Leman, L. E. Orgel and M. R. Ghadiri, *Science*, 2009, 325, 73–77 .
22. C.-H. Wong and S. C. Zimmerman, *Chem. Commun.*, 2013, 49, 1679–1695 .
23. M. Schmittel and K. Mahata, *Angew. Chem., Int. Ed.*, 2008, 47, 5284–5286 .
24. M. L. Saha, S. De, S. Pramanik and M. Schmittel, *Chem.*

Soc. Rev., 2013, 42, 6860–6909 .

25. V. Goral, M. I. Nelen, A. V. Eliseev and J.-M. Lehn, *Proc. Natl. Acad. Sci. U. S. A.*, 2001, 98, 1347–1352 .
26. J. Leclaire, L. Vial, S. Otto and J. K. M. Sanders, *Chem. Commun.*, 2005, 1959–1961 .
27. Z. Rodriguez-Docampo and S. Otto, *Chem. Commun.*, 2008, 5301–5303 .
28. A. G. Orrillo, A. M. Escalante and R. L. E. Furlan, *Chem. Commun.*, 2008, 5298–5300 .
29. A. M. Escalante, A. G. Orrillo and R. L. E. Furlan, *J. Comb. Chem.*, 2010, 12, 410–413 .
30. Y. Perez-Fuertes, A. M. Kelly, A. L. Johnson, S. Arimori, S. D. Bull and T. D. James, *Org. Lett.*, 2006, 8, 609–612 .
31. A. M. Kelly, Y. Perez-Fuertes, S. Arimori, S. D. Bull and T. D. James, *Org. Lett.*, 2006, 8, 1971–1974 .
32. R. J. Sarma, S. Otto and J. R. Nitschke, *Chem.–Eur. J.*, 2007, 13, 9542–9546 .
33. M. Hutin, G. Bernardinelli and J. R. Nitschke, *Chem.–Eur. J.*, 2008, 14, 4585–4593 .
34. S. Hagihara, H. Tanaka and S. Matile, *J. Am. Chem. Soc.*, 2008, 130, 5656–5657 .
35. N. Christinat, R. Scopelliti and K. Severin, *Angew. Chem., Int. Ed.*, 2008, 47, 1848–1852 .
36. B. Icli, N. Christinat, J. Tönnemann, C. Schüttler, R. Scopelliti and K. Severin, *J. Am. Chem. Soc.*, 2009, 131, 3154–3155 .
37. B. Icli, E. Solari, B. Kilbas, R. Scopelliti and K. Severin, *Chem.–Eur. J.*, 2012, 18, 14867–14874 .
38. K. D. Okochi, Y. Jinz and W. Zhang, *Chem. Commun.*, 2013, 49, 4418–4420 .
39. M. J. Barrell, A. G. Campaña, M. von Delius, E. M. Geertsema and D. A. Leigh, *Angew. Chem., Int. Ed.*, 2011, 50, 285–290 .
40. J.-S. Shin and N. A. Pierce, *J. Am. Chem. Soc.*, 2004, 126, 10834–10835 .
41. K. Lund, A. J. Manzo, N. Dabby, N. Michelotti, A. Johnson-Buck, J. Nangreave, S. Taylor, R. Pei, M. N. Stojanovic, N. G. Walter, E. Winfree and H. Yan, *Nature*, 2010, 465, 206–210 .
42. M. von Delius, E. M. Geertsema, D. A. Leigh and A. M. Z. Slawin, *Org. Biomol. Chem.*, 2010, 8, 4617–4624.
43. M. von Delius, E. M. Geertsema and D. A. Leigh, *Nat. Chem.*, 2010, 2, 96–101 .
44. M. von Delius, E. M. Geertsema, D. A. Leigh and D.-T. D. Tang, *J. Am. Chem. Soc.*, 2010, 132, 16134–16145 .
45. A. G. Campaña, A. Carlone, K. Chen, D. T. F. Dryden, D. A. Leigh, U. Lewandowska and K. M. Mullen, *Angew. Chem., Int. Ed.*, 2012, 51, 5480–5483 .
46. A. G. Campaña, D. A. Leigh and U. Lewandowska, *J. Am. Chem. Soc.*, 2013, 135, 8639–8645 .
47. G. Deng, F. Li, H. Yu, F. Liu, C. Liu, W. Sun, H. Jiang and Y. Chen, *ACS Macro Lett.*, 2012, 1, 275–279.
48. X. Li and D. G. Lynn, *Angew. Chem., Int. Ed.*, 2002, 41, 4567–4569.
49. L. Tauk, A. P. Schroder, G. Decher and N. Giuseppone, *Nat. Chem.*, 2009, 1, 649–656 .
50. M. D. Yilmaz and J. Huskens, *Soft Matter*, 2012, 8, 11768–11780 .
51. N. Sakai, M. Lista, O. Kel, S. Sakurai, D. Emery, J. Mareda, E. Vauthey and S. Matile, *J. Am. Chem. Soc.*, 2011, 133, 15224–15227 .
52. M. Lista, J. Areephong, N. Sakai and S. Matile, *J. Am. Chem. Soc.*, 2011, 133, 15228–15230 .
53. N. Sakai and S. Matile, *J. Am. Chem. Soc.*, 2011, 133, 18542–18545 .
54. T. Y. Luh, *Acc. Chem. Res.*, 2013, 46, 378–389 .
55. E. Orentas, M. Lista, N.-T. Lin, N. Sakai and S. Matile, *Nat. Chem.*, 2012, 4, 746–750 .
56. J. Areephong, E. Orentas, N. Sakai and S. Matile, *Chem. Commun.*, 2012, 48, 10618–10620 .
57. G. Sforazzini, R. Turdean, N. Sakai and S. Matile, *Chem. Sci.*, 2013, 4, 1847–1851 .
58. G. Sforazzini, E. Orentas, A. Bolag, N. Sakai and S. Matile, *J. Am. Chem. Soc.*, 2013, 135, 12082–12090 .
59. P. C. Gareiss, K. Sobczak, B. R. McNaughton, P. B. Palde, C. A. Thornton and B. L. Miller, *J. Am. Chem. Soc.*, 2008, 130, 16254–16261 .
60. A. V. Gromova, J. M. Ciszewski and B. L. Miller, *Chem. Commun.*, 2012, 48, 2131–2133 .
61. A. M. Escalante, A. G. Orrillo, I. Cabezudo and R. L. E. Furlan, *Org. Lett.*, 2012, 14, 5816–5819 .
62. P. Vongvilai and O. Ramström, *J. Am. Chem. Soc.*, 2009, 131, 14419–14425 .
63. J. W. Sadownik and D. Philp, *Angew. Chem., Int. Ed.*, 2008, 47, 9965–9970 .
64. N. Ollivier, J. Vicogne, A. Vallin, H. Drobecq, R. Desmet, O. El Mahdi, B. Leclercq, G. Goormachtigh, V. Fafeur and O. Melnyk, *Angew. Chem., Int. Ed.*, 2012, 51, 209–213 .
65. S. L. Wiskur and E. V. Anslyn, *J. Am. Chem. Soc.*, 2001, 123, 10109–10110.
66. M. He and J. W. Bode, *Org. Biomol. Chem.*, 2013, 11, 1306–1317.
67. A. R. Lippert, V. L. Keleshian and J. W. Bode, *Org. Biomol. Chem.*, 2009, 7, 1529–1532.
68. A. R. Lippert, A. Naganawa, V. L. Keleshian and J. W. Bode, *J. Am. Chem. Soc.*, 2010, 132, 15790–15799 .
69. K. K. Larson, M. He, J. F. Teichert, A. Naganawa and J. W. Bode, *Chem. Sci.*, 2012, 3, 1825–1828 .
70. S. Litvinchuk, H. Tanaka, T. Miyatake, D. Pasini, T. Tanaka, G. Bollot, J. Mareda and S. Matile, *Nat. Mater.*, 2007, 6, 576–580 .
71. A. Hennig, S. Hagihara and S. Matile, *Chirality*, 2009, 21, 826–835 .
72. S. Hagihara, H. Tanaka and S. Matile, *Org. Biomol. Chem.*, 2008, 6, 2259–2262.
73. T. Takeuchi, J. Montenegro, A. Hennig and S. Matile, *Chem. Sci.*, 2011, 2, 303–307 .
74. J. Montenegro, E.-K. Bang, N. Sakai and S. Matile, *Chem.–Eur. J.*, 2012, 18, 10436–10443 .
75. C. Gehin, J. Montenegro, E.-K. Bang, S. Takayama, H. Hirose, S. Futaki, A. Cajaraville, S. Matile and H.

Riezman, *J. Am. Chem. Soc.*, 2013, 135, 9295–9298 .

76. D. A. Tickell, M. F. Mahon, S. D. Bull and T. D. James, *Org. Lett.*, 2013, 15, 860–863 .
77. G. Mirri, S. D. Bull, P. N. Horton, T. D. James, L. Male and J. H. R. Tucker, *J. Am. Chem. Soc.*, 2010, 132, 8903–8905 .
78. P. M. S. D. Cal, J. B. Vicente, E. Pires, A. V. Coelho, L. F. Veiros, C. Cordeiro and P. M. P. Gois, *J. Am. Chem. Soc.*, 2012, 134, 10299–10305 .
79. E.-K. Bang, G. Gasparini, G. Molinard, A. Roux, N. Sakai and S. Matile, *J. Am. Chem. Soc.*, 2013, 135, 2088–2091 .
80. M. Piest and J. F. J. Engbersen, *J. Controlled Release*, 2011, 155, 331–340 .
81. G. A. Ellis, M. J. Palte and R. T. Raines, *J. Am. Chem. Soc.*, 2012, 134, 3631–3634 .
82. A. G. Torres and M. J. Gait, *Trends Biotechnol.*, 2012, 30, 185–190 .
83. C. B. Minkenberg, L. Florusse, R. Eelkema, G. J. M. Koper and J. H. van Esch, *J. Am. Chem. Soc.*, 2009, 131, 11274–11275 .
84. C. B. Minkenberg, F. Li, P. van Rijn, L. Florusse, J. Boekhoven, M. C. A. Stuart, G. J. M. Koper, R. Eelkema and J. H. van Esch, *Angew. Chem., Int. Ed.*, 2011, 50, 3421–3424 .
85. A. Aissaoui, B. Martin, E. Kan, N. Oudrhiri, M. Hauchecorne, J.-P. Vigneron, J.-M. Lehn and P. Lehn, *J. Med. Chem.*, 2004, 47, 5210–5223 .
86. W. Xia and P. S. Low, *J. Med. Chem.*, 2010, 53, 6811–6824 .
87. A. Herrmann, *Chem.–Eur. J.*, 2012, 18, 8568–8577 .
88. A. W. Jackson and D. A. Fulton, *Macromolecules*, 2012, 45, 2699–2708.
89. J. Grimm and W. Grimm, *Household Tales by the Brothers Grimm*, George Bell and Sons, London, 1884

CHAPTER 9 HYDROGEN BONDING PENALTY UPON LIGAND BINDING

Hongtao Zhao and Danzhi Huang

INTRODUCTION

Hydrogen bonding is an exchange reaction whereby the hydrogen bond donors and acceptors of the free protein and ligand break their hydrogen bonds with water and form new ones in the protein-ligand complex [1], [2], [3]. About thirty years ago, Wilkinson and coworkers found mutation of Cys-35 in Tyrosyl-tRNA synthetase to Ser-35 causes poorer ATP binding and catalysis although the hydroxyl group of serine forms far stronger hydrogen bonds than does the thiol group of cysteine [1]. Analysis of the hydrogen bonding geometry revealed that a hydrogen bond of Ser-35 is at least 0.5 Å longer than the optimum. Accordingly, Ser-35 would have to lose a good hydrogen bond with a bound water molecule to form this weak hydrogen bond with ATP in the enzyme-substrate complex, and thus the mutant shows poorer binding and catalysis. Therefore, enthalpic loss in hydrogen bonding could take place upon ligand binding if not compensated by formation of good hydrogen bonds between the protein and ligand.

Virtual screening has emerged as an efficient tool in drug discovery from lead identification to optimization and beyond [4], [5]. However, scoring functions that model the solvent environment as a continuum [6], [7] are still grossly inaccurate [8]. The role of individual waters can be critical in predication of binding affinities, and continuum models often provide poor results in treating bound waters in a confined cavity [9]. Glide docks explicit waters into the binding site and measures the exposure of polar/charged groups to the explicit waters. When a polar/charged ligand or protein group is judged to be inadequately solvated, a desolvation penalty is assessed [9], [10]. By contrast, most other scoring functions [11] do not properly take into account the enthalpic loss of hydrogen bonding upon ligand binding. Incorporation of bound water molecules into molecular docking was suggested for improvement of accuracy [12]. On the other hand, in high-throughput molecular docking campaigns a significant part of binding poses are rather unrealistic, e.g. burial of polar atoms in hydrophobic sites, and thus discarding them at an early stage is desirable. Filters such as van der Waals efficiency based on arbitrary cutoff are often used to remove poses that unlikely bind [13]. However, it seems lack of a reliable and efficient filter with transferable cutoff among different proteins.

Protein kinases play an important role in cell-signaling pathways regulating a variety of cellular functions. Dysregulation of kinase activity has been implicated in pathological conditions ranging from neuronal disorders to cellular transformation in leukemia [14]. The tyrosine kinase erythropoietin producing human hepatocellular carcinoma receptor B4 (EphB4) is involved in cancer related angiogenesis [15]. So far, two high-throughput virtual screening campaigns have been reported, with two scaffolds identified in the low micromolar range [13], [16]. Highly potent EphB4 inhibitors have been developed via chemical synthesis [17], [18], [19]. The marketed drug dasatinib, with Abl1 and Src as primary targets, also shows a very high affinity to Eph kinases [20].

Here, we report a new approach to calculate hydrogen bonding penalty (HBP) associated with ligand binding. HBP is further integrated into a binding energy calculation, and the fitted parameter of 1.7 kcal/mol is consistent with the estimate of contribution by formation of one neutral hydrogen bond ranging from 0.5 to 1.5 kcal/mol [21]. Moreover, statistics of HBP in kinase crystal structures and an application in a high-throughput docking campaign is presented.

METHODS

Binding of a ligand to a protein involves the breakage of hydrogen bonds with water molecules and formation of new

Citation: Zhao H, Huang D (2011) Hydrogen Bonding Penalty upon Ligand Binding. PLoS ONE 6(6): e19923. doi:10.1371/journal.pone.0019923.

hydrogen bonds between the protein and ligand, which can be described by the following equation [21] by using one pair of donor (D) and acceptor (A):

$$D\cdots OH_2 + A\cdots HOH \leftrightarrow D\cdots A + HOH\cdots OH_2 \quad (1)$$

Based on hydrogen bonding being an exchange reaction [1], [21], [22], its energy can be described using normalized weights:

$$E_{HB-unbound} = [(w_D + w_{O-H_2O}) + (w_A + w_{H-H_2O})] * E_{HB} \quad (2)$$

$$E_{HB-bound} = [f_{hb} * (w_D + w_A) + (w_{O-H_2O} + w_{H-H_2O})] * E_{HB} \quad (3)$$

wherein, w_D and w_A is the hydrogen bonding weight of a donor or acceptor, respectively, f_{hb} stands for the fraction of hydrogen bonding relative to that of an optimum geometry, and E_{HB} is unit hydrogen bonding energy. Hydrogen bonds with water are assumed to be in the optimum geometry. HBP (p_{HB}) associates with ligand binding is then described as

$$p_{HB} = (1 - f_{hb}) * (w_D + w_A) \quad (4)$$

Probing Hydrogen Bonding Status

Oxygen and nitrogen atoms in double or triple bonds are regarded as hydrogen bond acceptors, and hydrogen atoms bonded to oxygen, nitrogen or sulfur atoms are regarded as hydrogen bond donors. The existence of C—H...O hydrogen bonds has been confirmed by neutron diffraction data on organic compounds [23]. Analysis of 100 kinase crystal structures complexed with small molecule inhibitors at a resolution of at least 2.5 Å gives 64 short C—H...O interactions, showing typical hydrogen bonding features.

Each hydrogen bond donor or acceptor at the binding interface is firstly checked whether it forms hydrogen bond with water molecules. For this purpose, an optimum solvation radius (r_{sol}) is defined for each donor/acceptor and if a water molecule can be placed within 0.15 Å of the r_{sol} no penalty is applied. Here, 2.8 and 2.9 Å are used as r_{sol} for any oxygen and nitrogen, respectively, which were derived from an analysis of 397 crystal structures with X-ray resolutions below 1.0 Å. The r_{sol} of polar hydrogen is 1.9 Å (except 2.15 Å for H bonded to sulfur), which is the difference between the r_{sol} of nitrogen and the bond length [24]of N—H. The r_{sol} of other atom types are listed in Figure 1 and the values are mainly adapted based on the van der Waals radii of Bondi [25]. Details of probing hydrogen bonds with water were described. In case of not forming hydrogen bonds with water, the possibility of forming hydrogen bonds between the protein and ligand (including intra-molecular hydrogen bonds) is further checked and penalty (p_{HB}) is then calculated.

	Atom types	Hydrogen bonding weight	r_{sol} (Å)
O	$=O$, $-O^-$	1.0	2.8
	$-OH$	0.5	
	$-O-$	0.2	
N	$=N-$, $\equiv N$	1.0	2.9
H	$-NH^+$	1.0	1.88
	$-OH$, $-NH$, $-NH_3^+$, $-NH_2^+$	0.5	
	$-NH_2$	0.3	
	$CH=N$, $HC=O$	0	
	$-SH$	0.2	2.15
S	$=S$, $-S-$	-	3.2
C	Unsaturated	-	3.0
	Saturated	-	3.2
H	Apolar	-	2.28
F	-	-	2.99
Cl	-	-	3.27
Br	-	-	3.37
I	-	-	3.50

Figure 1. Hydrogen bonding weights and solvation radii of different atom types.

Fraction of hydrogen bonding

Similar to the strategy of evaluating hydrogen bonding energy in LUDI [26], the following equations are used to calculate the fraction of hydrogen bonding (f_{hb}) to that of an optimum geometry.

$$f_{hb}(r,\theta) = f(r)\cdot f(\theta) \quad (5)$$

$$f(r) = \begin{cases} 1 & if \quad r \le 2.0 \\ 1-0.5*(r-2.0) & if \quad 2.0 < r \le 2.8 \\ 0 & if \quad r > 2.8 \end{cases} \quad (6)$$

$$f(\theta) = \begin{cases} 1 & if \quad \theta \ge 150^o \\ 4.0*(-\cos\theta)^{0.25}-2.86 & if \quad 110^o \le \theta < 150^o \\ 0 & if \quad \theta < 110^o \end{cases} \quad (7)$$

wherein, r is the distance between the hydrogen atom and the acceptor and θ is the angle centered at hydrogen among donor, hydrogen and acceptor. The equation to calculate f(r) and f(θ) as well as the upper and lower limit in r and θ are derived from the calculation using density functional theory [27]. In case of one hydrogen atom is shared by two acceptors or one acceptor interacting with two donors, the f_{hb} for the corresponding donor/acceptor is additive but with 1 as the upper limit.

Hydrogen Bonding Penalty

The HBP at the protein-ligand interface is summarized over each donor/acceptor as

$$P_{HB} = \sum_{pro,lig} w*(1-\sum f_{hb}) \quad (8)$$

However, no penalty is applied for protein atoms which are not water accessible before ligand binding or participate in intra-molecular hydrogen bonds. Initial guess of hydrogen bonding weights (w) is based on chemical intuition by considering atomic partial charge and water solubility of a few small molecules. Empirical weights as proof-of-principle are then optimized with a trial-and-error procedure according to the fitted parameter in the binding free energy calibration.

Evaluation of binding free Energy

The equation used for fitting the calculated energies to the experimental free energies of binding ($\Delta G = RT\ln(K_d)$) is a three-parameter model

$$\Delta G = \alpha\Delta E_{ff} + \beta P_{HB} + \gamma \quad \textbf{(9)}$$

where, ΔE_{ff} is the interaction energy between the ligand and the protein calculated by the CHARMm force filed [28] and P_{HB} stands for HBP. Three parameters α, β, and γ are generated with fitting. ΔE_{ff} is calculated by the following equation:

$$\Delta E_{ff} = \Delta E_{vdW} + \Delta E_{coul} + \Delta G_{solv} + \Delta E_{strain} \quad \textbf{(10)}$$

where, ΔE_{vdW} is the intermolecular van der Waals energy, ΔE_{coul} is the intermolecular Coulombic energy in vacuo, ΔE_{strain} is the strain energy of ligand upon binding, and ΔG_{sol} is the change in solvation energy of ligand and protein upon binding.

The van der Waals and Coulombic interaction energy are calculated by subtracting the values of the isolated components from the energy of the complex with CHARMM [29] and the CHARMm22 force filed [28]. The van der Waals energy is calculated using the default nonbonding cutoff of 14 Å. Coulombic energy is calculated using infinite cutoff and a dielectric constant of 2.0. The electrostatic solvation energy was calculated by the finite-difference Poisson approach (FDP) [30] using PBEQ module [31] in CHARMM and a focusing procedure with a final grid spacing of 0.25 Å. The size of the initial grid is determined by considering a layer of at least 12.5 Å around the solute. The dielectric discontinuity surface was delimited by the van der Waals surface. The ionic strength is set to zero and the temperature to 300 K. Two finite-difference Poisson calculations are performed for each of the three systems (protein, ligand, and protein/ligand complex). The exterior dielectric constant was set to 78.5 and 2.0 for the first and second calculation, respectively, while the solute dielectric constant is 2.0 to take polar fluctuations into account. The solvation energy is the difference between the two calculations. The strain energy of the ligand is the energy difference between the bound and global minimum. Here, the global minimum is the one showing the lowest $E_{vdW}+E_{coul}+E_{bonded}+G_{sol}$ among all the poses that have been minimized outside of the protein.

Twenty-three inhibitors [32] of CDK2 (1H0V), 24 inhibitors [18] (**8** to **32**, excluding **30**) of EphB4 (2VWX), and 27 uncharged inhibitors [33] of p38 alpha MAP kinase (3GC7) are used as the training set. Thirty type II inhibitors [34] of Braf (3II5), 14 charged inhibitors [33] of p38 alpha and another 20 p38 alpha inhibitors [35] (1YWR) are used as the test set. Protein structures were taken from the X-ray structure (PDB code indicated in the brackets) and prepared as described below. Some key physiochemical properties of inhibitors are summarized in.

Version 4 of AutoDock [36] was used to generate the binding poses over the conformational search space using the Lamarckian genetic algorithm. The binding site was determined by 4.0 Å away from any atom of the ligand complexed in the respective protein structure. The number of energy evaluations was 2,750,000 and the number of poses was 50. Poses were further clustered using all atom RMSD cutoff of 0.3 Å to remove redundancy and in average 20 cluster representatives were kept. All other parameters were set as default. A few poses for each inhibitor were also generated by manual modification of the scaffold present in the respective crystal structure. All poses were further minimized by CHARMM in the respective proteins. The protein structure was kept rigid in all steps.

Preparation of Protein-Ligand Complexes

One hundred kinase crystal structures complexed with small molecule inhibitors at a resolution of at least 2.5 Å were downloaded from Protein Data Bank for analysis of HBP. Hydrogen atoms were added according to the protonation states of chemical groups at pH 7. Partial charges were then assigned using MPEOE method [37], [38]. The added hydrogen atoms were minimized by the conjugate gradient algorithm to a RMS of the energy gradient of 0.01 kcal mol^{-1} $Å^{-1}$. During minimization, the electrostatic energy term was screened by a distance-dependent dielectric of 4r to prevent artificial deviations due to vacuum effects, and the default nonbonding cutoff of 14 Å was used. Furthermore, the positions of all heavy atoms were fixed.

Preparation of the Compounds Library for Virtual Screening

The compounds were selected from Zinc library [39]. Preparation included the assignment of CHARMm atom types, force field parameters [28], and partial charges [37], [38], and energy minimization with a distance dependent dielectric function using the program CHARMM [29].

Enzymatic Assay

In vitro kinase activity was measured using the Panvera Z'lyte Tyr2 kinase assay PV3191 (Invitrogen) according to the manufacturer's instructions. The reaction assay (10 μL) contained 7.5 ng of EphB4 kinase (Proqinase, Germany), 30 μM ATP, and 5% DMSO. The reaction was performed at room temperature for 1 h.

RESULTS AND DISCUSSION

Statistics of Hydrogen Bonding Penalty in Kinase Complexes

Small HBPs can be observed for the binding modes of inhibitors in the X-ray structures. One example is c-Kit tyrosine kinase with its apo and holo form in complex with Imatinib (PDB codes 1T45 and 1T46). In the apo conformation, donors/acceptors at the ATP binding site form hydrogen bonds with bound water molecules. While upon ligand binding, as shown in the holo conformation, some water molecules are displaced by Imatinib. HBP on the protein part is close to zero because new hydrogen bonds to the protein are formed to compensate for the replacement of the water molecules. However, one nitrogen atom of the Imatinib pyrimidine ring becomes water inaccessible and does not form a new hydrogen bond, leading to a penalty of 1. By contrast, the other nitrogen atom remains hydrogen bonding with a nearby bound water molecule and thus has no penalty.

To check the distribution of HBP values in crystal structures, 100 kinase-ligand complexes are investigated. In this data set, all the small molecule inhibitors have molecular weights from 200 to 700 g/mol and number of donors or acceptors from 2 to 11. The HBP has been calculated for each of them and the values are in general small, with 62% smaller than 1 and 36% and 2% in the range from 1 to 2 and 2.0 to 2.1, respectively (Figure 2). It has also been observed that larger HBPs appear in some X-ray structures, e.g., the structures of PDB code 3KVX and 1JSV, and the large values actually originate from poor fitting of small molecules to the density, a common problem in crystallography [40] which can be manifested by clash of atoms.

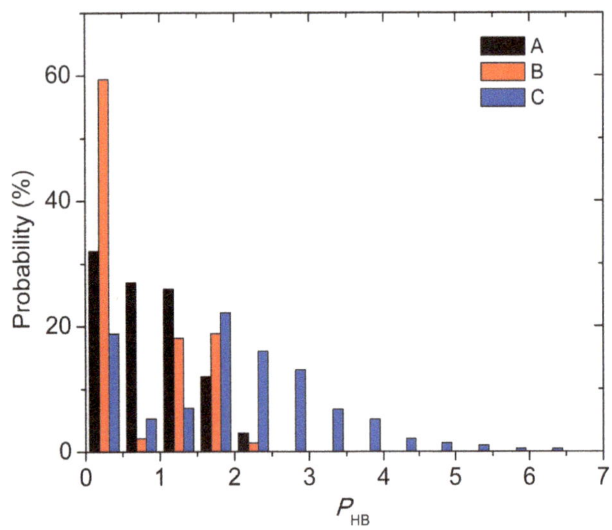

Figure 2. Distribution of hydrogen bonding penalties for: A) the binding modes in crystal structures of the 100 kinase complexes; B) poses with the most favorable calculated binding energies of the 138 molecules used in binding free energy calibration; C) all poses of the 138 molecules.

Distribution of HBPs for docked poses of small-molecule inhibitors is also evaluated. Here, the 138 molecules used in the binding free energy calibration are docked into the corresponding protein binding sites with Auto Dock. For each molecule, about 20 poses in average are generated. Then the HBPs and binding energies are calculated for all the poses. Firstly, the binding pose with the most favorable binding energy for each molecule is selected and the distribution of HBPs is plotted. As can be observed from **B** of Figure 2, the distribution is similar to that of the 100 kinase complex structures (**A**). On the other hand, the distribution of all poses (**C**) spreads more widely with the largest HBP being 6.5. Compared with the HBPs in the crystal structures (**A**), 2 is a reasonable threshold, and about 50% of poses with unrealistic binding modes can be filtered out from further evaluations.

Hydrogen Bonding Penalty Improves the Accuracy of Binding Energies Calculation

Binding energies can be calculated using equation 9 with the parameters obtained by least-squares fitting on the training data set of the 74 CDK2, EphB4, and p38α inhibitors as following:

$$\Delta G = 0.207 * \Delta E_{ff} + 1.72 * P_{HB} - 1.17 \quad \textbf{(11)}$$

The calculated binding energies show high correlation with the experimental values (R-square of 0.92) and a small RMS error of 0.69 kcal/mol (Figure 3A). Here, the parameter β-corresponds to the unit hydrogen bonding energy. Notably, the fitted value 1.72 kcal/mol is in agreement with the experimental value, e.g., breakage of a neutral hydrogen bond resulting in loss of energy from 0.5 to 1.5 kcal/mol [21]. Moreover, a charged primary amine or carboxyl group has a hydrogen bonding weight of 1.5 or 2.0, which can lead to a maximal penalty of 2.58 or 3.44 kcal/mol upon loss of the hydrogen bond/salt bridge. This value also agrees well with the experimental data (up to 4 kcal/mol) [21]. Hydrogen bonding weights were further used to rank the strength of individual hydrogen bonds in DNA base pairs, exhibiting good compatibility with the previously reported results.

The fitted model has been validated on a test set including 14 charged p38α inhibitors and 30 type II Braf inhibitors, with an RMS error of 1.12 kcal/mol (Figure 3B). Moreover, validation with different kinases shows general transferability of this model (Table 1). Transferability can be also seen for aspartic protease, e.g., HIV-1 protease and β-secretase, although a shift of 2.0 kcal/mol can be observed for the latter. Previously, we reported a two-parameter LIECE model for kinase inhibitors [13], which is not transferable for type II kinase inhibitors, HIV-protease or β-secretase inhibitors. The binding affinities predicted by the two-parameter LIECE on the 24 type I EphB4 inhibitors show about −5.0 kcal/mol shift compared with the experimental values. Clearly, the incorporation of HBP into the scoring function improves the general transferability besides the role of ligand reorganization energy [41].

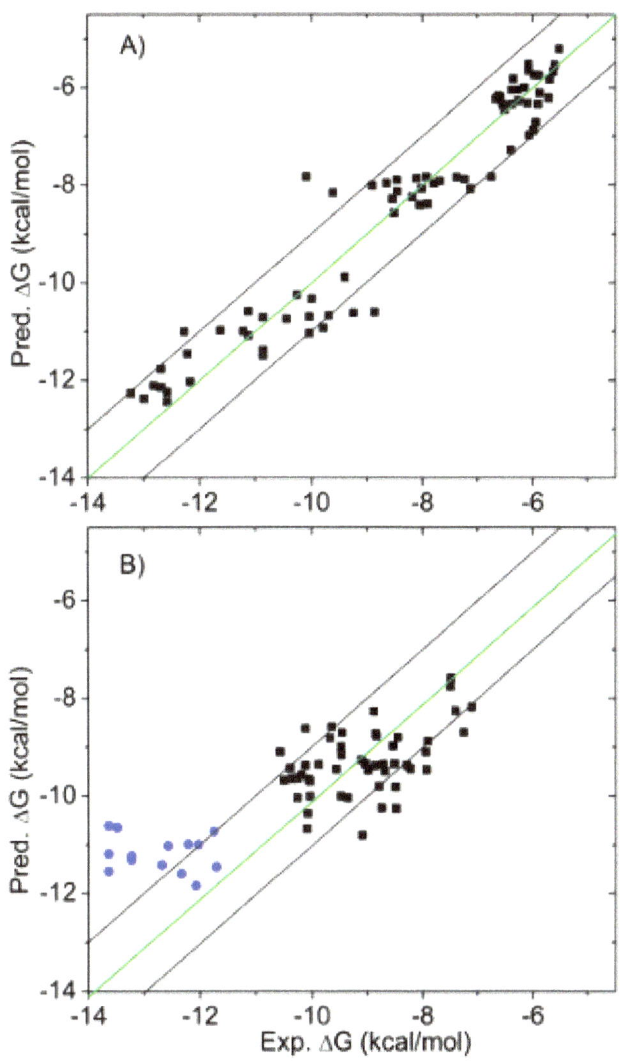

Figure 3. Comparison of the calculated versus experimental binding energies.

A) Training set of 74 inhibitors. R^2=0.92 and RMS error=0.69 kcal/mol; B) Validation set of 64 inhibitors. RMS error=1.12 kcal/mol. The blue dots indicated the 14 p38α inhibitors with one formal charge. The green diagonal line is the ideal line of perfect prediction. The black diagonals delimit the 1 kcal/mol error region.

Table 1. Further validation of the three-parameter model with kinases and aspartic protease

Protein	PDB code	ΔE_{ff} (kcal/mol)	P_{HB}	ΔG_{pred} (kcal/mol)	ΔG_{exp} (kcal/mol)
Abl	1OPJ	−64.80	1.24	−12.45	−10.81
Braf	1UWH	−57.61	1.27	−10.91	−10.45
JAK2	3E63	−30.18	0.00	−7.41	−7.91
Lck	2OFV	−59.13	0.53	−12.51	−13.23
JNK3	1PMV	−30.16	0.17	−7.12	−9.31
Ret	2X2L	−26.67	0.07	−6.58	−7.20
EGFR	1XKK	−66.60	2.30	−11.00	−10.91
CSrc	3G5D	−52.34	1.64	−9.19	−12.82
HIV-1 protease	1HIH	−65.71	1.49	−12.21	−11.01
	1HPX	−65.44	1.43	−12.26	−12.46
	1HXB	−61.24	0.95	−12.21	−13.49
	1HXW	−72.66	1.41	−13.78	−14.71
BACE-1	2QMF	−73.62	1.56	−13.72	−11.63
	2QP8	−68.76	0.47	−14.59	−11.05
	2XFI	−71.10	2.36	−11.83	−10.67

The derived model includes calculation of solvation energy by FDP which requires about 6 min on a single Intel 2.8 GHz CPU. Replacing the FDP approach with a distance-dependent dielectric model for solvation energy calculation gives similar accuracy for the neutral inhibitors at a much fast speed (10 seconds). However, distance-dependent dielectric model can only apply for non-charged compounds due to inaccurate treatment of the solvation effect, and also more false positives in a high-throughput virtual screening are observed. This comparison indicates that accurate calculation of solvation energies in prediction of binding affinities is necessary.

Virtual Screening for EphB4 Inhibitors

In a recent high throughput docking study for EphB4 inhibitors, ZINC "leads-now" library of about 20 million compounds (Mw≤350 and cLogP≤3.5) was first tailored by a pharmacophore model to generate a focused library of 103,177 compounds. This pharmacophore model was specifically designed for EphB4 type I inhibitors, consisting of a bi-dentate hydrogen bonding pattern and a conjugate hydrophobic group to be located in the deep ATP back pocket as well as geometric constraints thereof (H. Zhao, unpublished results). To our best knowledge, all known type I EphB4 inhibitors [13], [16], [17], [18] can fulfill this model.

The focused library was docked by AutoDock 4 and about 1 million poses were generated by clustering with a RMSD cutoff of 1.0 Å. The cluster representatives which do not form a hydrogen bond to NH of Met696 were further filtered out. The HBP (≤2) was then used to remove unrealistic poses (about 40%). The remaining poses were further ranked by the predicted binding energy, and the top about 30% compounds (22,517) with calculated binding energy smaller than −6 kcal/mol (~50 μM) were kept. Among them, 1381 compounds forming a hydrogen bond to Glu694 were selected and can be classified into 80 structural scaffolds. Finally, 7 scaffolds (9 compounds) of them were purchased for experimental measurements based on visual inspection of the binding modes, commercial availability and structural novelty. The procedures used in the virtual screening are shown in Figure 4. Comparison of the performances between the proposed and Auto Dock 4 scoring function is shown in.

Figure 4. Schematic picture of the high throughput docking approach.

HB stands for hydrogen bond. Met696 and Glu694 are the two key residues of the hinge loop (see also Figure 6).

Notably, 4 of the 9 tested compounds show inhibitory activity at micro-molar to high nano-molar range, with the most active compound showing IC_{50} at 300 nM (Figure 5). Interestingly, the two compound also show a high ligand efficiency [42] of −0.35 kcal/mol per non-hydrogen atom. The predicted binding mode of compound **3** (Figure 6) is further confirmed by the preliminary X-ray crystallography (J. Dong, unpublished results).

	Compound	Mw (g/mol)	P_{HB}	Pred. ΔG (kcal/mol)	IC_{50} (μM)[a]
1	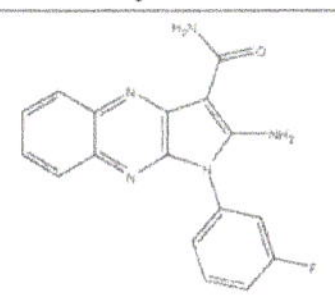	321	0.22	-7.9	0.38
2	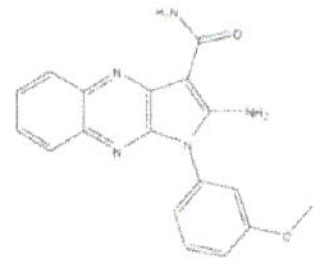	333	0.74	-8.0	0.30
3	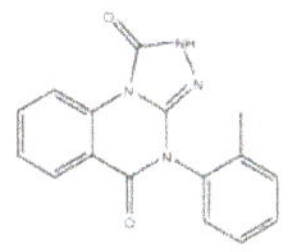	292	1.06	-6.6	15.7
4	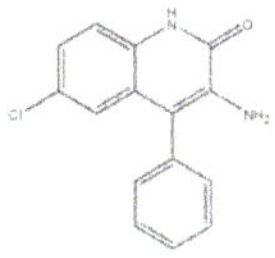	270	0.61	-6.4	42% at 20 μM

Figure 5. Identified EphB4 inhibitors by high throughput docking.

[a]All IC_{50} values are means of two to four dose-response measurements.

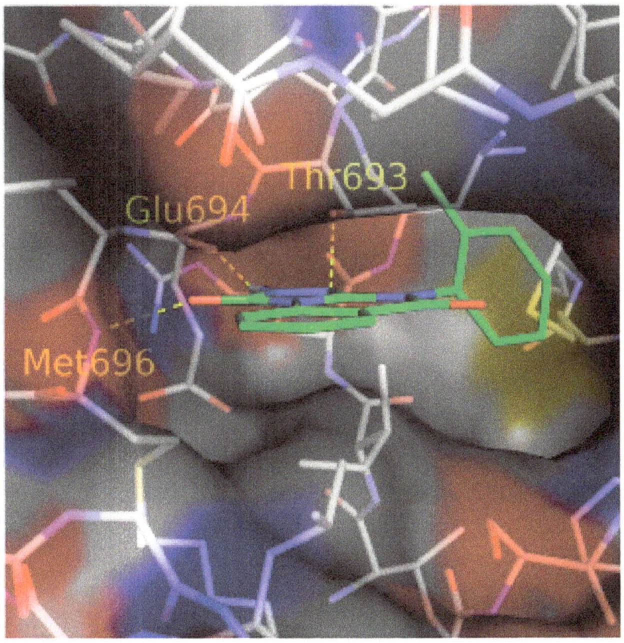

Figure 6. Binding mode of compound 3 (carbon atoms in green) predicted by docking.

The intermolecular hydrogen bonds to the residues at the hinge loop (Glu694 and Met696) and the gatekeeper (Thr693) are shown by yellow dashed lines. The protein surface is colored based on atom types with carbon in white, oxygen in red, and nitrogen in blue. This figure was prepared using PyMOL (Delano Scientific, San Carlos, CA).

CONCLUSION

Hydrogen bonding in biological system is a complex phenomenon as water competes with ligand for the hydrogen bonding sites. Removal of a group that forms a hydrogen bond in unfavorable geometry actually improves binding [21]. In view of hydrogen bonding being an exchange reaction [1], [21], [22], a new approach is proposed to evaluate the HBP upon ligand binding. Analysis of the 100 crystal structures indicates the penalty in general is low, predominantly smaller than 2 for inhibitors. A high throughput docking case shows HBP can function as an efficient filter to remove poses that unlikely bind. Incorporation of HBP into binding free energy calculation can significantly improve the predictive accuracy and transferability. The fitted parameter of 1.72 kcal/mol means loss of a neutral hydrogen bond would result in a penalty of from 0.34 to 1.72 kcal/mol in binding energy, consistent with the experimental data from 0.5 to 1.5 kcal/mol [21]. Four inhibitors of three scaffolds were discovered out of nine tested, and the binding affinity and ligand efficiency of the most potent compound is about 300 nM and 0.35 kcal/mol per non-hydrogen atom, respectively.

ACKNOWLEDGMENTS

We thank Dr. Amedeo Caflisch for useful discussions and comments on the manuscript. We thank Dr. Jing Dong for the preliminary X-ray structure. We are grateful to Armin Widmer (Novartis Basel) for continuous support with the program WITNOTP, which was used for visual analysis. Calculations were performed on the Schroedinger cluster at the Informatikdienste, University of Zurich.

REFERENCES

1. Wilkinson AJ, Fersht AR, Blow DM, Winter G (1983) Site-directed mutagenesis as a probe of enzyme structure and catalysis: tyrosyl-tRNA synthetase cysteine-35 to glycine-35 mutation. Biochemistry 22: 3581–3586.
2. Winter G, Fersht AR, Wilkinson AJ, Zoller M, Smith M (1982) Redesigning enzyme structure by site-directed mutagenesis: tyrosyl tRNA synthetase and ATP binding. Nature 299: 756–758.
3. Wilkinson AJ, Fersht AR, Blow DM, Carter P, Winter G (1984) A large increase in enzyme-substrate affinity by protein engineering. Nature 307: 187–188.
4. Bajorath J (2002) Integration of virtual and high-throughput screening. Nat Rev Drug Discov 1: 882–894.
5. Langer T, Hoffmann RD (2001) Virtual screening: an effective tool for lead structure discovery? Curr Pharm Des 7: 509–527.
6. Honig B, Nicholls A (1995) Classical electrostatics in biology and chemistry. Science 268: 1144–1149.
7. Feig M, Onufriev A, Lee MS, Im W, Case DA, et al. (2004) Performance comparison of generalized born and Poisson methods in the calculation of electrostatic solvation energies for protein structures. J Comput Chem 25: 265–284.
8. Schneider G (2010) Virtual screening: an endless staircase? Nat Rev Drug Discov 9: 273–276.
9. Friesner RA, Murphy RB, Repasky MP, Frye LL, Greenwood JR, et al. (2006) Extra precision glide: docking and scoring incorporating a model of hydrophobic enclosure for protein-ligand complexes. J Med Chem 49: 6177–6196.

10. Friesner RA, Banks JL, Murphy RB, Halgren TA, Klicic JJ, et al. (2004) Glide: a new approach for rapid, accurate docking and scoring. 1. Method and assessment of docking accuracy. J Med Chem 47: 1739–1749.
11. Wang R, Lu Y, Wang S (2003) Comparative evaluation of 11 scoring functions for molecular docking. J Med Chem 46: 2287–2303.
12. Thilagavathi R, Mancera RL (2010) Ligand-protein cross-docking with water molecules. J Chem Inf Model 50: 415–421.
13. Kolb P, Huang D, Dey F, Caflisch A (2008) Discovery of kinase inhibitors by high-throughput docking and scoring based on a transferable linear interaction energy model. J Med Chem 51: 1179–1188.
14. Hunter T (1998) The role of tyrosine phosphorylation in cell growth and disease. Harvey Lect 94: 81–119.
15. Pennisi A, Ling W, Li X, Khan S, Shaughnessy JD Jr, et al. (2009) The ephrinB2/EphB4 axis is dysregulated in osteoprogenitors from myeloma patients and its activation affects myeloma bone disease and tumor growth. Blood 114: 1803–1812.
16. Zhou T, Caflisch A (2010) High-throughput virtual screening using quantum mechanical probes: discovery of selective kinase inhibitors. ChemMedChem 5: 1007–1014.
17. Miyazaki Y, Nakano M, Sato H, Truesdale AT, Stuart JD, et al. (2007) Design and effective synthesis of novel templates, 3,7-diphenyl-4-amino-thieno and furo-[3,2-c]pyridines as protein kinase inhibitors and in vitro evaluation targeting angiogenetic kinases. Bioorg Med Chem Lett 17: 250–254.
18. Bardelle C, Cross D, Davenport S, Kettle JG, Ko EJ, et al. (2008) Inhibitors of the tyrosine kinase EphB4. Part 1: Structure-based design and optimization of a series of 2,4-bis-anilinopyrimidines. Bioorg Med Chem Lett 18: 2776–2780.
19. Lafleur K, Huang D, Zhou T, Caflisch A, Nevado C (2009) Structure-based optimization of potent and selective inhibitors of the tyrosine kinase erythropoietin producing human hepatocellular carcinoma receptor B4 (EphB4). J Med Chem 52: 6433–6446.
20. Karaman MW, Herrgard S, Treiber DK, Gallant P, Atteridge CE, et al. (2008) A quantitative analysis of kinase inhibitor selectivity. Nat Biotechnol 26: 127–132.
21. Fersht AR, Shi JP, Knill-Jones J, Lowe DM, Wilkinson AJ, et al. (1985) Hydrogen bonding and biological specificity analysed by protein engineering. Nature 314: 235–238.
22. Hine J (1972) Structural Effects on Rates and Equilibria .15. Hydrogen-Bonded Intermediates and Stepwise Mechanisms for Proton-Exchange Reactions between Oxygen-Atoms in Hydroxylic Solvents. J Am Chem Soc 94: 5766–&.
23. Steiner T, Saenger W (1993) Role of C-H…O Hydrogen-Bonds in the Coordination of Water-Molecules - Analysis of Neutron-Diffraction Data. J Am Chem Soc 115: 4540–4547.
24. Cordero B, Gomez V, Platero-Prats AE, Reves M, Echeverria J, et al. (2008) Covalent radii revisited. Dalton Trans. pp. 2832–2838.
25. Bondi A (1964) Van Der Waals Volumes+Radii. Journal of Physical Chemistry 68: 441–&.
26. Bohm HJ (1994) The development of a simple empirical scoring function to estimate the binding constant for a protein-ligand complex of known three-dimensional structure. J Comput Aided Mol Des 8: 243–256.
27. Morozov AV, Kortemme T, Tsemekhman K, Baker D (2004) Close agreement between the orientation dependence of hydrogen bonds observed in protein structures and quantum mechanical calculations. Proc Natl Acad Sci U S A 101: 6946–6951.
28. Momany FA, Rone R (1992) Validation of the General-Purpose Quanta(R)3.2/Charmm(R) Force-Field. J Comput Chem 13: 888–900.
29. Brooks BR, Bruccoleri RE, Olafson BD, States DJ, Swaminathan S, et al. (1983) Charmm - a Program for Macromolecular Energy, Minimization, and Dynamics Calculations. J Comput Chem 4: 187–217.
30. Warwicker J, Watson HC (1982) Calculation of the electric potential in the active site cleft due to alpha-helix dipoles. J Mol Biol 157: 671–679.
31. Im W, Beglov D, Roux B (1998) Continuum Solvation Model: computation of electrostatic forces from numerical solutions to the Poisson-Boltzmann equation. Computer Physics Communications 111: 59–75.
32. Gibson AE, Arris CE, Bentley J, Boyle FT, Curtin NJ, et al. (2002) Probing the ATP ribose-binding domain of cyclin-dependent kinases 1 and 2 with O(6)-substituted guanine derivatives. J Med Chem 45: 3381–3393.
33. Stelmach JE, Liu L, Patel SB, Pivnichny JV, Scapin G, et al. (2003) Design and synthesis of potent, orally bioavailable dihydroquinazolinone inhibitors of p38 MAP kinase. Bioorg Med Chem Lett 13: 277–280.
34. Berger DM, Torres N, Dutia M, Powell D, Ciszewski G, et al. (2009) Non-hinge-binding pyrazolo[1,5-a] pyrimidines as potent B-Raf kinase inhibitors. Bioorg Med Chem Lett 19: 6519–6523.
35. Koch P, Jahns H, Schattel V, Goettert M, Laufer S (2010) Pyridinylquinoxalines and pyridinylpyridopyrazines as lead compounds for novel p38 alpha mitogen-activated protein kinase inhibitors. J Med Chem 53: 1128–1137.
36. Goodsell DS, Olson AJ (1990) Automated docking of substrates to proteins by simulated annealing. Proteins 8: 195–202.
37. No KT, Grant JA, Scheraga HA (1990) Determination of Net Atomic Charges Using a Modified Partial Equalization of Orbital Electronegativity Method .1. Application to Neutral Molecules as Models for Polypeptides. Journal of Physical Chemistry 94: 4732–4739.
38. No KT, Grant JA, Jhon MS, Scheraga HA (1990) Determination of Net Atomic Charges Using a Modified Partial Equalization of Orbital Electronegativity Method .2. Application to Ionic and Aromatic-Molecules as

Models for Polypeptides. Journal of Physical Chemistry 94: 4740–4746.

39. Irwin JJ, Shoichet BK (2005) ZINC–a free database of commercially available compounds for virtual screening. J Chem Inf Model 45: 177–182.

40. Hawkins PC, Warren GL, Skillman AG, Nicholls A (2008) How to do an evaluation: pitfalls and traps. J Comput Aided Mol Des 22: 179–190.

41. Yang CY, Sun H, Chen J, Nikolovska-Coleska Z, Wang S (2009) Importance of ligand reorganization free energy in protein-ligand binding-affinity prediction. J Am Chem Soc 131: 13709–13721.

42. Hopkins AL, Groom CR, Alex A (2004) Ligand efficiency: a useful metric for lead selection. Drug Discov Today 9: 430–431.

CHAPTER **10**

VAN DER WAALS DISPERSION INTERACTIONS IN MOLECULAR MATERIALS: BEYOND PAIRWISE ADDITIVITY

Anthony M. Reilly and Alexandre Tkatchenko

INTRODUCTION

The study of intermolecular or non-covalent interactions has long been key to our understanding of complex molecular systems, particularly when considering the assembly of molecules into supramolecular systems or condensed matter. Predicting and studying cohesion in molecular materials relies heavily on accurate and physically realistic treatments for the different intermolecular forces that arise in such materials, with those that result from instantaneous fluctuations of electrons, which are referred to as dispersion or van der Waals (vdW) interactions,† being particularly important. Although they are typically weaker than electrostatic interactions or hydrogen bonding, their long-ranged nature and ubiquity in all molecular systems means they can always play an important role in non-covalent binding.[1,2] Indeed, as they scale with system size, in some cases strongly non-linearly,[3] their importance can grow for larger molecules and molecular assemblies.

Dispersion interactions are quantum-mechanical in nature and form part of the correlation energy of a system. They stem from the inherent zero-point fluctuations of electrons on an atom, which give rise to instantaneous multipole moments. These moments induce multipole moments on other atoms, which then interact with the original moment. The leading order contribution is from dipole–dipole interactions, with a $1/R^6$ dependence on distance.

The ability of an atom to form such moments (both instantaneous and induced) depends on its polarisability. The polarisability of an atom is influenced by its chemical bonding but also by its environment through polarisation and induction effects, where it is changed by coupling to the electric field of its surrounding atoms. Such effects are known to be non-additive in nature,[1] and are important for going beyond simple point charges in the modelling of electrostatics in force fields,[4,5] and modelling molecular polarisabilities.[6]

However, despite the central role of polarisability in vdW interactions, the vast majority of models ignore its non-additive and non-local nature, using simple pairwise-additive models with dispersion coefficients that do not capture any dependence on non-local environment. In neglecting many-body effects, these methods often require empirical fitting of parameters to a given molecule and environment.[2,7]

Pairwise-additive descriptions of vdW interactions have proven qualitatively successful in numerous applications ranging from bio-molecular simulations to modelling crystallisation. However, in recent years there has been a surge in interest in developing more sophisticated and quantitatively accurate approaches. The challenge of structure prediction, including polymorphism in molecular crystals,[8,9] and the availability of good-quality benchmark datasets[10–12] have driven developments for more realistic and less empirical approaches to vdW interactions.

Many of the developments in our understanding of vdW interactions in molecular systems have taken place in the context of vdW-inclusive density-functional theory (DFT). Workhorse semi-local and hybrid density functionals neglect long-range correlation[13] and considerable progress has been made in augmenting semi-local density functionals with non-local vdW contributions.[14–16] However, many of these vdW-inclusive DFT approaches are still founded on pairwise models of dispersion interactions, as use of fully ab initio approaches, such the random-phase approximation,[15] is not feasible due to their far higher computational cost than standard semi-local DFT.

Despite their successful application in many areas, including occasional successes for crystal-structure prediction,[17] quantitative and qualitative failures remain

---

Citation: Anthony M. Reilly and Alexandre Tkatchenko, Van der Waals dispersion interactions in molecular materials: beyond pairwise additivity, DOI: 10.1039/C5SC00410A.

when employing pairwise models of vdW interactions.[18–21] As a result, in recent years there has been considerable interest in studying and modelling non-additive and many-body contributions to vdW interactions, which can take a number of forms.[22] In this perspective article we highlight collective effects in vdW interactions in molecular materials, focusing on non-additivity in polarisabilities and methods that can capture these effects, in particular, the recently developed many-body dispersion (MBD) method.[23,24] Using a number of applications, we illustrate new insights into the role of collective vdW effects in the stability and properties of molecule materials. For a more in-depth discussion of the theory and physics of many-body vdW interactions the reader is referred to recent papers concerning a number of different approaches to the problem.[15, 24–27]

In the following section (Section 2) we briefly overview the theory of widely used pairwise descriptions of vdW interactions, before discussing the role of polarisation effects in vdW interactions and introducing a number of approaches to many-body van der Waals in Section 3. We illustrate the application of these methods to different systems in Section 4 before finally discussing the prospects and challenges for understanding and modelling collective vdW interactions in Section 5.

PAIRWISE-ADDITIVE DESCRIPTIONS OF VAN DER WAALS INTERACTIONS

One of the most widespread pictures and models of vdW interactions is the pairwise-additive one, whereby the vdW energy of a collection of atoms or molecules can be expressed as summation of attractive contributions from localised multipoles:

$$E_{\mathrm{vdW}} = \sum_{ij,i>j}^{N}\left(-\frac{C_6{}^{ij}}{R_{ij}{}^6}-\frac{C_8{}^{ij}}{R_{ij}{}^8}-\frac{C_{10}{}^{ij}}{R_{ij}{}^{10}}-\ldots\right) \tag{1}$$

where R_{ij} is the interatomic or fragment separation and C_6^{ij} is dipole–dipole dispersion coefficient for atoms or fragments i and j, with C_8 representing dipole–quadrupole and C_{10} representing quadrupole–quadrupole and dipole–octupole dispersion coefficients. The dispersion coefficients are determined by the corresponding polarisabilities of the atoms or fragments, relating the extent to which valence electrons can respond to local electric-field fluctuations. For the leading-order C_6 term the dispersion coefficient can be expressed as[1,28]

$$C_6{}^{ij} = \frac{3}{\pi}\int_0^{\infty}\alpha_i(i\omega)\alpha_j(i\omega)\mathrm{d}\omega, \tag{2}$$

where $\alpha(i\omega)$ is the frequency-dependent dipole polarisability. Eqn (1) has its origin in second-order perturbation theory, where the vdW interactions of two well-separated atoms can be shown to have a $1/R^6$ dependence. At small interatomic separations eqn (1) is clearly not physical, as the vdW energy diverges to $-\infty$. This arises from its derivation for well-separated fragments at long ranges. As a practical solution, in nearly all applications of the pairwise model damping is required at short distances. The foundation of many empirical force fields is the Lennard-Jones potential,[7] $E_{\mathrm{LJ}}(R_{ij}) = \sum_{ij,i\neq j}^{N} 4\epsilon\left[(\sigma/R_{ij})^{12}-(\sigma/R_{ij})^{6}\right]$ where the parameter σ controls the balance between the attractive vdW term and a short-range repulsion term, which models effects such as Pauli repulsion, or direct overlap of electron clouds. In vdW-inclusive DFT, damping functions are commonly used to switch off the vdW interaction at short distances,[14,16] not only due to this divergence but also because widely employed density functionals include short-range correlation, of which the vdW energy forms a part. These damping functions often contain one or two functional-specific parameters, which may control the distance at which the damping takes effect,[14,16] or even scale some or all of the coefficients to "match" better with DFT energies.[29,30]

Within the pairwise description of van der Waals interactions the accuracy obtainable (i.e. ignoring potential non-additive or high-order contributions) is governed by the dispersion coefficients used, whether to include higher than dipole contributions, and the damping or short-range attenuation of the interaction. Empirical potentials typically employ sets of parameters fitted to broad classes of molecules, e.g. bio-molecules,[31] organic molecules in a specific phase,[32,33] or small groups of molecules.[34] The fitting of such potentials can be highly empirical, with different terms in the potential compensating for deficiencies in other terms. Indeed, transferability of empirical potentials is a serious issue for this reason.[2]

In vdW-inclusive DFT, early approaches employed empirical parameter sets as well,[29,35] but more recent approaches aimed at more physically grounded schemes for obtaining dispersion coefficients, depending on hybridisation of atoms or the local electron density.[30,36,37] This is the first type of collective effect in vdW interactions,[22] where crowding of an atom by its neighbours results in reduction of polarisabilities from ideal free-atom values. The Tkatchenko–Scheffler (TS) model (also referred to as the DFT + vdW approach) uses the local Hirshfeld volume of an atom (V_{eff}) to reweigh accurate free-atom polarisabilities, C_6 coefficients, and van der Waals radii for this effect:

$$C_{6ii}^{\mathrm{eff}} = \left(\frac{V_i^{\mathrm{eff}}}{V_i^{\mathrm{free}}}\right)^2 C_{6ii}^{\mathrm{free}}. \tag{3}$$

$$R_{0i}^{\mathrm{eff}} = \left(\frac{V_i^{\mathrm{eff}}}{V_i^{\mathrm{free}}}\right)^{\frac{1}{3}} R_{0i}^{\mathrm{free}}. \tag{4}$$

$$\alpha_i^{\mathrm{eff}} = \left(\frac{V_i^{\mathrm{eff}}}{V_i^{\mathrm{free}}}\right)\alpha_i^{\mathrm{free}}. \tag{5}$$

The resulting C_6 coefficients are functionals of the local electron density and capture hybridisation effects well, with an accuracy of 5.5% for small molecules.[36] When used in DFT a damping function is required, with a single parameter fitted to the functional employed. Given its small-molecule accuracy and its ability to capture local effects on vdW interactions, we will use TS as a benchmark to illustrate the importance of many-body and collective effects. Other approaches, such as atom-centred potentials[38] and self-consistent vdW density functionals[39] are also employed in DFT: the interested reader is referred to a number of reviews on vdW-inclusive DFT.[14,16,40]

For sufficiently small or weakly polarisable systems, the pairwise-additive approximation is of enormous use and can be remarkably accurate. However, most molecular systems of interest are neither small nor weakly polarisable. Furthermore, in many instances it is differences in vdW energies that will govern important phenomena such as polymorphism and self-assembly, magnifying the importance of even small contributions. Such differences often depend on comparing molecules in very different environments. For example, the lattice energy of a molecular crystal is the balance between the energy of the solid lattice and that of a single isolated molecule. The vast difference in environments make collective effects that depend on an atom's or molecule's surroundings all the more important. Pairwise-additive models of vdW interaction will generally favour denser clusters and crystals, ignoring the contributions of specific arrangements of atoms and molecules in a molecular material. As such, pairwise-additive models can fail to model molecular materials both quantitatively and qualitatively. Quantitatively, ignoring collective effects leads to many pairwise methods overestimating cohesive energies of molecular crystals[19,41] and host–guest systems,[21,42] while also failing to capture the scaling of vdW interactions in different nano-materials, such as fullerenes.[3,43] Qualitatively, many approaches fail to predict the correct polymorphic ordering of molecular crystals of even relatively small molecules.[18,20] In the following section we will introduce a number of approaches to capturing many-body or collective effects in dispersion interactions. In Section 4 we will highlight the application of these methods to a number of the examples given above, showing how collective models of vdW interactions can overcome the limitations of pairwise methods.

BEYOND PAIRWISE-ADDITIVE MODELS FOR VAN DER WAALS INTERACTIONS

Despite the widespread and often successful application of pairwise-additive models of van der Waals interactions, it has long been known that they represent only an approximation to the true many-body nature of dispersion interactions. For this reason, a number of approaches have been developed to capture many-body contributions to vdW interactions, ranging from simple three-body terms added to the pairwise expression to treatments that consider the collective nature of vdW correlations in a more fundamental manner.

Triple-dipole Contributions

Given the origin of the pairwise approaches in second-order perturbation theory, the natural next step is to consider higher-order perturbative contributions. The Axilrod–Teller–Muto (ATM) three-body term considers triple-dipole energy contributions:[1,44,45]

$$E_{\mathrm{ATM}}(\boldsymbol{R}_i, \boldsymbol{R}_j, \boldsymbol{R}_k) = \frac{C_9^{ijk}(3\cos\theta_{\mathrm{a}}\cos\theta_{\mathrm{b}}\cos\theta_{\mathrm{c}} + 1)}{(R_{ij}R_{jk}R_{ik})^3}, \quad (6)$$

where the three angles θ_a, θ_b and θ_c are the internal angles of the triangle formed by the three atoms i, j and k, and C_9^{ijk} is the triple-dipole coefficient, which can also be obtained by integration of dipole polarisabilities over imaginary frequency,[1] but are also estimated from C_6 coefficients.[30] Unlike the pairwise term, the ATM expression can give both positive and negative contributions depending on the triangle formed by the atoms, although the major contribution is a repulsive one for an equilateral triangle, which is a common occurrence in close-packed solids.[1,44] The repulsion for an equilateral triangle arises because any interaction (e.g. formation of instantaneous and induced dipoles) between any two atoms of the triangle is frustrated by simultaneous interactions with the third, overall reducing the interaction compared to three distinct pairs of atoms.

The application of the ATM term has varied in different fields, as therefore has the understanding of its role and contribution. Its application to rare-gas solids yields a qualitatively correct picture of stability and elastic properties where pairwise methods fail.[46] However, as evaluating eqn (6) for a collection of atoms necessitates a triple sum over the atoms, at least in former times its used widespread use was prohibitively expensive for force-field simulations.[7] In the field of vdW-inclusive DFT such costs are negligible compared to the self-consistent cycle of the DFT calculation and a number of groups have investigated their use.[26,30,47] As we shall see below, the performance and importance of the ATM expression has been mixed. Damping the ATM contribution and matching it with pairwise contributions or a density functional are likely to be part of the reason for this,[26] but it is also the case that the ATM expression is just the first in a series of perturbative many-body contributions to vdW energy. Each of these terms can have alternating signs and analytical expressions of higher-order interactions become increasingly more cumbersome.[1,48] A deeper understanding of collective vdW effects requires going back to the definition of polarisability of molecules with many atoms, and this is what we do in the next section.

Non-additive Effects in the Polarisability

To better understand how collective behaviour can affect vdW interactions we need to step back and consider the definition of polarisability of molecular materials. The polarisability of a molecule or material relates the extent to which its electronic structure is distorted by an external electric field. Specifically, the dipole polarisability, $\boldsymbol{\alpha}$ relates an induced dipole moment to the applied electric field:

$$\boldsymbol{\mu} = \boldsymbol{\alpha E} \quad (7)$$

where $\boldsymbol{\mu}$ is the dipole moment of the atom and $\mathbf{E}$ is the local field. The dipole polarisability is a 3 × 3 tensor by construction. Its non-local nature is critical to modelling polarisation and induction effects in molecular materials, which has been an active area of research and method development for many years.[1,4,6,49,50] However, this aspect of polarisability and the importance of polarisation have largely been ignored in models of vdW interactions, despite the central role of polarisability in such interactions [cf.eqn (2)]. Accurately modelling vdW interactions requires considering polarisability and polarisation consistently.

Such a collective model of polarisability and polarisation can be achieved using classical electrodynamics, where the interaction between a collection of dipole polarisabilities can be modelled as:[1,6,51]

$$\boldsymbol{\mu}_p = \boldsymbol{\alpha}_p\left(\boldsymbol{E}_p - \sum_{q \neq p}^{N} \mathbf{T}_{pq}\boldsymbol{\mu}_q\right), \quad (8)$$

where $\boldsymbol{\mu}_p$ is the dipole moment of the pth atom, $\mathbf{E}_p$ is the local field at p, and the term in brackets denotes the overall electric field due to coupling with the dipole of the qth atom, through the dipole–dipole coupling tensor $\mathbf{T}_{pq}$. Although this model is classical, the input polarisabilities and definition of the coupling tensor can be based on a quantum-mechanical description. The coupling tensor is normally defined as:

$$\mathbf{T}_{pq}^{ij} = \nabla_i \nabla_j V(\mathbf{r}_{pq}), \quad (9)$$

where i and j represent the Cartesian components of the vector joining p and q, and V is the Coulomb potential.

It is possible to rearrange this expression and cast it in a matrix form so that an effective self-consistently screened (SCS) polarisability tensor can be defined that captures the effect of coupling between all the dipoles:[6,51]

$$\begin{pmatrix} \mu_1 \\ \mu_2 \\ \cdots \\ \mu_N \end{pmatrix} = \begin{pmatrix} \alpha_1^{-1} & \mathbf{T}_{12} & \cdots & \mathbf{T}_{1N} \\ \mathbf{T}_{21} & \alpha_2^{-1} & \cdots & \mathbf{T}_{2N} \\ \cdots & \cdots & \cdots & \cdots \\ \mathbf{T}_{N1} & \mathbf{T}_{N2} & \cdots & \alpha_N^{-1} \end{pmatrix}^{-1} \begin{pmatrix} E_1 \\ E_2 \\ \cdots \\ E_N \end{pmatrix}, \quad \begin{pmatrix} \mu_1 \\ \mu_2 \\ \cdots \\ \mu_N \end{pmatrix} = \mathbf{B} \begin{pmatrix} E_1 \\ E_2 \\ \cdots \\ E_N \end{pmatrix} \quad (10)$$

The effective polarisability matrix, **B**, is often referred to as the relay matrix.[6,51] The effective "screened" polarisability of e.g. the molecule or crystal, is obtained by fully contracting **B**. Alternatively, the polarisability can be partitioned among the atoms, with one practical scheme being to row or column contract **B**:[23]

$$\alpha_{p,\mathrm{SCS}} = \sum_{q}^{N} B_{pq}, \quad (11)$$

where B_{pq} is a 3 × 3 block of the 3N × 3N relay matrix. The result is an effective polarisability for each atom or fragment, which even if the input polarisabilities are isotropic is in general anisotropic due to the environment of each atom. Some care must be taken in the definition of the $\mathbf{T}_{pq}$, as the bare Coulomb potential ($1/|\mathbf{r}_q - \mathbf{r}_p|$) can lead to non-physical interactions between atoms at short separations, where the distributed nature of the atomic dipoles must be properly accounted for.[1] Damping of $\mathbf{T}_{pq}$ can yield reasonable results,[6] but modelling each dipole as having an isotropic Gaussian distribution (corresponding to an isotopic quantum harmonic dipole oscillator) gives a simple but remarkably useful model.[23,24,49] The above treatment can be performed using frequency-dependent polarisabilities, leading to frequency-dependent screened polarisabilities and hence screened C_6 coefficients viaeqn (2).[23] Typically, this is done by first obtaining the average of the eigenvalues of the screened polarisability tensor for each atom. As the TS polarisabilities are functionals of the local electron density and already contain the important short-ranged effect of hybridisation,[22,36] they are an ideal choice for the input polarisabilities in eqn (10), giving rise to the TS + SCS method for obtaining screened polarisabilities and C_6 coefficients, which can be used in a pairwise energy expression.

The screened polarisabilities and C_6 coefficients readily illustrate deviations from the isotropic model of polarisability and C_6 often used in force fields and vdW-inclusive DFT methods. The polarisability tensor should be a positive semi-definite 3 × 3 matrix, which can be represented by an ellipsoid, in the same fashion as anisotropic displacement tensors are depicted in crystallography.[52]Fig. 1 shows the screened polarisability tensors of benzene and aspirin as ellipsoids. In benzene, strong coupling in the plane of the molecule leads to an increase of in-plane polarisability and a corresponding reduction of out-of-plane polarisability. This is observed experimentally in the molecular polarisability with the out-of-plane polarisability being 55% of the in-plane value,[6] which is correctly captured by TS + SCS polarisabilities but not by the TS ones.[25]

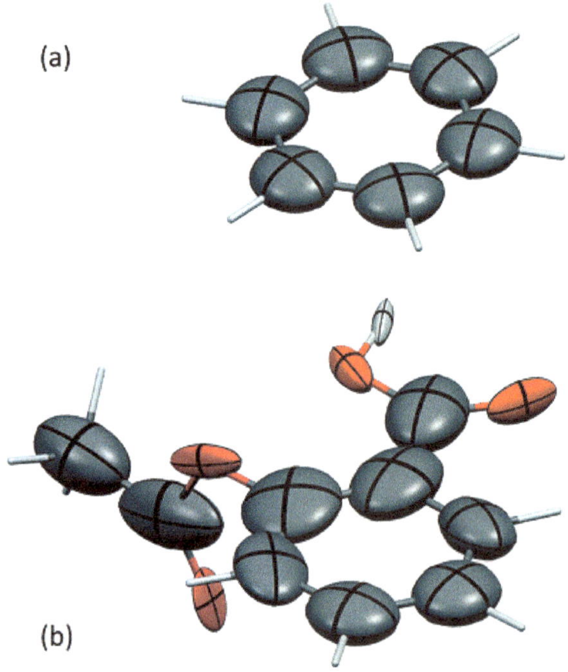

Figure 1. The anisotropic, screened polarisability tensors of (a) benzene and (b) aspirin. The tensors are represented as ellipsoids. Aliphatic H atoms are very strongly polarised towards their neighbouring C atom and are not shown for clarity. Mercury 3.4 has been used to produce these figures.[55]

For aspirin the same in-plane polarisation within its benzene ring can be observed, although now the different groups surrounding the ring distort the polarisabilities to different extents. Aspirin is a useful illustration of two competing polarisation effects of electrodynamic screening. Within a molecule the effect of collective vdW interactions on C_6 coefficients is to increase them, as atoms polarise strongly towards neighbours, with the average C_6 coefficient of the C atoms in aspirin being 43.3 a.u. compared to 31.0 a.u. for TS. However, in the solid state, coupling with the surrounding molecules limits an atom's ability to polarise with its neighbours, resulting in net depolarisation. This leads to an average C-atom C_6 of 39.0 a.u. in aspirin, a reduction of around 10% relative to the gas phase.[53] Comparing these coefficients to values from condensed-phase force fields is difficult, as they typically involve balancing dispersion, repulsion and electrostatic contributions and frequently have very different values. For example, an aromatic C···C interaction in the W99 force field has a dispersion coefficient of around 30 a.u.,[33] whereas in the UNI force field all C···C dispersion interactions have a coefficient of 42 a.u.[54]

Fig. 2(a) shows the screened polarisability tensors of a "buckyball catcher" host–guest system ($C_{60}@C_{60}H_{28}$).[56] The host molecule exhibits dramatic departures from a model of spherical and near identical polarisabilities for each C atom. In some cases the polarisation is so strong and deviates so much from the isotropic model that the polarisability tensor is not positive semi-definite, although it is still possible to calculate C_6 coefficients for these atoms. Fig. 2(b) shows the change in

TS + SCS C_6 coefficients when the complex is formed. The blue spheres denote atoms that depolarise when the complex is formed, leading to a smaller C_6 coefficient. This depolarisation is strongest at the interface between the two molecules and there is a clear asymmetry to the C_{60} molecule. Using the TS + SCS C_6 coefficients in a pairwise expression for the dispersion energy for this complex this yields a reduction of the binding energy of the order of 30 kJ mol^{-1} compared to TS,[56] illustrating the importance of capturing these depolarisation effects in cohesion. The TS + SCS approach to pairwise vdW energies has been applied to a number of systems by Bučko and co-workers,[57] and has shown that electrodynamic screening is important in a variety of situations.

(a)

(b)

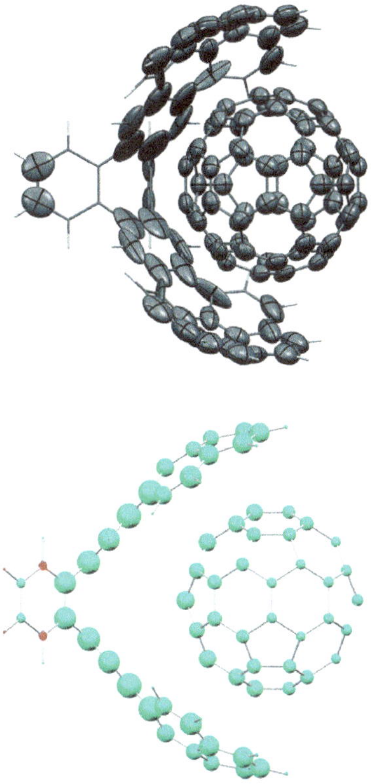

Figure 2 (a) The anisotropic, screened polarisability tensors of a "buckyball catcher" host–guest system and (b) a plot of the difference between C_6 coefficients for the whole complex and the separate parts (in the same geometry).[56] Some atoms polarise so strongly that they cannot be represented by an ellipsoid in (a). The size of the spheres in (b) indicates the magnitude of the difference, with blue meaning a decrease in C_6 upon forming the complex and red denoting an increase in the complex.

Van der Waals Correlations from the ACFD theorem

The self-consistently screened polarisabilities and C_6 coefficients discussed above capture important collective effects in vdW interactions, as does the ATM expression for triple-dipole contributions. However, the accurate treatment of vdW interactions requires methods than can capture both screening and many-body effects in a systematic manner.[22]

Within the framework of the adiabatic connection fluctuation-dissipation (ACFD) theorem there is an exact expression for the correlation energy in terms of density–density response functions $\chi(\mathbf{r},\mathbf{r}',i\omega)$, which describes the density response at $\mathbf{r}'$ as a result of a perturbation at $\mathbf{r}$.[15] The correlation energy of a system can be written in terms of the response functions of a corresponding mean-field system (χ_0), and the response functions of a corresponding interacting system, where the Coulomb interaction v is scaled by a parameter λ:[58]

$$E_c = -\frac{1}{2\pi}\int_0^\infty d\omega \int_0^1 d\lambda \; \mathrm{Tr}[\chi_\lambda v - \chi_0 v]. \tag{12}$$

The bare response function (χ_0) can be determined from the eigenfunctions and eigenvalues of an electronic-structure calculation,[59] but some approximation is required for the interacting response functions. Within the popular random-phase approximation (RPA) the interacting response functions are approximated in terms of the non-interacting ones as follows:[15,58]

$$\chi_\lambda^{\mathrm{RPA}} = \frac{\chi_0}{1-\chi_0 \lambda v}. \tag{13}$$

RPA is commonly employed with DFT or HF orbitals and contains the ring diagrams of coupled-cluster theory with double excitations. RPA can have remarkable performance for many properties in a variety of situations,[15,60] particularly for long-range correlation and dispersion, but has a substantial computational cost, scaling normally as N^4 or higher,[61] which has limited its widespread use. However, a recent N^3 algorithm may increase its use.[62]

The many-body Dispersion (MBD) Method

The ACFD theorem gives a route to the correlation energy of a system in terms of response functions. While it is natural to consider the full response of a system based on a full and accurate representation of its electronic structure, model response functions could also be used to capture non-local correlation energy.

It is possible therefore to obtain an accurate but efficient description of the long-range many-body van der Waals interactions by combining the ACFD expression for the correlation energy with response functions that model only the dipole response of an atom. This is the approach adopted by the many-body dispersion (MBD) method,[23] where the correlation or dispersion energy is given by:[24]

$$E_{c,\mathrm{MBD}} = \frac{1}{2\pi}\int_0^\infty d\omega \mathrm{Tr}[\ln(1-\mathbf{AT})], \tag{14}$$

with $\mathbf{T}$ being an appropriate dipole–dipole coupling tensor and $\mathbf{A}$ is a diagonal 3N × 3N matrix, with each block derived from an appropriate isotropic distributed polarisability: $\mathbf{A}_{lm} = -\delta_{lm}\alpha_l(i\omega)$. In this expression the dipole response of each atom is modelled using a single localised, isotropic quantum harmonic dipole oscillator. The RPA treatment of this model considers all orders of dipole coupling at long range, including triple and higher dipole energies and collective screening simultaneously and systematically.[23,24,63] Due to the localised description of the response or polarisability, which is centred on atoms, the MBD method is ideally suited to materials or systems with a finite band gap. The ACFD theorem expression can be recast as a straightforward Hamiltonian expression,[63] which can then be directly diagonalised using simple and efficient matrix algebra. The contribution of the MBD energy to atomic forces is also straightforward to derive and implement.[24]

The MBD method has to-date largely been coupled with density functional theory,[23,64,65] where the underlying functional accounts for short-range correlations. Therefore, some form of range separation or damping of the MBD correlation energy is required to avoid double counting, so that the MBD energy is curtailed at short distances. This is reflected in the input polarisabilities and the definition of the dipole–dipole coupling tensor. The polarisabilities are typically obtained in a two-step process where short-range hybridisation effects are considered by using TS polarisabilities and screening contributions are evaluated using the self-consistent screening equations [eqn (8)]. More rigorous range separation can be achieved by considering only short-range screening contributions, as in the MBD@rsSCS method,[24] and in general this improves the performance of the method. A more in-depth discussion of the theory can be found in a number of recent publications.[24,25]

MBD has been combined with DFT largely due to the balance of accuracy and computational cost that this combination achieves, as we shall discuss below. However, MBD could be used to provide a model of long-range correlation or dispersion in other methods including empirical approaches such as force fields. Bereau and von Lilienfeld have recently introduced a scheme using MBD with polarisabilities derived from Voronoi tessellation.[66]

COLLECTIVE VDW INTERACTIONS IN DISPERSION COEFFICIENTS, ENERGIES AND RESPONSE PROPERTIES

Pairwise and collective models of vdW interactions have been applied in a wide variety of situations. Taking examples from small molecules, supramolecular systems and molecular crystals, in this section we highlight the importance of collective vdW methods in accurately modelling molecular materials.

Collective effects in Polarisabilities and Dispersion Coefficients

We begin by briefly discussing the role of collective effects in polarisabilities and dispersion coefficients of atoms and molecules in different environments. For small molecules, pairwise-additive methods can capture C_6 coefficients quite well, with the TS method has an accuracy of 5.5% for a set of 1225 small-molecule C_6 values,[67] and other pairwise methods and vdW density functionals have values in the range of 8–30% using the same or comparable test sets.[30,68,69] However, the underlying polarisabilities of small molecules can still deviate substantially from the isotropic model employed by pairwise-additive approaches, as seen above for benzene and aspirin in Section 3.2. For 18 small molecules the TS method has a mean absolute relative error of 76.3% for the fractional anisotropy of the molecular polarisability with respect to experiment.[25] Including self-consistent screening, as in the TS + SCS method, captures a substantial portion of this anisotropy, reducing the error to 33.5%.

In molecular crystals the role of collective effects in molecules is more complex. As noted for aspirin above (Section 3.2), short-range polarisation within molecules increases C_6coefficients while long-range depolarisation due to surrounding molecules decreases them. The acene crystals are an excellent example of this, with average C-atom C_6 coefficient increasing from naphthalene to pentacene for the isolated molecule but decreasing in the solid as the molecule grows in size.[70] Capturing these effects is important for predicting the dielectric constant and optical properties of the acene crystals. The TS method overestimates dielectric constants substantially, while TS + SCS is much closer to experiment and benchmark calculations.[70,71] Screened C_6 coefficients of a set of semiconductors obtained from time-dependent DFT (TDDFT) have also been shown to differ significantly from pairwise values, with a pairwise vdW term based on screened C_6 coefficients noticeably improving the cohesive properties of the semiconductors.[72] The importance of defects and vacancies has also been studied in semiconductors (using TS + SCS), showing that C_6 coefficients are significantly altered by their presence, affecting stability and defect migration barriers.[73]

Collective effects in nanostructures lead not only to changes in C_6 coefficients but their fundamental scaling with system size. For a pair of fullerenes pairwise-additive models predict a C_6 coefficient between the two fullerenes that scales as n^2, where n is the number of C atoms in the fullerene. A number of recent studies have shown larger power laws. TS + SCS gives scaling of $n^{2.35}$,[3] while modelling the polarisability using spherical shells of valence electron density yields a power law in the asymptotic limit of $n^{2.75}$.[43] TDDFT calculations of smaller fullerenes shows a power law of $n^{2.2}$.[74] C_6 coefficients in other nanomaterials, such as multilayer graphene and carbon nanotubes also feature large and non-linear changes with system size that are not captured by simple pairwise methods.[3] This behaviour stems from the high symmetry and reduced dimensionality of these systems (e.g. "0-D" fullerenes; "2-D" graphene), which lead to very strong polarisation effects. For layers of graphene sheets, such effects lead to stronger interactions within sheets, which then reduces binding between sheets. TS overestimates the binding between sheets substantially [87 meV per C atom versus an experimental estimate of 52(2) meV per C atom]. In contrast, PBE0 + MBD and RPA calculations agree much better with experiment, with values of 50 meV per C atom and 48 meV per C atom, respectively.[24,75]

While the role of non-additive behaviour in C_6 coefficients is minor in small molecules, it is clear that as we move to solids and nanomaterials these effects become much more pronounced, changing not only absolute values but also altering fundamental scaling and behaviour. In the following subsections we will discuss how this manifests itself in cohesive energies and properties.

Collective vdW Interactions in Molecular Assemblies

Isolated molecular assemblies, ranging from small molecular dimers to supramolecular host–guest complexes, are amongst some of the most widely used benchmarks used to evaluate and understand vdW interactions and their contribution to cohesion. For small molecular dimers there are a number of databases of coupled-cluster singles, doubles and perturbative triples [CCSD(T)] binding energies, representing a "gold-standard" benchmark with which to compare different approaches. The S22 and S66 databases of Hobza and co-workers contain 22 and 66 dimers,[10,11] respectively, comprised of hydrogen-bonded interactions, vdW interactions and mixed bonding. These databases have been used to assess pairwise C_6/R^6 methods,[53,76] many-body dispersion (MBD)[24,53] and also vdW density functionals.[77] In general, many pairwise methods perform well for these databases, yielding mean absolute errors (MAEs) of 1–2 kJ mol^{-1}, better than the chemical-accuracy target of 4.2 kJ mol^{-1}. In comparison, semi-local functionals without pairwise vdW contributions can yield errors greater than 8 kJ mol^{-1},[76] highlighting the importance of these interactions, even for simple systems. Considering collective vdW contributions can further improve upon the performance of pairwise methods,[23,53] but in general their role in these small systems is not dominant.

In larger supramolecular systems, collective effects are clearly more important. The S12L dataset of Grimme[12] comprises 12 supramolecular host–guest complexes. Risthaus and Grimme found that including at least three-body vdW contributions (from the ATM expression) was necessary for these systems, with contributions of the order of 10 kJ mol^{-1}. The "buckyball catcher" in Fig. 2 has a contribution of 13.4 kJ mol^{-1} from the ATM expression.[21] Three-body contributions generally improved pairwise methods, although not consistently, with D3 coupled to the PBE functional[78] having a mean absolute error for binding energies of 8.8 kJ mol^{-1} but PBE + D3 + E_{ATM} having an MAE of 10 kJ mol^{-1}. The pairwise Tkatchenko–Scheffler vdW method yields an MAE of the order of 30 kJ mol^{-1},[24] despite easily reaching chemical accuracy for small molecules. Including collective vdW contributions using the MBD method reduces this error dramatically to around 6.7 kJ mol^{-1}, approaching close to the statistical error of benchmark diffusion Monte Carlo binding energies for a subset of the S12L.[42]

Decomposition of the many-body contributions in the S12L shows a significant contribution from three-body interactions, in line with the contribution seen with the ATM expression,[21]but the series only starts to converge with six-body contributions.[42] These are not short-ranged interactions of the same nature as the ATM-type three-body term, but rather the effect on the dispersion energy of coupling and polarisation of groups of six atoms via more long-ranged $1/R^3$ terms, as in eqn (10). In the $C_{60}@C_{60}H_{28}$ system, even atoms in the C_{60} molecule far from the host molecule are affected by screening, as can be seen from the change in C_6 coefficients in Fig. 2(b). As the many-body expansion typically alternates between positive and negative energy contributions, truncating at a lower order could lead to spurious over or under binding. Similar high-order contributions in DNA helices, where pairwise methods can omit more than 20% of the full many-body vdW energy.[79] Many-body contributions have also been found to be important for describing the conformational preference of the Ac-Phe-Ala_5-$LysH^+$ peptide chain.[80] It is clear from these studies that modelling the binding of supramolecular systems requires some model of collective and many-body vdW interactions.

Collective van der Waals effects in Molecular Solids

Molecular solids represent a diverse and challenging class of systems for applying different descriptions of vdW interactions to different properties, ranging from lattice energy to elastic properties and phonons.

Benzene Crystal

Benzene is naturally one of the most widely studied molecular crystals. Recently, several groups have studied the lattice energy of benzene and have found it important to include many-body contributions.[47,81–84] One approach is to expand the lattice energy as a mean-field contribution (e.g. a HF energy) plus a summation of contributions from dimers, trimers etc. within the lattice, calculated at a high level of theory. As MP2 cannot capture three-body dispersion contributions, Kennedy et al. have been able to estimated the three-body dispersion contribution by comparing MP2 energies for over 350 trimers from the crystal with CCSD(T) energies and using the ATM expression for further more-distant trimers.[81] The contribution is of the order of +3.2 kJ mol^{-1} (i.e. reducing the magnitude of the lattice energy), around 6% of the lattice energy, but smaller than previous estimates based solely on the ATM expression with different C_9 coefficients, which range from 3.8–7.0 kJ mol^{-1}.[47,81,82]

Yang et al. have used a similar expansion (including dimers, trimers, tetramers and additional correlation contributions) to estimate the benzene lattice energy as −55.9(9) kJ mol^{-1}, which compares very well with an experimental estimate of −55.3(22) kJ mol^{-1} obtained from sublimation enthalpies and theoretical zero-point contributions.[84] This value involved not only a significant three-body contribution to the lattice energy but a four-body one as well. It is worth noting that "body" is a complete benzene molecule in these approaches, and thus these many-body contributions are not directly comparable to the three-atom contributions of the ATM expression.

The MBD method has also been applied to benzene, where the difference between the TS pairwise lattice energy and the MBD lattice energy is around +11 kJ mol^{-1} when both are coupled to either the PBE or PBE0 (ref. 85) density functionals, reducing the lattice energy to −55.0 kJ mol^{-1} for PBE + MBD or −51.2 kJ mol^{-1} with PBE0 + MBD.[41] The pairwise TS method, which is accurate to 1.5 kJ mol^{-1} for small-molecule binding energies and 5.5% for C_6 coefficients,[36,53] significantly overestimates the lattice energy in the absence of collective screening and many-body energy contributions. The role of these effects is even more pronounced when considering larger oligoacenes such as naphthalene and anthracene.[41,70]

X23 dataset

Recent efforts to develop benchmark dataset such as the C21 (ref. 19) and the X23 (ref. 41 and 53) have permitted a broader assessment of absolute lattice energies with various methods. The X23 is based on the C21 database and contains a mixture of hydrogen-bonded solids, vdW-bound solids and solids with a mixture of the two interactions. Experimental lattice energies were determined from experimental sublimation enthalpies,[86] theoretical vibrational contributions (including vdW contributions) and, where available, experimental heat capacities, with an expected accuracy just better than chemical accuracy (4.2 kJ mol^{-1}), which is largely limited by experimental uncertainties in sublimation enthalpies.[41]

The TS pairwise method has a mean absolute error (MAE) for the X23 of 10.0 kJ mol^{-1} when coupled with the PBE0 functional, systematically overestimating lattice energies with an accuracy nearly an order of magnitude worse than that found for small dimers in the S22. Including many-body contributions with MBD yields an MAE of 3.9 kJ mol^{-1}, within the chemical-accuracy limit. The reduction in lattice energies is particularly pronounced for vdW-bound systems, as can be seen in Fig. 3, but is still important for hydrogen-bonded systems. The reduction in lattice energies arises from long-range depolarisation of vdW interactions, as seen for C_6 coefficients (Section 3.2) and repulsive many-body contributions, as would be partly obtained with the ATM expression (Section 3.1).

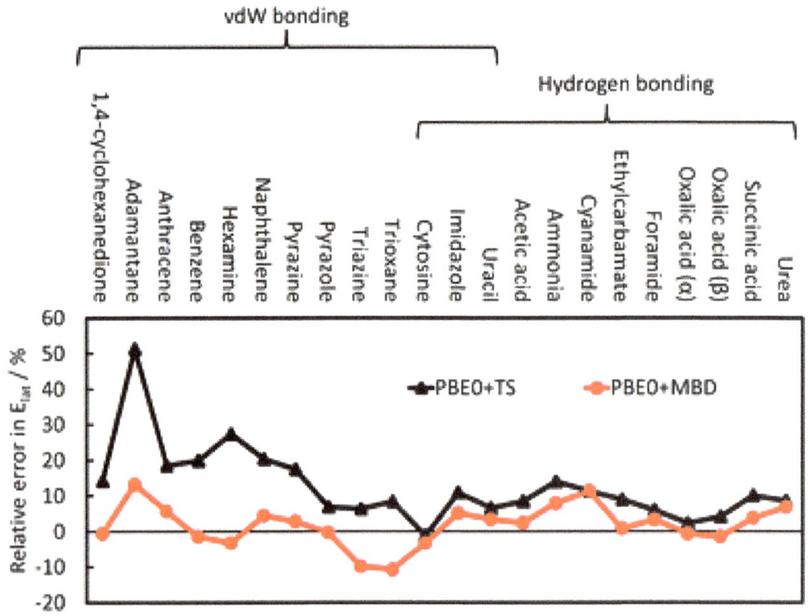

Figure 3. The relative error in the calculate lattice energies for 22 vdW-bound and hydrogen-bonded systems. A positive value indicates over-estimation of the experimental lattice energies; see ref. 41 for more details and a discussion on CO_2.

A number of other pairwise approaches have been tested with the C21 and X23 databases, including XDM, D2 and D3.[19,41,87] These methods have better MAEs than the TS pairwise approach, with D3 coupled with the TPSS meta-GGA functional[88] being capable of reaching chemical accuracy as well. However, these methods typically have a more flexible and empirical approach to fitting and determining the C_6 coefficients and damping of the short-range vdW interactions.[19,29,76] For example, in the D2 approach there is a global and empirical scaling of C_6 coefficients as part of the fitting process for the density functional employed. D3 and XDM both use the Becke–Johnson damping function,[89] which features an additional fitting parameter compared to TS and does not damp the vdW contribution to zero as $R \rightarrow 0$. The TS and MBD approaches are, in part, fitted to the S22 and S66 sets of small dimers to avoid including effects from larger molecules. D3 employs a wider set of benchmark data,[30] while in XDM these parameters are often re-fitted for a specific basis set,[90] potentially factoring in other contributions into the parameters.

Many of the empirical adjustments used in pairwise approaches would improve the performance of the TS method for molecular crystals, which we have noted before in the case of a global scaling of the TS C_6 coefficients for solids.[41] However, such adjustments limit the transferability of methods, creating specific approaches for different materials. As noted above, Risthaus and Grimme advocate the use of the ATM expression in host–guest systems. For molecular solids, using the ATM expression has proven difficult though. It requires an alternative form of the XDM method's damping function, worsening the underlying pairwise MAE and not yielding any additional improvement.[26] For D3, the three-body contributions worsens the MAE for the TPSS functional by about 1.3 kJ mol^{-1} but improves the HSE06 functional[91] result by 0.8 kJ mol^{-1}.[87] These inconsistent results with D3 and XDM methods stress the need for methods that model vdW interactions seamlessly and systematically across a range of different sizes and forms of molecular materials. By including collective vdW effects, MBD is able to systematically improve upon the performance of the TS method, providing an accurate method with minimal empiricism.

The non-local vdW-DF and vdW-DF2 functionals[39,92] have also been applied to the C21 set and show better performance than TS but worse than MBD or XDM, with MAEs of 6.1 kJ mol^{-1} and 6.4 kJ mol^{-1}, respectively.[19,41] Such functionals model the long-range correlation by performing a double integral over the whole electron density, summing contributions to the vdW using an interaction kernel that has a $1/R^6$ dependence at long range. Much of the work in improving these functionals has focussed not on the non-local part but on the semi-local functional used alongside it, as the performance of these methods is sensitive to the coupling between the non-local correlation and the semi-local functional used.[14,92] This can be exploited to tune these functionals to certain types of systems.[93] However, this leads to a broad spectrum of vdW-DF-based methods with their performance wildly varying across a range of molecular materials.

Molecular-crystal Polymorphism

The importance of many-body contributions extends from lattice energies to relative lattice energies and the challenge of polymorphism, where energy differences in the range of 0.1–1.0 kJ mol^{-1} need to be resolved.[8,94] On such scales even small many-body or collective vdW contributions can yield different predictions for the most stable polymorph or the low-energy portion of the polymorphic landscape.

TS and MBD have been applied to a number of polymorphic systems, demonstrating the importance of collective effects in these systems. Oxalic acid has two known forms, α and β [Fig. 4(a)]. Experimentally, the α form is found to be slightly more stable by around 0.2 kJ mol^{-1} in terms of lattice energy.[19,41,95] A number of pairwise vdW methods used in conjunction with DFT (including TS, XDM and D2)[19] systematically overestimate the stability of the β form by up to 4 kJ mol^{-1} [Fig. 4(b)]. Accounting for collective effects using MBD and employing the non-empirical hybrid functional PBE0 yields the correct qualitative ordering. The D3 approach also yields the correct ordering in conjunction with PBE0 but has a much larger energy difference than observed experimentally,[87] while vdW-DF2 is also able to reproduce the correct ordering (note however that this result will depend on the employed semi-local approximation to the exchange–correlation functional).[19] It is worth noting here that the best result in Fig. 4(b) in terms of absolute difference (PBE + MBD) represents the wrong ordering of the polymorphs. As methods often aim to minimise MAEs it is important that trends are still captured correctly. TPSS + D3 yields a smaller MAE for the X23 than PBE0 + D3 but incorrectly predicts β oxalic acid to be more stable.[87]

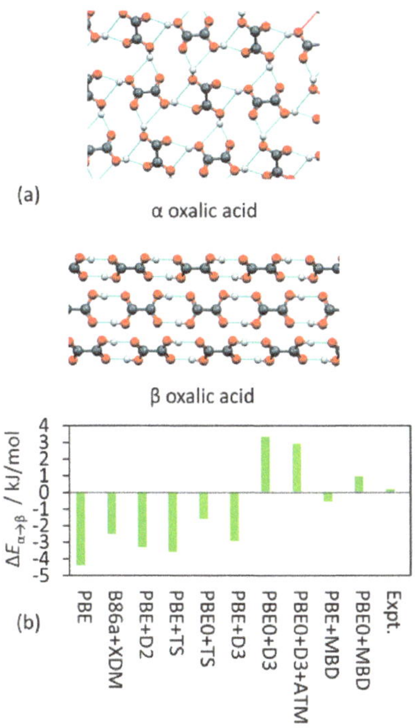

Figure 4 (a) The hydrogen-bonding interaction in α and β oxalic acid. (b) The relative energies between the polymorphs from experiment and as obtained with a number of different DFT methods.

Glycine is a challenging polymorphic system where computational methods often have difficulties. An experimental ordering has been established for three known polymorphs of glycine, with the γ one being be the most stable, followed by α and then β. The TS pairwise method systematically overestimates the stability of the α and β forms, while again including collective effects the correct experimental ordering is obtained with PBE0 + MBD.[20] The potential-energy surface and geometry of glycine is also markedly affected by many-body contributions.

Finally, oxalyl dihydrazide is another system where many-body contributions have been explored, although only in the three-body ATM form. Two separate studies have shown the ATM expression affecting relative energies by more than 1 kJ mol^{-1},[18] which in one case leads to a different ordering of the polymorphs.[18,96] The difference may arise in the damping of the three-body interaction or the manner in which the C_9 coefficients are determined.

The importance of collective vdW interactions in these systems is interesting, as they would normally be thought of as small hydrogen-bonded systems, where vdW interactions are very much a secondary concern. In larger systems the importance of vdW interactions can grow and with it the role of collective vdW interactions. In Section 3.2 the significant effect of screening on the C_6 coefficients of aspirin was highlighted. Overall, the MBD lattice energy for form-I aspirin is reduced by nearly 12 kJ mol^{-1} relative to the TS pairwise value.[53] Wen and Beran have estimated the 0 K lattice energy difference between form-I and form-II aspirin to be of the order of ±0.1 kJ mol^{-1} using MP2-based methods.[97] MBD agrees with this, yielding an essentially degenerate lattice energy difference for the two forms. TS gives an energy difference of −0.2 kJ mol^{-1} in favour of form II ($\Delta E = E_I - E_{II}$). Despite the lattice energy change of nearly 12 kJ mol^{-1}, the relative energies are very similar with the pairwise and many-body treatment. However, as we shall see below, going beyond the 0 K lattice energy reveals substantial differences between the two vdW treatments.

Beyond 0 K Lattice Energies

Many studies of polymorphism focus on lattice potential energies, devoid of any zero-point or vibrational contributions. However, the finite-temperature stability of a given structure is dependent on lattice free-energy differences at a given temperature. In the harmonic limit these can be readily determined from phonon or solid-state vibrational spectra of the different polymorphs.[98,99]

As noted above, the two known forms of aspirin[100] are degenerate in terms of lattice energy when calculated with a variety of different methods. However, while intergrowth of the two forms occurs,[101] form I appears to be the more thermodynamically stable and abundant form.[102] The free-energy difference between the two forms at 298 K is relatively small (−0.7 kJ mol^{-1}, favouring form II) when calculated with the TS pairwise method. However, when the MBD method is used there is significant stabilisation of form I by 2.6 kJ mol^{-1},[99] rationalising experimental observations. This arises due to softening of low-frequency vibrational modes in form I, making it entropically stabilised. The different arrangements of methyl groups and hydrogen bonds in the lattice may explain the ability of many-body interactions to influence the vibrations differently in the forms. Given the fact that the two forms are quite similar structurally (differing in weak C–H···O hydrogen bonds),[100] such effects are likely to be found in other polymorphic systems where more significant differences can occur.

Dispersion interactions also play an important role in the elastic or mechanical properties of molecular materials, which are important for their processing and practical application, e.g.tabletting of pharmaceuticals.[103] Recently, the importance of pairwise vdW interactions has been highlighted for peptide nanostructures.[104] Aspirin also highlights the role of collective vdW interactions in elastic behaviour. With the TS pairwise method the bulk modulus of form-I aspirin is 12.40 GPa,[99] compared to a value of 7.77 GPa at 300 K.[105] Collective vdW contributions reduce the calculated bulk modulus by over 20% to 9.58 GPa, with a similar reduction for form II. Much of the remaining difference is likely to arise from vibrational contributions.[106] The shear modulus is also reduced by over 20% by many-body contributions in both forms. The reduction in the strength of vdW interactions that arises to collective effects is more apparent in derived properties such as the elastic constants. This phenomenon has also been seen using the TS + SCS method, which has been used to determine the bulk modulus of rare-gas molecular crystals by Bučko et al.[57] For rare-gas crystals the screened pairwise method reduces bulk moduli by around 6–7%. For benzene, PBE with D2, TS, and TS + SCS all yield a bulk modulus of 10 GPa,[57,107,108] overestimating a 0 K experimental estimate of 8.0 GPa by 25%.[109] In contrast, RPA based on PBE orbitals (E_{xx}/RPA@PBE) yields a value of 7.5 GPa,[110] which is much closer to experiment.

While the focus of research in vdW interactions has largely been on cohesive energies and structure, much of the importance of molecular materials derive from their properties. The study of elastic properties and vibrations already suggests that collective vdW effects play an important role in governing the behaviour of molecular materials beyond their fundamental stability.

FUTURE DIRECTIONS AND CHALLENGES

In recent years our understanding of the importance and role of van der Waals interactions, especially those of a collective nature, has substantially increased. Here, we highlight some of the different aspects and areas of vdW interactions that still pose significant challenges, while simultaneously offering significant opportunities for new insights.

Considerable effort has been put into creating a number of benchmark datasets for assessing different approaches to non-covalent interactions,[10–12,19,41] which have spurred development of new methods and our understanding of collective vdW interactions. These have largely focused on energetics and to a lesser extent geometry. For solids, obtaining suitable benchmark energies is challenging in the absence of a tractable high-level computational method. Benchmark data for derived properties, such as phonons and elastic properties, are even harder to obtain from theory alone and experimental data requires accounting for thermal and zero-point contributions. However, further understanding of the importance of vdW interactions hinges on new benchmark data to push and challenge computational methods. Gaining acceptance for new general-purpose benchmark datasets, as the S22 and S66 already are for small-molecule dimers, is also crucial.

The role of many-body and collective vdW interactions in liquids and solutions, which play a critical role in so much of chemistry, has not been explored to the same extent as for solids or isolated systems. Probing such systems requires sampling of their dynamic behaviour, which is computationally demanding from first principles. However, as the recent work shows, it is possible to decouple models of collective vdW interactions from DFT, allowing their use in empirical potentials.[66] The Drude oscillator model has been used to model induction and vdW interactions in rare-gas liquids[27] and water.[111] For general use, correlation between models of vdW interactions and e.g. repulsion terms in force fields may make initial parameterisation and applications challenging. However, new force fields may lead to new insights into collective effects on scales not encountered in "perfect" molecular crystals or isolated supramolecular systems. For example, recent experimental work has shown that the role of dispersion in molecular recognition of apolar alkyl chains in solution appears to be much smaller than anticipated from pairwise models of vdW interactions,[112] which might well stem from screening of vdW interactions.

Expanding the application of vdW-enabled methods will require extending and developing these collective vdW methods further. The MBD method at present considers only dipolar responses to the vdW energy, ignoring higher multipole contributions. While these contributions are short ranged compared to the dipole term, which appears to be dominant in many cases,[41,113,114] higher multipole contributions may well be more important for molecular materials at high pressures, where atoms approach closer together. The model response functions used in MBD could be extended to capture such higher-order responses.[24,25] Currently, the use of MBD requires a single empirically fitted range-separation parameter to couple it with a chosen density functional, just as pairwise methods require one or more parameters in their damping functions. Defining the range separation in terms of the functional itself or developing new functionals implicitly alongside MBD would remove this empiricism and may enhance its transferability. Another area of on-going development is the application of MBD to metallic systems. Due to its localised and atom-centred nature, MBD is ideally suited to systems with a band gap. Delocalised metallic states can lead to strong screening effects within the metal but also between surfaces and adsorbing molecules, as captured by the vdW^{surf} method.[115] MBD could be extended to capture these effects by modelling delocalised states alongside localised atom-centred contributions, with Wannier functions being one possible route to achieve this.[64]

The importance of developing these methods and concepts further lies in the need to model, and ultimately design, hybrid materials that may contain different manifestations of collective vdW interactions. Such manifestations will need to be captured in a systematic and seamless manner to achieve this: an ongoing challenge for the vdW research community.

CONCLUSIONS

Dispersion or vdW interactions are critical for the stability of numerous molecular materials. However, many studies consider such interactions only in a pairwise-additive form (mainly consisting of C_6/R^6 terms), neglecting collective effects that originate from the quantum-mechanical many-body electronic interactions they result from.

Considering these collective effects in the dipole limit can yield screened polarisabilities and C_6 coefficients for atoms that can dramatically depend on the local and non-local environment of an atom or molecule, varying significantly from "one size fits all" C_6 coefficients used in many pairwise approaches. Such environment-aware interactions could play an important role in molecular recognition and assembly. Beyond capturing such effects in the coefficients used in the pairwise expression, there are additional many-body energy contributions that need to be captured. The many-body dispersion method,[23] captures both effects in an efficient RPA-like treatment of dispersion and non-local correlation, and has been used here to illustrate their role.

In small-molecule dimers collective effects are not essential for achieving accurate binding energies. However, in supramolecular systems even the simplest triple-dipole contribution (from the ATM expression) can yield corrections to binding energies of the order of 10 kJ mol^{-1}.[21] However, analysing the many-body energy suggests that contributions as high as sixth order are needed for converged interaction energies, as the collective effect of atoms on polarisabilities is much more long ranged than the three-body ATM energy contribution.

Many-body effects are also important in molecular crystals. Recent work by a number of groups has highlighted the need for capturing three- and even four-molecule interactions to accurately model the stability of the comparatively "simple" and symmetric benzene crystal.[81,84] The MBD method corrects systematic overestimation of lattice energies found in the TS pairwise method and yields the correct polymorphic ordering for a number of challenging polymorphic systems. Adding three-body contributions to other approaches has yielded mixed results though, highlighting the need for methods that treat vdW interactions in a systematic and seamless manner.

More recent explorations of vdW interactions in response properties highlights the role of collective vdW interactions not just in stability but in the properties and function of molecular materials. More benchmark data are needed to fully explore these aspects of vdW interactions and there is also a need to move beyond static lattices and isolated systems to consider liquids, solutions, and disordered and dynamic systems.

ACKNOWLEDGEMENTS

The authors acknowledge funding support from the European Research Council (ERC Starting Grant VDW-CMAT) and thank A. Ambrosetti, R. A. DiStasio Jr, V. V. Gobre, O. A. von Lilienfeld, N. Marom, M. Scheffler and G.-X. Zhang for numerous stimulating discussions and collaborations in this field.

REFERENCES

1. A. J. Stone, The Theory of Intermolecular Forces, Oxford University Press, UK, 2nd edn, 2013.
2. A. J. Stone, Science, 2008, 321, 787–789 .
3. V. V. Gobre and A. Tkatchenko, Nat. Commun., 2013, 4, 2341.
4. P. Ren and J. W. Ponder, J. Comput. Chem., 2002, 23, 1497–1506 .
5. P. Ren, C. Wu and J. W. Ponder, J. Chem. Theory Comput., 2011, 7, 3143–3161 .
6. B. T. Thole, Chem. Phys., 1981, 59, 341–350 .
7. M. P. Allen and D. J. Tildesley, Computer Simulation of Liquids, Oxford University Press, Oxford, 1989.
8. S. L. Price, Acc. Chem. Res., 2009, 42, 117–126 .
9. J. Bernstein, Polymorphism in Molecular Crystals, Oxford University Press, USA, 2007.
10. P. Jurěcka, J. Šponer, J. Černý and P. Hobza, Phys. Chem. Chem. Phys., 2006, 8, 1985–1993.
11. J. Řezáč, K. E. Riley and P. Hobza, J. Chem. Theory Comput., 2011, 7, 2427–2438 .
12. S. Grimme, Chem.-Eur. J., 2012, 18, 9955–9964 .
13. K. Burke, J. Chem. Phys., 2012, 136, 150901 .
14. J. Klimeš and A. Michaelides, J. Chem. Phys., 2012, 137, 120901 .
15. J. F. Dobson and T. Gould, J. Phys.: Condens. Matter, 2012, 24, 073201 .
16. S. Grimme, Wiley Interdiscip. Rev.: Comput. Mol. Sci., 2011, 1, 211–228 .
17. M. A. Neumann, F. J. J. Leusen and J. Kendrick, Angew. Chem., Int. Ed., 2008, 47, 2427–2430 .
18. S. Wen and G. J. O. Beran, J. Chem. Theory Comput., 2012, 8, 2698–2705 .
19. A. Otero-de-la-Roza and E. R. Johnson, J. Chem. Phys., 2012, 137, 054103 .
20. N. Marom, R. A. DiStasio Jr, V. Atalla, S. Levchenko, A. M. Reilly, J. R. Chelikowsky, L. Leiserowitz and A. Tkatchenko, Angew. Chem., Int. Ed., 2013, 52, 6629–6632 .
21. T. Risthaus and S. Grimme, J. Chem. Theory Comput., 2013, 9, 1580–1591 .
22. J. F. Dobson, Int. J. Quantum Chem., 2014, 114, 1157–1161 .
23. A. Tkatchenko, R. A. DiStasio Jr, R. Car and M. Scheffler, Phys. Rev. Lett., 2012, 108, 236402 .
24. A. Ambrosetti, A. M. Reilly, R. A. DiStasio Jr and A. Tkatchenko, J. Chem. Phys., 2014, 140, 18A508 .
25. R. A. DiStasio Jr, V. V. Gobre and A. Tkatchenko, J. Phys.: Condens. Matter, 2014, 26, 213202 .
26. A. Otero-de-la-Roza and E. R. Johnson, J. Chem. Phys., 2013, 138, 054103 .
27. A. P. Jones, J. Crain, V. P. Sokhan, T. W. Whitfield and G. J. Martyna, Phys. Rev. B: Condens. Matter Mater. Phys., 2013, 87, 144103 .
28. H. B. G. Casimir and D. Polder, Phys. Rev., 1948, 73, 360–372 .
29. S. Grimme, J. Comput. Chem., 2006, 27, 1787–1799 .
30. S. Grimme, J. Antony, S. Ehrlich and H. Krieg, J. Chem. Phys., 2010, 132, 154104 .
31. C. Oostenbrink, A. Villa, A. E. Mark and W. F. Van Gunsteren, J. Comput. Chem., 2004, 25, 1656–1676 .
32. W. L. Jorgensen, D. S. Maxwell and J. Tirado-Rives, J. Am. Chem. Soc., 1996, 118, 11225–11236 .
33. D. E. Williams, J. Comput. Chem., 2001, 22, 1154–1166 .
34. D. C. Sorescu, B. M. Rice and D. L. Thompson, J. Phys. Chem. B, 2000, 104, 8406–8419 .
35. M. A. Neumann and M.-A. Perrin, J. Phys. Chem. B, 2005, 109, 15531–15541 .
36. A. Tkatchenko and M. Scheffler, Phys. Rev. Lett., 2009,

102, 073005 .

37. A. D. Becke and E. R. Johnson, J. Chem. Phys., 2007, 127, 154108 .

38. O. A. von Lilienfeld, I. Tavernelli, U. Rothlisberger and D. Sebastiani, Phys. Rev. Lett., 2004, 93, 153004 .

39. M. Dion, H. Rydberg, E. Schröder, D. C. Langreth and B. I. Lundqvist, Phys. Rev. Lett., 2004, 92, 246401 .

40. A. Tkatchenko, L. Romaner, O. T. Hofmann, E. Zojer, C. Ambrosch-Draxl and M. Scheffler, MRS Bull., 2010, 35, 435–442 .

41. A. M. Reilly and A. Tkatchenko, J. Chem. Phys., 2013, 139, 024705 .

42. A. Ambrosetti, D. Alfé, R. A. DiStasio and A. Tkatchenko, J. Phys. Chem. Lett., 2014, 5, 849–855 .

43. A. Ruzsinszky, J. P. Perdew, J. Tao, G. I. Csonka and J. M. Pitarke, Phys. Rev. Lett., 2012, 109, 233203 .

44. B. M. Axilrod and E. Teller, J. Chem. Phys., 1943, 11, 299–300 .

45. Y. Muto, Proc. Phys.-Math. Soc. Jpn., 1943, 17, 629–631.

46. G. J. Keeler and D. N. Batchelder, J. Phys. C: Solid State Phys., 1970, 3, 510 .

47. O. A. von Lilienfeld and A. Tkatchenko, J. Chem. Phys., 2010, 132, 234109 .

48. W. L. Bade, J. Chem. Phys., 1958, 28, 282–284 .

49. A. Mayer, Phys. Rev. B: Condens. Matter Mater. Phys., 2007, 75, 045407 .

50. M. W. Cole, D. Velegol, H.-Y. Kim and A. A. Lucas, Mol. Simul., 2009, 35, 849–866 .

51. J. Applequist, J. R. Carl and K.-K. Fung, J. Am. Chem. Soc., 1972, 94, 2952–2960 .

52. K. N. Trueblood, H.-B. Bürgi, H. Burzlaff, J. D. Dunitz, C. M. Gramaccioli, H. H. Schulz, U. Shmueli and S. C. Abrahams, Acta Crystallogr., Sect. A: Found. Crystallogr., 1996, 52, 770–781 .

53. A. M. Reilly and A. Tkatchenko, J. Phys. Chem. Lett., 2013, 4, 1028–1033 .

54. A. Gavezzotti and G. Filippini, J. Phys. Chem., 1994, 98, 4831–4837 .

55. C. F. Macrae, I. J. Bruno, J. A. Chisholm, P. R. Edgington, P. McCabe, E. Pidcock, L. Rodriguez-Monge, R. Taylor, J. van de Streek and P. A. Wood, J. Appl. Crystallogr., 2008, 41, 466–470 .

56. A. Tkatchenko, D. Alfé and K. S. Kim, J. Chem. Theory Comput., 2012, 8, 4317–4322 .

57. T. Bučko, S. Lebègue, J. Hafner and J. G. Ángyán, Phys. Rev. B: Condens. Matter Mater. Phys., 2013, 87, 064110 .

58. H.-V. Nguyen and S. de Gironcoli, Phys. Rev. B: Condens. Matter Mater. Phys., 2009, 79, 205114 .

59. (a) S. L. Adler, Phys. Rev., 1962, 126, 413–420 ; (b) N. Wiser, Phys. Rev., 1963, 129, 62–69 .

60. X. Ren, P. Rinke, C. Joas and M. Scheffler, J. Mater. Sci., 2012, 47, 7447–7471 .

61. X. Ren, P. Rinke, V. Blum, J. Wieferink, A. Tkatchenko, A. Sanfilippo, K. Reuter and M. Scheffler, New J. Phys., 2012, 14, 053020 .

62. M. Kaltak, J. Klimeš and G. Kresse, Phys. Rev. B: Condens. Matter Mater. Phys., 2014, 90, 054115 .

63. A. Tkatchenko, A. Ambrosetti and R. A. DiStasio Jr, J. Chem. Phys., 2013, 138, 074106 .

64. P. L. Silvestrelli, J. Chem. Phys., 2013, 139, 054106 .

65. M. Modrzejewski, G. Chalasinski and M. M. Szczesniak, J. Chem. Theory Comput., 2014, 10, 4297–4306 .

66. T. Bereau and O. A. von Lilienfeld, J. Chem. Phys., 2014, 141, 034101 .

67. A. Tkatchenko and O. A. von Lilienfeld, Phys. Rev. B: Condens. Matter Mater. Phys., 2008, 78, 045116 .

68. E. R. Johnson and A. D. Becke, J. Chem. Phys., 2005, 123, 024101 .

69. O. A. Vydrov and T. Van Voorhis, Phys. Rev. A, 2010, 81, 062708 .

70. B. Schatschneider, J.-J. Liang, A. M. Reilly, N. Marom, G.-X. Zhang and A. Tkatchenko, Phys. Rev. B: Condens. Matter Mater. Phys., 2013, 87, 060104 .

71. S. Sharifzadeh, A. Biller, L. Kronik and J. B. Neaton, Phys. Rev. B: Condens. Matter Mater. Phys., 2012, 85, 125307 .

72. G.-X. Zhang, A. Tkatchenko, J. Paier, H. Appel and M. Scheffler, Phys. Rev. Lett., 2011, 107, 245501 .

73. W. Gao and A. Tkatchenko, Phys. Rev. Lett., 2013, 111, 045501 .

74. J. Kauczor, P. Norman and W. A. Saidi, J. Chem. Phys., 2013, 138, 114107 .

75. S. Lebègue, J. Harl, T. Gould, J. G. Ángyán, G. Kresse and J. F. Dobson, Phys. Rev. Lett., 2010, 105, 196401 .

76. L. Goerigk, H. Kruse and S. Grimme, ChemPhysChem, 2011, 12, 3421–3433 .

77. O. A. Vydrov and T. Van Voorhis, J. Chem. Phys., 2010, 133, 244103 .

78. J. P. Perdew, K. Burke and M. Ernzerhof, Phys. Rev. Lett., 1996, 77, 3865–3868 .

79. R. A. DiStasio Jr, O. A. von Lilienfeld and A. Tkatchenko, Proc. Natl. Acad. Sci. U. S. A., 2012, 109, 14791–14795 .

80. M. Rossi, S. Chutia, M. Scheffler and V. Blum, J. Phys. Chem. A, 2014, 118, 7349–7359 .

81. M. R. Kennedy, A. R. McDonald, A. E. DePrince, M. S. Marshall, R. Podeszwa and C. D. Sherrill, J. Chem. Phys., 2014, 140, 121104 .

82. S. Wen and G. J. O. Beran, J. Chem. Theory Comput., 2011, 7, 3733–3742 .

83. R. Podeszwa, B. M. Rice and K. Szalewicz, Phys. Rev. Lett., 2008, 101, 115503 .

84. J. Yang, W. Hu, D. Usvyat, D. Matthews, M. Schütz and G. K.-L. Chan, Science, 2014, 345, 640–643 .

85. C. Adamo and V. Barone, J. Chem. Phys., 1999, 110, 6158–6170 .

86. NIST WebBook and references within, http://webbook.nist.gov/chemistry/name-ser.html, Accessed September, 2014.

87. J. Moellmann and S. Grimme, J. Phys. Chem. C, 2014, 118, 7615–7621.

88. J. Tao, J. P. Perdew, V. N. Staroverov and G. E. Scuseria, Phys. Rev. Lett., 2003, 91, 146401 .
89. E. R. Johnson and A. D. Becke, J. Chem. Phys., 2006, 124, 174104 .
90. A. Otero-de-la-Roza and E. R. Johnson, J. Chem. Phys., 2012, 136, 174109.
91. J. Heyd, G. E. Scuseria and M. Ernzerhof, J. Chem. Phys., 2003, 118, 8207–8215 ; (b) J. Heyd, G. E. Scuseria and M. Ernzerhof, J. Chem. Phys., 2006, 124, 219906 .
92. K. Lee, E. D. Murray, L. Kong, B. I. Lundqvist and D. C. Langreth, Phys. Rev. B: Condens. Matter Mater. Phys., 2010, 82, 081101 .
93. J. Klimeš, D. R. Bowler and A. Michaelides, J. Phys.: Condens. Matter, 2010, 22, 022201 .
94. S. L. Price, Acta Crystallogr., Sect. B: Struct. Sci., Cryst. Eng. Mater., 2013, 69, 313–328 .
95. H. G. M. de Wit, J. A. Bouwstra, J. G. Blok and C. G. de Kruif, J. Chem. Phys., 1983, 78, 1470–1475 .
96. D. Presti, A. Pedone, M. C. Menziani, B. Civalleri and L. Maschio, CrystEngComm, 2014, 16, 102–109.
97. (a) S. Wen and G. J. O. Beran, Cryst. Growth Des., 2012, 12, 2169–2172 ; (b) Y. Huang, Y. Shao and G. J. O. Beran, J. Chem. Phys., 2013, 138, 224112 .
98. S. Baroni, S. de Gironcoli, A. Dal Corso and P. Giannozzi, Rev. Mod. Phys., 2001, 73, 515–562 .
99. A. M. Reilly and A. Tkatchenko, Phys. Rev. Lett., 2014, 113, 055701 .
100. A. D. Bond, R. Boese and G. R. Desiraju, Angew. Chem., Int. Ed., 2007, 46, 615–617 .
101. A. D. Bond, R. Boese and G. R. Desiraju, Angew. Chem., Int. Ed., 2007, 46, 618–622 .
102. S. Varughese, M. S. R. N. Kiran, K. A. Solanko, A. D. Bond, U. Ramamurty and G. R. Desiraju, Chem. Sci., 2011, 2, 2236–2242.
103. S. Karki, T. Friščić, L. Fábián, P. R. Laity, G. M. Day and W. Jones, Adv. Mater., 2009, 21, 3905–3909 .
104. I. Azuri, L. Adler-Abramovich, E. Gazit, O. Hod and L. Kronik, J. Am. Chem. Soc., 2014, 136, 963–969 .
105. J. D. Bauer, E. Haussühl, B. Winkler, D. Arbeck, V. Milman and S. Robertson, Cryst. Growth Des., 2010, 10, 3132–3140.
106. Y. P. Varshni, Phys. Rev. B: Solid State, 1970, 2, 3952–3958 .
107. T. Bučko, J. Hafner, S. Lebègue and J. G. Ángyán, J. Phys. Chem. A, 2010, 114, 11814–11824 .
108. W. A. Al-Saidi, V. K. Voora and K. D. Jordan, J. Chem. Theory Comput., 2012, 8, 1503–1513 .
109. E. J. Meijer and M. Sprik, J. Chem. Phys., 1996, 105, 8684–8689 .
110. D. Lu, Y. Li, D. Rocca and G. Galli, Phys. Rev. Lett., 2009, 102, 206411 .
111. A. Jones, F. Cipcigan, V. P. Sokhan, J. Crain and G. J. Martyna, Phys. Rev. Lett., 2013, 110, 227801 .
112. L. Yang, C. Adam, G. S. Nichol and S. L. Cockroft, Nat. Chem., 2013, 5, 1006–1010 .
113. N. Marom, A. Tkatchenko, M. Rossi, V. V. Gobre, O. Hod, M. Scheffler and L. Kronik, J. Chem. Theory Comput., 2011, 7, 3944–3951 .
114. S. N. Steinmann and C. Corminboeuf, J. Chem. Theory Comput., 2012, 8, 4305–4316 .
115. V. G. Ruiz, W. Liu, E. Zojer, M. Scheffler and A. Tkatchenko, Phys. Rev. Lett., 2012, 108, 146103

CHAPTER 11 INTERMOLECULAR INTERACTIONS

Andrew G. P. Maloney Peter A. Wood and Simon Parsons

INTRODUCTION

Methods for interpretation of molecular crystal structures have advanced considerably in the past decade. While analy-sis of intermolecular interactions using graphical tools such as Mercury[1] can be achieved in a matter of seconds using fast algorithms based on geometry, it is increasingly common to evaluate intermolecular interaction energies using ab initio methods,[2–7] PIXEL calculations,[8] symmetry-adapted perturbation theory[9] or force fields.[10] The results can be used to visualise contributing energy terms using Hirshfeld surfaces[11] or energy vectors or frameworks.[12–14] This progress has been applied to areas such as polymorphism[15] and energy landscapes,[16] cocrystals and solvates,[17] crystal engineering,[18] molecular recognition[19] and extreme conditions research.[20]

The techniques listed above have been applied extensively in work on organic materials.[21] There is, nevertheless, substantial interest in intermolecular interactions in metal-containing systems.[22] Crystal engineering and supramolecular chemistry frequently make use of strategies involving metals. For example, Orpen and co-workers have shown that H-bond acceptors based on metal halides and oxalates can be used to form more reliable and reproducible supramolecular building blocks than those based on purely organic ligands.[23,24] Use of diplatinum thiocarboxylate complexes in bottom-up assembly of conductive one-dimensional nano-structures has been suggested on the basis of the strong (~50 kJ mol^{-1}) intermolecular Pt···Pt interactions that occur in these systems.[25] Density functional theory (DFT) has been used to study the different intermolecular energies of alterna-tive RuIJII) hydrogenation catalysis pathways,[26–28] while more recently Li et al. performed calculations to investigate the performance of cobalt and copper analogues of a pre-existing nickel catalyst for olefin purification.[29] A great deal of com-putational effort has been invested in the study of the bind-ing affinities and selectivities of metalloprotein–ligand inter-actions in these systems,[30–32] and the desire to find more efficient methods of drug design means that computational analysis of metal-based pharmaceuticals is an ever expanding field,[33] exemplified by the analysis of the interactions of zinc ions with several anti-inflammatory drugs.[34] Metal–organic frameworks, large porous structures consisting of metal ions linked by organic ligands, are increasingly being studied as potential gas storage and separation materials, and a variety of computational methods are used to study the adsorption of small molecules in these systems.[35–37]

The development of the PIXEL method,[8,21,38,39] a semi-empirical technique for evaluating intermolecular interactions based on integrations over calculated electron densities of molecules, has allowed energetic analysis in organic crystal structures to be carried out quickly with an accuracy comparable to high level quantum mechanical methods.[40,41] PIXEL calculations yield the total lattice energy partitioned into individual molecule–molecule energies, which are themselves partitioned into four terms: coulombic, polarisation, dispersion and repulsion. The separation of contributions allows for the character of individual interactions and overall crystal packing to be inferred from the dominant terms, providing chemical insight. The PIXEL method requires definition of cer-tain atomic parameters, but these are mostly physically mea-surable quantities such as ionisation potentials, and one of the most appealing features of the method is the transferability of parameters across many different chemical systems. The PIXEL method is thus a potentially valuable addition to established techniques applied to metal-containing systems, such as ab initio methods and calculations based on force fields.

The aims of this paper are (i) to define a validated PIXEL parameter set for use with the d-block metals, and (ii) to

Citation: Andrew G. P. Maloney, Peter A. Wood and Simon Parsons, Intermolecular interactions, DOI: 10.1039/c5ce01522g.

illustrate possible applications of the method in interpreting intermolecular interactions in metal-containing structures. This expands the potential applicability of the PIXEL method from organic materials (~40% of the CSD) to the majority of organic and organometallic structures (~85% of the CSD).

PARAMETERISATION AND VALIDATION CRITERIA

Definition of Metal Parameters

The four energy terms evaluated during a PIXEL calculation depend on a small number of fundamental atomic parameters. Values of these parameters for atoms common in organic chemistry are embedded in the PIXEL code. In the present work we have defined new values for transition metals and validated them against experimental sublimation enthalpies (H_{sub}) using the convention that lattice energies are approximately equal to $-H_{sub}$. This Procedure assumes that there are no intramolecular structural changes on pass-ing from the solid state to the gas phase. The names used to refer to the parameters in the following sections are those used in the PIXEL program documentation and Gavezzotti's publications, where full details of their definition, use and significance can be found.[21,38,39]

Definition of some parameters is straightforward. ZTOT and ZVAL are the total number of electrons and the number of valence electrons in the neutral atom; POTIO is the first ionisation energy (in atomic units) and WEIGHT is the atomic weight.[42] For other parameters a choice among several possibilities needs to be made. Unless otherwise specified, values used for non-metallic elements in the compounds were the program defaults.

Dispersion energies are calculated in a London-type expression in which the ionisation energy of a pixel is used to approximate its 'oscillator strength'. DIFA (also given the symbol β in Gavezzotti's papers) is a "variable ionisation" parameter which controls the diminution of the ionisation energy of a pixel as the distance from the nucleus increases.[21] Gavezzotti's recommended value of 0.4 $Å^{-1}$ was used throughout. Variation of DIFA(Ti) between 0.1 and 1.0 $Å^{-1}$ yielded values of the lattice energy of $TiCl_4$ between −46.4 and −57.7 kJ mol^{-1}. A value of −53.0 kJ mol^{-1} was obtained‡ with the standard value 0.4 $Å^{-1}$, compared to the experimental sublimation enthalpy of 51.9 kJ mol^{-1} (calculated from the enthalpies of formation of the gas and solid taken from the NIST Chemis-try WebBook, http://webbook.nist.gov/chemistry/).

The covalent radii of atoms, RINTER, are used in PIXEL to check for short internuclear distances, and not used to calculate energies. Values were taken from Emsley's compilation.[42] RAVDW, the van der Waals radius, is used to assign pixels of electron density to atomic basins. The sets of values reported by Nag et al.[43] and Batsanov[44] were tested.

The atomic polarisability, POLZE or α in $Å^3$, appears in the calculation of both the polarisation and dispersion terms. For non-metallic species the PIXEL method makes use of the Slater– Kirkwood approximation to estimate α according to eqn (1),

$$\alpha = \frac{1}{a_0}\left(\frac{R_{\mathrm{vdW}}}{1.05\sqrt{3}}\right)^4 \qquad (1)$$

where a_0 is the Bohr radius and R_{vdW} is the van der Waals radius. The Nag and Batsanov radii were tested. The Clausius–Mossotti relation (eqn (2)) is another simple method for estimating atomic polarisabilities:

$$\alpha = \frac{\varepsilon - \varepsilon_0}{\varepsilon + 2\varepsilon_0}\frac{3V_{\mathrm{m}}}{4\pi} \qquad (2)$$

where V_m is the atomic volume, ε is the dielectric constant of the species and ε_0 is the permittivity of free space. Atomic vol-umes were obtained from the crystal structures of the elemental metals at room temperature and pressure. For pure metals, the dielectric constant $\varepsilon \to \infty$, so that the first term of this equation tends to unity, giving $\alpha = 3V_m/4\pi$. Variation of α(Ti) between 3.5 and 5.0 $Å^3$ yielded values of the $TiCl_4$ lattice energy between −45.9 and −57.1 kJ mol^{-1}.

The electronegativity, ELNEG, is used in the calculation of the repulsion energy. Both Pauling and Allred–Rochow values[42] were investigated.

PIXEL Calculations and the Treatment of Ligand Parameters

PIXEL calculations can be carried out in one of two ways using the programs PIXELc or PIXELd of Gavezzotti's CLP package of modelling and structure analysis software.[45] PIXELd is intended for the calculation of interaction energies of discrete dimers. PIXELc is used for calculations on a complete crystal structure, yielding a lattice energy which is further broken-down into the contributions from individual molecule–molecule pairs out to a specified cluster cut-off distance.

PIXELc calculations can be carried out straightforwardly on crystal structures with one or two molecules in the asymmetric unit. More complex cases can be handled with some user intervention. For example, where a molecule lies on a special position the space group symmetry should be lowered and the structure specified using whole molecules. Calcula-tions on disordered structures can be carried out using per-mutations of PIXELc calculations considering different molecular pairs.[46]

We thank a referee to this paper for pointing out that some structures may in fact be more readily amenable for processing using a series of PIXELd dimer calculations, while PIXELc can also be used to calculate dimer energies for a user-specified list of symmetry operations if the cluster cut-off distance is set to zero. Although both of these procedures would fail to reflect the many-body nature of the polarisation contribution to the lattice energy, they could be applied to the interpretation of crystal packing. We also note that a cor-rection is applied to PIXELc lattice energies of structures in polar space groups.[47] Details of calculations along with an extensive set of tutorial examples are provided with the CLP documentation.

All PIXEL calculations in the present study were carried out using a version of PIXELc[48] which had been modified to read a table of transition metal parameters described in section 2.1. For each structure OH, NH and CH distances were normalised to 0.993, 1.015 and 1.089 Å, respectively.[49] This procedure involves moving H positions to values typically obtained by neutron diffraction studies and corrects approximately for the effects of asphericity of H-atom electron densities which lead to systematic shortening of distances involving hydrogen atoms when determined by X-ray diffraction. The electron density was obtained in a single-point calculation with a B3LYP functional and a 6-31G** basis-set (Gaussian09)[50] for main group elements and first-row transition metals. Second-row transition metal species were treated with the LanL2DZ basis-set, and third-row metals used the LanL2DZ basis-set with pseudopotentials to model the core orbitals of the metal atom. The "cube" format electron den-sity files were then used in PIXEL calculations. Unless other-wise specified, the pixel size for all calculations was 0.16 × 0.16 × 0.16 Å (corresponding to 'condensation level' 4), and the cluster cut-off distance was kept at the default value gen-erated by PIXEL for each structure.[21]

While atomic parameters in PIXEL calculations are intended to be transferable between different compounds, values of atomic polarisabilities may be varied depending on chemical bonding. For instance, three different atomic polarisabilities are used for carbon, depending on whether it is aliphatic, aromatic or 'bridging aromatic' as in naphtha-lene. While such differences might ideally be taken into account for other species (e.g. carbonyl and ether oxygen atoms), the dominance of carbon in organic compounds means that it is much more important to account for varia-tion in its different chemical environments than it is for less abundant atomic species. While in practice only α(C) is usu-ally varied, modification of the atomic polarisabilities of other species has been applied previously by Gavezzotti,[51] for example for chloride ions in ionic organic crystals.

Although carbon is a common constituent of many ligands, it may be necessary to consider alternative values of polarisabilities of non-carbon atoms in cases such as homoleptic carbonyl complexes where the molecular surface is composed of exposed oxygen atoms. PIXEL analysis of molecular carbon monoxide using the default parameters in the program yields a lattice energy of −7.9 kJ mol^{-1}. The experimental sublimation enthalpy is 7.9(2) kJ mol^{-1} (average value from three determinations). However, when carbon monoxide acts as a ligand, PIXEL results were found to be around 20 kJ mol^{-1} lower than the literature value when the default value of α(O) (0.75 Å^3) was used (e.g. CrIJCO)$_6$, literature sublimation enthalpy 69.6 kJ mol^{-1}, calculated lattice energy −47.8 kJ mol^{-1}). Carbonyl oxygen therefore, like carbon, seems to require its own value of α(O) depending on whether the CO is ligating or not. By testing different values of atomic polarisability of O, a value of α(O) = 1.0 Å^3 was chosen for this species when carbon monoxide is acting as a ligand, yielding a lattice energy of −70.5 kJ mol^{-1} for CrIJCO)$_6$. Sup-port for these adjustments was obtained by calculation (AIMAll)[52] of atomic polarisabilities in CO and CrIJCO)$_6$ by the Atoms In Molecules method[53] as described by Keith[52,54] at the B3LYP/6-31G** level, using structures optimised at the same level of theory. The values of α(O) obtained for CO and CrIJCO)$_6$ were 0.57 and 0.86 Å^3 respectively, a similar relative increase to the one proposed above. Ligating sulfur atoms were given a polarisability of 3.6 Å^3. Program defaults were otherwise used, Table 1 showing the full set of values used for ligand atoms in all calculations.

Construction of a Validation set of Experimental Sublimation Enthalpies

A compilation of experimental sublimation enthalpies[55] was cross-referenced with the Cambridge Structural Database[56,57] to obtain a set of transition metal complexes for which both crystal structure and experimental sublimation data are avail-able. Sublimation enthalpies are notoriously difficult to mea-sure, and different determinations may yield wildly disparate results: for example two different measurements of the sublimation enthalpy of ferrocene give values of 64.6 kJ mol^{-1} and 84.0 kJ mol^{-1}.[58,59] For this reason, only compounds with a minimum of two independent sublimation enthalpy determinations were used for validation. The enthalpy values were arithmetically averaged with no weighting after elimination of any egregious outliers. The full validation data set, which contains 43 different compounds, is given in Table 2, and chemical structures are given in the ESI.† Also listed in Table 2 are the CSD refcodes, along with average experimental sublimation enthalpies calculated from data in ref. 55 and PIXEL calculated lattice energies. All complexes investigated had centrosymmetric crystal structures, and no polarisation corrections[47] were necessary.

Table 1. Atomic polarisabilities (Å3) for non-metal atoms used in the PIXEL calculations.

Atom	Atomic polarisability
H	0.39
C aliphatic	1.05
C aromatic	1.35
C aromatic bridge	1.90
N	0.95
O	0.75
O carbonyl	1.00
F	0.40
Cl	2.30
S	3.00
S ligating	3.60

Table 2. The 43 compounds used for validation and parameterisation. Compound numbers correspond to a scheme contained within the supplemen-tary information. PIXEL energies shown here were calculated with parameter set 5. All energies are in kJ mol^{-1}

Compound number	CSD refcode	Metal	Experimental sublimation enthalpy	−PIXEL calculated lattice energy	|Difference| (%)
1	ACACCR07	Cr	119.5 ± 8.7	124.5	4.2
2	ACACCS	Cr	120.4 ± 13.3	118.1	1.9
3	ACACCU02	Cu	118.7 ± 9.9	113.4	4.5
4	ACACMN21	Mn	119.3 ± 6.0	126.7	6.2
5	ACACVO12	V	140.6 ± 0.4	143.7	2.2
6	BZCRCO14	Cr	94.6 ± 4.7	93.8	0.8
7	CCRTOL01	Cr	94.0 ± 1.4	97.7	3.9
8	CDCPTI04	Ti	122.5 ± 3.0	131.8	7.6
9	CEHPIO01	Ti	94.6 ± 9.0	93.5	1.2
10	COACAC10	Co	138.7 ± 4.0	136.2	1.8
11	CPNDYV07	V	58.2 ± 1.1	76.8	32.0
12	CUBEAC01	Cu	156.6 ± 4.0	157.0	0.3
13	CUQUIN05	Cu	166.3 ± 5.3	159.6	4.0
14	DBENCR11	Cr	83.6 ± 6.1	93.7	12.1
15	DCYPCO04	Co	71.1 ± 1.5	75.8	6.6
16	DERNOD05	Cu	124.9 ± 2.9	129.5	3.7
17	DMTCCU	Cu	150.8 ± 4.6	170.9	13.3
18	DPIMNI	Ni	128.1 ± 24.2	119.2	7.5
19	DURHEE	Ni	120.4 ± 11.5	99.2	17.6
20	FEACAC03	Fe	121.92 ± 9.0	120.0	1.6
21	FEROCE27	Fe	73.6 ± 4.2	71.5	2.9
22	FOHCOU02	Cr	69.3 ± 2.6	70.5	1.7
23	IGAGEC	Cr	117.5 ± 7.8	81.6	30.6
24	IPEZOS	Cu	129.0 ± 1.9	124.0	3.9
25	IPTCNI10	Ni	145.5 ± 3.5	152.8	5.0
26	LIYLIO	Co	132.6 ± 15.3	134.2	1.2
27	MACACU10	Cu	133.0 ± 2.5	124.7	6.2
28	NCKLCN01	Ni	71.2 ± 1.0	74.1	4.1
29	NIDCAR06	Ni	153.7 ± 2.9	151.6	1.4
30	NISALO01	Ni	109.5 ± 3.5	115.4	5.4
31	QQQBWP03	Cu	113.5 ± 2.4	115.6	1.9
32	TCBMNI	Ni	146.0 ± 8.5	148.8	1.9
33	ACACPD01	Pd	128.1 ± 4.0	138.6	8.2
34	CYCPRU06	Ru	78.8 ± 3.4	83.0	5.3
35	FUBYIK01	Mo	72.6 ± 3.5	65.7	9.5
36	HCYPMO02	Mo	87.0 ± 7.8	82.9	4.7
37	HQUIPD	Pd	163.3 ± 6.7	154.4	5.5
38	KOKPEF	Hf	104.4 ± 4.9	94.9	9.1
39	KOVSOD02	W	74.9 ± 2.9	55.9	25.4
40	QQQCXJ02	Ir	112.0 ± 21.6	115.3	2.9
41	REGSAY	Hf	127.2 ± 3.5	109.1	14.2
42	REPKIH	W	90.4 ± 8.2	101.2	11.9
43	SINWER	Os	76.2 ± 3.9	84.6	11.0

Calculations of Individual Intermolecular Interactions

Dimers displaying a variety of intermolecular interaction types were selected to compare the PIXEL results with those calculated using higher level computational methods. A range of interactions was investigated involving chromium hexacarbonyl, vanadyl stacking and metal hydrides participating in dihydrogen bonding.[60–62] For each system a combination of Mercury 3.5 and Materials Studio V5[63] was used to obtain a structural model which was then optimised using Gaussian09. In cases where calculations had been previously reported, we used the same level of theory and basis-set as in the literature study. The optimised structures were then used for PIXELc calculations as described above. Further computational details are given in the relevant sections below, but Hirshfeld surface analysis was carried out with Crystal Explorer 3.1,[64] where the required wavefunctions were calculated with the program TONTO 3.2 rev. 4048,[65] while analysis of PIXEL results was accomplished using processPIXEL.[12]

RESULTS

Parameter set Selection and Reproduction of Experimental Sublimation Enthalpy Data

Five different parameter sets were constructed using different combinations of methods for estimation of the van der Waals radii, polarisability and electronegativity, as defined in section 2.1. The different combinations and values for each parameter set are given in Tables S2 and S3 in the ESI.† Following Gavezzotti,[40] the performance of the sets was quantified using the gradients and correlation coefficients of straight-line fits of the experimental sublimation enthalpies of Table 2 to the calculated lattice energies. The straight-line fitting statistics are listed in Table S2;† the fits used unit weights and were constrained to intercept at the origin.

The data in Table S2† show that there is little difference between the performance of the different parameter sets, demonstrating the robustness of the PIXEL method to different choices of reasonable parameters. We recommend parameter set 5 (Table 3) because both the volume (and therefore the polarisability) of an elemental metal and its Allred–Rochow electronegativity are more unambiguously defined in terms of readily accessible experimental data than the quantities used in other parameter sets. Furthermore, the gradient for 2nd and 3rd row transition metals for set 5 is nearer unity than the other sets.

For parameter set 5 the overall gradient of the straight line fit is 0.99(1) with a correlation coefficient of 0.92 (Fig. 1). These data compare favourably with respective values of 0.96 and 0.89 obtained by Gavezzotti for 172 organic crystal structures.[21] The apparent improvement over organic materials obtained for the metal complexes is probably ascribable to limitation of the validation data-set to compounds for which multiple sublimation enthalpy determinations are available. If all available data (105 complexes containing first second and third row metals) are used the gradient and correlation coefficient are 0.96(2) and 0.69, respectively (Fig. 2).

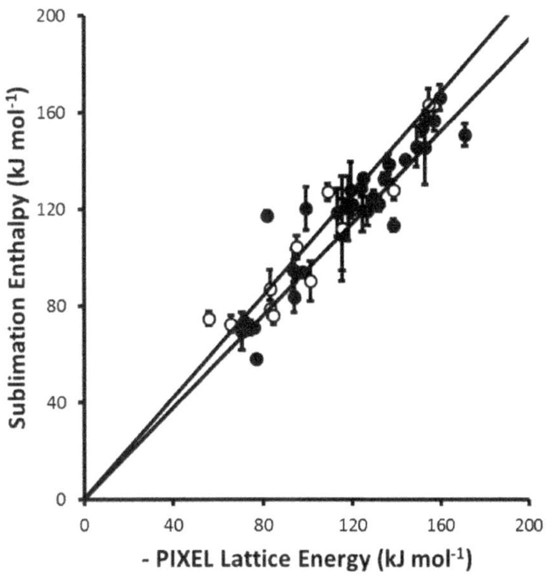

Figure 1. Comparison of experimental sublimation enthalpies with values calculated by PIXEL (parameter set 5) for 1st row (closed circles) and 2nd and 3rd row (open circles) transition metal complexes for which the sublimation enthalpy has been determined multiple times. The black lines are ±5% of the experimental sublimation enthalpy. The least-squares straight line through the data points is y = 0.99IJ1)x (R = 0.92).

Table 3. Transition metal parameters for set 5, this set was used to generate the calculated lattice energies in Table 2 and for the examples studied in section 4. Abbreviations and units are given in section 2.1

Atom	*DIFA*	*RAVDW*	*RINTER*	*ZTOT*	*ZVAL*	*POLZE*	*ELNEG*	*POTIO*	*WEIGHT*
Ti	0.4	2.15	1.32	22	4	4.18	1.32	0.251	47.88
V	0.4	2.05	1.22	23	5	3.31	1.45	0.248	50.94
Cr	0.4	2.05	1.17	24	6	2.86	1.56	0.249	52.00
Mn	0.4	2.05	1.17	25	7	2.93	1.60	0.273	54.94
Fe	0.4	2.05	1.16	26	8	2.81	1.64	0.289	55.85
Co	0.4	2.00	1.16	27	9	2.62	1.70	0.289	58.93
Ni	0.4	2.00	1.15	28	10	2.61	1.75	0.281	58.69
Cu	0.4	2.00	1.35	29	11	2.81	1.75	0.284	63.55
Zn	0.4	2.10	1.31	30	12	3.63	1.66	0.345	65.38
Zr	0.4	2.30	1.45	40	12	5.56	1.33	0.251	91.22
Nb	0.4	2.15	1.34	41	13	4.30	1.60	0.253	92.91
Mo	0.4	2.10	1.29	42	14	3.72	2.16	0.261	95.94
Tc	0.4	2.05	1.23	43	15	3.41	1.90	0.267	98.91
Ru	0.4	2.05	1.24	44	16	3.23	2.20	0.271	101.07
Rh	0.4	2.00	1.25	45	17	3.29	2.28	0.274	102.91
Pd	0.4	2.05	1.28	46	18	3.51	2.20	0.307	106.42
Ag	0.4	2.10	1.34	47	19	4.07	1.93	0.278	107.87
Cd	0.4	2.20	1.41	48	20	5.15	1.69	0.330	112.41
Hf	0.4	2.35	1.44	72	12	5.32	1.23	0.251	178.49
Ta	0.4	2.20	1.34	73	13	4.31	1.33	0.277	180.95
W	0.4	2.10	1.30	74	14	3.78	1.40	0.289	183.85
Re	0.4	2.05	1.28	75	15	3.51	1.46	0.288	186.21
Os	0.4	2.00	1.26	76	16	3.34	1.52	0.310	190.20
Ir	0.4	2.00	1.26	77	17	3.40	1.55	0.330	192.22
Pt	0.4	2.05	1.29	78	18	3.61	1.44	0.329	195.08
Au	0.4	2.10	1.34	79	19	4.04	1.42	0.339	196.97
Hg	0.4	2.05	1.44	80	20	5.87	1.44	0.384	200.59

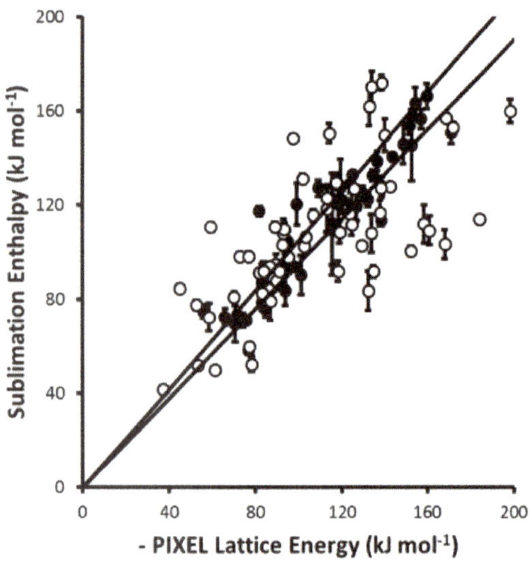

Figure 2. A similar comparison to that shown in Fig. 1 but including complexes (shown as open circles) for which the sublimation enthalpy has been determined only once. All points from Fig. 1 are shown as closed circles. The least-squares straight line through the data points is y = 0.96IJ2)x (R = 0.69).

The data presented in Fig. 1 are dominated by first-row transition metal complexes, these data alone yielding a gradient and correlation coefficient equal to 0.99(2) and 0.92, respectively. The data for second and third row complexes are more limited, and the fitting statistics are 1.02(3) and 0.92. Some elements, such as technetium and gold, are not represented at all.

Overall, while the data presented above indicate that the PIXEL method can be applied with some confidence to first-row metal complexes; more data are needed to establish this for compounds containing heavier metals.

It is important to note that the assumption made in calculating lattice energy by the PIXEL method is that no change in molecular structure occurs on passing from the crystalline state to the gas phase. PIXEL calculations of the lattice energies of amino acids, for example, are in poor agreement with experimental values because amino acids exist as zwitterions in the solid state but as neutral molecules in the gas phase. The energy associated with the transfer of a proton from the ammonium to the carboxylate group does not form part of the PIXEL analysis, and would need to be calculated separately in any calculation aiming to reproduce the experimental sublimation enthalpy. This consideration may account for some of the differences between observed and calculated data presented in Table 2.

The experimental sublimation enthalpies of the biscyclopentadienyl complexes of Cr, Fe, Ni, Ru and Os all lie between 70 and 80 kJ mol^{-1}, and are reproduced to within 10 kJ mol^{-1} in the PIXEL calculations. By contrast the relative difference between the experimental and PIXEL values for Cp_2V is the poorest in Table 2 (entry 11), but it is notable that the experimental value seems anomalously low at 58.2(11) kJ mol^{-1}. The lattice energy of trisIJhexafluoroacetylacetonato-O, O′)-chromiumIJIII) (entry 23) is under-estimated [calc. 81.6, expt. 117.5(78) kJ mol^{-1}], a finding consistent with previous PIXEL calculations on fluorinated polycyclic hydrocarbons.[66] Decreasing the damping parameter used to calculate dispersion energies from 3.0 to 2.4 together with a small increase in α(F) from 0.4 to 0.5 Å^3 to account for the effect of negative charge distribution in the (anionic) ligand yields a lattice energy of 113.4 kJ mol^{-1}, illustrating the sensitivity of the calculation to the parameterisation used for fluorine.

EXAMPLES

Chromium Hexacarbonyl

Chromium hexacarbonyl (CSD refcode FOHCOU02)[67] crystallises in space group Pnma with the molecules forming layers in the mirror planes at y = 1/4 and 3/4. The sublimation enthalpy is 69.3(26) kJ mol^{-1} (an average of twelve determina-tions),[55] and the lattice energy calculated by PIXEL is −70.5 kJ mol^{-1} (see Table 2). The three principal intermolecular inter-actions are to the molecules labelled A, B and C in Fig. 3; Table 4 shows the breakdown of the component energy terms and the Cr···Cr distance for each interaction.

The data in Table 4 confirm that the interactions are predominantly dispersion based. The influence of the Cr-atom on the dispersion term in interaction A can be demonstrated by setting its polarizability close to zero§ (0.0001 Å^3), which changes the dispersion energy from −18.7 to −16.6 kJ mol^{-1}. Setting both α(Cr) and α(O) to zero gives E_{disp} = −3.5 kJ mol^{-1} showing that the bulk of the interaction energy derives from the oxygen atoms. The role of electrostatics in the three inter-actions is illustrated in Fig. 4, where the electrostatic potential of each molecule is mapped onto its Hirshfeld surface.[11,68] Favourable overlap occurs when the negatively-charged (red) regions of one molecule are in contact with the positively-charged (blue) regions of a neighbouring molecule, and the white lines separating these regions are contiguous. This arrangement is seen for interactions A and C, but in the interaction with molecule B there is some overlap between negative regions. This is the source of the more negative (stabilising) coulombic terms for interactions A and C, which is not apparent from geometric analysis alone.

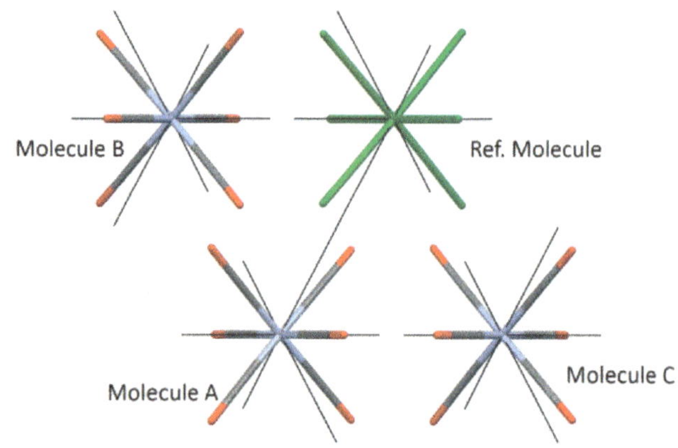

Figure 3. Energy vectors (obtained using process PIXEL) for the total intermolecular energies of chromium hexacarbonyl as viewed down the crystallographic c-axis. The energy vectors show the interaction energies scaled to that of the strongest interaction. The reference molecule is shown in green, and the strongest interaction is between this and molecule A. The interaction to molecule B is the second strongest, and that to molecule C is the weakest

Table 4. PIXEL component terms and Cr···Cr distances for the three principal intermolecular interactions in CrIJCO)$_6$. Values are in kJ mol^{-1} unless otherwise indicated

Interacting molecule	E_{coul}	E_{disp}	E_{pol}	E_{rep}	E_{tot}	Cr···Cr (Å)
A	−5.8	−18.7	−1.8	13.0	−13.3	6.236
B	−2.6	−16.9	−1.2	8.9	−11.9	6.213
C	−4.3	−11.2	−1.1	7.4	−9.1	6.888

BisIJacetylacetonato)-oxo-vanadium IJIV)

In the crystal structure of bisIJacetylacetonato)-oxo-vanadium-IJIV) (VOIJacac)$_2$, CSD refcode ACACVO12),[69] the molecules are arranged in offset stacks disposed about inversion centres. The spacing between the two planes defined by the four ligating acac oxygen atoms in each molecule is 3.46 Å. A similar motif occurs in the crystal structures of other vanadyl complexes.[70–72] The sublimation enthalpy is 140.6(4) kJ mol^{-1} (the average of two measurements),[55] and the lattice energy estimated by PIXEL is −143.7 kJ mol^{-1}. Based on the PIXEL results, the stacking interaction is the strongest intermolecular contact with an energy of −65.0 kJ mol^{-1}, which is an order of magnitude higher than in typical π-stacking interactions involving phenyl groups,[41] and similar to a strong hydrogen bond. A series of single-point ab initio calculations¶ performed at the crystal structure geometry with 'normalised' H-atom positions confirms the value obtained by PIXEL: B3LYP-D/6-31G*: −65.9 kJ mol^{-1}; B2PLYP-D/6-31G*: −52.9 kJ mol^{-1}; MO5-2X/6-311++G**: −53.5 kJ mol^{-1}; SCS MP2: −67.7 kJ mol^{-1}.

The component terms (in kJ mol^{-1}) of the stacking energy are: E_{coul} = −28.8, E_{disp} = −67.9, E_{pol} = −9.8 and E_{rep} = +41.5, showing that, though the dispersion term is the largest, there is also substantial electrostatic character (Fig. 5). The contribution of the metal atom to the dispersion energy can be estimated by running the PIXEL calculation with α(V) set to approximately zero. This procedure reduces the magnitude of the interac-tion energy by 16.4 kJ mol^{-1} to −51.5 kJ mol^{-1}, showing that the relatively high polarizability of vanadium (3.31 Å^3) has an important influence on the intermolecular interaction ener-gies. The magnitude of the coulombic term results from the efficient overlap of positive and negative regions of the electrostatic potentials of the two molecules (Fig. 6).

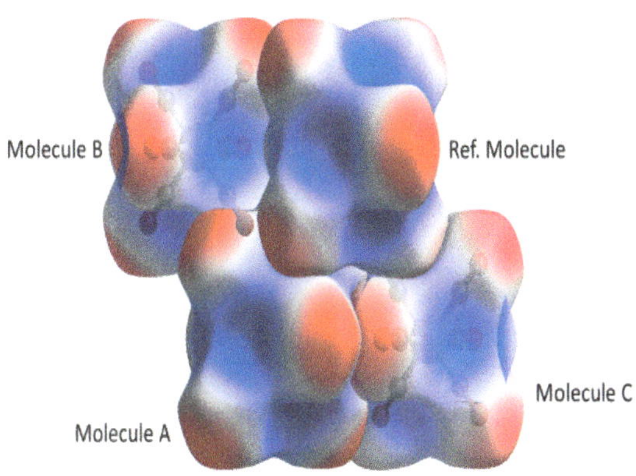

Figure 4. Coulombic interactions in CrIJCO)$_6$ visualised using the electrostatic potential (ESP) mapped onto Hirshfeld surfaces. The sur-face is mapped from −0.036 au (red) to 0.047 au (blue). Wave func-tions were obtained at the HF/STO-3G level

Dihydrogen Bonding

A metal-bound hydride may be sufficiently negatively charged such that it can act as a hydrogen bond acceptor, forming a dihydrogen bond.[74] Theoretical analysis of such interactions has shown that they are attractive and are predominantly electrostatic in character.[75] Interaction energies for a series of dihydrogen-bonded dimers comprising a Ru-hydride com-plex and carboxylic acids or alcohols with different levels of fluorination have been evaluated using either DFT (B3LYP) or Hartree–Fock (HF) calculations with the LanL2DZ basis-set.[60–62] PIXEL calculations were carried out using electron densities calculated with the same geometries, functionals and basis-sets as reported for the DFT calculations, and the results are compared in Fig. 7 and Table 5.

The level of agreement between the PIXEL and DFT results is similar to that found for the sublimation enthalpies: the gradient of the straight line in Fig. 7 is 0.91 and the correlation coefficient 0.96; differences are of the order of 5 kJ mol^{-1}. The data in Table 5 show that the interactions are predominantly electrostatic in character, in accordance to observations made by Liu and Hoffmann, and that the coulombic term increases with increasing fluorination of the donor as expected on the basis of induction effects. The increase in the dispersion inter-action which also occurs on fluorination is a result of a change in the orientation of the alcohol (Fig. 8) which enables forma-tion of a secondary dispersion interaction between the fluorine atoms on the donor and the hydrogen atoms of the Ru complex.[76]

CONCLUSIONS

A straight-line fit of experimental sublimation enthalpies of a series of first-row transition metal complexes to the lattice energy magnitudes calculated using the PIXEL method has a gradient of 0.99(2) and a correlation coefficient of 0.92. These figures indicate that the performance of the PIXEL parameterisation for molecules containing first-row metals is similar to that described by Gavezzotti for organic materials. The use of the method has been illustrated using interactions in the crystal structures of chromium hexacarbonyl and bisIJacetylacetonato)-oxo-vanadiumIJIV). Intermolecular interactions in CrIJCO)^6 are dominated by dispersion, as expected, but the shortest interaction is not the strongest on account of the second shortest interaction having a more complementary match of electrostatic potentials. Stacking interactions in VOIJacac)^2 have an energy in the region of 60 kJ mol^{-1} , a figure confirmed by a variety of flavours of DFT calculation.

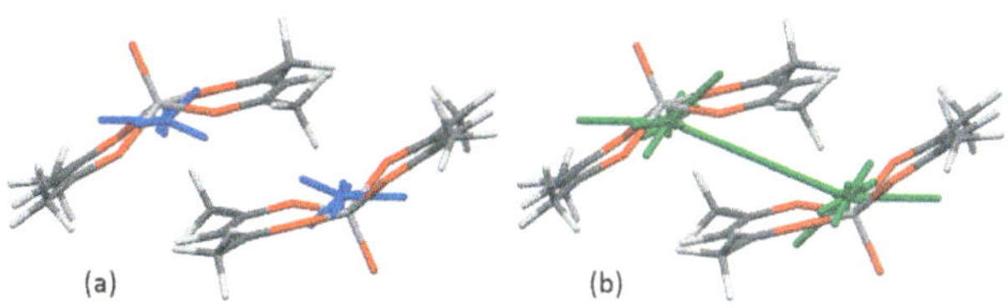

Figure 5. Energy vectors for the coulombic (a) and dispersion (b) components of the stacking interaction between bisIJacetylacetonato)-oxovanadiumIJIV) molecules.

This energy is similar to a strong acid–base hydrogen bond such as benozoic acid–imidazole,[21] and explains why stacking interactions are a recurring feature of vanadyl crystal structures. Application to guest-binding in metal–organic frameworks will be described in a future paper. The gradient and correlation coefficient of the fit between sublimation enthalpies and PIXEL lattice energies for second and third-row complexes are 1.02(3) and 0.92, respectively, but these figures are based on a much less extensive set of data than those for the first-row systems. The parameters for 2nd and 3rd row transition metals should be used with caution, but they are consistent with DFT results for dihydrogen bonds involving ruthenium hydride complexes and intermolecular embraces.

The PIXEL method assumes that an interaction is truly non-covalent, and it cannot be applied to interactions involving, for example, partial bond formation or other electron sharing regimes for which full quantum mechanical treatments are necessary. Distinction between non-covalent and more complex interactions is often quite intuitive in organic chemistry, but this may not be the case in transition metal chemistry, and so caution is also necessary in this context. It would not be appropriate, for example, to apply PIXEL calculations to modelling the transition between long and short Jahn–Teller distortions or to aurophilic interactions, which both demand high-level quantum mechanical treatments even though the interatomic distances which characterise them extend to 3 Å and beyond. This said, comparison of PIXEL and quantum mechanical treatments for such systems might be used to detect the presence of complex behaviour.

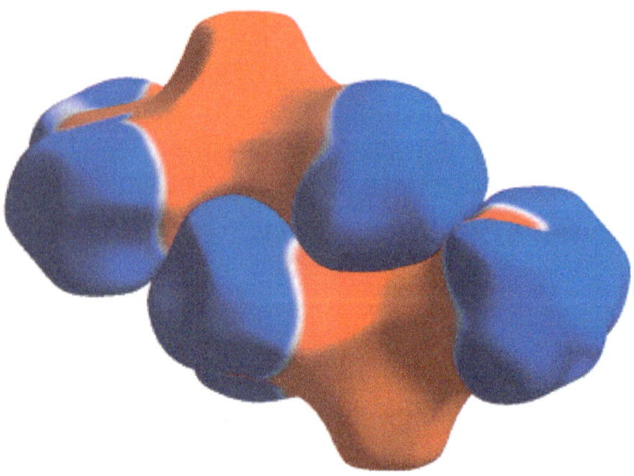

Figure 6. The stacking interaction in bis IJacetylacetonato)-oxovanadium IJIV) shown using the electrostatic potential mapped onto Hirshfeld surfaces. The surface is mapped from −0.005 au (red) to +0.005 au (blue). Wavefunctions were obtained using DFT with the Becke88 exchange potential and LYP correlation potential, and the 6-31G* basis-set

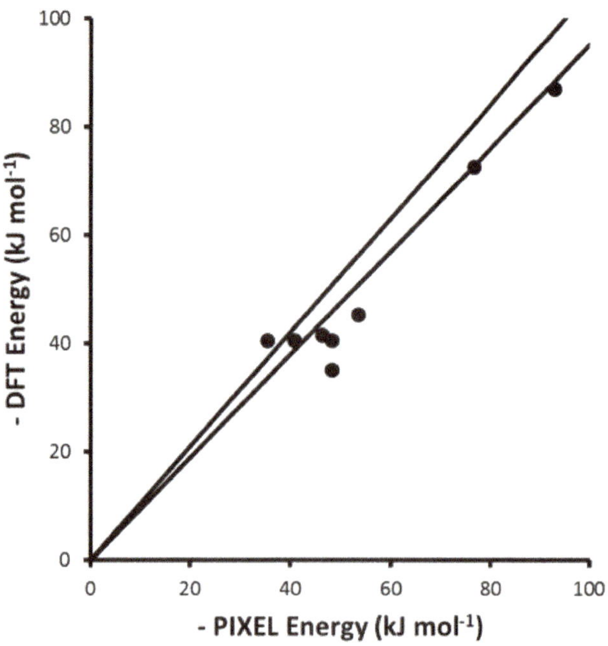

Figure 7. Comparison of PIXEL calculated energies with DFT reference energies for a selection of dihydrogen bonded complexes. The black lines are ±5% of the DFT calculated interaction energy. The leastsquares straight line through the data points is y = 0.91x (R = 0.96).

The accurate and efficient quantification of crystalpacking and intermolecular energies is having a transformative effect on solid-state organic chemistry. It is no longer necessary to base the interpretation of a structure or a crystal engineering strategy on the assumption that short interactions are strong interactions.[77] Systematic quantification of molecule–molecule energies further reveals important interactions that are easily missed on the basis of analysis of distances alone, a particular issue with electrostatic and van der Waals interactions which lack characteristic geometric signatures.78 While the PIXEL calculations do not replace quantum mechanical methods, they have proved to be an extremely valuable tool for interpretation of the thermodynamic stability of crystalline organic phases. The parameterisation described here should enable these advantages to be extended to systems containing metals, an extension which dramatically increases the domain of applicability for the PIXEL method.

Table 5. Breakdown of PIXEL values into component and total energy terms and reference DFT energies. dpe is $PH_2CH_2CH_2PH_2$ and PP_3 is PIJ$CH_2CH_2PH_2)_3$. All values are in kJ mol^{-1}

Complex	Coul	Disp	Rep	Pol	Tot	DFT	Ref.
$CpRu(CO)(PH_3)H\cdots HOOCCF_3$	-53.0	-22.4	63.6	-36.6	-48.4	-35.1	60
$CpRu(CO)(PH_3)H\cdots HOC(CF_3)_3$	-50.7	-27.3	65.0	-35.4	-48.4	-40.6	60
$CpMo(dpe)H_2H\cdots HOCH(CF_3)_2$	-62.3	-35.1	81.3	-30.3	-46.4	-41.6	61
$CpW(dpe)H_2H\cdots HOCH(CF_3)_2$	-69.6	-36.6	85.6	-32.9	-53.5	-45.3	61
$PP_3RuHH^{ax}\cdots HOCH_3$	-54.8	-15.4	57.1	-22.3	-35.4	-40.6	62
$PP_3RuHH^{eq}\cdots HOCH_3$	-51.8	-18.8	52.8	-23.1	-40.9	-40.5	62
$PP_3RuHH^{ax}\cdots HOCF_3$	-103.6	-20.9	92.6	-61.0	-92.9	-86.9	62
$PP_3RuHH^{eq}\cdots HOCF_3$	-92.6	-23.7	117.1	-77.4	-76.6	-72.6	62

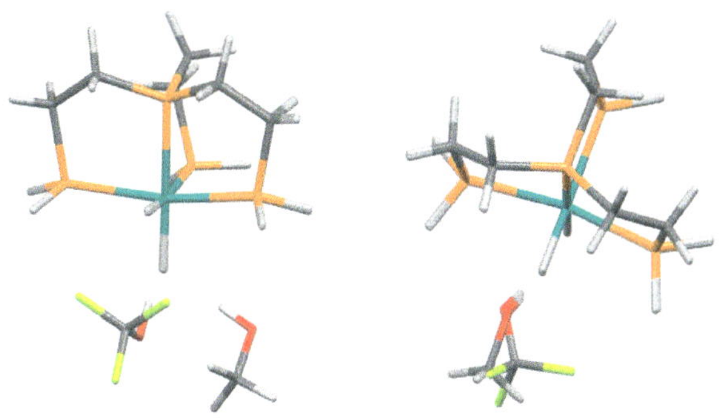

Figure 8. Binding geometries to the axial hydride (left) and equatorial hydride (right) in PP$_3$RuHH where PP$_3$ is PIJCH$_2$CH$_2$PH$_2$)$_3$. Changes in the disposition of the alcohol upon fluorination are shown by an overlay in each case.

CODE AVAILABILITY

A version of the CLP package incorporating the PIXEL parameterisation given here for the first-row metals is available from Professor Gavezzotti's web-site, http://users.unimi.it/gavezzot. PIXEL calculations using parameters for the second and third row metals can be carried out by editing the symbol, wei, ravrg, ravdw, zeta, zval, polat, elneg, potio and difa data statements in the file alldat.for and re-compiling the code. These arrays respectively correspond to the Atom, WEIGHT, RINTER, RAVDW, ZTOT, ZVAL, POLZE, ELNEG, POTIO and DIFA entries in Table 3.

ACKNOWLEDGEMENTS

We thank the Cambridge Crystallographic Data Centre and the EPSRC for studentship funding to AGPM. This work has made use of the resources provided by the EaStCHEM Research Computing Facility (http://www.eastchem.ac.uk/ facilities), which is partially supported by the eDIKT initiative. (http://www.edikt.org.uk). We also thank Professor Nik Kaltsoyannis (University College London) for helpful advice, Tanya van Mourik (University of St Andrews) for assistance with SCS MP$_2$ calculations, and Professor Andrew Bond for his assistance in the use of processPIXEL with transition metal complexes. We also thank Dr Neil Feeder and Dr Colin Groom (CCDC) for their comments on the manuscript and Professor Angelo Gavezzotti for his advice, insight and interest over many years.

REFERENCES

1. C. F. Macrae, I. J. Bruno, J. A. Chisholm, P. R. Edgington, P. McCabe, E. Pidcock, L. Rodriguez-Monge, 2. R. Taylor, J. van de Streek and P. A. Wood, J. Appl. Crystallogr., 2008, 41, 466–470.
2. G. J. O. Beran, Angew. Chem., Int. Ed., 2015, 54, 396–398.
3. R. Centore, M. Causa, S. Fusco and A. Carella, Cryst. Growth Des., 2013, 13, 3255–3260.
4. H. C. S. Chan, J. Kendrick, M. A. Neumann and F. J. J. Leusen, CrystEngComm, 2013, 15, 3799–3807.
5. J. A. Chisholm, S. Motherwell, P. R. Tulip, S. Parsons and J. Clark, Cryst. Growth Des., 2005, 5, 1437–1442.
6. O. V. Shishkin, S. V. Shishkina, A. V. Maleev, R. I. Zubatyuk, Vasylyeva and K.
7. Merz, ChemPhysChem, 2013, 14, 847–856.
8. J. Yang, W. Hu, D. Usvyat, D. Matthews, M. Schuetz and K.-L. Chan, Science, 2014, 345, 640–643.
9. A. Gavezzotti, Z. Kristallogr., 2005, 220, 499–510.
10. E. G. Hohenstein and C. D. Sherrill, Wiley Interdiscip. Rev.: Comput. Mol. Sci., 2012, 2, 304–326.
11. A. Gavezzotti, New J. Chem., 2011, 35, 1360–1368.
12. M. A. Spackman and D. Jayatilaka, CrystEngComm, 2009, 11, 19–32.
13. A. D. Bond, J. Appl. Crystallogr., 2014, 47, 1777–1780.
14. M. J. Turner, S. P. Thomas, M. W. Shi, D. Jayatilaka and A. Spackman, Chem. Commun., 2015, 51, 3735–3738.
15. O. V. Shishkin, V. V. Medvediev, R. I. Zubatyuk, O. O. Shyshkina, N. V. Kovalenkoc and J. K. Volovenko, CrystEngComm, 2012, 14, 8698–8707.
16. F. P. A. Fabbiani, C. R. Pulham and J. E. Warren, Z. Kristallogr., 2014, 229, 667–675.
17. R. M. Bhardwaj, L. S. Price, S. L. Price, S. M. Reutzel-Edens, J. Miller, I. D. H. Oswald, B. F. Johnston and A. J. Florence, Cryst. Growth Des., 2013, 13, 1602–1617.
18. A. O. Surov, K. A. Solanko, A. D. Bond, A. Bauer-Brandl and L. Perlovich, CrystEngComm, 2015, 17, 4089–4097.
19. J. D. Dunitz and A. Gavezzotti, Cryst. Growth Des., 2012, 12, 5873–5877.
20. A. Gavezzotti, Acta Crystallogr., Sect. D: Biol. Crystallogr., 2008, 64, 905–908. R. D. L. Johnstone, D. Francis, A. R. Lennie, W. G. Marshall, A. Moggach, S. Parsons, E. Pidcock and J. E. Warren, CrystEngComm, 2008, 10, 1758–1769.
21. A. Gavezzotti, Molecular Aggregation - Structure Analysis and Molecular Simulation of Crystals and Liquids, Oxford Univer-sity Press, New York, 1st edn, 2007.
22. I. Dance, CrystEngComm, 2003, 5, 208–221.
23. C. J. Adams, A. Angeloni, A. G. Orpen, T. J. Podesta and B. Shore, Cryst. Growth Des., 2006, 6, 411–422.
24. C. J. Adams, P. C. Crawford, A. G. Orpen and T. J. Podesta, Dalton Trans., 2006, 4078–4092.
25. A. Perez Paz, L. A. Espinosa Leal, M.-R. Azani, A. Guijarro, J. Sanz Miguel, G. Givaja, O. Castillo, R. Mas-Balleste, F. Zamora and A. Rubio, Chem. – Eur. J., 2012, 18, 13787–13799.
26. D. Di Tommaso, S. A. French and C. R. A. Catlow, J. Mol. Struct.: THEOCHEM, 2007, 812, 39–49.
27. D. Di Tommaso, S. A. French, A. Zanotti-Gerosa, F. Hancock, J. Palin and C. R. A. Catlow, Inorg. Chem., 2008, 47, 2674–2687.
28. S. A. French, D. Di Tommaso, A. Zanotti-Gerosa, F. Hancock and C. R. A. Catlow, Chem. Commun., 2007, 2381–2383.
29. H. Li, E. N. Brothers and M. B. Hall, Inorg. Chem., 2014, 53, 9679–9691.
30. T. Dudev and C. Lim, Annu. Rev. Biophys., 2008, 37, 97–116.

31. D. P. Martin, Z. S. Hann and S. M. Cohen, Inorg. Chem., 2013, 52, 12207–12215.
32. I. Yruela, Metallomics, 2013, 5, 1090–1109.
33. A. De Almeida, B. L. Oliveira, J. D. G. Correia, G. Soveral and Casini, Coord. Chem. Rev., 2013, 257, 2689–2704.
34. H. Chiniforoshan, L. Tabrizi, M. Hadizade, M. R. Sabzalian, N. Chermahini and M. Rezapour, Spectrochim. Acta, Part A, 2014, 128, 183–190.
35. R. Q. Snurr, A. O. Yazaydin, D. Dubbeldam and H. Frost, in Metal-Organic Frameworks: Design and Application, ed. L. R. MacGillivray, John Wiley & Sons, Inc., Hoboken, New Jersey, 2010, ch. 11, pp. 313–339.
36. S. Keskin, J. Liu, R. B. Rankin, J. K. Johnson and D. S. Sholl, Ind. Eng. Chem. Res., 2009, 48, 2355–2371.
37. T. Düren, Y.-S. Bae and R. Q. Snurr, Chem. Soc. Rev., 2009, 38, 1237–1247.
38. A. Gavezzotti, J. Phys. Chem. B, 2002, 106, 4145–4154.
39. A. Gavezzotti, J. Phys. Chem. B, 2003, 107, 2344–2353.
40. L. Maschio, B. Civalleri, P. Ugliengo and A. Gavezzotti, J. Phys. Chem. A, 2011, 115, 11179–11186.
41. W. B. Schweizer and J. D. Dunitz, J. Chem. Theory Comput., 2006, 2, 288–291.
42. J. Emsley, The Elements, Oxford University Press, New York, 1989.
43. S. Nag, K. Banerjee and D. Datta, New J. Chem., 2007, 31, 832–834.
44. S. S. Batsanov, Inorg. Mater., 2001, 37, 871–885.
45. A. Gavezzotti, New J. Chem., 2011, 35, 1360–1368.
46. K. A. Solanko and A. D. Bond, Acta Crystallogr., Sect. B: Struct. Sci., 2011, 67, 437–445.
47. B. P. van Eijck and J. Kroon, J. Phys. Chem. B, 1997, 101, 1096–1100.
48. A. Gavezzotti, A computer program package for the calculation of intermolecular interactions and crystal energies, University of Milan, Italy, 2003.
49. F. H. Allen and I. J. Bruno, Acta Crystallogr., Sect. B: Struct. Sci., 2010, 66, 380–386.
50. M. J. Frisch, G. W. Trucks, H. B. Schlegel, G. E. Scuseria, M. A. Robb, J. R. Cheeseman, G. Scalmani, V. Barone, B. Mennucci, G. A. Petersson, H. Nakatsuji, M. Caricato, X. Li, H. P. Hratchian, A. F. Izmaylov, J. Bloino, G. Zheng, J. L. Sonnenberg, M. Hada, M. Ehara, K. Toyota, R. Fukuda, J. Hasegawa, M. Ishida, T. Nakajima, Y. Honda, O. Kitao, H. Nakai, T. Vreven, J. A. Montgomery Jr., J. E. Peralta, F. Ogliaro, M. Bearpark, J. J. Heyd, E. Brothers, K. N. Kudin, V. N. Staroverov, R. Kobayashi, J. Normand, K. Raghavachari, A. Rendell, J. C. Burant, S. S. Iyengar, J. Tomasi, M. Cossi, N.
51. Rega, J. M. Millam, M. Klene, J. E. Knox, J. B. Cross, V. Bakken, C. Adamo, J. Jaramillo, R. Gomperts, R. E. Stratmann, O. Yazyev, A. J. Austin, R. Cammi, C. Pomelli, J. W. Ochterski, R. L. Martin, K. Morokuma, V. G. Zakrzewski, G. A. Voth, P. Salvador, J. J. Dannenberg, S. Dapprich, A. D. Daniels, Ö. Farkas, J. B. Foresman, J. V.
52. Ortiz, J. Cioslowski and D. J. Fox, Gaussian 09, Revision B.01, Gasussian, Inc., 2009.
53. A. Gavezzotti, Acta Crystallogr., Sect. B: Struct. Sci., 2010, 66, 396–406.
54. T. A. Keith, A software package for performing quantitative and visual QTAIM (Quantum Theory of Atoms in Molecules) analyses of molecular systems, TK Gristmill Software, Overland Park, KS, USA, 12.06.03 edn, 2012.
55. R. F. W. Bader, Atoms in Molecules: A Quantum Theory, Oxford University Press, Oxford, UK, 1990.
56. T. A. Keith, in The Quantum Theory of Atoms in Molecules: From Solid State to DNA and Drug Design, ed. C. F. Matta and R. J. Boyd, WILEY-VCH, Weinham, 2007, ch. 3, pp. 61–95
57. J. S. Chickos and W. E. Acree, J. Phys. Chem. Ref. Data, 2002, 31, 537–698
58. I. J. Bruno, J. C. Cole, P. R. Edgington, M. Kessler, C. F. Macrae, P. McCabe, J. Pearson and R. Taylor, Acta Crystallogr., Sect. B: Struct. Sci., 2002, 58, 389–397
59. F. Allen, Acta Crystallogr., Sect. B: Struct. Sci., 2002, 58, 380–388.
60. R. M. Stephenson and S. Malanowski, Handbook of the Thermodynamics of Organic Compounds, Elsevier, New York, 1987.
61. G. Beech and R. Lintonbon, Thermochim. Acta, 1971, 2, 86–88
62. N. V. Belkova, M. Besora, L. M. Epstein, A. Lledos, F. Maseras and E. S. Shubina, J. Am. Chem. Soc., 2003, 125, 7715–7725.
63. I. Andrieu, N. V. Belkova, M. Besora, E. Collange, L. M. Epstein, A. Lledos, R. Poli, P. O. Revin, E. S. Shubina and A. V. Vorontsov, Russ. Chem. Bull., 2003, 52, 2679–2682.
64. E. I. Gutsul, N. V. Belkova, M. S. Sverdlov, L. M. Epstein, E. S. Shubina, V. I. Bakhmutov, T. N. Gribanova, R. M. Minyaev, C. Bianchini, M. Peruzzini and F. Zanobini, Chem. Eur. J., 2003, 9, 2219–2228
65. Accelrys, Accelrys Software Inc., Cambridge, UK, Release 5.5.3 edn, 2010.
66. S. K. Wolff, D. J. Grimwood, J. J. McKinnon, M. J. Turner, D. Jayatilaka and M. A. Spackman, CrystalExplorer 3.1, University of Western Australia, 2012.
67. D. Jayatilaka and D. J. Grimwood, in Computational Science -ICCS 2003, Pt IV, Proceedings, ed. P. M. A. Sloot, D. Abramson, A. V. Bogdanov, J. J. Dongarra, A. Y. Zomaya and Y. E. Gorbachev, 2003, vol. 2660, pp. 142–151
68. F. Cozzi, S. Bacchi, G. Filippini, T. Pilati and A. Gavezzotti, Chem. – Eur. J., 2007, 13, 7177–7184.
69. A. Whitaker and J. W. Jeffery, Acta Crystallogr., 1967, 23, 977–984
70. M. A. Spackman, J. J. McKinnon and D. Jayatilaka, CrystEngComm, 2008, 10, 377–388.
71. M. Hoshino, A. Sekine, H. Uekusa and Y. Ohashi, Chem. Lett., 2005, 34, 1228–1229

72. D. I. Arnold, F. A. Cotton, J. H. Matonic and C. A. Murillo, Chem. Commun., 1996, 2113–2114
73. P. K. Hon, R. L. Belford and C. E. Pfluger, J. Chem. Phys., 1965, 43, 3111–3115.
74. U. Schilde, W. Bansse, E. Ludwig and E. Uhlemann, Z. Kristallogr., 1995, 210, 627–628
75. F. Neese, Wiley Interdiscip. Rev.: Comput. Mol. Sci., 2012, 2, 73–78.
76. V. I. Bakhmutov, Dihydrogen Bonds: Principles, Experiments and Applications, John Wiley & Sons, Inc., Hoboken, NJ, 2008
77. Q. Liu and R. Hoffmann, J. Am. Chem. Soc., 1995, 117, 10108–10112.
78. H. Jacobsen, Phys. Chem. Chem. Phys., 2009, 11, 7231–7240.
79. J. Dunitz, IUCrJ, 2015, 2, 157–158.
80. A. Gavezzotti, CrystEngComm, 2013, 15, 4027–4035.

CHAPTER 12

CATION–π INTERACTIONS

Matthew R. Davis and Dennis A. Dougherty

INTRODUCTION

The importance of noncovalent interactions in determining macromolecular structure and molecular associations in biology is well appreciated. Nevertheless, it remains a challenging task to confirm the importance of any particular noncovalent interaction, and even more difficult to quantify one. Different issues arise in attempting to evaluate hydrophobic effects, hydrogen bonds, or ion-pairing interactions, as each has its own unique characteristics. For some time we have been evaluating a particular noncovalent interaction – the cation–π interaction – in a range of proteins, most typically neuroreceptors and ion channels.[1–4] Only in recent years has structural information become available for such systems. Remarkably, the available structures point to a general structural motif for cation–π interactions, the so-called aromatic box.[2,5–7] This is formed by 3 to 5 aromatic side chains that can encapsulate the positive charge of a cationic molecule. This molecular cage has been seen in many receptors and in other binding sites for cationic guests. It is not known what the advantage of this motif is, nor is it clear to what extent 3–5 aromatics can simultaneously contribute to the binding of a cation.

Even if a structure is available and it shows a close contact between an ion and a π system, it cannot be assumed that a functionally significant noncovalent interaction exists. Over the past 20 years we have addressed this issue using non-canonical amino acid mutagenesis.[4,8] The aromatic of interest (the side chain of a phenylalanine (Phe), tyrosine (Tyr), or tryptophan (Trp)) is progressively fluorinated. Fluorine is well known to be deactivating in a cation–π interaction, and its effects are typically additive. One thus expects a correlation between protein function and/or ligand binding and degree of fluorination if a cation–π interaction is important. In a number of systems we have found a linear trend between the activation of a receptor by a cationic ligand and the calculated binding of a sodium ion to a series of fluorinated aromatic rings (indoles to mimic the side chain of Trp or benzenes to mimic Phe/Tyr). We considered this compelling evidence for a cation–π interaction.

This "fluorination strategy" is surprisingly general. Linear plots have been seen in over 30 cases, spanning a range of proteins and ligand types. Drug-like molecules with widely differing structures have been studied, including quaternary ammonium ions (acetylcholine) and protonated amines, including primary (glycine, GABA, serotonin), secondary (epibatidine, cytidine, varenicline) and tertiary (nicotine). In addition, more complex cations such as granisetron, ondansetron,[9] and the guanidinium toxin tetrodotoxin (TTX)[10] have shown linear fluorination plots. In contrast, a study of another guanidinium compound, *meta*-chlorophenyl biguanide (*m*CPBG) binding to the 5-HT_3 (serotonin) receptor showed behavior that was difficult to interpret.[11] In all cases we compared experimental data to the binding of Na^+ to the appropriate aromatics. While it may be reasonable to assume that a primary ammonium ion (RNH_3^+) is well modeled by Na^+, more complex ions such as a quaternary ammonium or a guanidinium show much different charge distributions (Fig. 1) and so may display different binding behaviors.

Citation: Matthew R. Davis and Dennis A. Dougherty, cation–π interactions, DOI: 10.1039/C5CP04668H.

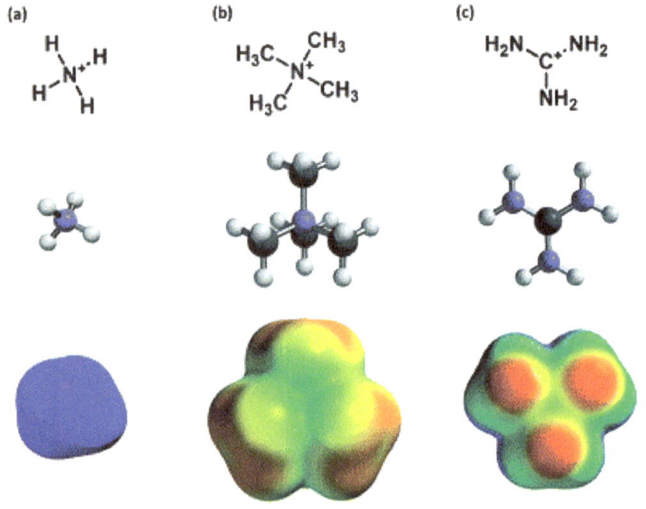

Figure 1 Cations examined in this study. (a) The ammonium ion, (b) the tetramethylammonium ion, and (c) the guanidinium ion. Pictured are molecular structures and potential energy surfaces. (Geometry optimized M06/6-31G(d,p) ranging from +400 (red) to +700 (blue) kJ mol^{-1}.)

To address this issue we have computationally evaluated fluorination effects on cation–π interactions involving the more complex cations ammonium (NH_4^+), tetramethylammonium (NMe_4^+), and guanidinium (Fig. 1). Substituent effects on cation–π interactions and related noncovalent interactions involving benzene have been the subject of several recent investigations, including some with very high levels of theory.[12–14] These studies have revealed some unanticipated effects in such noncovalent interactions. The more modest goals of the present work involve the trends in cation–π binding energies in response to progressive fluorination for several combinations of cation and aromatic. When constrained to a cation–π binding geometry, these larger cations mimic the trends seen with Na^+ as probe ion.

METHODS

All calculations were performed using Spartan 14[15] unless otherwise stated.

Calculating cation–ϖ energies:

Cation–π interactions to benzene and derivatives were evaluated with full geometry optimization at M06/6-31G(d,p)[16] with energies calculated using eqn (1):

$$BE = (E_+ + E_\pi) - E_{TOT} \qquad (1)$$

where E_{TOT} is the total complex energy, E_+ is the energy of the cation alone, and E_π is the energy of the aromatic system, whether it is a single aromatic or an aromatic box bereft of a cation. In this form, a positive BE signifies a favorable interaction.

Distance Dependence

Distance dependence data were obtained by constraining the distance of the ion in question to benzene and performing a constrained geometry-optimized M06/6-31G(d,p) calculation. This distance constraint was increased iteratively from the starting (optimal) geometry, and the binding energy calculated using eqn 1. These data were fit using the KaleidaGraph™ software (Synergy Software).

Aromatic Box Calculations

Ab initio aromatic boxes with a complexed ion were generated using Spartan 14. Geometry-minimized (M06/6-31G(d,p) ≡ M06/6-31G**) structures were obtained for ammonium bound to 3 or 4 benzene molecules and for tetramethylammonium binding to 3, 4, or 5 benzene molecules. The binding energies were obtained using eqn (1), where E_π is the energy of the aromatic box without the tetramethylammonium ion.

Existing aromatic box motifs were imported from the Protein Data Bank using Gaussview 5,[17] and protons were added to the structures. Cation–π binding energies were calculated using Gaussian09.[18] Aromatic boxes were generated by deleting backbone and β atoms of the amino acids as well as the noncationic portion of acetylcholine or trimethyllysine (to generate NMe_4^+in both cases). The cation–π binding energy was computed by taking the single-point energy (M06/6-31G(d,p)) with and without the tetramethylammonium ion. Individual cation–π interactions between tetra-methylammonium and aromatic residues were found by deleting all but the cation and aromatic group of interest and computing a single point energy.

Fluorination Series

In most cases, full geometry optimization was performed for complexes of cations to fluorinated aromatics. In some cases, however, this resulted in geometries that would not be considered cation–π interactions. In such cases, fluorines were appended to the benzene ring with C–F bond distances set to values determined by optimization of the isolated aromatic system. This allowed determination of a cation–π binding energy with the cation at the position appropriate to the benzene–ion complex. We will refer to this as the single-point binding energy. The structures were then allowed to relax to the lowest-energy geometry in the gas phase. The binding energy itself was calculated using eqn (1).

Several cation–π structures represented relatively shallow local minima, and were challenging to isolate. For the three separate orientations of the tetramethylammonium ion (one, two and three methyl groups facing the benzene), starting structures were the molecular mechanics-minimized structure. This enforced a symmetry on the structure that was maintained while optimizing at a higher level of theory. To achieve a stacked conformation for guanidinium–benzene, the distance from the center of the guanidinium molecule to three *meta* carbons of benzene was constrained to 3.4 Å and the (constrained) equilibrium geometry was calculated at M06/6-31G(d,p). The final stacked geometry was obtained using this restricted geometry as a starting point for full geometry optimization.

RESULTS

Choice of computational method

We considered several different levels of theory to find which offers the best combination of accuracy and computational efficiency. Since one goal of the present work is to enable evaluation of larger systems related to protein crystal structures, we did not consider more computationally expensive correlational methods such as CCSD, nor did we consider overly large basis sets of the sort that have often been used to evaluate various methods. We considered four prototype cation–π interactions for which experimental gas phase data are available (Table 1).[19–22] Some time ago, we computed the binding of Na^+ to a number of simple aromatic systems.[23] We found that a relatively simple level of theory – HF/6-31G(d,p) – performed remarkably well. As shown in Table 1, it remains the most accurate level of theory considered for both Na^+ and K^+ binding to benzene. HF does, however, perform poorly on the larger ions. MP2 performed inconsistently depending on the ion studied. Also, several years ago we evaluated the most popular DFT method at the time – B3LYP – and found the results to be disappointing. Subsequent studies have confirmed that B3LYP is not especially effective for noncovalent interactions without complex dispersion correcting potentials.[24–29]

Table 1 Binding energies to benzene (kcal mol^{-1})[a]

Ion	HF 6-31G(d,p)	HF 6-311+G(d,p)	MP2 6-31G(d,p)	MP2 6-311+G(d,p)	B3LYP 6-31G(d,p)	B3LYP 6-311+G(d,p)	M06 6-316(d,p)	M06 6-311+G(d,p)	Expt.
Na^+	27.1	23.2	29.9	24.9	28.4	23.8	26.8	22.2	28.0
K^+	19.3	15.6	23.8	19.5	20.8	16.1	20.3	16.3	19.2
NH_4^+	15.3	14.5	19.0	19.2	17.5	15.8	19.5	18.2	19.3
NMe_4^+	6.6	5.6	11.3	11.7	7.7	6.0	10.8	10.0	9.4
Error[b]	1.9	4.3	2.2	1.5	1.4	3.6	0.98	2.6	—

a Computed values are ΔE; experimental values are ΔH. All data in the gas phase. *b* Mean absolute error of the four values.

In recent years, a wide array of DFT methods has been developed, each targeting a specific computational challenge. Several methods have been specifically parameterized for noncovalent interactions, including M06-L, M06, M06-2X and others. A previous comparison of over a dozen DFT methods showed that M06 and M06-2X both performed well.[16] For cation–π interactions, we find that M06 performs very well (Table 1), even with a modest basis set. The most significant effect is a substantial improvement in the binding to tetramethylammonium. Surprisingly, M06-2X overestimated the binding energies to benzene of Na^+ and K^+ by 2 and 4 kcal mol^{-1}, respectively, and so we did not consider it further. Clearly, when considering DFT methods one must make a choice on a case-by-case basis. Interestingly, for all the methods considered except MP2, the larger 6-311+G(d,p) basis set gave poorer results than 6-31G(d,p). This likely reflects a cancellation of effects to some extent, but the results are consistent across the various cations considered. Given the results of Table 1, we have settled on M06/6-31G(d,p) as the optimal level of theory. It gives quite good results for all the systems considered. We note that the simpler HF/6-31G(d,p) method reproduces all the trends seen with M06; the largest differences are seen in the quantitative results for tetramethylammonium and guanidinium ions.

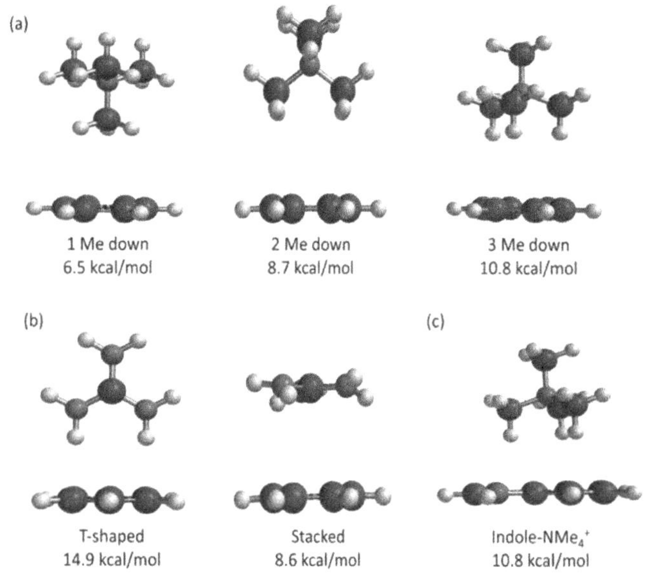

Figure 2 Examples of large cations binding to benzene. (a) Tetramethylammonium ions binding with one, two and three methyl groups towards the face of benzene. (b) Guanidinium ions binding in T-shaped and stacked conformations. (c) Tetramethylammonium binding to indole. Cation–π binding energies computed at M06/6-31G(d,p).

Cation–ϖ Interactions with Benzene

The cation–π binding energies and optimized geometries were calculated at M06/6-31G(d,p) for a number of different ions. For the simple metal ions Na^+ and K^+ as well as NH_4^+, the cation binds to benzene in one conformation: directly centered in the middle of the benzene ring (*en face*). To evaluate the tetramethylammonium–benzene interaction, the complex was minimized in three separate cation–π geometries: with one, two, and three methyl groups *en face* (Fig. 2). Geometries with more methyl groups *en face* had higher binding energies, consistent with previously reported work.[30]

Cation–π energies of guanidinium ions to benzene were calculated in two separate conformations: T-shaped and stacked (see Fig. 2). As has been seen in other studies,[31,32] T-shaped interactions have a much stronger binding energy. Interestingly, according to M06/6-31G(d,p) the guanidinium ion adopts a propeller shape in the stacked conformation, with the amine groups adopting a 25 degree torsion angle from the plane of the cation. All cation–π binding energies are listed in Table 2.

Table 2 Cation–π binding energies (kcal mol^{-1})^a

	Benzene	F-benzene	F_2-benzene		F_3-benzene	
			1,3	1,4	1,2,3	1,3,5
a **M06/6-31G(d,p) calculations; full geometry optimization unless otherwise noted. *b* Binding energy obtained *via* single point method.**						
Sodium	26.8	22.7	19.2	18.9	16.1	15.6
Potassium	20.3	17.7	14.6	14.4	12.0	11.7
Ammonium	19.5	16.2	12.8	12.9	10.1	10.1
Tetramethylammonium (1)	6.5	5.3^b	4.3^b	—	2.9^b	3.0
Tetramethylammonium (2)	8.7	7.2^b	5.5^b	5.7	4.0^b	—
Tetramethylammonium (3)	10.8	9.0^b	7.1^b	7.0	5.3^b	5.4
Guanidinium (T-shaped)	14.9	12.3	10.4	9.8	7.0^b	7.9
Guanidinium (stacked)	8.6	7.2^b	5.8^b	—	4.6^b	5.1

Distance dependence

In biological systems cations might bind to aromatic rings at non-optimal geometries (the optimum being essentially van der Waals contact), due to other constraints imposed by the protein and/or ligand. This might be especially true for the aromatic box motif, as discussed below. As such, it is important to probe non-optimal binding geometries. To evaluate this situation, we have determined the distance dependence of a cation–π interaction to benzene for Na^+, K^+, NH_4^+, and tetramethylammonium (1-methyl down). As shown in Fig. 3, similar trends are seen, and the cation–π interaction distance dependence is not overly steep. For all three cations, displacement by a full Å from the optimal distance results in roughly a 40% drop in binding energy. When fit to a $1/r^n$ function, n values of 1.9, 2.0, 2.8, and 2.8 are obtained for Na^+, K^+, NH_4^+, and tetramethylammonium, respectively. These results indicate that even a non-optimal cation–π interaction can contribute significantly to ligand binding or protein stability.

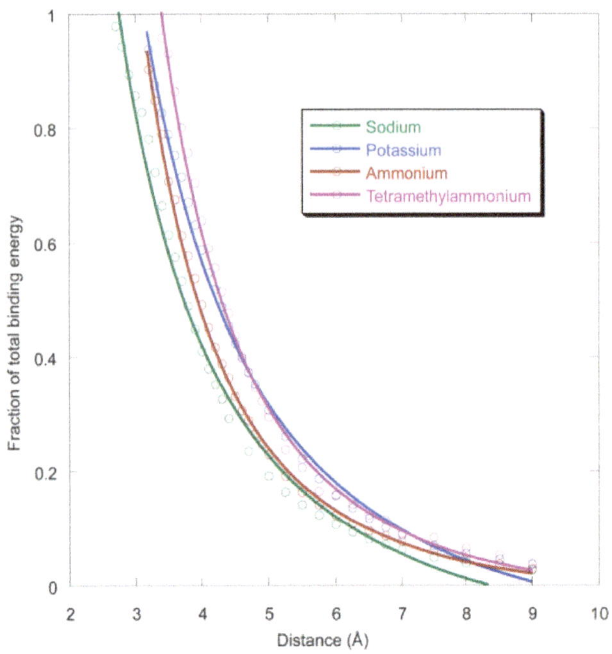

Figure 3 Distance dependence of cation–π binding for sodium, potassium, ammonium, and NMe_4^+ (one methyl down) ions. Binding energy is relative to the optimal binding, and distance is from the center of the benzene to Na^+ or K^+; N atom of ammonium; or methyl C of tetramethylammonium. Complexes with very short distances deviate from this electrostatic plot and were not included in the fitting.

The Aromatic Box

In recent years, it has become apparent that there is a common motif for a biological cation–π binding site – the so-called aromatic box.[2,5–7] This is formed by 3–5 aromatic side chains that encapsulate the positive charge of a cationic molecule (Fig. 4 and 5). An early and especially compelling example was seen in the ACh binding proteins (AChBP), which provide excellent models for the nicotinic acetylcholine receptor (nAChR) binding site.[33,34] Five aromatics form a box around the quaternary ammonium of ACh and around cationic sites of other ligands in this much discussed system (Fig. 5a). Another frequently seen box is found in proteins that bind a trimethyllysine (KMe_3^+) side chain of a histone protein.[35–39] The KMe_3^+ structure is a common and critical motif as part of the "histone code". A number of crystal structures of transcriptional regulators bound to KMe_3^+ have appeared, and all show an "aromatic box" that binds the quaternary ammonium ion through cation–π interactions (Fig. 5b).

We first probed the aromatic box by studying the simultaneous binding of 4 benzene rings to a tetramethylammonium.[40] As seen in Table 3 and Fig. 4a, a tetramethylammonium can easily bind to multiple benzenes simultaneously. The total binding energy is 39.2 kcal mol^{-1} (Table 3). Each pairwise interaction has three methyl groups *en face*, and so four optimal interactions of this sort would produce a binding energy of 43.2 kcal mol^{-1} (4 × 10.8; listed as "Theoretical" in Table 3). The individual binding energies are nearly optimal, as evidenced by the "Sum" column in Table 3. Similar results are seen for 2 or 3 or 5 benzenes (ESI†), although the 5th benzene binds less tightly than the others.

Table 3 Binding energies (kcal mol^{-1}) to a 4-benzene aromatic box

Table 3 Binding energies (kcal mol^{-1}) to a 4-benzene aromatic box

	Binding energy	Theoretical[a]	Sum[b]	Discrepancy[c]
NH_4^+	59.6	78.0	71.1	18.4
NMe_4^+	39.2	43.2	42.3	4.0

[a] Four times the binding energy for a single ion–benzene complex. [b] Four times the binding energy of a single cation-box aromatic. [c] The difference between the theoretical binding energy and the calculated binding energy.

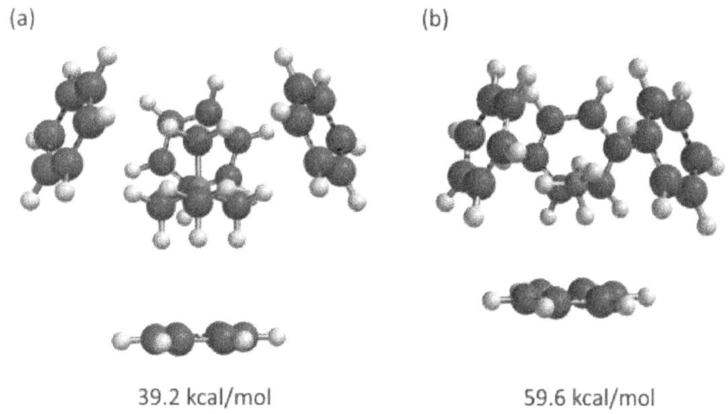

Figure 4 Geometry-optimized complexes of four benzenes to (a) one tetramethylammonium and (b) one ammonium ion. Binding energies computed at M06/6-31G(d,p).

Receptors that bind primary ammonium ions (RNH_3^+) such as GABA or serotonin also have a cluster of aromatic amino acids at the agonist binding site, so we considered the binding of ammonium to an aromatic box. Again, four benzenes can fit comfortably around an ammonium ion with no obvious steric conflicts (Fig. 4b). Experimental studies of K^+, which is similar in size to ammonium, have shown that multiple benzenes can complex the ion.[21,41,42] The total binding energy of ammonium to four benzenes is 59.6 kcal mol^{-1}, but now the individual interactions are not optimal; the sum value (four times the binding energy of a single cation-box aromatic) differs considerably from the true binding energy. The even larger difference between the theoretical value and the binding energy could reflect a small negative cooperativity due to conflicts between induced dipoles generated by the binding.

Such induced dipoles are expected to be smaller when tetramethylammonium is the ion, consistent with the smaller deviation seen in Table 3.

The aromatic box motif found in biological systems uses Phe, Tyr, and Trp, although there is a general bias toward Tyr and Trp.[32] Based on studies using Na^+ as a probe ion,[44] Trp is expected to provide the strongest cation–π binding. Indeed, calculated binding energies for tetramethylammonium (3 methyls down) to benzene (Phe), phenol (Tyr), and indole (Trp) are 10.8, 12.5, and 15.1 kcal mol^{-1}, respectively. As in previous studies, the cation is significantly offset from the center of the phenol,[45] and it lies over the six-membered ring of the indole.[32]

We replaced the ACh of AChBP with tetramethylammonium and evaluated the total binding energy as well as the interaction to each individual ring (Fig. 5a and Table 4). We find a total binding energy of 28.9 kcal mol^{-1}, with four of the five aromatics contributing significantly. The largest contribution comes from TrpB, the residue that has been shown experimentally to make a functionally important cation–π interaction in most nAChRs.[4]

Table 4 Binding energies (kcal mol^{-1}) of tetramethylammonium to protein aromatic boxes[a]

AChBP		**Histone binding protein**	
***a* M06/6-31G(d,p) calculations; single-point energies. *b* Binding energy if all interactions were optimal and independent.**			
Aromatic box	28.9	Aromatic box	32.3
Predicted[b]	67.7	Predicted[b]	42.7
TrpB	8.3	Trp50	13.9
TyrC2	8.0	Trp47	12.0
TyrC1	7.5	Tyr26	9.0
TrpD	4.8		
TyrA	2.3		
Sum	30.9	Sum	34.9

We applied a similar analysis to the binding of KMe_3^+ to a histone binding protein (Fig. 5b).[43] We find a total binding energy of 32.3 kcal mol^{-1} for the tetramethylammonium. Trp47 and Trp50 contribute the bulk of the binding energy, but Tyr26 is still significant (Table 4). Based on the distance dependence of tetramethylammonium binding from Fig. 3, it is evident that all eight potential methyl···aromatic interactions of Fig. 5b contribute to the total binding energy. For both protein structures, the calculated total binding energy is slightly less than the sum of the individual interactions, as was seen in the model structures. Also for both systems, the total binding energy is much less than what would be expected if all cation–π interactions were optimal; deviations of 39 and 10 kcal mol^{-1} are seen for AChBP and the histone binding protein, respectively.

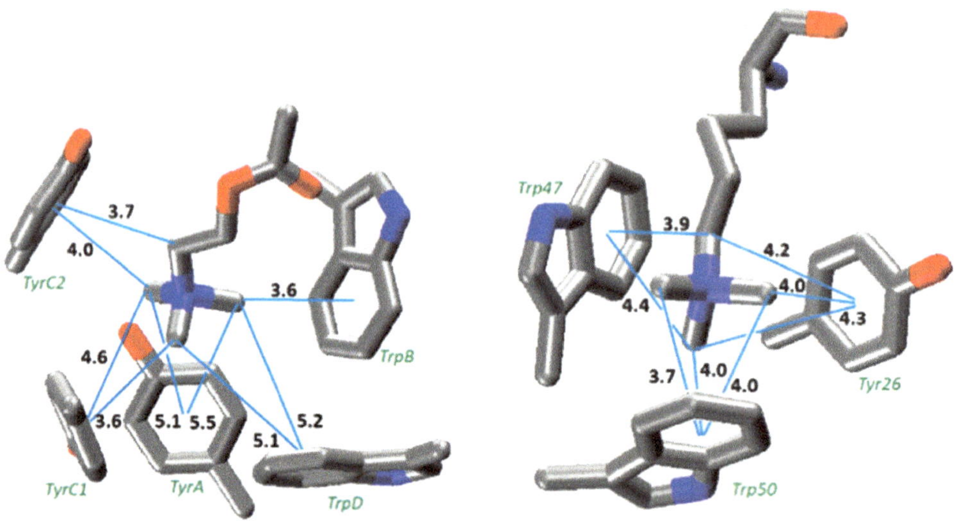

Figure 5 The aromatic box motif binding sites of (left) ACh binding to AChBP (PDB: 3WIP).[34] Residues are labeled according to their conventional designation in the nAChR. (right) Trimethyllysine binding to the polycomb chromodomain from *Drosophila* histone H3 protein (PDB: 1PFB).[43] Distances (Å) are from a carbon to the ring centroid; the calculated optimal distance for such an interaction is 3.6 Å.

Fluorination Studies

In our experimental studies of a possible cation–π interaction to a Phe or a Tyr,[4] we have used 4-F-Phe, 3,5-F_2-Phe, and 3,4,5-F_3-Phe as our cation–π interaction probes, while for Trp we use the 5-, 5,7-, 5,6,7-, and 4,5,6,7-fluorotryptophans. As such, we have focused on the analogous benzenes and indoles for our computational studies (Tables 2 and 5). For the most part these studies were straightforward: the ion remained roughly centered over the ring and the cation–π binding energy progressively decreased as fluorines were introduced.

Table 5 Cation–π binding energies (kcal mol^{-1})[a]

	Indole	F-indole 5	F2-indole 5,7	F3-indole 5,6,7	F4-indole 4,5,6,7
Sodium	32.7	28.8	25.7	22.5	18.9
Tetramethyl-ammonium (3)	15.2	13.5	11.9	10.1	9.1[b]

[a] M06/6-31G(d,p) calculations; full geometry optimization unless otherwise noted. [b] Binding energy obtained *via* single point method.

The exceptions occurred with the larger ions tetramethylammonium and guanidinium binding to the fluorobenzenes and highly fluorinated indoles. In these cases full geometry optimization led to edge-on binding, with the cation attracted to the fluorine(s). Full details of these results are presented in the ESI.†

While these edge-on structures are interesting with regard to possible gas phase studies, the focus of the present work is the binding of cationic ligands to proteins, where structural constraints are likely to discourage edge-on geometries. We developed two strategies to determine the effect of fluorination on such constrained systems. First, we simply added fluorines to the ion–benzene complex (which is in an *en face* cation–π geometry) and performed single point calculations. Alternatively, we could use the symmetrical 1,4- and 1,3,5-fluorobenzenes, for which the ions stayed centered over the ring. The two strategies produced similar results, with the expected trends on progressive fluorination.

Comparison of Fluorination Calculations with *in vivo* data

As noted above, we previously compared experimental results for a fluorination study with binding energies for Na^+ to the analogous aromatic.[4] We can now perform more realistic comparisons. For primary amines, the previously described cation–π interaction between glycine and fluorinated phenylalanine residues in the glycine receptor[46] was re-evaluated using the results for the binding of an ammonium ion to fluorinated benzenes. As shown in Fig. 6, a linear trend between agonist potency and cation–π binding energy is still obtained.

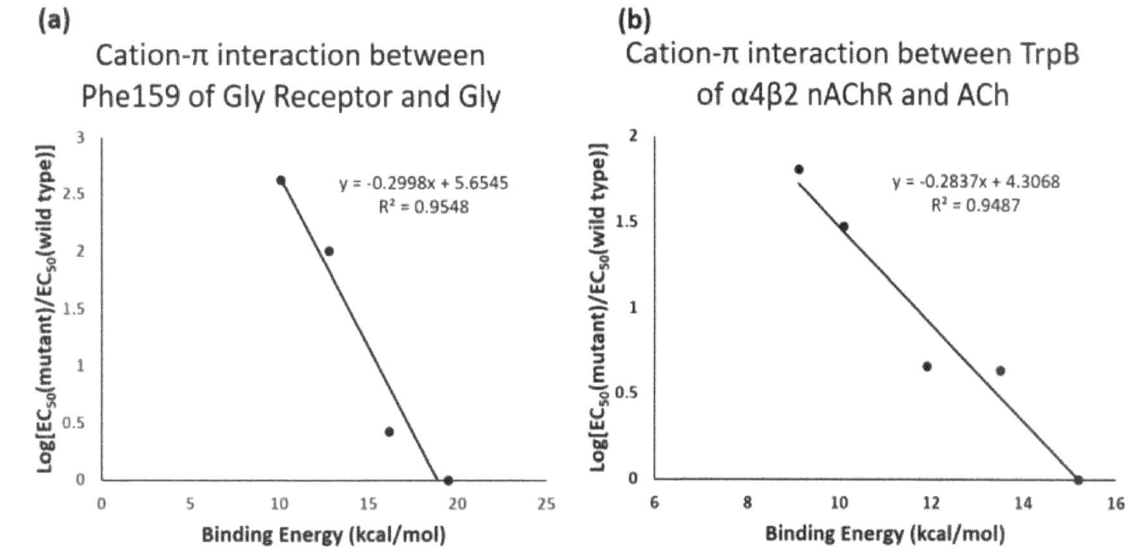

Figure 6 Cation–π binding plots for (a) glycine at the glycine receptor using computed cation–π binding energies for ammonium ions to fluorinated benzenes using *in vivo* data collected for activation of glycine receptor mutants[46] and (b) acetylcholine at the α4β2 nicotinic acetylcholine receptor (nAChR) using *in vivo* data collected for activation of α4β2 nAChR mutants[47] using cation–π energies for tetramethylammonium ions binding to fluorinated indoles.

We have several demonstrations of strong cation–π interactions between ACh and nicotinic ACh receptors.[47] The relevant calculation then is tetramethylammonium binding to fluorinated indoles. Fig. 6 shows such an analysis, considering tetramethylammonium binding to the indole ring with three methyl groups *en face* (Fig. 6). In each case, the calculated cation–π binding energies line up well with experimental results, confirming that the results of such an analysis of receptor binding are independent of the nature of the probe cation.

DISCUSSION

The importance of the cation–π interaction in stabilizing protein structures and mediating protein–ligand interactions is well-established, and computational methods have always played an important role in evaluating cation–π interactions. Here we evaluated several modern computational protocols to find one that could be conveniently applied by experimentalists to evaluate potential cation–π interactions in biological structures. Gratifyingly, we find that a not overly complex method – M06/6-31G(d,p) – provides very good results on systems for which experimental data are available. For larger cations such as tetramethylammonium it provides a significant improvement over HF methods. Note that we did not employ counterpoise corrections or other methods to account for likely basis set superposition error (BSSE). It may well be that there is a compensation effect, where the relatively small basis set and the neglect of BSSE roughly compensate for each other. In the end, we feel the method espoused here is quite adequate for evaluating cation–π interactions.

The aromatic box motif is common in protein cation–π binding sites. This suggests that multiple aromatics can simultaneously have favorable interactions with a cation, although it seems likely that not all interactions would be optimal in a protein structure. This prompted us to revisit the issue of the distance dependence of the cation–π interaction. As in an earlier, preliminary study,[1] we find that the distance dependence is not steep (Fig. 3). For example, it could be argued that the cation–π interaction should be treated as an ion–quadrupole interaction. Ideally, an ion–quadrupole interaction should show a $1/r^3$ distance dependence for the binding energy. However, the cation–π interaction is better modeled by a 1/

r^2 distance dependence for the prototype Na^+ and K^+ ions. For larger ions a roughly $1/r^3$ distance dependence is seen, likely reflecting the increased importance of polarizability for the larger ions. In any case, displacement of the ion by a full angstrom from van der Waals contact (the optimal binding arrangement) diminishes the interaction by less than half.

We have taken two approaches to evaluating the aromatic box motif. First, we evaluated the binding of multiple benzenes to both tetramethylammonium and ammonium ions under optimal conditions. We find that four aromatics can easily bind simultaneously to either ion. For tetramethylammonium, the total binding energy is close to what would be expected if all four individual benzene tetramethylammonium interactions were optimal and independent. For ammonium there is a significant deviation from expectation based on four optimal ammonium···benzene interactions. Mostly, this is because the individual interactions are not optimal; clearly the ammonium ion is a bit too small to perfectly bind four benzenes simultaneously. There is a second, non-additivity factor that we attribute to a clash between induced dipoles in the four benzenes. Nevertheless, the global conclusion is that multiple aromatics can readily contribute to the binding of a cation through cation–π interactions.

We next considered two exemplary aromatic boxes from protein structures: the acetylcholine binding protein (AChBP) and a polycomb chromodomain binding to a peptide model of trimethyllysine from the histone H3 protein. In each case we truncated the ligand (ACh or trimethyllysine) to a tetramethylammonium. We anticipated that all contacts between tetramethylammonium and an aromatic of the box would not be optimal, and indeed that is the case. As shown in Table 4, the "predicted" binding energy – that which would arise if each cation–π interaction was optimal – is substantially larger than the actual binding energy. Recalling that an optimal cation–π interaction for tetramethylammonium to Tyr or Trp would be 12.5 or 15.1 kcal mol^{-1}, it is clear that each interaction evaluated individually is weaker than optimal. The histone binding protein has three interactions approaching the *ab initio*-generated geometries and interaction energies, whereas some components of the AChBP box make only relatively minor contributions to the overall binding energy. Nevertheless, it is clear that a ligand can benefit simultaneously from multiple cation–π interactions to components of an aromatic box. Gratifyingly, the largest interaction in the AChBP is to TrpB, which is the residue most commonly involved in a functionally significant cation–π interaction in studies of nAChRs.

We also considered the extent to which the total binding energy deviates from the sum of the individual interactions. In each case, the deviation is not large, suggesting a relative lack of cooperativity (positive or negative) in the biological aromatic boxes.

We conclude that while the positioning of the aromatics in the biological structures is not optimal, a cation can still make energetically significant cation–π interactions to multiple amino acid side chains. The shallow distance dependence of the cation–π interaction presaged this observation.

While not as large as is theoretically possible, total binding energies for the two aromatic boxes are quite large. Like most noncovalent interactions, the cation–π interaction is attenuated on moving from the gas phase to a condensed medium, although the attenuation is not as large as is typically associated with a hydrogen bond or an ion pair, since the desolvation penalty is less severe.[48] Additionally, these calculations do not include any adjustments that the protein structure must make to adapt to the ligand; here we are considering perfectly preorganized binding sites.

It is also interesting that, despite the presence of 5 cation–π interactions, the AChBP shows a smaller binding energy than the histone binding protein with 3 cation–π interactions, because the cation–π interactions in the latter are individually much stronger. Experimentally, AChBP binds ACh with a ~4 mM affinity,[33] while affinities on the order of 0.3 mM have been determined for the binding of KMe_3^+ to a histone binding protein.[49]

An additional goal of the present work was to determine whether previous experimental studies of complex biological receptors – in which cation–π binding to a receptor was correlated to the affinity of a Na^+ to a substituted aromatic ring – would be altered if more realistic models of the biological ligand were used. The ligands studied included simple protonated amines, (RNH_3^+), ACh and related compounds ($RNMe_3^+$), and derivatives of guanidinium. As such, we have evaluated the binding of NH_4^+, NMe_4^+, and guanidinium to a series of fluorinated benzenes and indoles.

As noted above, several highly detailed analyses of noncovalent interactions involving aromatics have appeared in recent years.[49] These have emphasized direct interactions between the ion and the substituent, an interaction that might be expected to favor geometries in which the ion migrates away from the center of the ring toward the substituent(s). We find that for the simpler ions (Na^+, K^+, and ammonium) the ion stays over the center of the ring as fluorines are introduced. In contrast, for tetramethylammonium and guanidinium, edge-on geometries are competitive with and in some cases superior to the cation–π geometries for multiply fluorinated aromatics (ESI†). However, when in a cation–π binding geometry – achieved either through constraints or by using symmetrically substituted benzenes – tetramethylammonium and guanidinium follow the same patterns as the smaller ions.

Our results allow more reasonable modeling of the experimental studies. We find, perhaps not surprisingly, that the linear trends between receptor activation and cation–π binding energy arise regardless of the precise nature of the cation used in the calculations. These more realistic calculations do, however, allow some interesting quantitative comparisons to be made. We have several studies of ACh binding to a series of fluorinated Trp derivatives. The largest effect we have seen is a 540-fold drop in affinity on going from Trp to F_4-Trp, corresponding to 3.7 kcal mol^{-1}.[4] Our calculations for tetramethylammonium show a corresponding drop of 6.1 kcal mol^{-1}. For a simple ammonium ion, our biggest effect has been seen for GABA binding to the $GABA_A$ receptor, with a 16 500-fold drop in affinity on going from Phe to 3,4,5-F_3-Phe, corresponding to 5.75 kcal mol^{-1}.[4] Our calculations on ammonium show a corresponding drop of 8.4 kcal mol^{-1}. We consider these results to be quite encouraging. As noted above, one might expect some attenuation of a cation–π interaction on going from the gas phase to a receptor binding site. The fact that the trend between the two studies is in the right direction, with the protein binding being roughly 60% of the gas phase binding in both cases, suggests that the present computational model may be able to provide semi-quantitative analyses of protein cation–π binding sites.

CONCLUSIONS

The cation–π interaction, common in protein structures, can be modeled well using the recently-developed density functional M06 and the modest 6-31G(d,p) basis set. Calculations at this level of theory show that in the gas phase, there is a relatively shallow distance dependence for the cation–π interaction. This allows multiple interactions to contribute to ligand binding, providing a rationale for the aromatic box motif commonly seen in protein cation–π interaction binding sites. In addition, we examined cation–π binding to the AChBP and histone binding protein aromatic boxes, and reproduce *in vivo* observations.

All the ions studied here show a linear fluorination trend when in a cation–π *(en face)* binding geometry, enforced either by computational constraints or symmetrical substitution patterns. This supports the use of fluorination plots as a means to evaluate potential cation–π interactions.

ACKNOWLEDGEMENTS

We thank the NIH (NS 34407) for support of this work.

REFERENCES

1. D. A. Dougherty, *Science*, 1996, **271**, 163–168.
2. D. A. Dougherty, *Acc. Chem. Res.*, 2013, **46**, 885–893 .
3. J. C. Ma and D. A. Dougherty, *Chem. Rev.*, 1997, **97**, 1303–1324 .
4. E. B. Van Arnam and D. A. Dougherty, *J. Med. Chem.*, 2014, **57**, 6289–6300 .
5. E. A. Meyer, R. K. Castellano and F. Diederich, *Angew. Chem., Int. Ed.*, 2003, **42**, 1210–1250 .
6. L. M. Salonen, M. Ellermann and F. Diederich, *Angew. Chem., Int. Ed. Engl.*, 2011, **50**, 4808–4842 .
7. J. Cheng, R. Goldstein, A. Gershenson, B. Stec and M. F. Roberts, *J. Biol. Chem.*, 2013, **288**, 14863–14873 .
8. D. A. Dougherty and E. B. Van Arnam, *ChemBioChem*, 2014, **15**, 1710–1720 .
9. N. H. Duffy, H. A. Lester and D. A. Dougherty, *ACS Chem. Biol.*, 2012, **7**, 1738–1745 .
10. V. P. Santarelli, A. L. Eastwood, D. A. Dougherty, R. Horn and C. A. Ahern, *J. Biol. Chem.*, 2007, **282**, 8044–8051 .
11. T. F. MIles, PhD thesis, California Institute of Technology, 2015.
12. R. M. Parrish and C. D. Sherrill, *J. Am. Chem. Soc.*, 2014, **136**, 17386–17389 .
13. S. E. Wheeler and J. W. G. Bloom, *J. Phys. Chem. A*, 2014, **118**, 6133–6147 .
14. S. E. Wheeler and K. N. Houk, *J. Am. Chem. Soc.*, 2009, **131**, 3126–3127 .
15. I. C. Wavefunction Inc.
16. Y. Zhao and D. Truhlar, *Theor. Chem. Acc.*, 2008, **120**, 215–241.
17. T. K. Roy Dennington and J. Millam, *Gaussview 5*, Semichem, Inc, Shawnee Mission, KS, 2009
18. M. J. Frisch, G. W. Trucks, H. B. Schlegel, G. E. Scuseria, M. A. Robb, J. R. Cheeseman, G. Scalmani, V. Barone, B. Mennucci, G. A. Petersson, H. Nakatsuji, M. Caricato, X. Li, H. P. Hratchian, A. F. Izmaylov, J. Bloino, G. Zheng, J. L. Sonnenberg, M. Hada, M. Ehara, K. Toyota, R. Fukuda, J. Hasegawa, M. Ishida, T. Nakajima, Y. Honda, O. Kitao, H. Nakai, T. Vreven, J. A. Montgomery Jr, J. E. Peralta, F. Ogliaro, M. J. Bearpark, J. Heyd, E. N. Brothers, K. N. Kudin, V. N. Staroverov, R. Kobayashi, J. Normand, K. Raghavachari, A. P. Rendell, J. C. Burant, S. S. Iyengar, J. Tomasi, M. Cossi, N. Rega, N. J. Millam, M. Klene, J. E. Knox, J. B. Cross, V. Bakken, C. Adamo, J. Jaramillo, R. Gomperts, R. E. Stratmann, O. Yazyev, A. J. Austin, R. Cammi, C. Pomelli, J. W. Ochterski, R. L. Martin, K. Morokuma, V. G. Zakrzewski, G. A. Voth, P. Salvador, J. J. Dannenberg, S. Dapprich, A. D. Daniels, Ö. Farkas, J. B. Foresman, J. V. Ortiz, J. Cioslowski and D. J. Fox, *Gaussian 09*, Gaussian, Inc, Wallingford, CT, USA, 2009.
19. M. Meot-Ner (Mautner) and C. A. Deakyne, *J. Am. Chem. Soc.*, 1985, **107**, 469–474 .
20. M. Meot-Ner and C. A. Deakyne, *J. Am. Chem. Soc.*, 1985, **107**, 474–479 .
21. J. Sunner, K. Nishizawa and P. Kebarle, *J. Phys. Chem.*, 1981, **85**, 1814–1820 .
22. B. C. Guo, J. W. Purnell and A. W. Castleman, *Chem. Phys. Lett.*, 1990, **168**, 155–160 .
23. S. Mecozzi, A. P. West and D. A. Dougherty, *J. Am. Chem. Soc.*, 1996, **118**, 2307–2308 .
24. S. Grimme, *J. Comput. Chem.*, 2004, **25**, 1463–1473 .
25. E. R. Johnson and G. A. DiLabio, *Chem. Phys. Lett.*, 2006, **419**, 333–339 .
26. E. R. Johnson, R. A. Wolkow and G. A. DiLabio, *Chem. Phys. Lett.*, 2004, **394**, 334–338 .
27. S. Kristyán and P. Pulay, *Chem. Phys. Lett.*, 1994, **229**, 175–180 .
28. E. Torres and G. A. DiLabio, *J. Phys. Chem. Lett.*, 2012, **3**, 1738–1744 .
29. S. Tsuzuki and H. P. Lüthi, *J. Chem. Phys.*, 2001, **114**, 3949–3957 .
30. A. Pullman, G. Berthier and R. Savinelli, *J. Comput. Chem.*, 1997, **18**, 2012–2022 .
31. E. M. Duffy, P. J. Kowalczyk and W. L. Jorgensen, *J. Am. Chem. Soc.*, 1993, **115**, 9271–9275 .
32. J. P. Gallivan and D. A. Dougherty, *Proc. Natl. Acad. Sci. U. S. A.*, 1999, **96**, 9459–9464 .
33. T. K. Sixma and A. B. Smit, *Annu. Rev. Biophys. Biomol. Struct.*, 2003, **32**, 311–334 .
34. J. A. Olsen, T. Balle, M. Gajhede, P. K. Ahring and J. S. Kastrup, *PLoS One*, 2014, **9**, e91232 .
35. R. M. Hughes, K. R. Wiggins, S. Khorasanizadeh and M. L. Waters, *Proc. Natl. Acad. Sci. U. S. A.*, 2007, **104**, 11184–11188 .
36. S. A. Jacobs and S. Khorasanizadeh, *Science*, 2002, **295**, 2080–2083 .

37. H. Li, S. Ilin, W. Wang, E. M. Duncan, J. Wysocka, C. D. Allis and D. J. Patel, *Nature*, 2006, **442**, 91–95 .

38. P. V. Peña, F. Davrazou, X. Shi, K. L. Walter, V. V. Verkhusha, O. Gozani, R. Zhao and T. G. Kutateladze, *Nature*, 2006, **442**, 100–103.

39. S. D. Taverna, H. Li, A. J. Ruthenburg, C. D. Allis and D. J. Patel, *Nat. Struct. Mol. Biol.*, 2007, **14**, 1025–1040.

40. J. Cheng, Z. Gong, W. Zhu, Y. Tang, W. Li, Z. Li and H. Jiang, *J. Phys. Org. Chem.*, 2007, **20**, 448–453 .

41. O. M. Cabarcos, C. J. Weinheimer and J. M. Lisy, *J. Chem. Phys.*, 1998, **108**, 5151–5154 .

42. G. K. Koyanagi and D. K. Bohme, *Int. J. Mass Spectrom.*, 2003, **227**, 563–575 .

43. J. Min, Y. Zhang and R. M. Xu, *Genes Dev.*, 2003, **17**, 1823–1828 .

44. S. Mecozzi, A. P. West Jr. and D. A. Dougherty, *Proc. Natl. Acad. Sci. U. S. A.*, 1996, **93**, 10566–10571 .

45. A. Campo-Cacharrón, E. M. Cabaleiro-Lago, J. A. Carrazana-García and J. Rodríguez-Otero, *J. Comput. Chem.*, 2014, **35**, 1290–1301

46. S. A. Pless, A. P. Hanek, K. L. Price, J. W. Lynch, H. A. Lester, D. A. Dougherty and S. C. R. Lummis, *Mol. Pharmacol.*, 2011, **79**, 742–748 .

47. X. Xiu, N. L. Puskar, J. A. Shanata, H. A. Lester and D. A. Dougherty, *Nature*, 2009, **458**, 534–537 .

48. J. P. Gallivan and D. A. Dougherty, *J. Am. Chem. Soc.*, 2000, **122**, 870–874 .

49. H. van Ingen, F. M. A. van Schaik, H. Wienk, J. Ballering, H. Rehmann, A. C. Dechesne, J. A. W. Kruijzer, R. M. J. Liskamp, H. T. M. Timmers and R. Boelens, *Structure*, 2008, **16**, 1245–1256

CHAPTER 13

TOWARD THE IDEAL SYNTHESIS AND MOLECULAR FUNCTION THROUGH SYNTHESIS-INFORMED DESIGN

Paul A. Wender

"The great book, always open and which we should make an effort to read, is that of Nature". Antoni Gaudi 3.8 billion years of chemical evolution has produced a molecular library of unsurpassed size, structural diversity and functional value – our planet's chemome. Only in the last few decades have we started to develop the analytical tools needed to "access and read" this vast treasure trove of molecularly encoded information. The lessons learned thus far have been transformative, revolutionizing chemistry and most other scientific disciplines from molecular anthropology to zoology. New drugs (*e.g.*, taxol), imaging agents (*e.g.*, green fluorescent protein), materials (*e.g.*, biopolymers), research tools (*e.g.*, RNAi), catalysts (*e.g.*, ribozymes), reactants (*e.g.*, enediynes) and reaction processes (*e.g.*, self-replication, bond activation, photosynthesis) have been uncovered, providing a wealth of fundamentally new knowledge and inspiring new innovations of great societal, health and economic benefit.

Notwithstanding the enormous potential value of Nature's library, tapping this vast body of information and inspiration has had and continues to have its challenges. Amongst these is our varied ability to not only make but also supply Nature's molecules.[1] Clearly, synthesis has made remarkable strides forward in addressing this challenge by enabling access to many natural products once considered difficult if not impossible to make. Woodward's view of erythromycin as a "hopelessly complex" synthetic target in the 1950s, for example, gave way with the emergence of new methodology and analytical tools and the benefit of brilliant and dedicated coworkers, to his group's impressive synthesis of this very same target some 25 years later.[2]Imaginative solutions to comparably, and even more complex targets have appeared with increasing frequency, collectively demonstrating the emerging firepower of contemporary synthesis and the impact of those who have creatively advanced its frontiers. Most natural product targets can now be made in the laboratory if suitable resources are provided. The challenge now, and for many targets it is formidable, is to do so in a green, step- and time-economical, if not ideal, way.[3] Addressing this challenge will take more than improvements in reaction and synthesis efficiency as a long synthesis, even if it proceeds in 100% overall yield and thus 100% selectivity, is still a long synthesis with associated costs in time, effort, resources, and environmental impact.

There are three principal ways to more commonly achieve supply-impacting syntheses (Fig. 1: 1–3). While varying in stages of advancement, all involve a more holistic approach to the problem of synthesis design, *i.e.*, one involving not only optimization of single reactions but also of the entire strategy, its implementation and its downstream consequences (*e.g.*, waste and environmental impact). The commonly used and often effective approach to a supply-impacting synthesis is to optimally sequence reactions selected from the current reaction lexicon (Fig.1: 1). For a given synthetic problem, some sequences work, others work even better and some do not impact supply (for metrics read on). Given that most reactions, including the venerable Diels–Alder cycloaddition, allow for only two target bonds to be formed per synthetic operation and many targets require numerous bonds to be formed to connect with starting materials, this approach has its mathematical limitations as longer syntheses generally produce less final product due to the arithmetic of overall yield attrition. Other economies are similarly adversely impacted by increased step count. To address this problem, there is a growing interest in and use of processes that allow for greater target-relevant complexity increases per step, *i.e.*, for more bonds to be made per synthetic operation. The use of multicomponent reactions, serialized sequences (a.k.a., domino or cascade processes) and strategies derived therefrom provide attractive ways to achieve

Citation: Paul A. Wender, Toward the ideal synthesis and molecular function through synthesis-informed design, DOI: 10.1039/C4N-P00013G.

greater per-operation complexity increases and therefore shorter synthetic sequences and better economies.

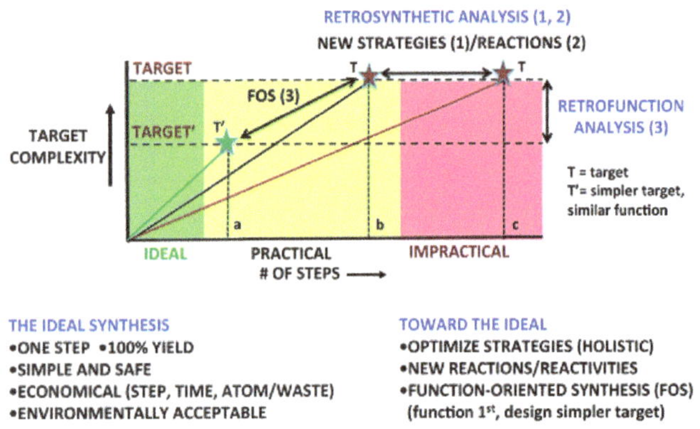

Figure 1. Toward the ideal synthesis. Optimized sequences, step-saving strategies (1): *e.g.*, cascade, domino and multicomponent processes) and related new reactions (2) can reduce the number of steps needed to access a target of complexity "T" from "c" to "b". Alternatively, if in addition to synthetic studies the goal is target function, synthesis-informed design of a simplified target "T'" with function equivalent to "T", *i.e.* function-oriented synthesis (3), could lead to a greatly shortened (a) and thus practical synthesis.

A related second approach, also based on achieving greater increases in target-relevant complexity, is obviously to expand the lexicon of reactions (Fig. 1(2)). New reactions and reactivities change the way one thinks about bond construction. They enable new and often greener synthetic strategies with more choices for route selection. Think about the synthesis of say 1-carbomethoxycyclohex-3-ene now and making the same target before the introduction of the Diels–Alder cycloaddition. The difference is transformative as the cycloaddition provides a new way to think about the problem and not just improvements over existing processes.

A third approach (Fig. 1(3)) to improved synthetic efficacy places an initial emphasis, not on target *structure* but on target *function* (*e.g.*, imaging agents, medicinal agents, materials, diagnostics, catalysts, probes, *etc.*). This change in orientation from structure to function has a dramatic impact on how one defines the synthetic problem and thus on approaches to its solution. It is based on the view that most syntheses are directed at targets with some desired function. Indeed the first section of most publications and proposals related to synthesis often justify a synthesis effort on the basis of the target presenting both challenging synthetic problems and great potential or demonstrated value associated with its function. A central tenet of this third approach, which we have referred to as function-oriented synthesis (FOS),[4] is that a given function can be derived from a variety of structures. It follows that one could design simpler and therefore synthetically more accessible targets with comparable or improved function. This does not eliminate the opportunity to address new synthetic challenges as one could also design targets that would serve that purpose as well. The FOS approach thus shifts the problem from an initial focus on structure and retrosynthetic analysis to an initial focus on function and "retrofunction" analysis. An especially attractive aspect of this approach is that one creates new targets – inspired by Nature or by *de novo* design, and then associated innovative strategies to address their supply. Synthetic chemists are superbly positioned to advance this field as it draws on translating modes of action, *i.e.*, structure-driven mechanism, to new candidate structures that might exhibit that same function and selecting from the latter the ones that could be the most effective and most readily synthesized. It clearly creates many exciting opportunities for chemists to design new, synthetically accessible structures with improved or totally new functions.

Common to all three approaches (Fig. 1(1–3)) is the goal of moving synthesis to a more effective level marked by improvements in quality, speed and value. To do this, metrics are needed to quantify "improvements". Comparisons with past achievements provide one means of measuring progress relative to the reaction and strategy lexicon of the past. Such comparisons generally favor newer work as newer science draws on the rich legacy of the past *and* new advances. Another approach is to start instead with a more aspirational measure of progress, comparing a plan against what might be an ideal solution to a problem. Such prospective comparisons do more to accelerate advancement of the field as they define and orient thinking toward ultimate goals and not just improvements over the past.

The "ideal synthesis" provides a template for measuring a synthetic plan against what might ultimately be possible. It has been defined in various ways and for various purposes over the years, often reflecting the intrinsic objectives of synthesis as well as the emerging priorities of the time. Jim Hendrickson – a friend and former collaborator, offered the following perspective drawn in part from his seminal contributions to computer-based synthesis design: *"The ideal synthesis creates a complex skeleton from simpler starting materials and so must link several such synthon molecules via construction reactions. Ideally, the synthesis would start from available small molecules so functionalized as to allow constructions linking them together directly, in a sequence only of successive construction reactions involving no intermediary refunctionalizations, and leading directly to the structure of the target, not only its skeleton but also its correctly placed functionality. If available, such a synthesis would be the most economical, and it would contain only construction reactions."*[5] There is great insight in this view especially as it relates to many aspects of computer-based design. There are also issues that are subject to interpretation and other unmentioned factors that go beyond connectivity analysis and are of major importance in contemporary synthesis design.

In the 1980s, sensitive to and emphasizing the importance of starting material availability (irrespective of its complexity), as well as operational ease, step economy, and most significantly *safety* – this last a matter of unique and pre-eminent importance in a comprehensive definition of an ideal synthesis, we offered the following: *"ideal syntheses [are] those in which the target molecule is assembled from readily available starting materials in one simple, safe, economical, and efficient operation."*[6] The motivation here was to chart progress by comparisons, numerical where possible, to what might ultimately be possible if not ideal and to integrate the importance of simple and safe operations into synthesis design plans.

There are points of significance in the above definition that merit emphasis. Safety, always a priority and indispensable determinant of a successful synthesis plan, clearly is a first priority in any comprehensive definition of an "ideal synthesis" as it increasingly influences if not drives decisions about synthesis design, route selection and execution. Indeed as we evolve the various "orientations" of synthesis (*e.g.*, diversity-oriented synthesis (DOS), function-oriented synthesis (FOS), target-oriented synthesis (TOS)), safety-oriented synthesis (SOS), an emphasis introduced herein, will assume a position of unique importance and generality. Given the role of safe practices to the field, it is expected that SOS will receive even greater attention in future efforts to design, evaluate, and execute synthetic plans. It clearly addresses the interdependence of design and execution. In the final analysis, the exciting and demanding challenge of synthesis is that one must address all factors that are associated with the design and execution of a plan and safety is a consideration of pre-eminent importance and an essential component of any definition.

Starting material availability, whether simple or complex, is another obvious consideration in planning how best to address a synthetic need. While *simple* starting materials are often sought in a plan, it is more generally the *availability* of a starting material, irrespective of whether it is simple or complex, that determines whether the synthesis will indeed prove to be the most step economical and green if not ideal route. Indeed semi-synthesis, based generally on complex starting materials, has saved the day in many efforts to achieve supply-impacting syntheses including, for example, the semi-syntheses of the medicinal agent taxol[7] and of prostratin,[8] a preclinical lead for HIV/AIDS eradication. It will continue to be a major impact area for synthesis. As science moves forward and manufacturers and fine chemical producers provide more complex, commercially available building blocks, the pool of readily available starting materials will also increasingly include more complex components, thus leading to shorter syntheses. In essence, suppliers, exploiting the brilliant contributions from many laboratories to feedstock enhancements, are moving the starting line of *simplicity* closer to the finish line of *complexity*. Synthetic biology and engineered biosynthesis represent other sources of synthetic building blocks that are impacting synthesis now and will surely influence synthesis in the future including semi-synthesis. Hybrid efforts, combining, for example, biosynthesis and semi-synthesis, have already produced stunning practical syntheses in many fields (*e.g.*, beta-lactam antibiotics).

Step economy is a major driver of synthetic success, a point we and others have emphasized for decades. It requires a holistic approach to design and execution rather than individual step optimizations. It serves to focus innovation not just on improvements but on seeking the best design (closest to the ideal) that accelerates the advancement and especially the broader impact of the field. While the ideal one-step synthesis is unlikely to be commonly achieved (although Wöhler and Reppe did so with urea and cyclooctatetraene, respectively, and many total syntheses have weighed in at under 5 steps), the importance of its inclusion in an "ideal" definition is that it provides a metric by which we could judge where we are and how close we come to an invariant ultimate goal. Such ideal goals drive innovation as they place emphasis on finding the best solution that our lexicon can deliver or defining what reactions might be needed to reach the ideal and not just better the past. Indeed it was this orientation that led many years ago, for example, to our search for more step economical ways to make triquinanes. That search for synthetic brevity approaching the ideal resulted in a three-step synthesis of silphinene, based on the arene-alkene metacycloaddition (Fig. 2).[6] This remarkable process is still one of the most complexity-increasing reactions in our reaction lexicon, allowing for the formation of three new rings and up to six stereocenters in one synthetic operation from relatively simple starting materials.[9]

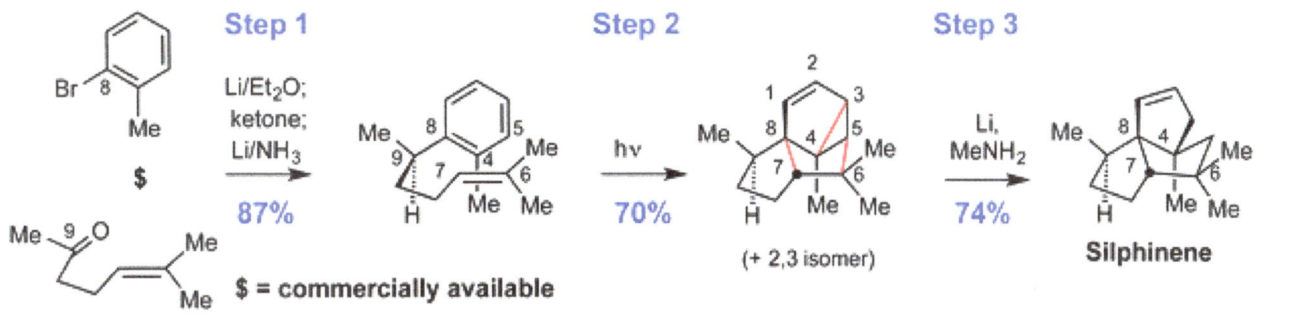

Figure 2 New reactions enable new strategies and more step- and time-economical access to natural and designed targets.

In 1993, sensitive to the growing importance of time and environmental issues associated with synthesis design, we noted in a theoretical overview directed at the "ideal synthesis" and ways to achieve it that *"while the best solution [to a synthetic problem] might be defined in various ways, most practitioners would agree that it will be the one that comes closest to the 'ideal synthesis'...defined as one in which the target molecule is prepared from readily available starting materials in one simple, safe, environmentally acceptable, and resource-effective operation that proceeds quickly and in quantitative yield."*[3] The terms "resource-effective" and "quickly" made their way into this definition as it is clear that even a short synthesis would be ineffective, *i.e.*, far from the ideal, if it were to require too much effort and time to execute. Along with safety, a special priority is thus placed in this definition on conserving the time of those skilled in synthesis as their gift and time represent our most important resources. Most synthesis efforts whether directed at training and research in academia or synthesis campaigns in industry are heavily influenced by what we have referred to as time economy, thus necessitating its inclusion in an ideal synthesis.

Many of the economies of synthesis design are interdependent. For example, step economy favors time economy as steps require development and execution time. Similarly, step economy favors atom economy and more significantly waste minimization, as reaction solvents and purifications are the largest contributors to the waste stream of any process even though solvents are not included in the otherwise useful concept of atom economy.[10] Mapping the fate of atoms in reactants into products is important but reaction and purification solvents often are the major contributors to atom loss in a synthesis because the solvent concentration in a reaction is generally significantly higher than that of reactants. Step reduction thus reduces solvent waste.[11] Solvents can of course be recycled but that requires energy and time and incurs cost. While less emphasized, "energy economy" is a metric of importance in many processes and figures in recycling costs. Simply put, greater step economy correlates with greater "solvent/energy economy". Finally, step economy is also logically at the core of redox economy and the importance of minimizing unnecessary protection steps as redox operations and protections and deprotections often add steps to a synthesis.[12] Step economy more generally emphasizes eliminating unnecessary steps and using reactions and strategies that maximize increases in target relevant complexity.

While discussions about the ideal synthesis will and should continue, and new metrics (*e.g.*, energy economy, solvent economy) will be added, the transition over the past 40 years from comparisons between current and previous syntheses to comparisons between current and ideal syntheses provides an important set of metrics (step count, atom loss, overall yield, overall time) to evaluate where we are and what more might ultimately be possible. A related aspirational set of metrics is what could be called minimally acceptable syntheses which in essence are those that supply what is minimally needed to solve the problem of material availability and route generality. The metrics here are hugely important to design and are context and problem dependent. Scalability, for example, plays a role. If one simply wants only to record a successful synthesis then one would only need to produce enough material to establish purity and to completely characterize what one has produced. If on the other hand one wishes to learn about the properties or activities of a target, then more material will be needed. Still more would be needed for more complex and compound demanding studies associated with, for example, *in vivo* work. Similarly, time economy will also influence design in varied ways. The difference in time constraints and thus synthesis design approaches between producing a creative synthesis of a molecule as part of a multi-year PhD program and producing a scale-up synthesis campaign of a compound cleared for first human dosing in say 6 months is huge and will dramatically influence what design is deemed acceptable. Metrics drive design and thus influence whether one seeks a plan that is minimally acceptable or closer to the ideal. Personnel, safety, development time, execution time, step count, scale, yield, cost, overall selectivity, purity, number of purifications, waste stream, environmental impact, energy requirements, equipment and toxicities are some of the measurable factors that figure in and influence synthesis design and thus the quality, timing and impact of a given plan and its execution.

The above analyses lead to an interesting question and problem and form the basis for the aforementioned third approach to design. What can one do when a target cannot be produced in a timely, safe, step-economical, and resource-effective manner? Obviously one approach would be to invent new reactions that would result in a practical synthesis. This is an important direction. However, reaction invention or discovery and development take time (even the metathesis reaction evolved over decades before reaching broad and practical use in the 1990s). If the target is needed now, a not uncommon situation for many molecules of say medicinal or similar importance, one has a problem principally dictated by time. Many synthetic targets could be produced in a practical fashion if given enough time and resources. That is often not the case. It was this problem (we need it now, but our reaction lexicon is not compatible with it being made now) that led to the genesis of what we have referred to as "function-oriented synthesis" (FOS). As we noted in our earlier overview of FOS: *"the central principle of FOS is that the function of a biologically active lead structure can be recapitulated, tuned, or greatly enhanced with simpler scaffolds designed for ease of synthesis and also synthetic innovation."*[4] In essence and as noted before, function is not the unique property of any one structure but could be exhibited in varying degrees by many structures. Simply put, one need not synthesize a bird to create a flying machine. Nature's library provides a lesson on flight (a bird) from which knowledge can be extracted (a wing's concavity provides lift) thereby enabling the design of many new structures (*e.g.*, gliders, planes and jets) that function even better for intended human use (aviation).

Nature's library is rich with structures whose complexity is determined in part by their natural functions that generally would be irrelevant to many intended human uses of those structures. For example, if one needs a methylating agent, methyl iodide would be far less complex and easier to prepare than Nature's methylating agent, S-adenosyl methionine. Similarly, many natural products are derived from restricted biosynthetic pathways evolved to achieve distribution, target association, metabolism and clearance objectives peculiar to their ecosystem and mostly irrelevant to their potential use in say human therapy. It follows that only a subset of the structural features and therefore the complexity of a natural lead might be needed to recapitulate or even exceed its desirable function for human therapy. 25 years ago this FOS analysis provided the starting point for our phorbol ester tumor promotion and subsequent bryostatin studies. Computer modeling and structure-function studies led to a hypothesis about structural features of phorbol esters required for binding and modulation of protein kinase C (PKC) activity, a highly important protein family implicated in phorbol ester activity. This analysis produced a design blueprint that led to the first rationally designed PKC modulators (Fig. 3).[13] The structural lead for this program, phorbol, was synthesized in 29 steps requiring years to execute.[14] The first designed PKC modulators required 7 steps (step-economy) and two weeks (time-economy) to synthesize and exhibited function not unlike the endogenous PKC modulator.[13]

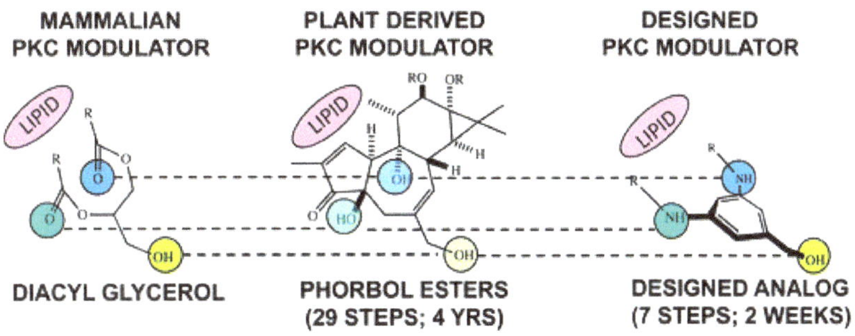

Figure 3 Step- and time-economy associated with the synthesis of structurally varied natural and designed compounds that exhibit a similar function, *i.e.*, binding to and activation of PKC.

In a subsequent FOS effort, bryostatin (Fig. 4), putatively produced in its ecosystem to deter predation of bryozoan larvae by fish, emerged as a significant medicinal lead and is now being studied in clinical trials for cancer and Alzheimer's disease. Our earlier FOS studies on phorbol suggested that bryostatin might associate with PKC through a similar array of H-bond donors and acceptors. This then gave way to similar computer based FOS studies on bryostatin. In 1998, the first designed bryostatin analog was produced.[15] It weighed in at under 30 steps when otherwise impressive syntheses of the natural product at the time required over 70 steps. Significantly, the first bryostatin analogs, bryologs, were found to exhibit comparable or superior activity to bryostatin when screened by the National Cancer Institute for growth inhibition against a panel of human cancer cell lines. We and others have since produced additional bryologs based on this FOS approach. There are also now several impressive total syntheses of natural bryostatins, with three requiring only around 40 steps.[16] In our case, the analog program actually preceded and paved the way for what eventually resulted in a concise total synthesis of bryostatin 9 using a then unexplored Prins macrolactonization strategy.[17]

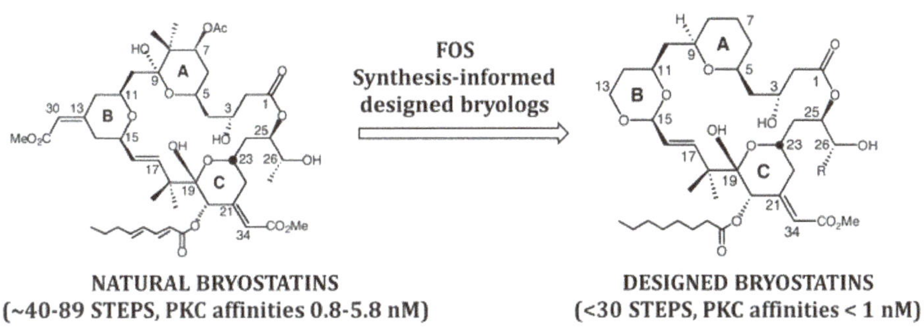

Figure 4 Simplified and more synthetically accessible bryologs inspired by the natural product bryostatin.

Such FOS efforts are not restricted to small molecule design and synthesis in biology and medicine. Materials, polymers, aggregates and other chemical entities with sought after function are also addressable through synthesis-informed design. Indeed, the field of FOS is likely to be creatively led by those skilled in synthesis with an interest in functional targets. For example, new functional probes, *e.g.*, *molecular* tweezers, *molecular* computers, *molecular* wires and on, representing the "*molecularization*" of functional materials, are possible through synthesis-informed design as exemplified by the rapidly growing list of amazing molecular devices.[18]

In another example of the interplay of function and synthesis design, 15 years ago we started a program on drug delivery based on the view that existing drugs and probes could be improved and others enabled through the introduction of better delivery technologies (function: molecules that breach biological barriers).[19] Many molecules of therapeutic interest suffer from an inability to cross biological barriers and enter cells. Some of the most important new research tools, potential therapeutics, and globally recognized scientific targets of current interest, for example, such as siRNA, mRNA, DNA, and therapeutic proteins and peptides do not enter cells well if at all. As noted previously, Nature›s library provides lessons on structure-function relationships and therein the basis to design biomimetics. With respect to drug delivery and barrier passage, one natural lead is found in the form of the protein HIV Tat (Fig. 5) that unlike most proteins enters cells. Research suggested that a sequence of 9 residues (RKKRRQRRR) might be responsible for cellular entry (function). Through a reverse-engineering effort involving systematic truncations, alanine scans and other modifications, we suggested for the first time that cellular entry of HIV Tat was a function of its arginine content and more specifically the number and spatial array of guanidinium groups in the so called Tat 9-mer.[20]

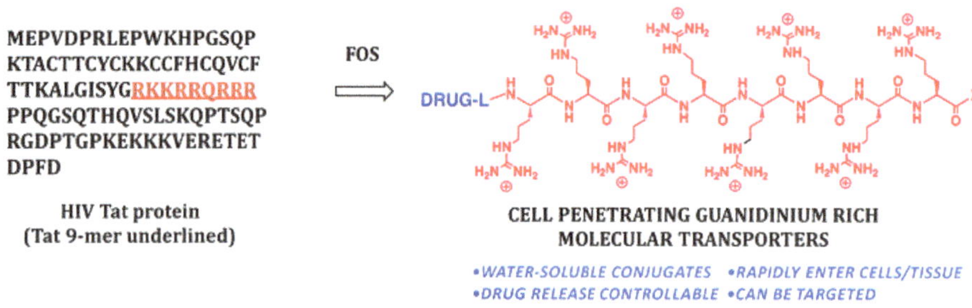

Figure 5 Cell penetrating guanidinium rich molecular transporters for general drug delivery inspired by the naturally occurring cell penetrating protein HIV Tat.

The beauty of design is that once one understands how a function can be achieved one can create synthetically more accessible systems designed to function more effectively. We first found that octaarginine (R8: RRRRRRRR) exhibits better cellular uptake than the Tat 9-mer. This "design" provided already a modest synthetic benefit as the Tat 9-mer requires 18 steps to synthesize while the octaarginine requires only 16 steps. More importantly, because R8 is a homooligomer of arginine, we proposed its assembly through a segment doubling strategy, reducing 18 steps to only 9 and showing how a focus on function could result in significant step economy with an actual gain in function. Unlike conventional solid phase synthesis for which two steps are required for each added subunit, in segment doubling three steps are needed for each doubling of length (starting monomers give in 3 steps a dimer which in 3 more steps give a 4-mer and 3 more give the 8-mer). Thus an 8-mer can be formed in three doubling events, *i.e.*, 9 steps.[21] This is a pure strategic reduction in steps arising from the doubling process and made possible by design of a more effective and more accessible homooligomer. This approach was enabled by knowledge of and focus on function (all arginines would work better) and synthesis-informed design with an eye on step- and time-economy.

It gets better. We have since used this information to make a variety of new guanidinium rich drug delivery systems that we refer to as cell penetrating, guanidinium rich molecular transporters (GRMoTrs). GRMoTrs have been shown to transport (function) small molecules, probes, imaging agents, metals, peptides, proteins, PNAs, RNAs and DNAs into cells in culture and across tissue barriers in animals.[22] They have also been advanced into human clinical trials. We have also shown that attachment of transporters to drugs like taxol that are rendered ineffective due to Pgp export resistance, provides drug-conjugates that overcome the resistance of the drug itself (Fig. 6).[23] In this case the step-economy is significant as one does not need to discover a new drug but rather use FOS to fix the functional problems of an existing drug. In all primary disease samples obtained from ovarian cancer patients, the taxol-transporter conjugate outperforms existing therapy presumably by evading Pgp export. The conjugates thus serve as molecular patches, rapidly getting into cells and releasing free drug only after cell entry at a rate determined by release design.

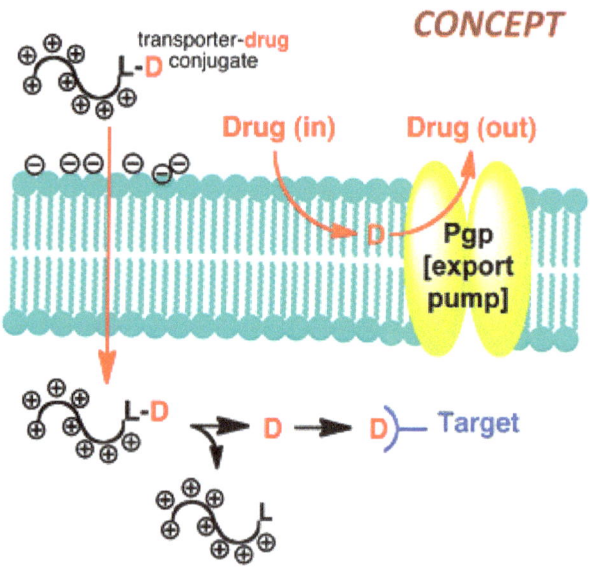

Figure 6 Free drug (D, *e.g.*, taxol; L, releasable linker), as is often the case for membrane soluble (non-polar) drugs, is exported by the membrane bound export pump, rendering the cell drug resistant. Drug-MoTr conjugate is highly polar, enters cells through a different mechanism evading Pgp export and thus overcomes the very resistance that the free drug elicits. Once inside the cell, the drug-MoTr conjugate releases free drug to associate with its intracellular target.

And it gets still better. Significantly, and related to the theme of synthesis-informed design directed at function, when we initiated efforts to study the use of GRMoTrs to transport RNAs, polyanions that do not cross the non-polar membrane of cells, we realized that access to even longer sequences of guanidinium rich (polycation) systems would be needed. To address the resultant need for a step economical strategy to make longer oligomers, we joined in a collaboration with Bob Waymouth, Jim Hedrick and their coworkers, and were able to show that one could produce guanidinium rich oligomers in one step (!) through an organocatalytic oligomerization process (Fig. 7).[24] Simply put in terms of comparative metrics, making a 32-mer by solid phase synthesis would require 64 steps (2 steps/unit added). Even segment doubling would require 15 steps (3 steps/segment doubling). With the oligomerization process, one can produce 32-mers with good polydispersity in only *one synthetic operation*. Focus on the seemingly unattainable "ideal" in this case paid off. Indeed this approach theoretically allows one to make any length of oligomer in only one step. The value of this strategy is that it allows one to rapidly make many different types of oligomers, including di-, tri- and poly-block oligomers in one synthetic operation followed by Boc removal to liberate the free guanidinium groups. This near-ideal, time- and step-economical strategy allowed for a rapid study of factors that contribute to siRNA complexation and the subsequent use of these transporters to complex and deliver siRNA into cells.[25] The amphipathic, guanidinium rich oligocarbonate-siRNA complexes were shown to enter HaCaT cells and selectively knocked down expression of a fluorescent protein (tdTomato) while not affecting expression of another fluorescent protein control (EGFP). This strategy is now driving development of new classes of biodegradable and non-biodegradable transporters for ferrying other oligonucleotides and a variety of other molecular cargos across biological barriers (*e.g.*, membranes, skin, blood brain, ocular and most recently cell wall[26] barriers).

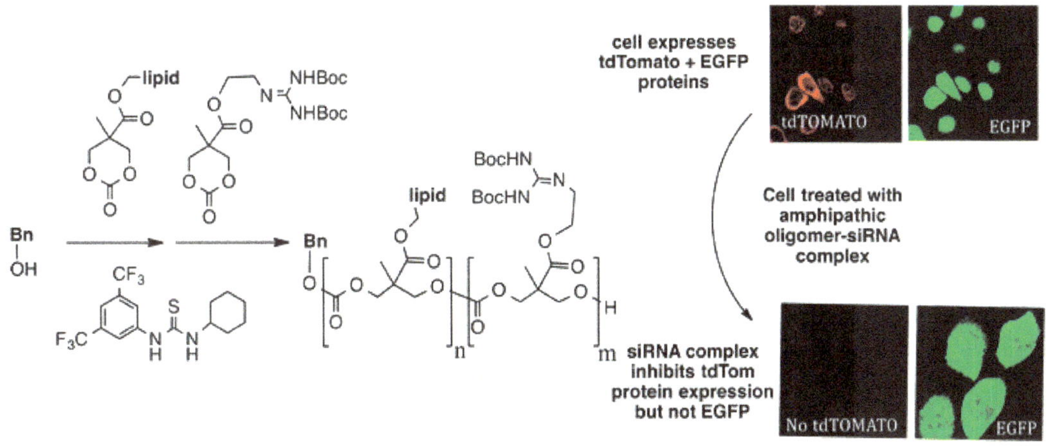

Figure 7 A one-step synthesis of amphipathic oligocarbonate molecular transporters that after Boc removal, complex, deliver and release siRNA in cells, suppressing the expression of only the target protein tdTomato and not expression of the control EGF protein.

The powerful and general view that *form follows function*, the essence of function-oriented synthesis (FOS), can change how one thinks about synthesis especially the synthesis of compounds that exhibit sought after function (*e.g.*, new medicinal agents; smart materials; devices; environmental sensors; imaging agents; energy collection, storage and conversion systems; and research tools) but whose structures are currently beyond the timely and practical reach of synthesis. Function-oriented synthesis is driven by design and it opens a field of opportunities for chemists to not only reproduce what Nature has made but to create new structures and associated functions, whether inspired by Nature or devised by *de novo* design, that might be better suited for their intended human use. FOS addresses the concern that some natural products are too complex and thus not likely to be supplied on scale in a timely fashion by focusing creative attention on the design of simpler and thus more accessible, often bio-inspired structures that can be accessed *now* through synthesis. It opens an exciting field of opportunities that synthetic chemists are uniquely able to lead and develop and with that to create new fields impacting all of molecular science. In short, through synthesis-informed design directed at function, one can achieve practical and creative syntheses in a time- and step-economical, simple, safe, resource effective, and environmentally acceptable fashion, of targets with improved or totally new functions that could transform science in ways that are of profound health, economic and societal benefit.

REFERENCES

1. P. A. Wender and B. L. Miller, *Nature*, 2009, **460**, 197–201 .
2. R. B. Woodward, E. Logusch, K. P. Nambiar, K. Sakan, D. E. Ward, P. Au-Yeung Balaram, L. J. Browne and P. J. Card, *et al.* , *J. Am. Chem. Soc.*, 1981, **103**, 3215–3217
3. (*a*) P. A. Wender and B. L. Miller, Toward the ideal synthesis: connectivity analysis and multi-bond forming processes, in *Organic synthesis: theory and applications*, ed. T. Hudlicky, JAI, Greenwich, CT, 1993, vol. 2, pp. 27–66 ; (*b*) P. A. Wender, D. Wright and S. Handy, *Chem. Ind.*, 1997, **19**, 765 .
4. P. A. Wender, V. A. Verma, T. J. Paxton and T. H. Pillow, *Acc. Chem. Res.*, 2008, **41**, 40–49 .
5. J. B. Hendrickson, *J. Am. Chem. Soc.*, 1975, **97**, 5784–5800
6. P. A. Wender and R. J. Ternansky, *Tetrahedron Lett.*, 1985, **26**, 2625–2628
7. (*a*) J. N. Denis, A. E. Greene, D. Guénard, F. Guéritte-Voegelein and P. Potier, *J. Am. Chem. Soc.*, 1988, **110**, 5917 (*b*) R. A. Holton; R. J. Biediger and P. D. Boatman, Semisynthesis of taxol and taoxtere, in *Taxol: science and applications*, ed. M. Suffness, CRC Press, Boc Raton, Fl, 1995, pp. 97–121 .
8. P. A. Wender, J.-M. Kee and J. M. Warrington, *Science*, 2008, **320**, 649–652 .

9. D. Chappell and A. T. Russell, *Org. Biomol. Chem.*, 2006, **4**, 4409–4430
10. B. M. Trost, *Science*, 1991, **254**, 1471–1477
11. (*a*) P. T. Anastas and J. C. Warner, *Green chemistry: theory and practice*, Oxford University Press, New York, NY, 1998, p. 30 ; (*b*) R. A. Sheldon; I. Arends and U. Hanfeld, *Green chemistry and catalysis*, Wiley-VCH, Weinheim, Germany, 2007 ; (*c*) R. Sheldon, Introduction to green chemistry, organic synthesis and pharmaceuticals, in *Green chemistry in the pharmaceutical industry*, ed. P. J. Dunn, A. S. Wells and M. T. Williams, Wiley-VCH GmbH & KGaA, Weinheim, Germany, 2010 ; (*d*) M. C. Bryan, B. Dillon, L. G. Hamann, G. J. Hughes, M. E. Kopach, E. A. Peterson, M. Pourashraf, I. Raheem, P. F. Richardson, D. T. Richter and H. F. Sneddon, *J. Med. Chem.*, 2013, **56**, 6007–6021 .
12. Noah Z. Burns, Phil S. Baran and Reinhard W. Hoffmann, *Angew. Chem., Int. Ed.*, 2009, **48**, 2854–2867 .
13. P. A. Wender, K. F. Koehler, N. A. Sharkey, M. L. Dell'Aquila and P. Blumberg, *Proc. Natl. Acad. Sci. U. S. A.*, 1986, **83**, 4214–4218
14. P. A. Wender, K. D. Rice and M. E. Schnute, *J. Am. Chem. Soc.*, 1997, **119**, 7897–7898
15. P. A. Wender, J. De Brabander, P. G. Harran, J.-M. Jimenez, M. F. T. Koehler, B. Lippa, C.-M. Park, C. Siedenbiedel and G. R. Pettit, *Proc. Natl. Acad. Sci. U. S. A.*, 1998, **95**, 6624
16. For total syntheses of natural bryostatins, see: (*a*) M. Kageyama, T. Tamura, M. H. Nantz, J. C. Roberts, P. Somfai, D. C. Whritenour and S. Masamune, *J. Am. Chem. Soc.*, 1990, **112**, 7407–7408 ; (*b*) D. A. Evans, P. H. Carter, E. M. Carreira, A. B. Charette, J. A. Prunet and M. Lautens, *J. Am. Chem. Soc.*, 1999, **121**, 7540–7552 ; (*c*) K. Ohmori, Y. Ogawa, T. Obitsu, Y. Ishikawa, S. Nishiyama and S. Yamamura, *Angew. Chem., Int. Ed.*, 2000, **39**, 2290–2294 ; (*d*) B. M. Trost and G. Dong, *Nature*, 2008, **456**, 485–488 ; (*e*) G. E. Keck, Y. B. Poudel, T. J. Cummins, A. Rudra and J. A. Covel, *J. Am. Chem. Soc.*, 2011, **133**, 744–747 ; (*f*) P. A. Wender and A. J. Schrier, *J. Am. Chem. Soc.*, 2011, **133**, 9228–9231 ; (*g*) Y. Lu, S. K. Woo and M. J. Krische, *J. Am. Chem. Soc.*, 2011, **133**, 13876–13879 . For a formal synthesis, see: ; (*h*) S. Manaviazar, M. Frigerio, G. S. Bhatia, M. G. Hummersone, A. E. Aliev and K. J. Hale, *Org. Lett.*, 2006, **8**, 4477–4480 and references therein.
17. P. A. Wender and A. J. Schrier, *J. Am. Chem. Soc.*, 2011, **133**, 9228–9231 .
18. As an example relating to work of the guest editor on "molecular switches", see: M. Stein, A. Breit, T. Fehrentz, T. Gudermann and D. Trauner, *Angew. Chem., Int. Ed.*, 2013, **52**, 9845–9848 .
19. E. Geihe Stanzl, B. M. Trantow, J. R. Vargas and P. A. Wender, *Acc. Chem. Res.*, 2013, **46**, 2944–2954
20. P. A. Wender, D. J. Mitchell, K. Pattabiraman, E. Pelkey, L. Steinman and J. B. Rothbard, *Proc. Natl. Acad. Sci. U. S. A.*, 2000, **97**, 13003–13008 .
21. P. A. Wender, T. C. Jessop, K. Pattabiraman, E. T. Pelkey and C. L. VanDeusen, *Org. Lett.*, 2001, **3**, 3229–3232 .
22. P. A. Wender, W. C. Galliher, E. A. Goun, L. R. Jones and T. H. Pillow, *Adv. Drug Delivery Rev.*, 2008, **60**, 452–472 .
23. E. A. Dubikovskaya, S. H. Thorne, Th. H. Pillow, C. H. Contag and P. A. Wender, *Proc. Natl. Acad. Sci. U. S. A.*, 2008, **105**, 12128–12133 ; P. A. Wender, W. C. Galliher, N. M. Bhat, T. H. Pillow, M. M. Bieber and N. H. Teng, *Gynecol. Oncol.*, 2012, **126**, 118–123 .
24. C. B. Cooley, B. M. Trantow, F. Nederberg, M. K. Kiesewetter, J. L. Hedrick, R. M. Waymouth and P. A. Wender, *J. Am. Chem. Soc.*, 2009, **131**, 16401–16403 .
25. E. I. Geihe, C. B. Cooley, J. R. Simon, M. K. Kiesewetter, J. A. Edward, R. P. Hickerson, R. L. Kaspar, J. L. Hedrick, R. M. Waymouth and P. A. Wender, *Proc. Natl. Acad. Sci. U. S. A.*, 2012, **109**, 13171–13176 .
26. J. M. Hyman, E. I. Geihe, B. M. Trantow, B. Parvin and P. A. Wender, *Proc. Natl. Acad. Sci. U. S. A.*, 2012, **109**, 13225–13230

CHAPTER 14 ARTIFICIAL SWITCHABLE CATALYSTS

Victor Blanco, David A. Leigh and Vanesa Marcos

INTRODUCTION

Different enzymatic syntheses within cells often occur in parallel. To ensure that the transformations proceed with the necessary spatial and temporal control, and without unwanted interference from other reaction pathways, enzyme activity is often modulated through feedback loops and a variety of trigger-induced effects.[1,2] In contrast, the reactions promoted by man-made catalysts usually take place according to the initially chosen reaction conditions.[3] Incorporating stimuli-responsive features into artificial catalysts can confer an additional 'bio-like' level of control over chemical transformations, potentially enabling such systems to perform tasks in synthesis that are difficult or impossible to accomplish in other ways. For example, switchable catalysts can be used to speed up or slow down the rate of a reaction according to the presence of a specific analyte or to change the stereochemical outcome of a reaction. In recent years their utility in performing more advanced tasks, such as turning 'on' and 'off' different catalysts to promote alternative reactions from a mixture of different building blocks, have started to be explored.

In this review we discuss the state-of-the-art in the emerging field of artificial switchable catalysis, including systems whose activity and/or selectivity can be controlled in different chemical reactions, grouping them by the type of external triggering stimulus or other changes in the experimental conditions. We conclude by considering the future outlook for this area.

LIGHT-DRIVEN SWITCHING

Photoresponsive processes, such as photosynthesis[4] and vision,[5] exemplify the effectiveness of light in initiating and regulating complex molecular and biochemical processes. The use of light to induce a change in catalyst state is particularly attractive as it is non-invasive, offers excellent temporal and spatial control, and can be precisely regulated with an appropriate light source. Artificial photoswitchable catalysts[6] are generally based on photoactive units incorporated within the molecular architecture or through intermolecular interactions between a photochemical additive and the catalyst active site. In this review we classify these systems according to the principal type of effect used to achieve control: cooperative, steric or electronic effects, or the aggregation/ dissociation of the catalyst.

Cooperative effects

Cooperative effects in switchable catalysis are often based on a large geometrical change, typically occurring during an isomerization process, that induces a change in the distance or orientation of key sites in the catalyst. A seminal example of a cooperative template for catalysis was reported by Würthner and Rebek (Scheme 1),[7] who used the isomerization of the azobenzene unit in *E*/*Z*-**1** to change the relative positions of two carbazole-based adenine receptors to control the rate of the amide coupling reaction between aminoadenosine (**2**) and an adenosine-derived ester (**3**) (Scheme 1). In the extended *E*-**1** form, the bound substrates (**2**, **3**) are held too far apart to react. Upon UV isomerization to the folded form of the receptor, *Z*-**1**, bound **2** and **3** are held in close proximity and the amidation reaction is accelerated. By determining the binding constants of the substrates to the switchable catalyst and calculating the concentrations of the intermolecular complex, the authors showed that *Z*-**1** is approximately 50 times more active than *E*-**1**. However, the system suffers from product inhibition: product **4**binds strongly to two imide groups of *Z*-**1** whereas each of the reactants, **2** and **3**, bind only to one imide group. Indeed, neither thermal nor photochemical reverse isomerization of the catalyst could be induced while **4** is bound to *Z*-**1**.[8]

Citation: Victor Blanco, David A. Leigh and Vanesa Marcos, Artificial switchable catalysts, DOI: 10.1039/C5CS00096C.

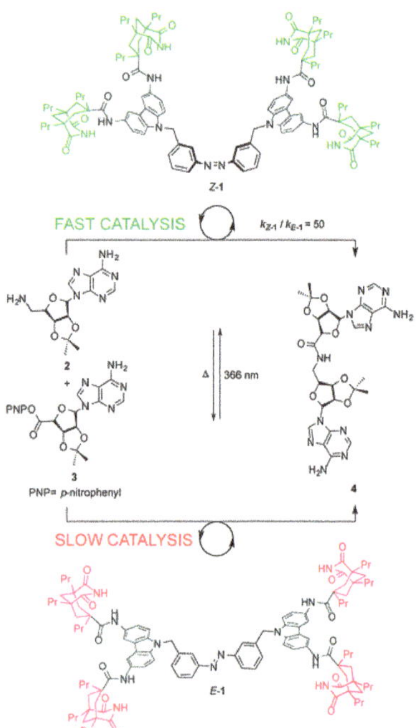

Scheme 1 Controlling the rate of amidation using photoswitchable template *E*/*Z*-**1**.[7]

An example of a photoswitchable catalyst based on cooperative interactions was reported by Cacciapaglia and co-workers in 2003 (Scheme 2).[9] The bis-barium(II) complex of azobis(benzo-18-crown-6) ether **5** was used to control the rate of basic ethanolysis of 4-carboxyacetanilide **6**, exploiting the reversible *E* to *Z* isomerization of the azobenzene spacer. Efficient ethanolysis requires the participation of two metal binding sites, which is only feasible for *Z*-**5**, formed upon UV irradiation of the less active *E*-**5** isomer ($k_{Z\text{-}5}/k_{E\text{-}5}$ up to 5). The authors demonstrated that the catalytic activity could be switched between 'faster' and 'slower' multiple times by alternating between exposure to UV or visible light, illustrating the reversibility of the photoisomerization reaction.

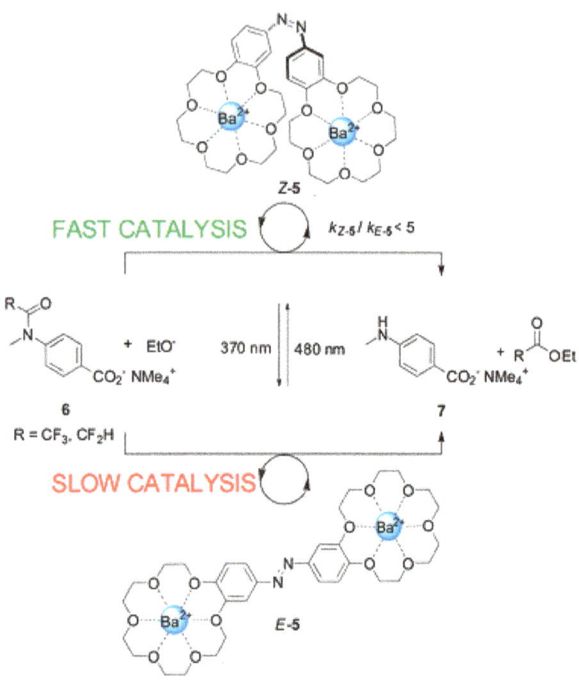

Scheme 2 Controlling the rate of anilide ethanolysis using photo-switchable catalyst E/Z-5.9

Linking two triaryl alcohol groups through an azobenzene moiety resulted in cooperative bifunctional switchable catalyst *E*/*Z*-**8**, which can turn 'on' or 'off' the Morita–Baylis–Hillman reaction of 3-phenylpropanal (**9**) and 2-cyclopente-1-one (**10**) upon exposure to light (Scheme 3).[10] In the *E*-**8** isomer the hydroxyl groups are too far apart to engage in the intramolecular hydrogen bonding needed to increase the acidity of the trityl alcohol, affording **11** in 37% yield, only a small increase over the background reaction (27%). In contrast the *Z*-**8** isomer, generated by UV irradiation of *E*-**8**, afforded **11** in 78% yield. The catalytic activity of *Z*-**8** is significantly higher than other bis(trityl alcohol) derivatives similarly activated by intramolecular hydrogen bonding, suggesting that there are other cooperative enhancements in the catalysis mechanism due to having the two trityl alcohols in close proximity in *Z*-**8**.

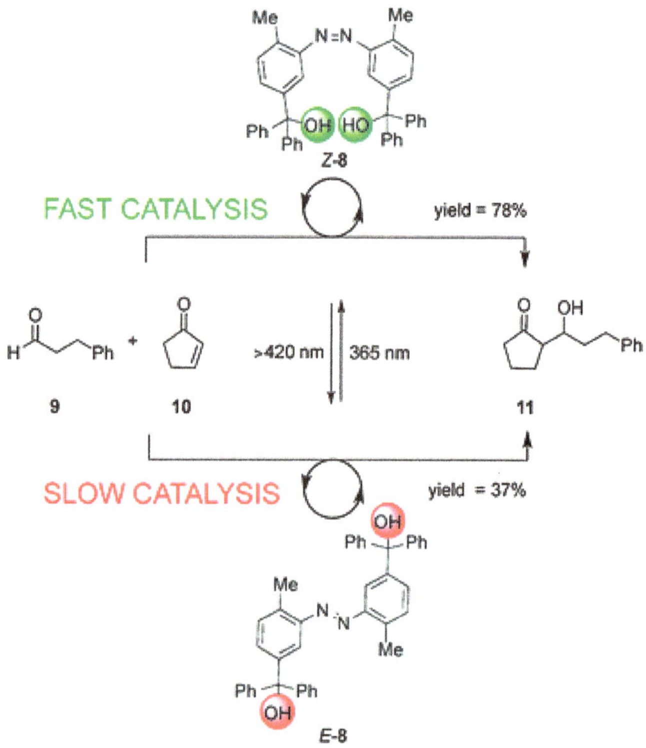

Scheme 3 Controlling the rate of the Morita–Baylis–Hillman reaction using photoswitchable catalyst *E*/*Z*-**8**.[10]

Recently, a 3,3′-dicarboxylic acid enzyme mimic (*E*/*Z*-**12**) that acts as switchable catalyst with glycosidase activity has been described (Scheme 4).[11] Isomers *Z*-**12** and *E*-**12** have significantly different pK_a values. In *E*-**12** both carboxylic acid groups ionize at roughly the same pH. In contrast, the deprotonation of *Z*-**12** occurs in a stepwise fashion, with the carboxylic acid and carboxylate forms coexisting at pH values between 4.7 and 6.5. *Z*-**12** showed a rate enhancement of six orders of magnitude for the hydrolysis of nitrophenyl glycoside **13** (giving **14**) compared to *E*-**12**. The catalytic activity of *E*/*Z*-**12** could be switched 'on' and 'off' multiple times over the course of the reaction by alternating exposure to UV or visible light.

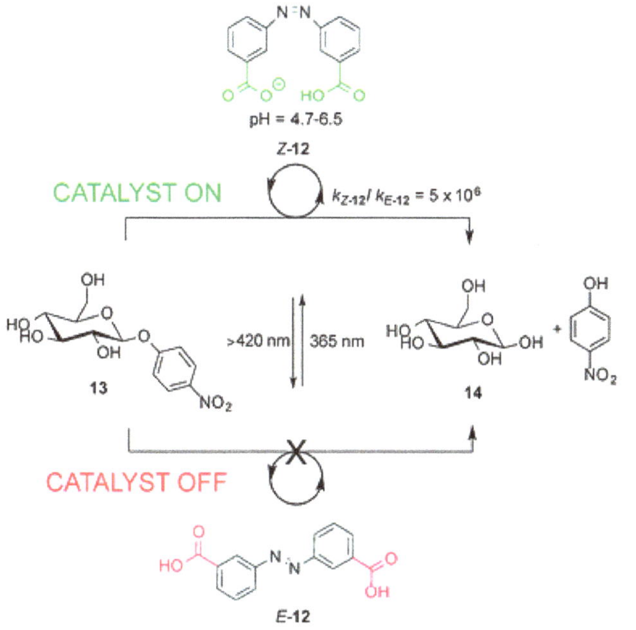

Scheme 4 Controlling the rate of hydrolysis of nitrophenyl β-D-glucose 13 using photoswitchable catalyst E/Z-12.[11]

Cooperative interactions can also be used to change the stereochemical outcome of switchable catalysed reactions. Branda and co-workers successfully controlled the stereochemical outcome of the cyclopropanation of styrene (**16**) using a photoswitchable catalyst (Scheme 5).[12] Ligand **15** consists of a dithienylethene-based chiral bis(oxazoline) which, in its open form (o-**15**), binds Cu(I) ions in a bidentate fashion. This results in a rigid chiral environment for the metal ion that subsequently catalyzes cyclopropanation of **16** with **17** to afford **18** as a *trans–cis* mixture with significant enantioselectivity (30–50% ee). In contrast, the closed form of the ligand (c-**15**; formed upon radiation at 313 nm), is only able to form monodentate complexes with Cu(I). The chiral environment is less well expressed in this complex and results in negligible enantioselectivity in the cyclopropanation reaction (5% ee). A limitation of this catalytic system is the low photocyclization efficiency in the presence of Cu(I), with only 23% of the closed form (c-**15**) present at the stationary state at 313 nm, meaning that switching the catalyst state only results in a small disruption in the stereoselectivity of the catalysed reaction.

The first example of a cooperative photoswitchable catalyst able to control both catalytic activity and the stereoselectivity of the reaction was reported by Wang and Feringa (Scheme 6).[13] A dimethylaminopyridine (DMAP) Brønsted base and a hydrogen bonding thiourea, which are known to cooperate in the catalysis of Michael additions, were attached to the rotor and stator of a light-driven unidirectional rotary motor (**19**). The resulting catalyst was able to alter both the rate and the stereochemical outcome of the Michael addition of aromatic thiol **21** to 2-cyclohexen-1-one (**20**). When the (*P*,*P*)-*E*-**19** isomer is used as catalyst, the reaction is very slow (7% yield) and afforded the racemic Michael adduct (*rac*-**22**) (bottom, Scheme 6).

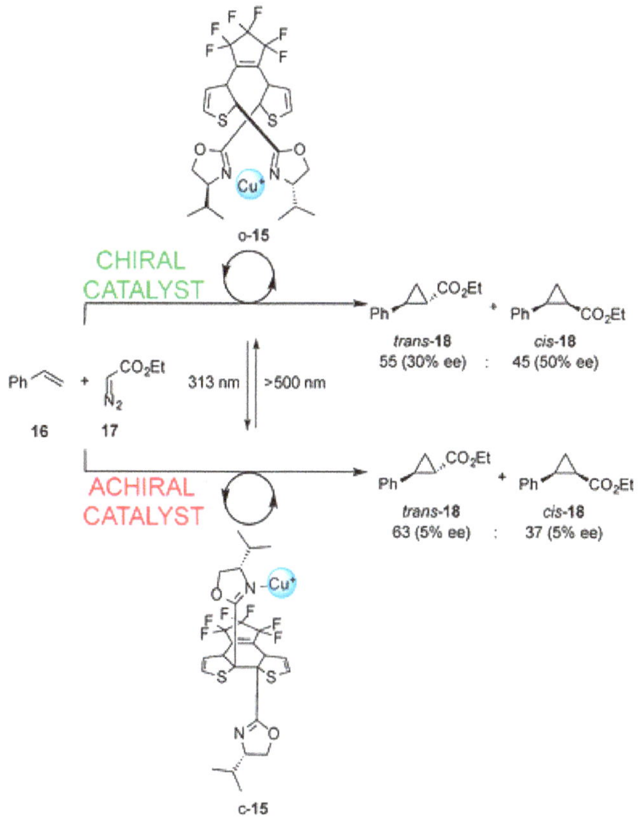

Scheme 5 Controlling the stereoselectivity of the cyclopropanation of styrene using photoswitchable catalyst o/c-**15**.[12]

Upon UV irradiation, *E–Z* isomerization takes place forming the (*M*,*M*)-*Z*-**19** isomer, which shows a higher catalytic activity and moderate enantioselectivity (50% yield, 50% ee) in favour of the (*S*)-**22** enantiomer (middle, Scheme 6). By thermally-activated helix inversion (*P*,*P*)-*Z*-**19** is formed, which is also able to catalyse the reaction giving a higher yield and a similar degree of stereoselectivity (83% yield, 54% ee) but in favour of the (*R*)-**22** enantiomer (top, Scheme 6). Subsequent photochemical and thermal isomerization can return the catalyst to its original state, (*P*,*P*)-*E*-**19**, *via* a fourth diastereoisomer (*M*,*M*)-*E*-**19**. This rotary motor organocatalyst allows for control over the rate and the selectivity in a reversible and sequential manner using light and temperature. The Feringa group later reported on a second generation molecular rotor where the phenyl spacers between the motor core and the catalytic units are absent.[14] The resulting switchable catalyst is able to switch from a *E*-state with low catalytic activity and no stereocontrol in the Henry reaction of nitromethane with α,α,α-trifluoroketones, to two *Z*-states, in which the catalytic units are in close proximity, affording the corresponding products with opposite stereoselectivity in excellent yields and good enantioselectivity.[14]

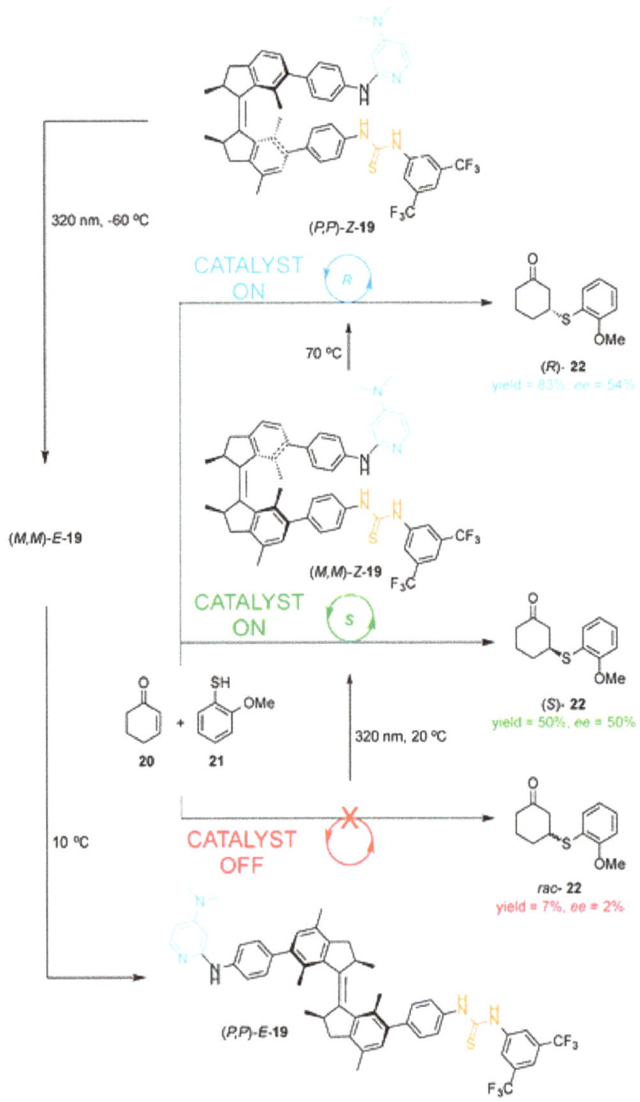

Scheme 6 Controlling the rate and the stereoselectivity of a Michael addition using photoswitchable catalyst **19**.[13]

The photoswitchable unit used in Feringa's rotary catalysts has also been exploited by Craig and colleagues in a switchable catalyst able to modulate the stereochemical outcome of asymmetric Heck arylations and Trost allylic alkylations.[15] This catalyst consists of a macrocycle containing a chiral bis(phosphine) ligand linked to the photoswitchable unit. Modulation of the ligand geometry upon exposure to UV-light afforded isomers of the catalyst with different enantioselectivities for the Heck arylation and Trost allylic alkylation reactions.[15] The Feringa group have also recently demonstrated dynamic control of chirality using phosphine ligands attached to their rotary molecular motor system.[16]

Steric Effects

These catalyst systems feature a group that blocks access of the substrate to the active site in one state of the catalyst, normally through a large geometrical change during the isomerization process. Different approaches have been used to achieve this, such as host–guest binding, hydrogen bonding or simple steric bulk to shield the catalytic site. An early example using host–guest interactions to modulate catalytic activity was described by Ueno and co-workers (top, Scheme 7).[17] Complexation of β-cyclodextrin (**23**) with 4-carboxyazobenzene (**24**) alters the rate of the hydrolysis of nitrophenol ester **25** in basic media. β-Cyclodextrin **23** acts as a host for the hydrophobic aryl substituent of **25**, facilitating attack of one of the peripheral hydroxyl groups on the ester moiety. However, *E*-**24** is a competitive inhibitor of the binding of **25** within **23**, decreasing the rate of catalysis, whereas the *Z* isomer does not fit inside the cyclodextrin cavity. The system does not turnover and the catalyst and the azobenzene must exceed the substrate concentration by one and two orders of magnitude, respectively to compensate for unfavourable product inhibition.

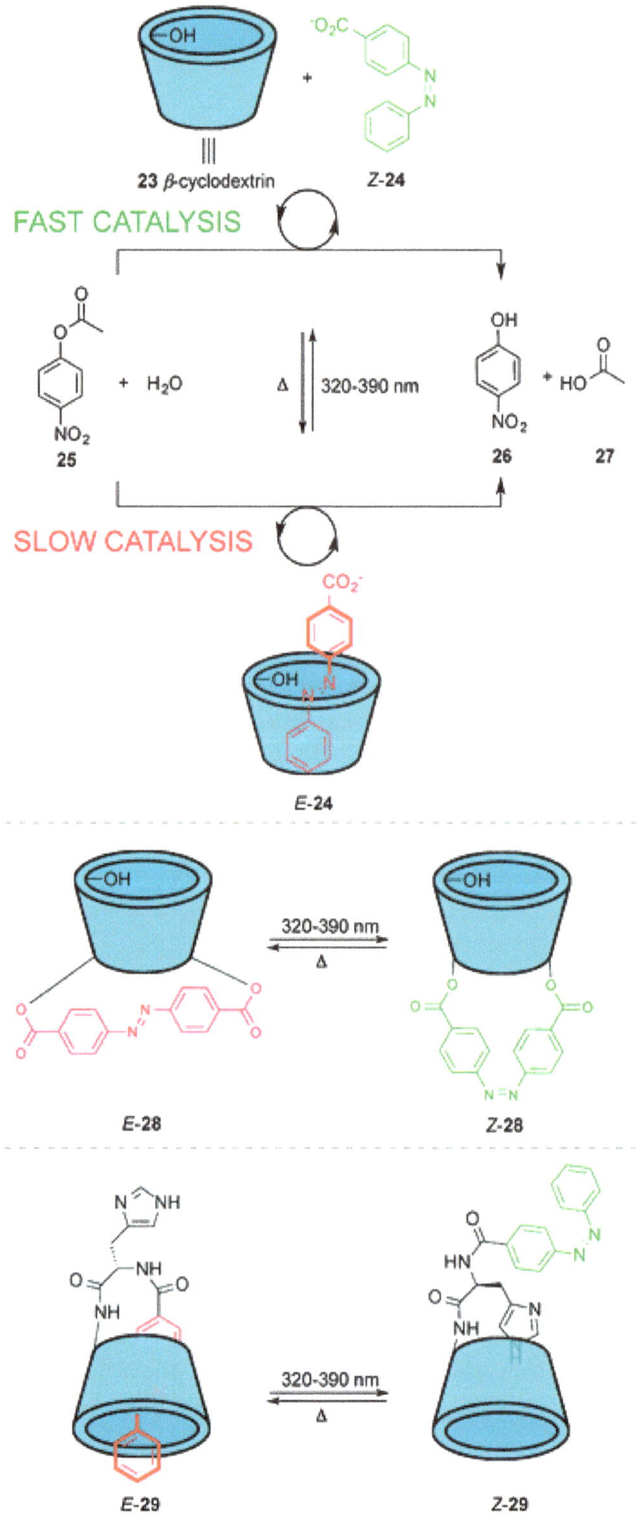

Scheme 7 Controlling the rate of ester hydrolysis using photoswitchable β-cyclodextrin catalysts.[17,18]

The same group developed photoswitchable catalysts based on β-cyclodextrins by covalently linking azobenzene moieties to the lower rim of β-cyclodextrin (middle and bottom, Scheme 7). The extended linker in *E*-**28** makes the internal cavity shallow, decreasing the binding of the ester substrate in the hydrophobic cavity (middle, Scheme 7).[17] Isomerization to *Z*-**28** creates a deeper cavity that offers stronger binding to the substrate resulting in a five-fold increase in the rate of catalysed ester hydrolysis. The group also investigated a switchable catalyst with an azobenzene unit attached to the β-cyclodextrin at a single point *via* a histidine residue (bottom, Scheme 7).[18] In the 'off' state, *E*-**29**, the azobenzene unit occupies the cavity and the system displayed no catalytic activity. In *Z*-**29** the azobenzene does not fit within the cavity and, therefore, the substrate can bind to the cyclodextrin turning 'on' the catalytic activity for the hydrolysis of various esters, including nitrophenyl acetate, Boc-D-alanine-*p*-nitrophenyl ester and Boc-L-alanine-*p*-nitrophenyl ester.

The azobenzene units were later tethered to Au nanoparticles (NPs) and used in conjunction with a Zn(II) coordinated β-cyclodextrin dimer as a heterogeneous photoswitchable catalyst to modulate the catalytic activity in ester hydrolysis.[19]

Following these strategies, in 2009 Harada and co-workers reported control over the polymerization of δ-valerolactone by photoisomerization of a *E*-cinnamoyl-α-cyclodextrin to *Z*-cinnamoyl-α-cyclodextrin, which turned 'off' the catalytic activity through the inhibition of the binding and inclusion of the δ-valerolactone monomer into the cyclodextrin cavity.[20]

Rebek *et al.* have described a light-responsive cavitand-piperidinium complex that acts a switchable catalyst for the Knoevenagel condensation of aromatic aldehydes with malonitrile (**33**) (Scheme 8).[21] Cavitand **30** is armed with an isopropyl-substituted azobenzene unit that modulates the rate of the reaction. In the *E* state the cavitand binds to the piperidinium cation (**31**) and accelerates the rate of the reaction, producing up to 3.5-fold increase in rate compared to the reaction carried out with piperidinium acetate alone. Photoisomerization of the catalyst generates *Z*-**30** for which the azobenzene unit can bind within the cavity, slightly inhibiting catalysis (*E*-**30** accelerates the reaction ~1.5× faster than *Z*-**30**). NOESY analysis indicates that the ammonium group of the piperidinium cation in the complex with *E*-**30** protrudes from the open end of the cavitand. As the substrates are not guests in the complex, a wide range of aldehydes is tolerated by the mechanism of catalysis.

Inoue and colleagues have exploited the reversible photoisomerization of stilbene derivatives in a switchable catalyst for the chemical fixation of carbon dioxide (Scheme 9).[22] Control over catalysis of the condensation of carbon dioxide with 1,2-epoxypropane (**37**) was achieved by reversible coordination of 3,5-di-*t*-butyl-2-stilbazole (**36**) to aluminium porphyrin **35**.

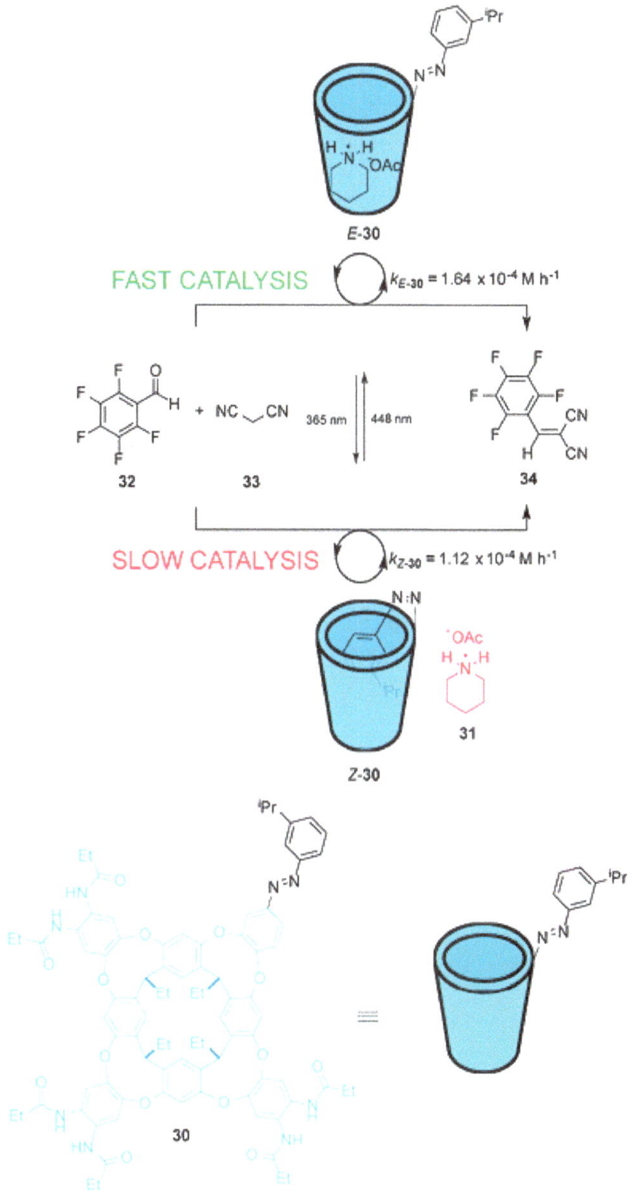

Scheme 8 Controlling the rate of a Knoevenagel condensation using photoswitchable catalyst *E*/*Z*-**30**.[21]

Due to unfavourable steric interactions between the porphyrin ring system and the bulky *tert*-butyl groups in the *E*-2-stilbazole (*E*-**36**), the pyridine group is unable to effectively coordinate to the Al(III) centre in **35**, resulting in a low yield of **38** (2% conversion over 18 h). Irradiation with UV light afforded the *Z*-2-stilbazole (*Z*-**36**), which bound to the aluminium porphyrin **35**enhancing the rate of the reaction (23% conversion over 18 h). Switching between UV and visible light multiple times allowed continuous modulation of the rate of the catalysed condensation reaction.

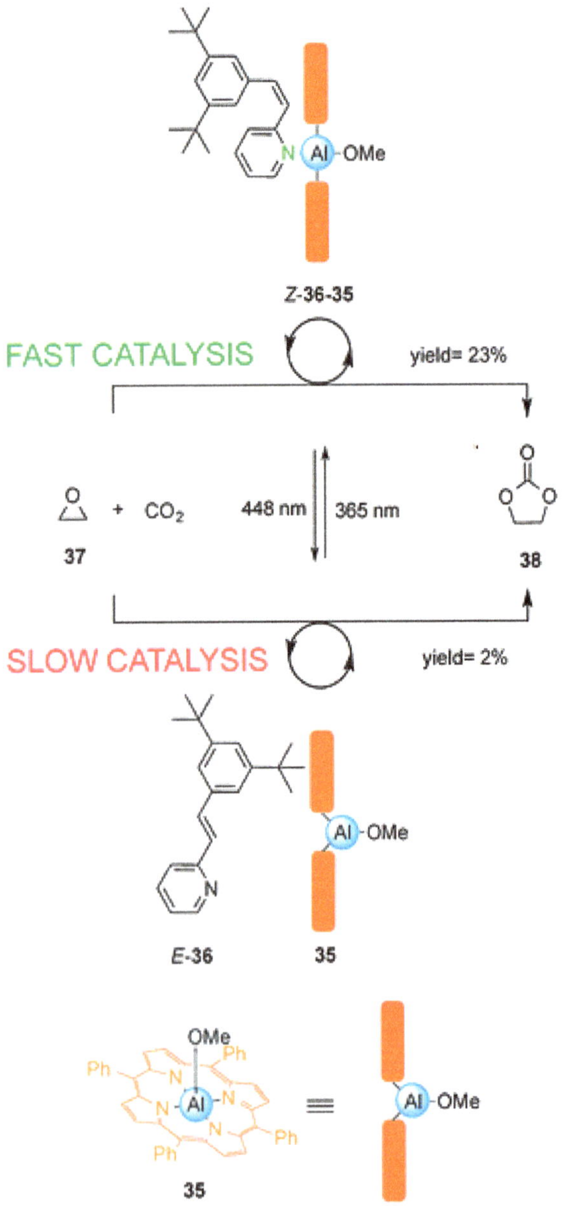

Scheme 9. Controlling the rate of chemical fixation of carbon dioxide using reversible coordination of stilbene derivative **36** to aluminum porphyrin **35**.[22]

Hecht and co-workers described a series of photoswitchable base catalysts (**39a–c**) able to control the rate of aza-Henry reactions (Scheme 10).[23] Catalysts **39a–c** comprise conformationally restricted tertiary piperidine bases as the catalytic unit and 3,5-disubstituted azobenzenes as a switchable steric shield. The *E*-**39a–c** catalysts have an inaccessible basic site and exhibit low basicity in acid–base titration experiments. UV irradiation affords the *Z*-**39a–c** isomers, in which the basic piperidine nitrogen atom is exposed, enhancing the basicity by almost one order of magnitude. Switchable catalyst **39c** was able to effectively control the rate of the aza-Henry reaction of nitroethane (**40**) to *p*-nitroanisaldehyde (**41**) ($k_{Z\text{-}39c}/k_{E\text{-}39c}$ = 35.5). Immobilization of the catalyst on silica gel particles enabled the system to be used as a heterogeneous switchable catalyst.[24]

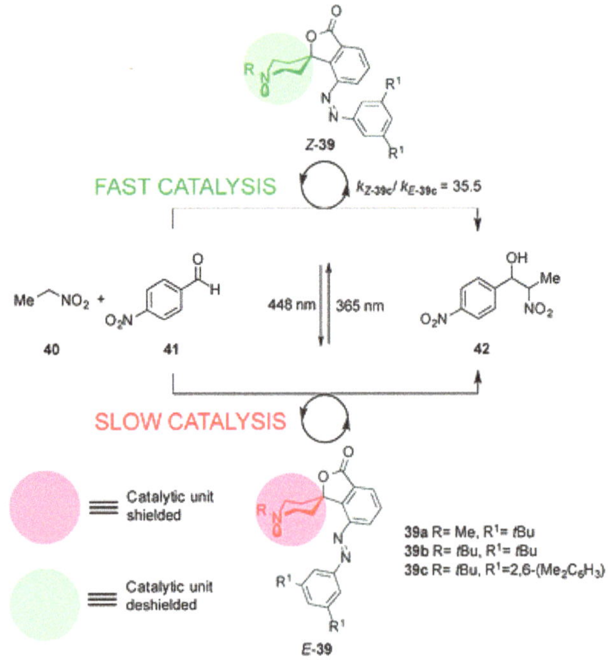

Scheme 10 Controlling the rate of the aza-Henry reaction using switchable catalyst *E*/*Z*-**39**.[23]

Pericás and co-workers developed a switchable thiourea organocatalyst to control Michael addition reactions through hydrogen-bond shielding of the catalytic unit (Scheme 11).[25]Catalyst **43** consists of a hydrogen bond donor (the thiourea unit) and a blocking moiety (a nitro group) located either side of an azobenzene unit. The thermodynamically stable *E*-**43**diastereomer effectively catalyses the Michael addition of acetylacetone (**44**) to *m*-bromonitrostyrene (**45**) (full conversion after 19 h). UV irradiation forms *Z*-**43**, in which the nitro group binds to the thiourea, leading to significantly slower catalysis (23% conversion after 20 h).

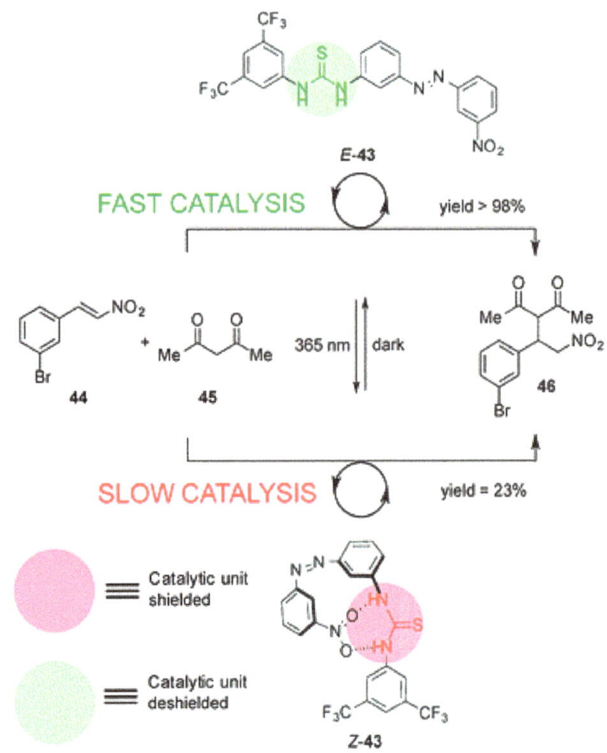

Scheme 11 Controlling the rate of Michael addition using switchable catalyst *E*/*Z*-**43**.[25]

Electronic Effects

Using photoinduced changes in electronic properties to alter the rate, chemo- and/or regioselectivity of catalysts has been less explored compared to steric and cooperative effects. The most popular concept is to use the breaking or forming of conjugation between the active site and an electronically activating group. The first example of photo-modulating catalytic activity *via*electronic changes was reported by Branda, using a photoactive diarylethene unit that undergoes ring-closing and ring opening reactions when exposed to UV and visible light, respectively (Scheme 12).[26] The researchers developed a mimic of the bioactive form of vitamin B6, pyridoxal 5′-phosphate (PLP), in which the catalytic activity is dependent upon electronic connectivity between the functional groups. In the ring-open form of the catalyst (o-**47**), the pyridinium and aldehyde units are electronically isolated from each other, switching 'off' the catalytic activity for the racemization of L-alanine (less than 3% over 140 hours). However, UV irradiation to form the ring-closed isomer (c-**47**) results in a fully conjugated structure, turning 'on' catalysis and resulting in 30% deuterium exchange after 140 h and near-complete exchange (95%) after an additional 90 h. The *in situ* switching of the catalyst between its inactive and active forms was successfully demonstrated through alternating exposure to UV and visible light.

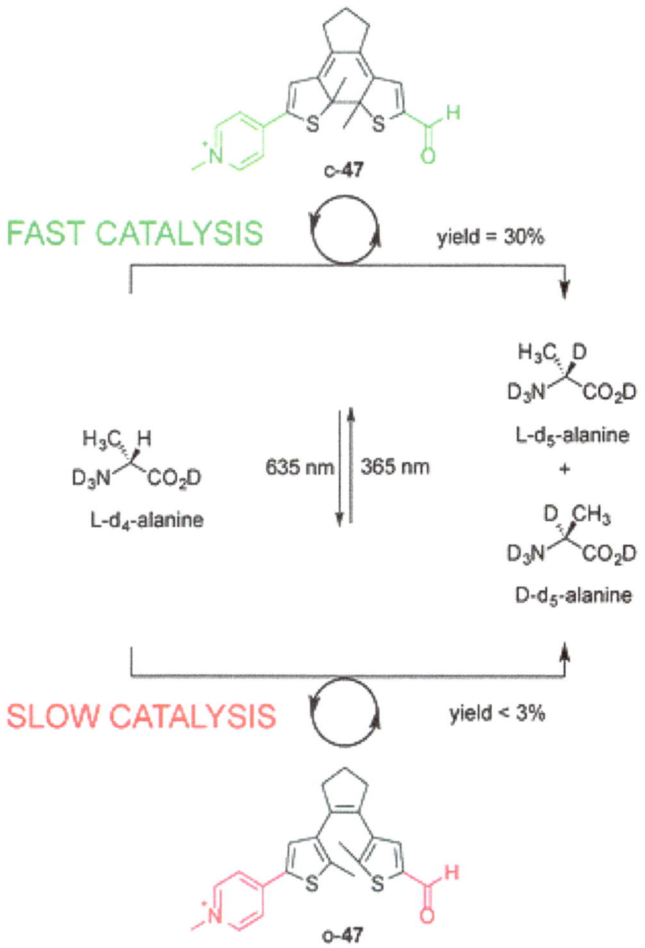

Scheme 12 Controlling the rate of the racemization of L-alanine using switchable catalyst o/c-**47**.[26]

Bielawski and Neilson described control of the rate of transesterification and amidation reactions promoted by an electronically-modulated N-heterocyclic carbene (NHC) organocatalyst (Scheme 13).[27] Switchable catalyst **48** is based on a NHC unit incorporated into a dithienylethene scaffold, where the length of the conjugated π-system modulates the electronic density at the carbenoid center of the imidazolium salt. In the presence of visible light and base the open form of the NHC catalyst, o-**48**, catalyzes transesterification and amidation reactions. Upon UV irradiation to the closed form, c-**48**, the rate of both transesterification and amidation reactions is significantly decreased ($k_{o\text{-}48}/k_{c\text{-}48}$ = 12.5 and 100, respectively). The difference in catalytic activity was rationalized on the basis of NMR experiments using an isotopic label at the C2 'carbene' carbon that showed that the ring-open form (o-**48**) exists as an imidazolium species while the ring-closed species c-**48** forms a less active alcohol adduct. The rate of the amidation reaction was successfully switched several times between the fast and slow states by alternating exposure to UV and visible light over the course of the reaction.

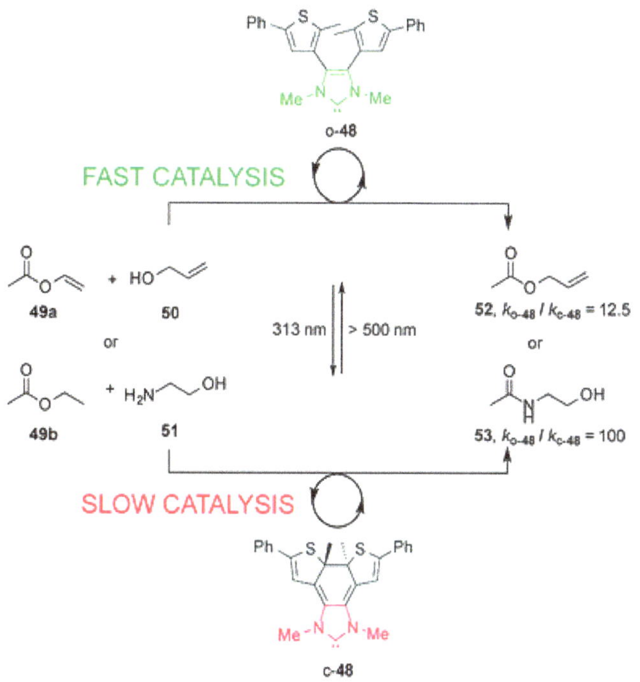

Scheme 13 Controlling the rate of transesterification and amidation reactions using switchable catalyst o/c-**48**.[27]

The same group expanded the scope of the NHC switchable organocatalyst **48** to include the ring-opening polymerization of cyclic esters, such as δ-valerolactone and ε-caprolactone, achieving good control over the catalytic activity.[28]

The ability to photomodulate the electronic density of the NHC was further extended to control the rate of the rhodium(I)-catalyzed hydroboration of styrene (Scheme 14).[29] Under visible light, open Rh(I) complex o-**54** efficiently promoted the reaction. Exposure to UV light to form the close form, c-**54**, attenuated the rate of catalysis by almost one order of magnitude, an effect attributed to the decrease in donor strength of the NHC ligand upon photocyclization. The rate of the hydroboration of styrene with pinacolborane was switched several times over the course of a single reaction *via* alternating UV and visible irradiation.

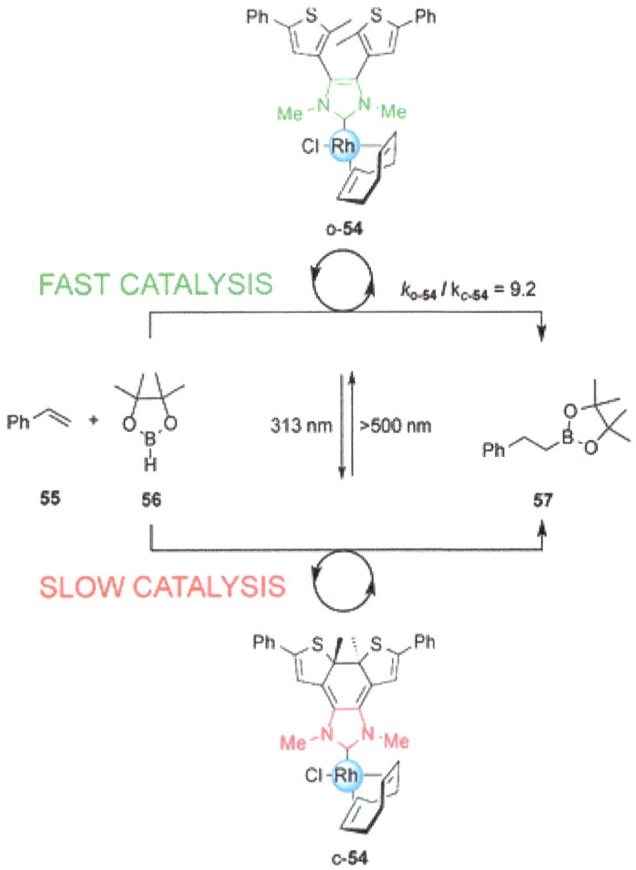

Scheme 14 Controlling the rate of a hydroboration reaction using switchable rhodium catalyst o/c-**54**.[29]

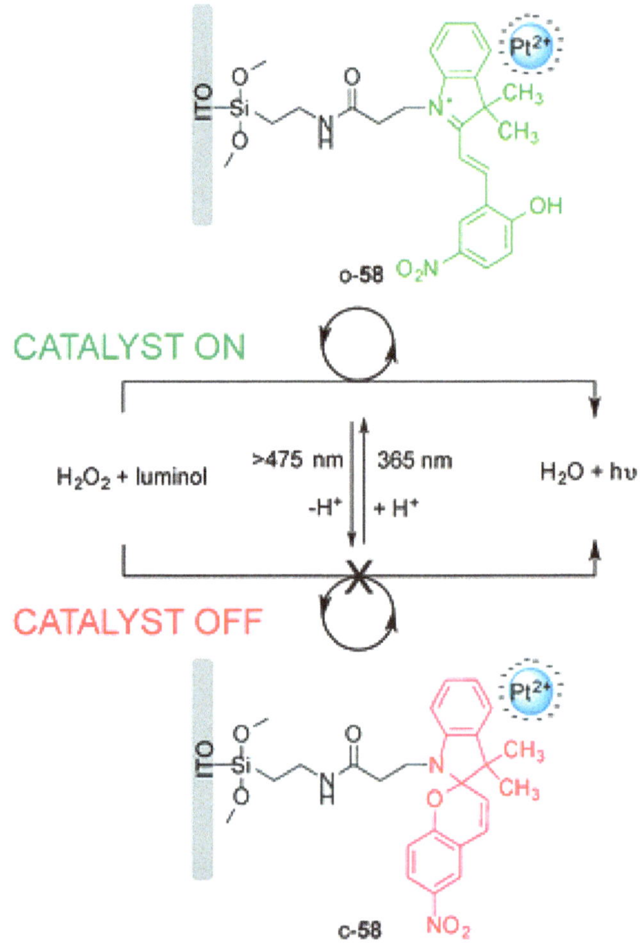

Scheme 15 Heterogeneous photoswitchable catalyst o/c-**58** for electrocatalysis and chemiluminescence.[30]

Willner and co-workers reported on a surface-confined photoisomerizable monolayer coupled to negatively charged Pt nanoparticles for photoswitchable electrocatalysis and chemiluminescence (Scheme 15).[30] The catalyst (**58**) is based on a photoswitchable nitrospiropyran linked to an indium tin oxide (ITO) electrode to form a monolayer that in closed form (c-**58**) lacks affinity for the negatively charged Pt nanoparticles and shows no catalytic activity towards H_2O_2 reduction. Upon UV irradiation the nitrospiropyran ring opens and picks up a proton to form a positively charged species (o-**58**), which attracts the Pt nanoparticles to the ITO electrode facilitating electrocatalytic H_2O_2 reduction. The reaction was monitored both electrochemically and *via* luminol addition which undergoes chemoluminescence upon UV irradiation in the presence of H_2O_2. The catalytic activity was successively switched 'on' and 'off' through alternating exposure to different wavelengths of light for up to ten cycles without detectable degradation of the heterogeneous catalyst.

Aggregation/Dissociation

The dynamic self-assembly of molecules into supramolecular ensembles is the basis of many functional systems in biology and has inspired the construction of synthetic systems with dynamic and responsive properties. Grzybowski's group reported the photoswitchable catalyzed hydrosilylation of *p*-anisaldehyde (**60**) and diphenylsilane (**61**), controlled by reversible aggregation/dispersion of catalytic Au nanoparticles (**59**) (Scheme 16).[31] Heterogeneous catalyst **59** consists of Au nanoparticles decorated with a mixture of alkyl amines and azobenzene-terminated alkyl thiol moieties. Under visible light, the Au nanoparticles (*E*-**59**) remain dispersed and exposed to a large solvent surface area, efficiently promoting the hydrosilylation reaction ($k_{E\text{-}59} = 9 \times 10^{-4}$ mM^{-1} min^{-1}). Irradiation with UV light isomerizes the *E*-azobenzene units of the thiol chains to the *Z* isomer, developing electric dipoles that cause aggregation of the Au nanoparticles (*Z*-**59**), decreasing their catalytic activity ($k_{Z\text{-}59} = 1 \times 10^{-5}$ mM^{-1} min^{-1}; $k_{E\text{-}59}/k_{Z\text{-}59} = 90$).

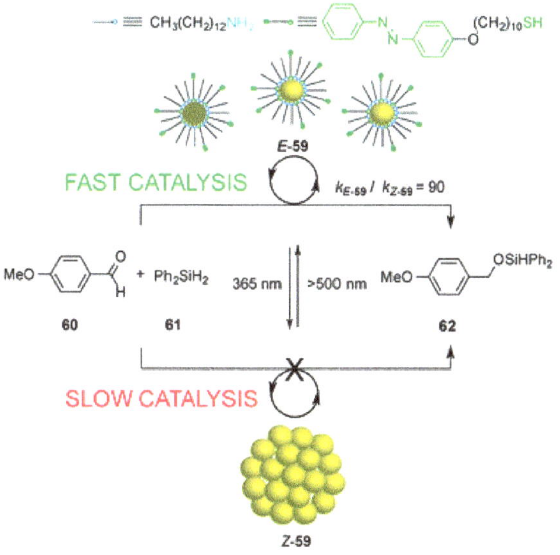

Scheme 16 Heterogeneous photoswitchable catalyst *E*/*Z*-**59** for the hydrosilylation of *p*-anisaldehyde (**60**).[31]

Shibasaki's group have developed a photoswitchable nucleophilic catalyst **63**, that controls catalytic activity through switching between aggregated/dissociated states of a dipeptide derivative (Scheme 17).[32] The switchable catalyst has an azobenzene unit linked through the peptide to a nucleophilic 4-aminopyridyl group. The aggregated/dissociated state varies according to the *E*/*Z* geometry of the azobenzene. The *E*-**63** isomer has limited solubility due to extensive aggregation, almost switching 'off' the catalysed-rearrangement of 2-acyloxybenzofuran (**64**) (14% conversion after 100 min). In contrast *Z*-**63** catalyzes the reaction effectively (80% conversion after 100 min). The switchable catalyst could also be used to promote the *O*-*tert*-butoxycarbonylation of 1-naphthol.

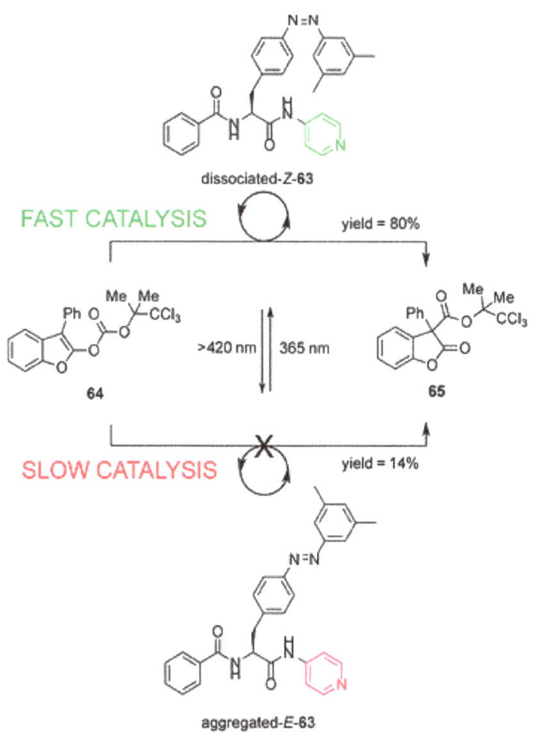

Scheme 17 Controlling the rate of a rearrangement using photo-switchable catalyst *E*/*Z*-**63**.[32]

PH-DRIVEN SWITCHING

Chemically-driven processes driven by changes in pH are common in both nature[2] and artificial molecular machines. However, in the context of controlling catalysis few examples have been reported thus far. An early use of a change in pH to control catalytic activity was reported by Schanz,[33] who exploited the change in the electron-donating ability of N-heterocyclic carbene (NHC) ligands upon protonation to moderate the activity of Ru(II)-based olefin metathesis catalysts (Scheme 18).

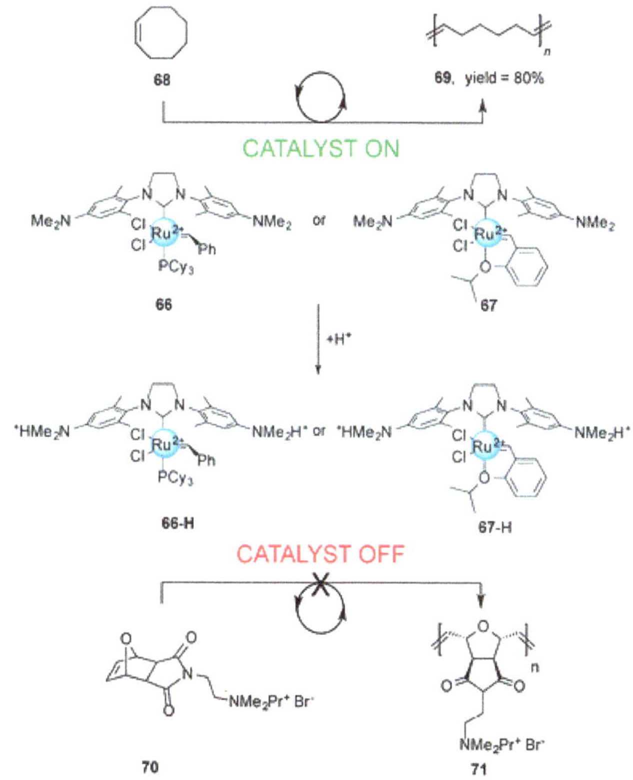

Scheme 18 Controlling the rate of ROMP using pH-switchable catalysts **66** and **67**.[33]

The first generation Grubbs-type **66** and Hoveyda–Grubbs-type **67** catalysts bear a pH-responsive H_2ITap ligand (H_2Tap = 1,3-bis(2′,6′-dimethylaminophenyl)-4,5-dihydroimidazol-2-ylidene). Control over the catalytic activity of these two pH-switchable ligands was investigated in ring-opening metathesis polymerization (ROMP) reactions. Under neutral conditions the NMe_2 groups in the catalyst backbones are uncharged and the system showed excellent catalytic activity in the ROMP of cyclooctene (>80% conversion in 7 min with **66** and in 28 min with **67**). Upon protonation with HCl, both catalysts become dicationic and the ROMP of cationic *exo*-7-oxanorbornene becomes negligible. The researchers also investigated the performance of these pH-responsive catalysts in the ring closing metathesis (RCM) of diallyl malonic acid. The non-protonated forms of **66** and **67** showed good conversion (45% and 59% after 60 min, respectively). In acid media, the RCM proceeded at slower rates and did not afford complete conversion.

In an extension of this work,[34] the group demonstrated that gradual addition of *p*-toluenesulfonic acid (TsOH) reduced the rate of the ROMP reaction with the *exo*-7-norbornene derivative. DFT calculations of the atomic charge at the metal center of catalyst **66** and its mono- and dicationic forms confirmed that the propagation rates should be significantly affected upon protonation.

Plenio and co-workers subsequently described related Grubbs–Hoveyda complexes related to **66** and **67** that showed significant changes in the double-bond geometry (*E*/*Z* ratio = 0.78 to 1.04) and microstructure of the product in the ROMP of norbornene upon protonation of the catalyst.[35]

Recently, synthetic rotaxane catalysts[36] in which the catalytic activity or the outcome of the reaction can be controlled by changes in pH have been reported.[37] Rotaxane **72** consists of a dibenzo-24-crown-8 macrocycle and a thread bearing a dibenzylamine/ammonium unit, that acts as catalyst, and triazolium rings as alternative binding sites for the macrocycle (Scheme 19).[38] The organocatalytic group can be concealed or revealed, and its activity turned 'on' and 'off', by the well-defined acid/base-promoted translocation of the macrocycle between the binding sites on the thread. In neutral conditions, the triazolium ring is the preferred binding site for the macrocycle in **72** and so the dibenzylamine residue is available for catalysis (Scheme 19, top). Upon protonation (**72**-H), the macrocycle preferentially binds the ammonium unit, switching 'off' the catalytic activity (Scheme 19, bottom). Switchable rotaxane catalyst **72**proved to very effectively control the rate of Michael addition of an aliphatic thiol (**74**) to *trans*-cinnamaldehyde (**73**), affording product **75** only when the reaction was catalyzed with the non-protonated rotaxane (**72**). The catalysis could be turned 'on' *in situ* through deprotonation of **72**-H with similar effectiveness (Scheme 19).

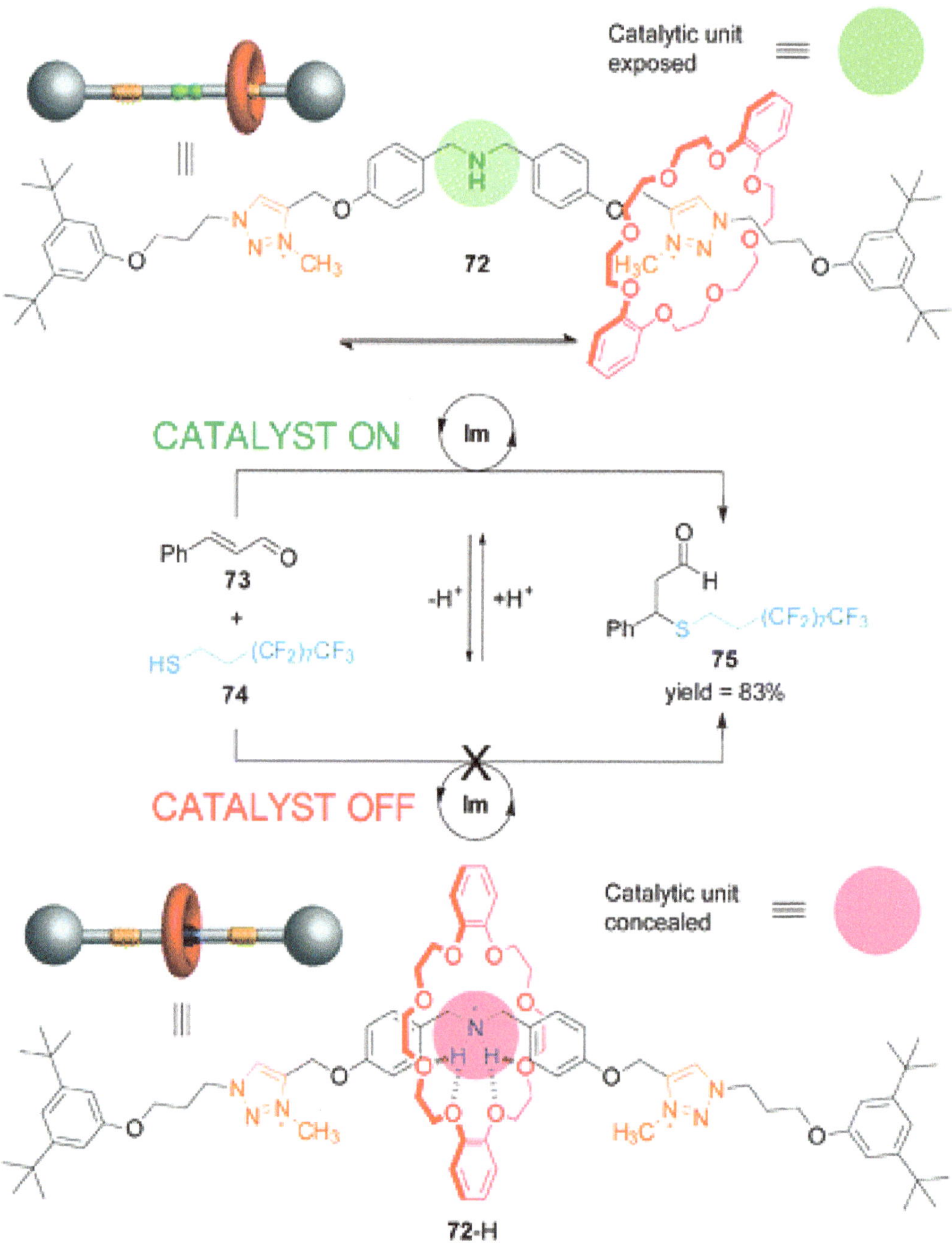

Scheme 19 Controlling the rate of Michael addition using a pH-switchable rotaxane catalyst (**72**).[38]

Rotaxane **72** catalyzes the Michael addition reaction shown in Scheme 19 through iminium ion activation. However, secondary amines can also catalyse reactions through other activation pathways. One goal of switchable catalysts is to be able to use them to react selectively with particular building blocks within a pool of potential reactants. By switching 'on' and 'off' various types of catalyst in different orders by applying different stimulus it may then prove possible to carry out different sequences of reactions, generating different products from a common pool of building blocks. To accomplish this it is necessary to establish the scope and limitations of the switchable catalysts involved in the processes. Accordingly rotaxane **72** was investigated for its ability to catalyze a variety of reaction types through different activation modes (Fig. 1).[39]

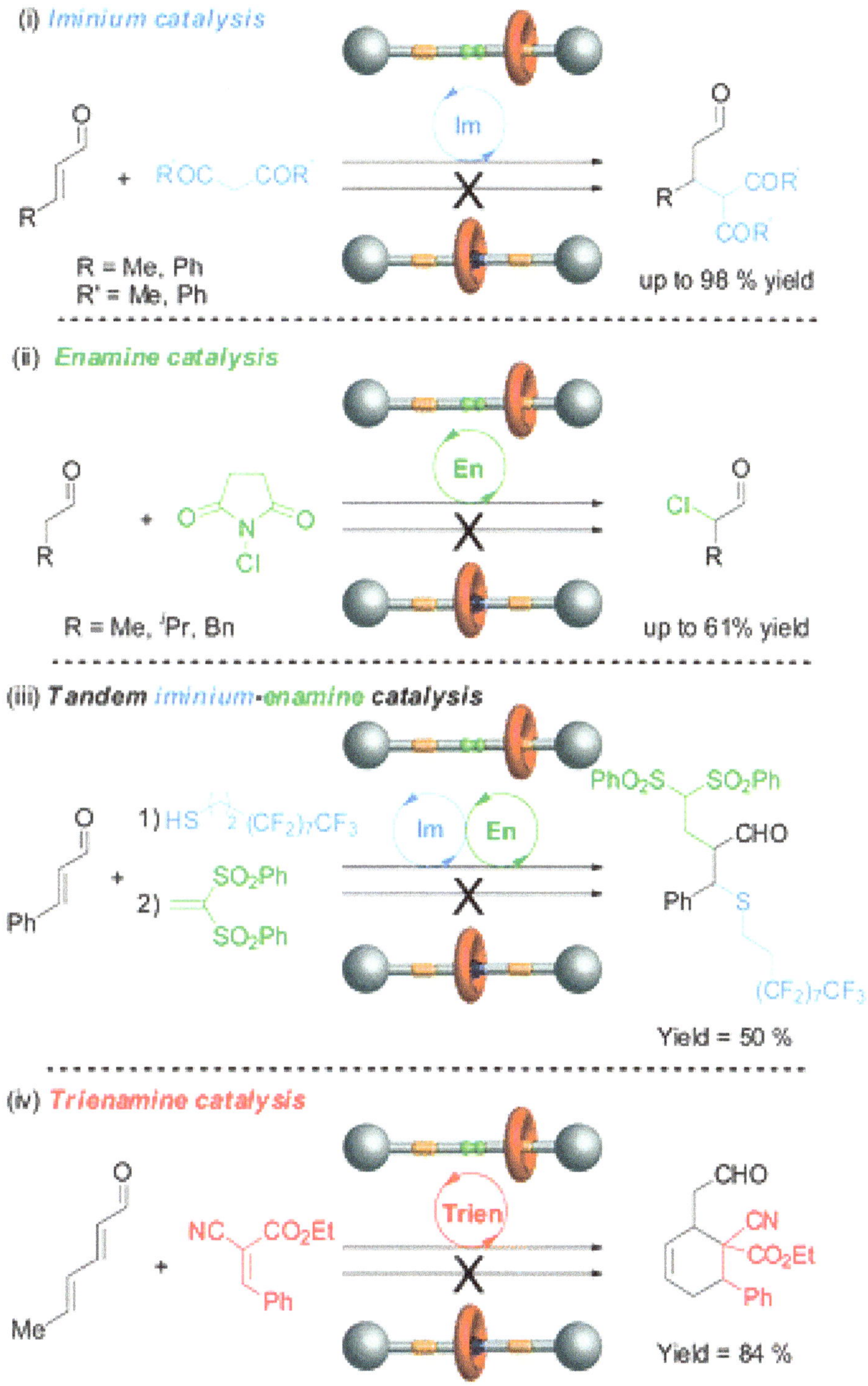

Figure 1 The scope of switchable rotaxane organocatalyst 72 through different activation mechanisms.39

The non-protonated rotaxane **72** ('on-state') very effectively catalyzed the β-functionalization of carbonyl compounds with C or S-nucleophiles through iminium activation (often 95–98% conversion), although the protonated form **72**-H ('off-state') also promotes the most facile of these transformations to a lesser degree (up to 14% conversion) (Fig. 1(i)). Rotaxane **72**also promoted nucleophilic addition or substitution reactions *via* enamine catalysis, although with less effectiveness (40–61% conversions). However, switching is more effective with enamine catalysis, with the 'off-state' of the catalyst showing no detectable catalytic activity in these reactions (Fig. 1(ii)). The rotaxane catalyst is even able to promote tandem iminium-enamine reaction sequences with high efficiency (Fig. 1(iii)) and the Diels–Alder reaction of a dienal through a trienamine activation pathway (Fig. 1(iv)).

This type of switchable catalysis was extended to an asymmetric organocatalytic rotaxane (**76**) that features a simple acyclic chiral secondary amine housed within a rotaxane framework (Scheme 20).[40] Switchable chiral rotaxane **76** was able to control the catalyzed Michael addition of 1,3-diphenylpropan-1,3-dione (**78**) to aliphatic α,β-unsaturated aldehydes (**77a–c**). The acyclic chiral secondary amine promotes the reaction with stereochemical control comparable to—or better than—commercial cyclic amine organocatalysts, albeit at the expense of slower rates of conversion. *In situ* switching between the catalyst states by alternating addition of acid and base afforded excellent control over both the rate and stereoselectivity of the reactions.

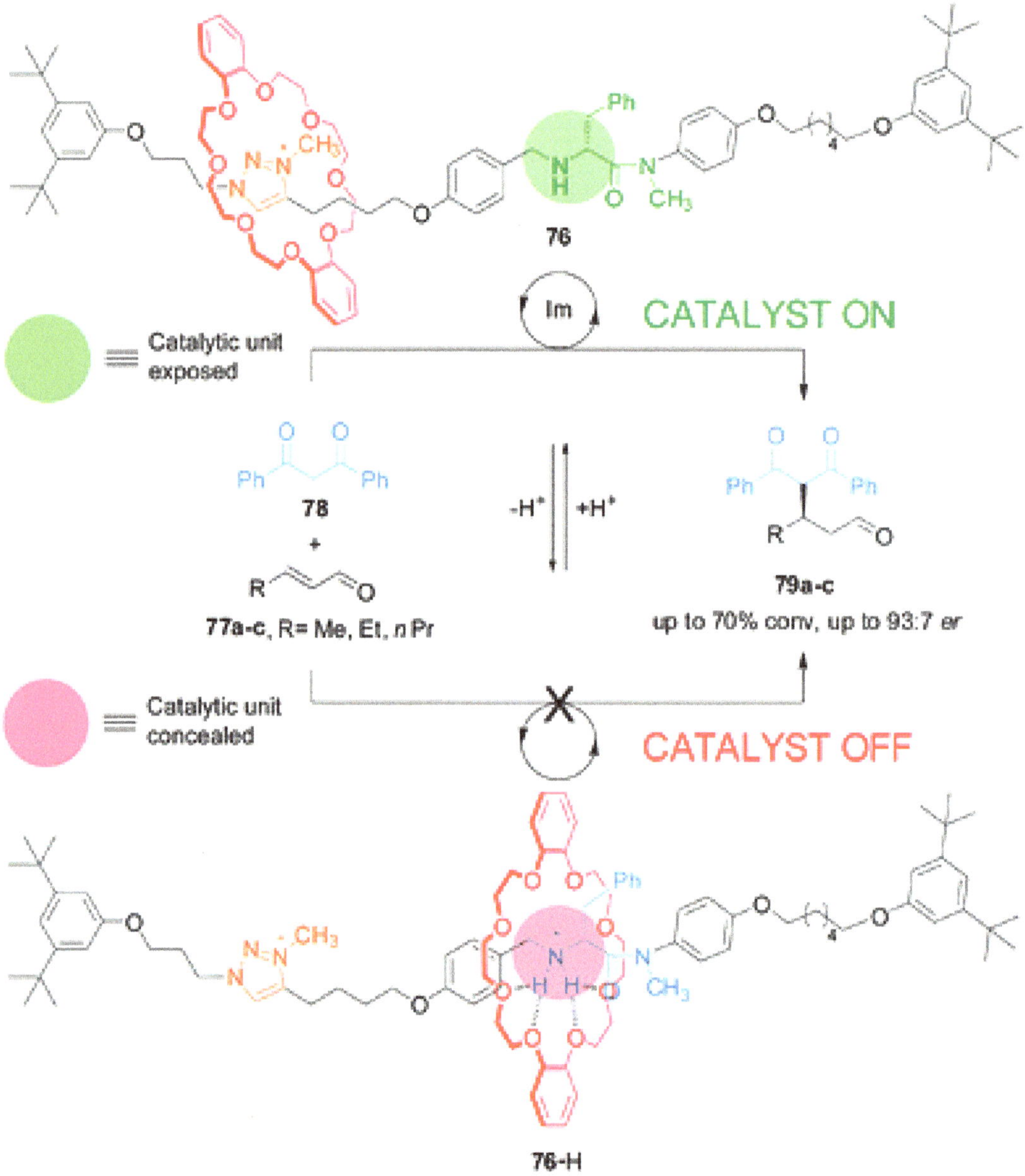

Scheme 20 Controlling the rate and stereoselectivity of Michael addition using a switchable rotaxane asymmetric organocatalyst **76**.[40]

Recently, a switchable rotaxane system was developed, **80**, featuring two different organocatalytic sites: a squaramide moiety and a dibenzylamine group (Scheme 21).[41] When the rotaxane is protonated (**80**-H), the macrocycle preferentially interacts with the ammonium unit revealing the squaramide unit, which can promote the Michael addition of 1,3-diphenylpropan-1,3-dione (**78**) to *trans*-β-nitrostyrene (**81**) through hydrogen bond catalysis (75% conversion after 18 h). In basic media the macrocycle preferentially resides over the squaramide, revealing the secondary amine which promotes the Michael addition of 1,3-diphenylpropan-1,3-dione (**78**) to crotonaldehyde (**77a**) *via* iminium ion catalysis (40% after 40 h). In this way the catalyst state controls which building blocks react together—and what product is formed—from a mixture of reactants.

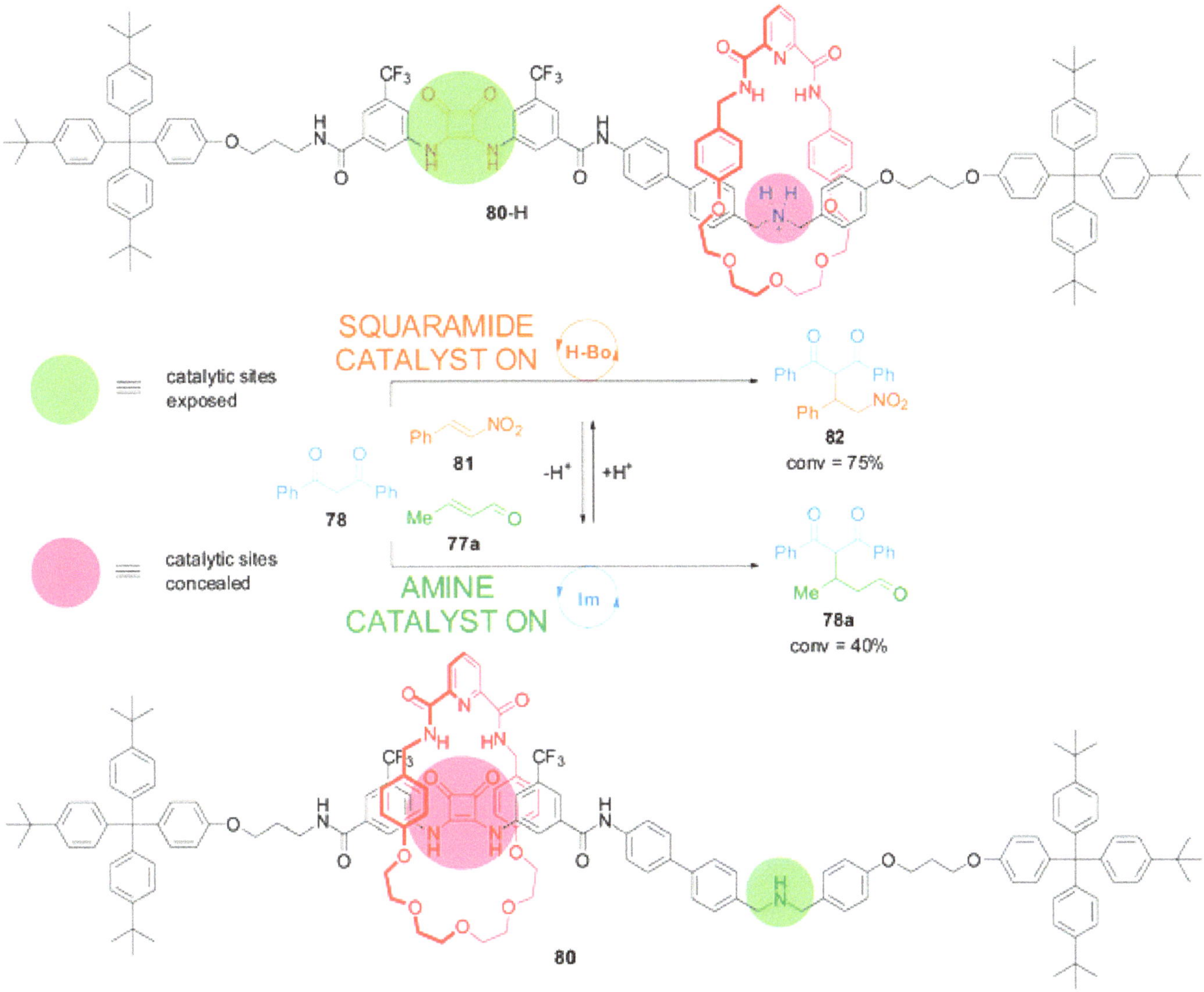

Scheme 21 Controlling the reaction outcome from a mixture of building blocks using a switchable rotaxane featuring two different organocatalytic sites, **80**.[41]

COORDINATION-DRIVEN SWITCHING

One of the first uses of metal-ion-coordination to enhance reactivity was reported by Shinkai and co-workers in 1983 (Scheme 22). They described a flavin derivatised with a crown ether (**83**) that reacted stoichiometrically with an NADH model (**84**) at an enhanced rate in the presence of alkali metal ions (up to 1.8-fold rate enhancement) or ammonium groups (2.4-fold acceleration).[42]

Many examples followed in which the activity of a catalytic centre could be influenced by tuning coordination binding events.[43] Two main approaches have been developed for this purpose, the addition/removal of external ligands that can complex to a metal ion present in the catalyst (Section 4.1), and the addition/removal or metal ions that can coordinate to binding motifs in the structure of the catalytic system (Section 4.2).

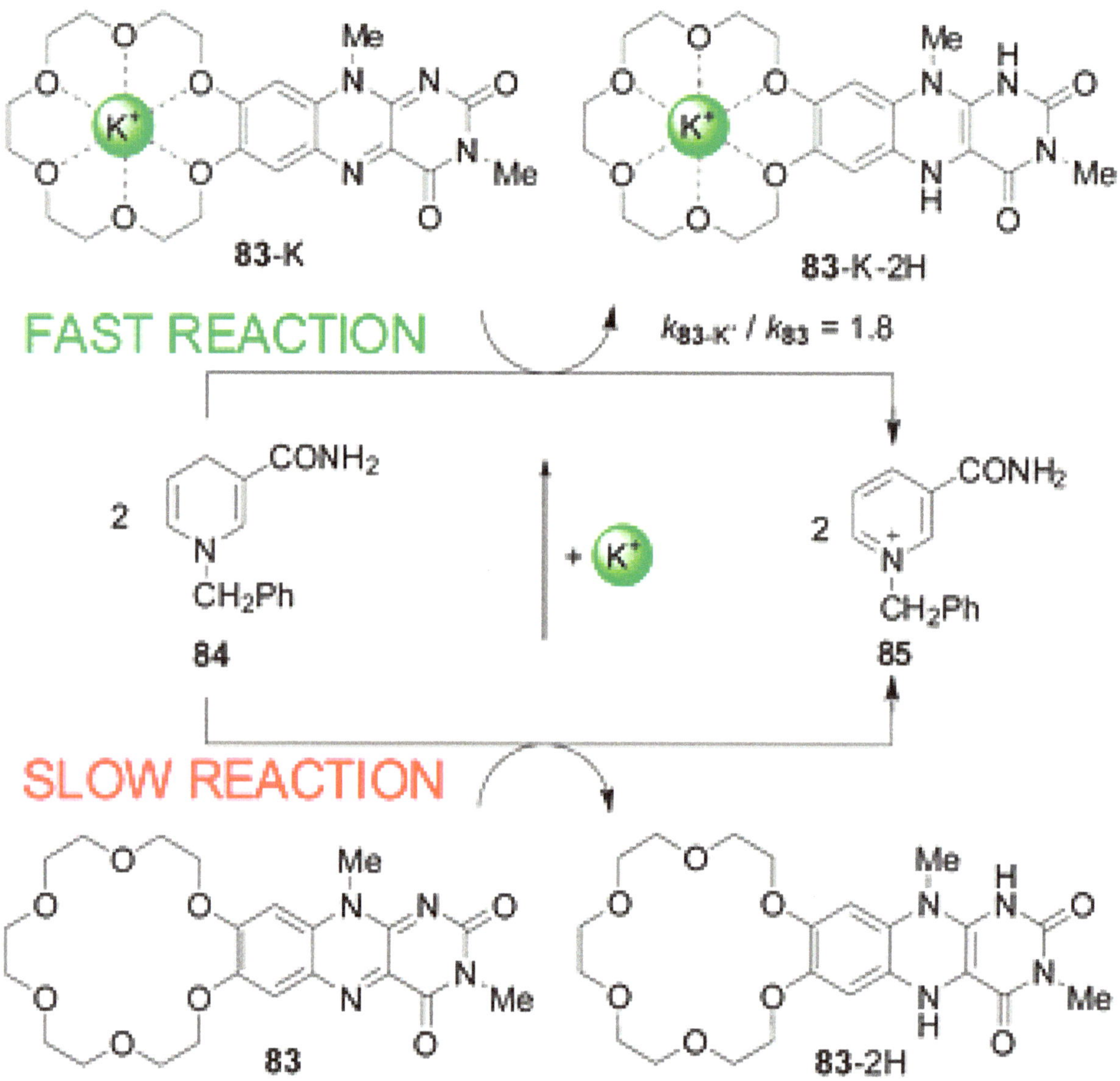

Scheme 22 Acceleration of the redox reaction between **83** and **84** by coordination of an alkali metal ion to the crown ether.[42]

Ligand-based switching

Mirkin and co-workers have developed a series of heterometallic macrocycles incorporating metal-based catalytic units whose activity can be regulated through a reversible coordination-induced conformational change in the molecule (*e.g.* Schemes 23–29).[43] The switching mechanism is based on what the authors term the 'weak link approach'.[44] It involves metal complexation in side-arms additional to the catalytic site, through hemilabile ligands that typically bind to the auxiliary metal centres in a bidentate fashion in one state of the catalyst and in a monodentate mode in the other state. The metal centre is usually Rh(I), Cu(I) or Pt(II). In bidentate mode the molecule adopts a compact or 'closed' conformation. Small monodentate ligands (*e.g.* Cl^- and/or CO) that bind to the metal centre more strongly than the 'weak link' (*e.g.* a thioether) of the bidentate ligand can displace it from the metal ion, inducing a conformational change in the molecule, which now adopts a more open conformation. The process can be reversed as removal of the small monodentate ligands allows the weak binding motif of the hemilabile ligand to re-coordinate to the metal centre, restoring the closed conformation.

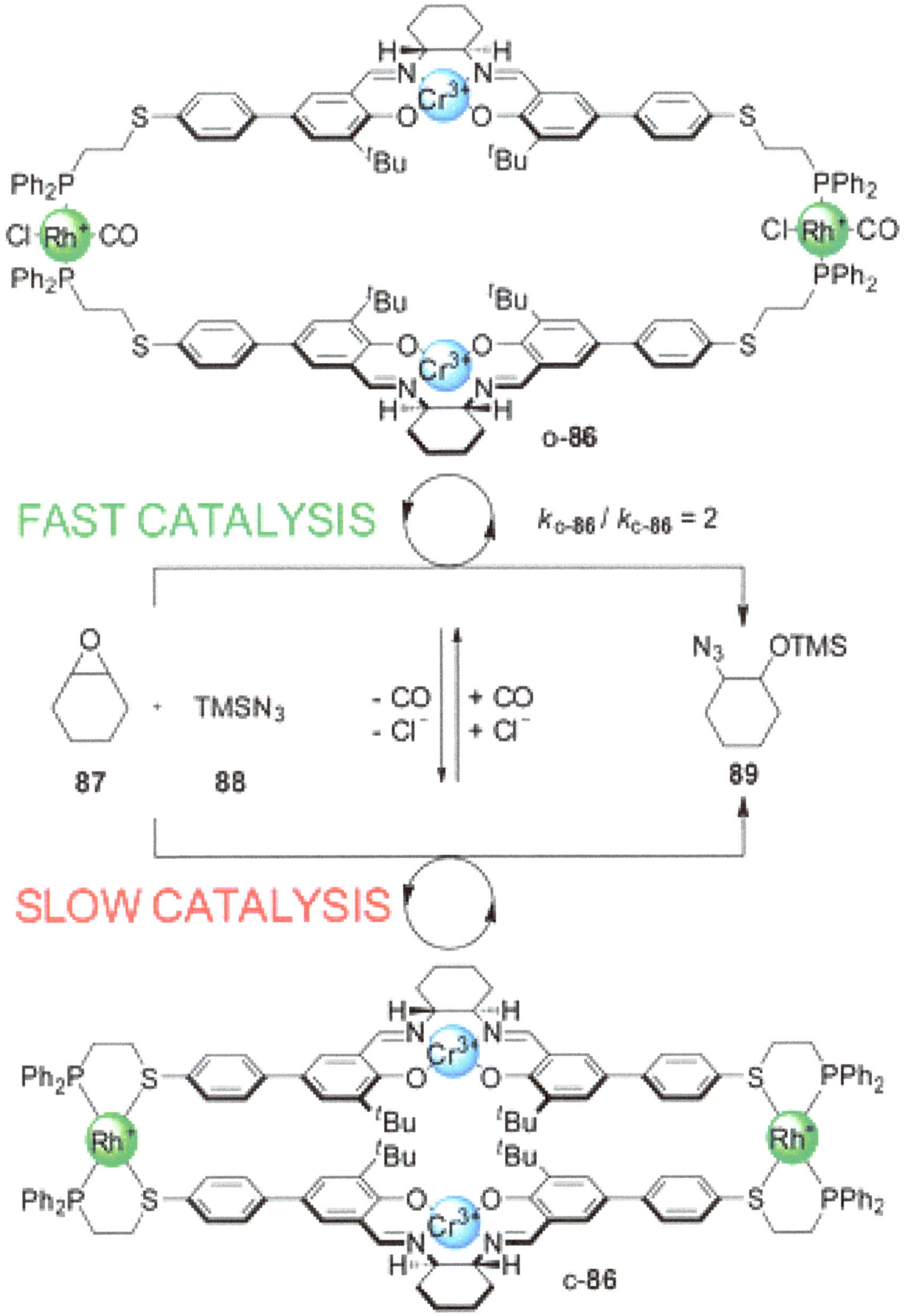

Scheme 23 Control of the rate of an epoxide-opening reaction with catalyst o/c-**86** through a change in catalyst conformation induced by coordination of small ligands to the side-arm metal complexes.[45]

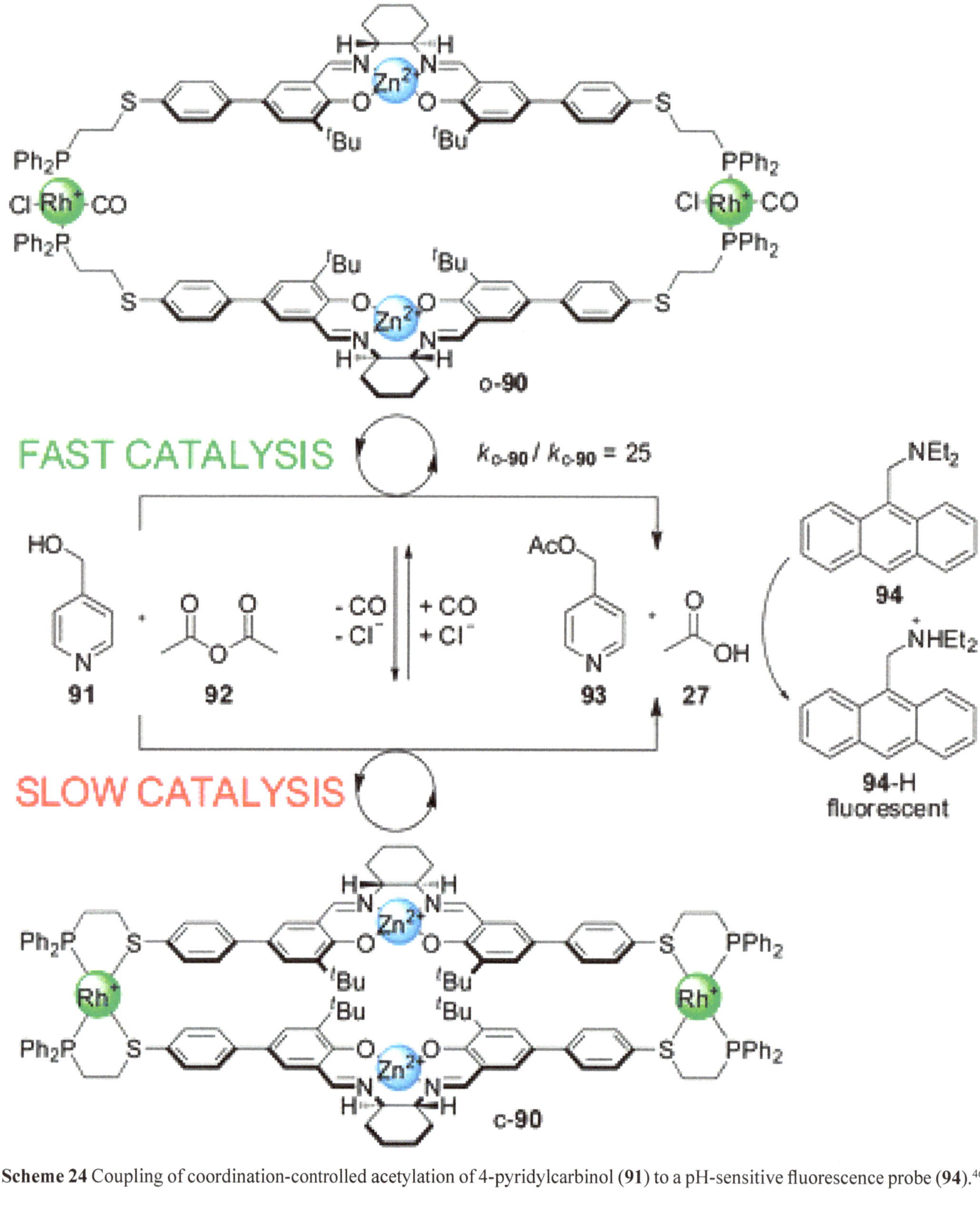

Scheme 24 Coupling of coordination-controlled acetylation of 4-pyridylcarbinol (**91**) to a pH-sensitive fluorescence probe (**94**).[46]

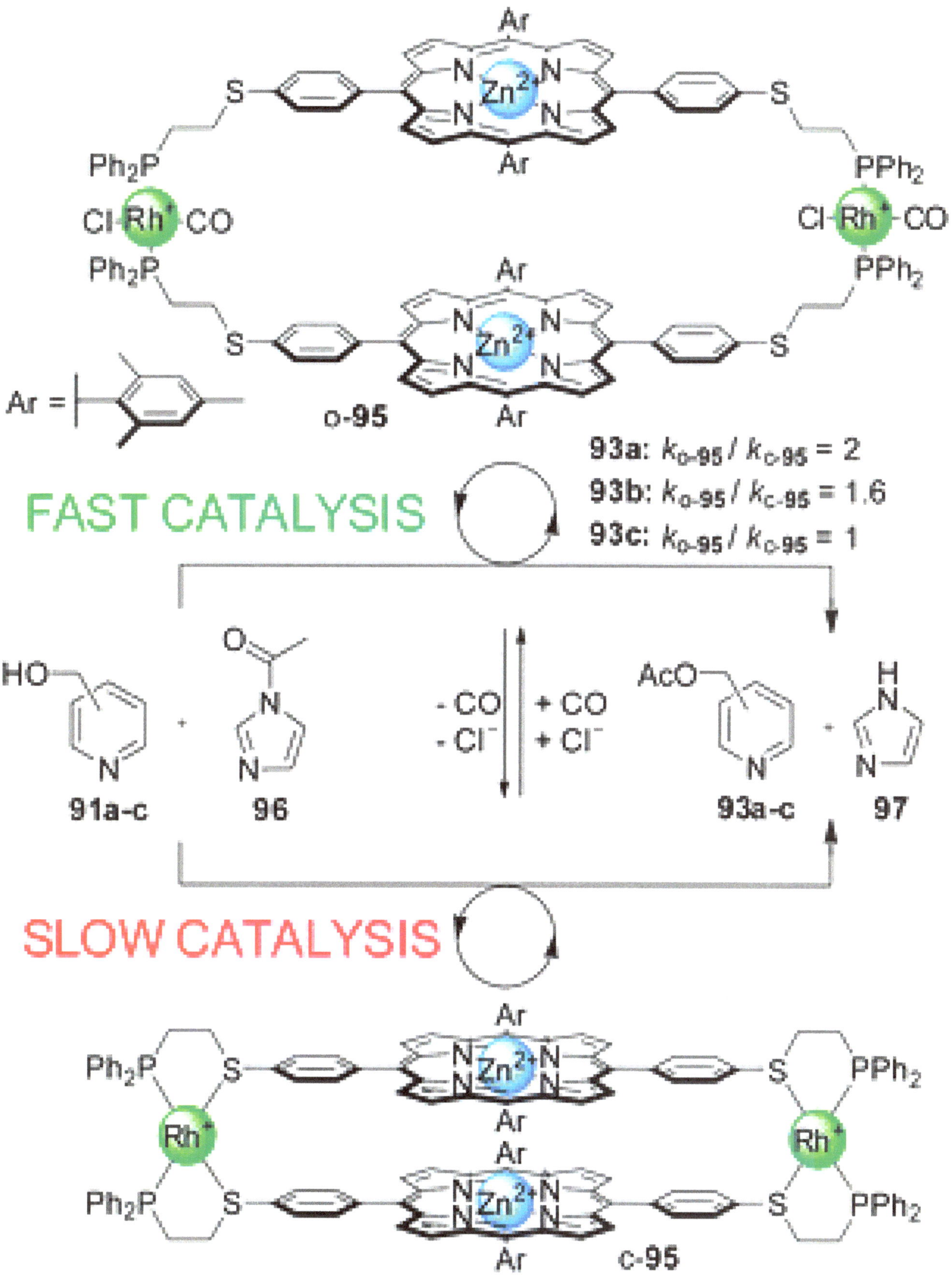

Scheme 25 Regioselective control over the rate of acetylation of pyridine carbinol isomers using porphyrin-based switchable catalyst o/c-**95**.[49]

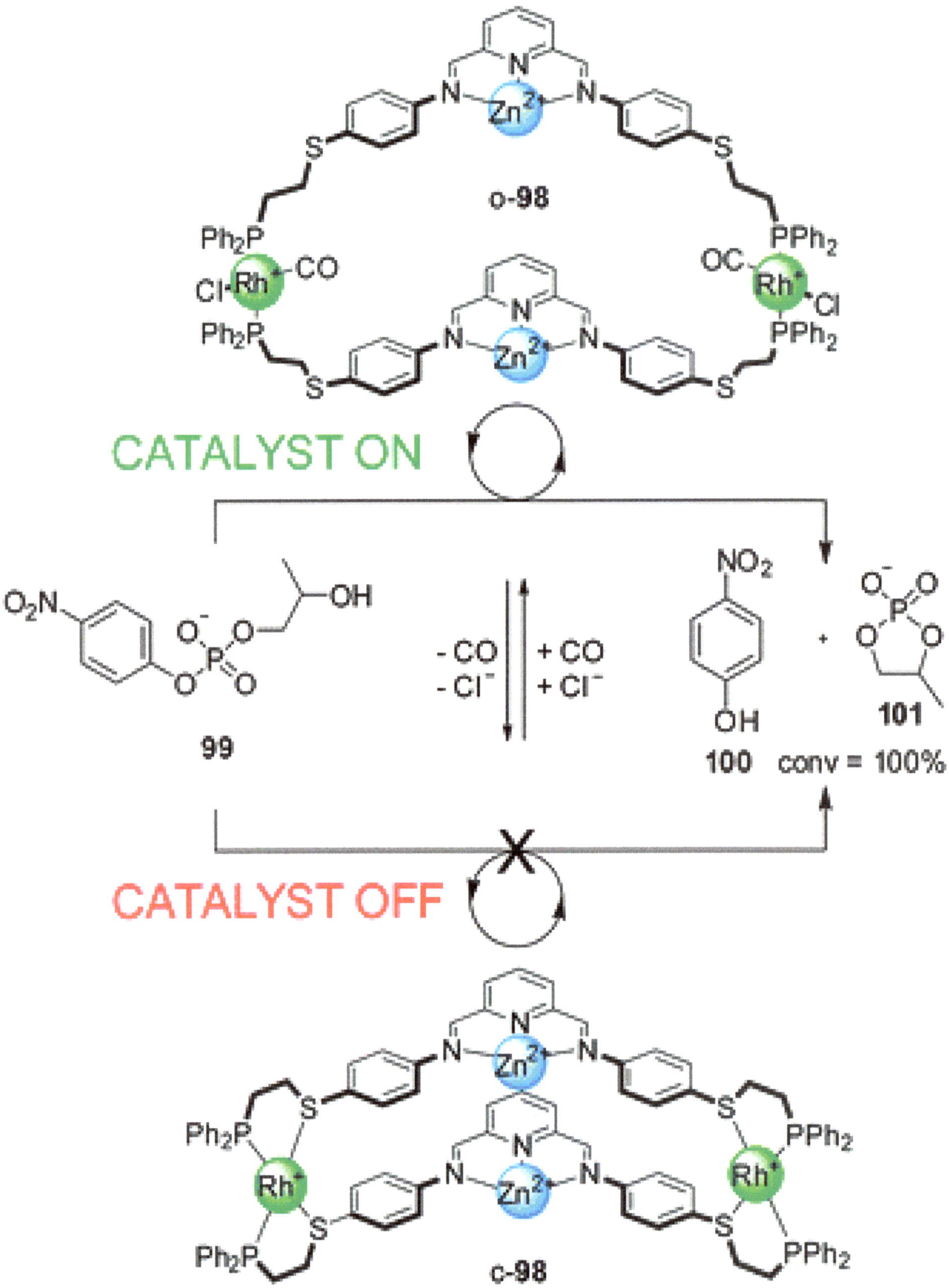

Scheme 26 Activation/deactivation by coordination-triggered switching between active 'open' and inactive 'closed' conformations of Zn(II) catalyst o/c-**98**.[50]

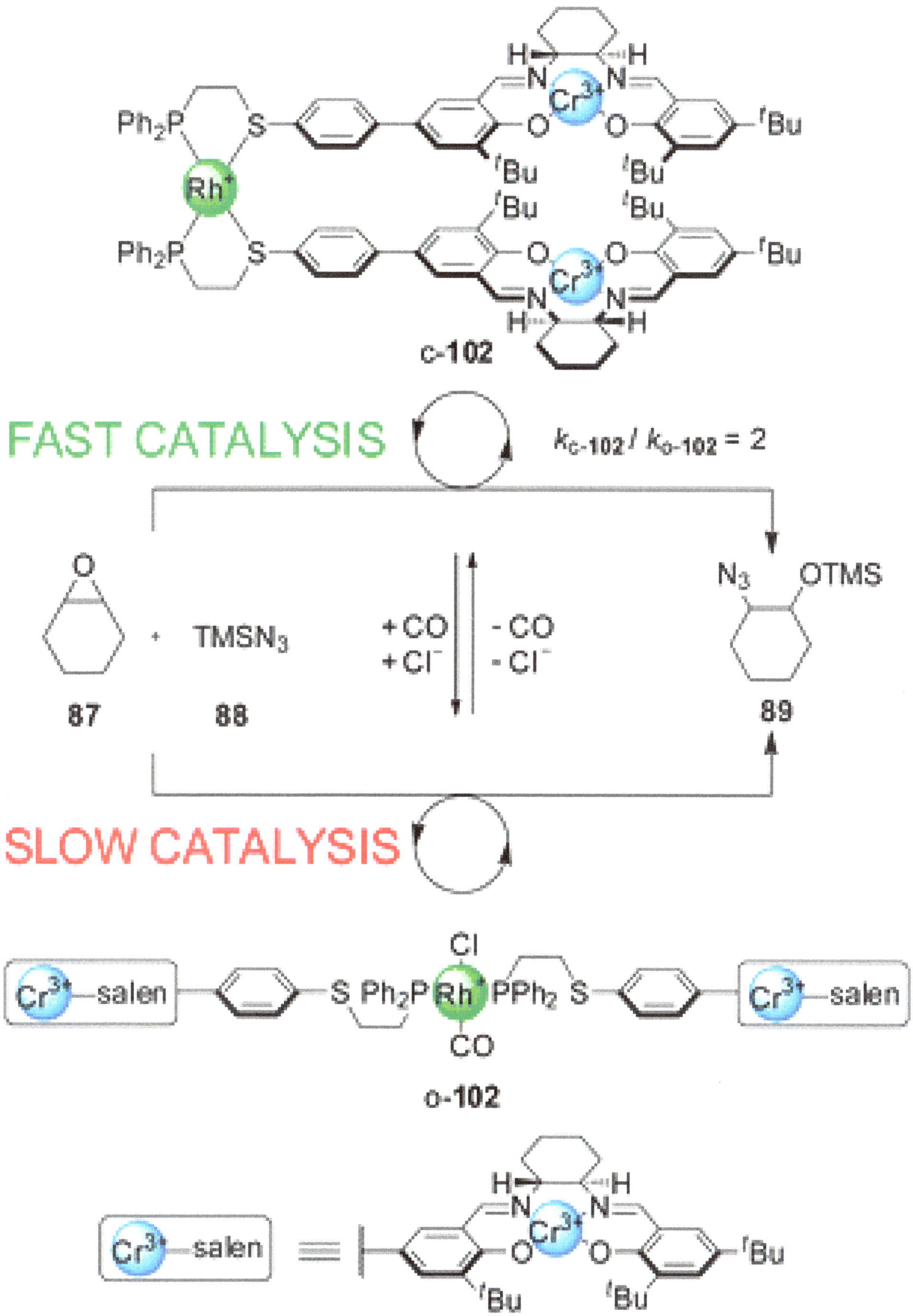

Scheme 27 Coordination-driven switching of the activity of tweezer catalyst o/c-**102**.[51]

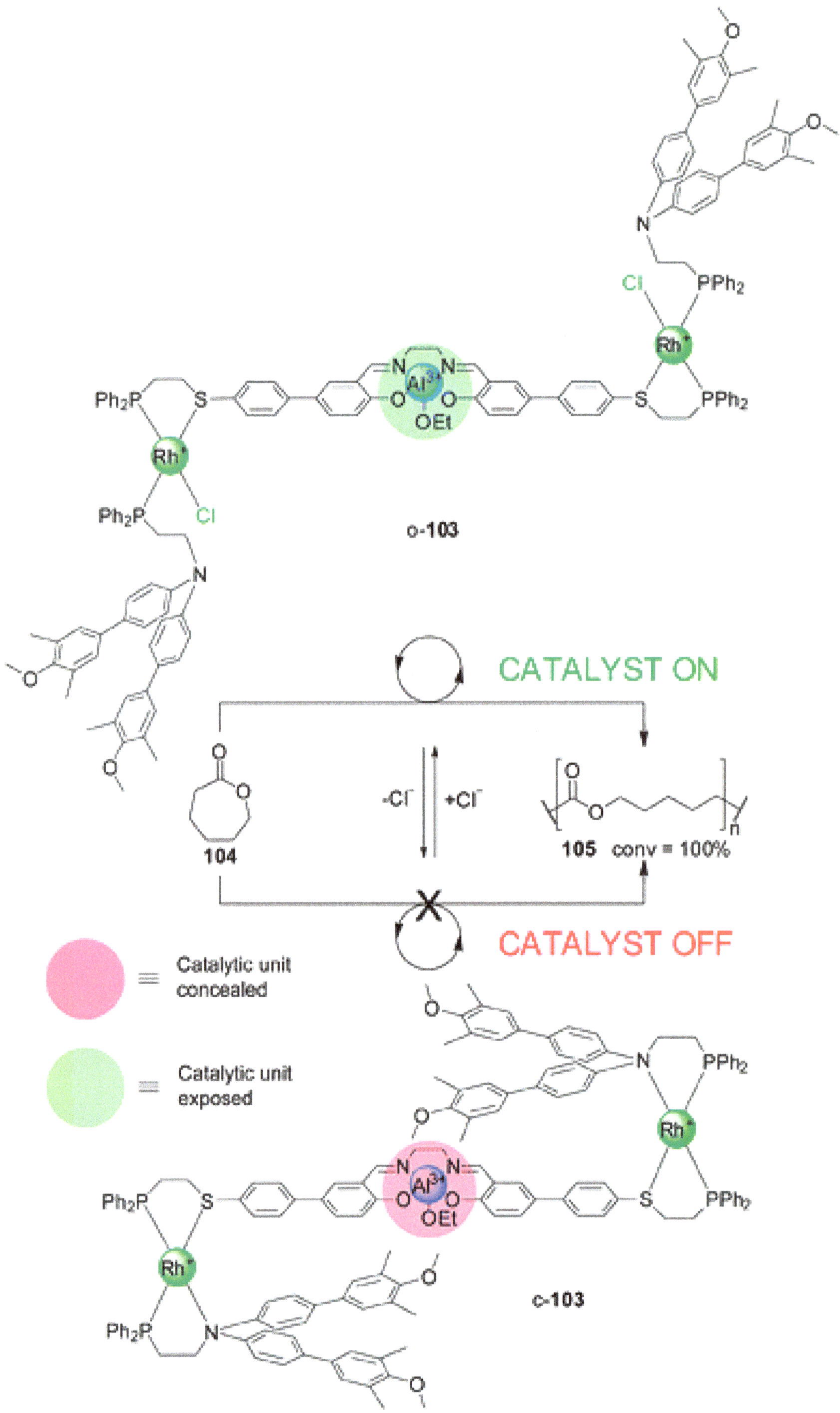

Scheme 28 Control of the activity of Al(III)–salen catalyst o/c-**103** through chloride binding.[52]

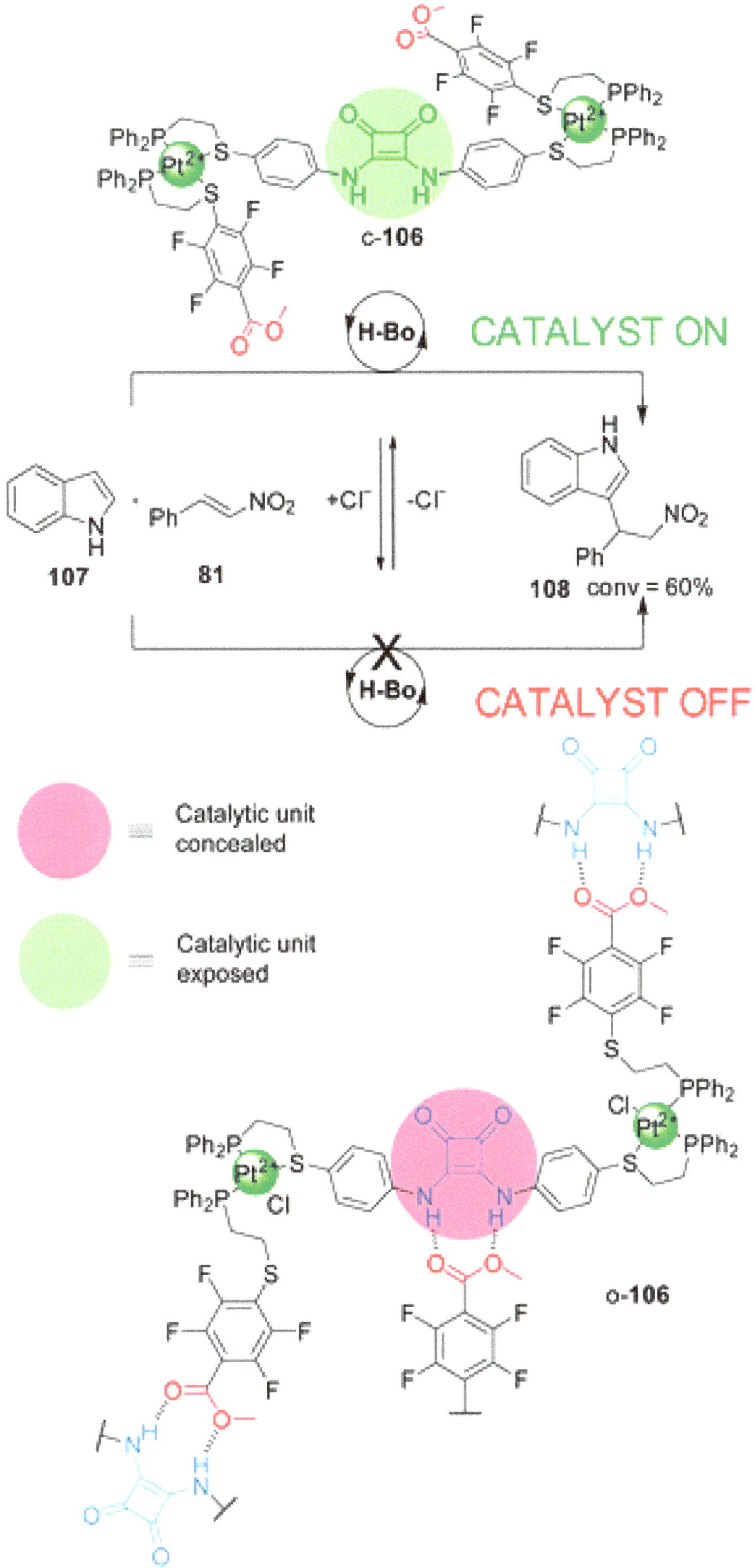

Scheme 29 Switching 'on' and 'off' squaramide-based hydrogen-bond organocatalyst **106** by chloride-mediated aggregation.[53]

The rationale behind the application of this concept in switchable catalysis is that the control of the distance between the catalytic sites could influence their activity. The first such system developed consisted of a tetrametallic macrocycle with two Cr(III)–salen complexes as the catalytic units and two Rh(I) centres with P,S-bidentate ligands as the side-arm complexes (Scheme 23).[45] When Cl^- and CO were added to the reaction media, they bound to the Rh(I) ions replacing the thioether ligand and the system adopted the 'open' conformation (o-**86**). Removal of the CO under reduced pressure or through a N_2 purge switched the catalyst back to the closed conformation (c-**86**), as both external ligands are required to displace Rh–S bonds. The catalytic activity of the two conformations was tested in the ring opening of cyclohexene oxide (**87**) with $TMSN_3$ (**88**), which is catalyzed by Cr(III)–salen catalysts through a bimetallic intermediate. The 'open' conformation of the catalyst showed a modest two-fold enhancement of the reaction rate compared to the 'closed' conformation.

When a Zn(II)–salen complex was used instead of Cr(III) in switchable catalyst **90**, the Cl^- and CO-induced open–closed switching was more effective at controlling reactivity (Scheme 24).[46] Acetylation of 4-pyridylcarbinol (**91**) with acetic anhydride (**92**) proceeded around 25× faster with the 'on' ('open') state than with the 'off' ('closed') state of the catalyst. The authors postulated that the closed state of the catalyst promoted the reaction through Lewis-acid-activation at a single site, while in the open state both metal centres participate in the catalysis, activating both the acetic anhydride and pyridylcarbinol substrates. The reaction was coupled to diethylaminomethylene anthracene (**94**), a pH-sensitive fluorophore, resulting in a fluorimetric sensor able to detect concentrations of Cl^- as low as 800 nM (Scheme 24).

It was found that the switching of catalyst **90** was also possible using acetate anions instead of chloride, which could be used to establish autocatalytic rate acceleration. Under the reaction conditions the 'off' state of **90** catalyzed the acetylation of pyridine carbinol at a very slow rate. Upon addition of 10 mol% of *tert*-butylammonium acetate, a percentage of the inactive catalyst switched to its active form, which started to catalyze the reaction generating as the products 4-acetoxymethylpyridine (**93**) and acetic acid (**27**). The acetic acid is deprotonated by a base present (diethylaminomethylene anthracene, **94**) and coordinates to additional Rh(I) centres, generating additional amounts of the active form of the catalyst, progressively accelerating the reaction until saturation at around 80% conversion. The autocatalytic sequence effectively amplifies detection of acetate ion.[47]

Cu(I) can also be used instead of Rh(I) in the switching mechanism. Bidentate ligands such as phenanthroline and 4,4′-bipyridine, and monodentate species such as *tert*-butylisonitrile, were found to promote the switching from the closed to the open conformation by coordination to the copper(I) centre.[48] However, with Cu(I) the switching is irreversible and the original off state cannot be restored. The corresponding Cr(III)–salen-based catalyst with the Cu(I) side-arm complex had a 12-fold rate enhancement in the 'on' state compared to the 'off' state for the reaction between 4-pyridylcarbinol (**91**) and acetic anhydride (**92**).

Switchable catalysts of this type based on catalytic units other that Cr(III)–salen complexes have also been explored. A system featuring Zn(II)–porphyrin catalytic centres (**95**) was tested in the acetylation of different isomers of pyridine carbinol (**91a–c**) with acetylimidazol (**96**), but did not afford better results than previous systems (only a 2-fold acceleration of the reaction, Scheme 25).[49] However, the system displayed interesting regioselectivity. The open conformation of the switchable catalyst (o-**95**) showed enhancement of the catalytic activity with the *para*- and *meta*-isomers of pyridylcarbinol (**91a** and **91b**, respectively), but not with *ortho*-pyridylcarbinol (**91c**). Therefore, discrimination of 2-pyridylcarbinol over the 3- and 4-isomers was achieved.

An unwelcome characteristic of most of these systems is that discrimination between the active and inactive states is limited by significant catalytic activity displayed by the 'off' closed state. This limitation was overcome in a system featuring Zn(II)–pyridine–bisimine catalytic motifs (**98**) with switching again triggered by the addition/removal of Cl^- and CO (Scheme 26).[50] In the hydrolysis of 2-(hydroxypropyl)-*p*-nitrophenyl phosphate (**99**) in a water/methanol 1 : 3 mixture the closed state of the catalyst (c-**98**) was completely inactive while the open state (o-**98**) led to full conversion within 40 min. The absence of activity in the 'off' state was attributed to the short Zn–Zn distance that precluded the bimetallic intramolecular reaction, and to the presence of an acetate ion bridging the two Zn(II) centres hindering access of the substrates. *In situ* switching between both forms of the catalyst was also demonstrated.

The same type of switching mechanism has been applied with acyclic catalytic assemblies, including a 'tweezer' catalyst, **102**, incorporating two units of the Cr(III)–salen catalytic moiety and a Rh(I) centre coordinated to two P,S-hemilabile ligands (Scheme 27).[51] Again, addition or removal of Cl^- and CO ligands triggers switching between an 'open' and a 'closed' form. However, the catalytic activity of the system in the ring-opening reaction of cyclohexene oxide (**87**) with trimethylsilylazide (**88**) is reversed compared to the macrocyclic systems shown in Scheme 23. Here, the closed state (c-**102**) is the active form of the catalyst, as the Cr(III) catalytic centres are preorganized favouring the bimetallic catalysis. The open form (o-**102**) shows lower catalytic activity as a result of the more flexible arrangement of the salen sites, in part due to the phosphine ligands occupying a *trans*-coordination relationship across the Rh(I) centre. The system could be switched *in situ* by CO saturation/desaturation, alternating between reaction rates of ~0.013 mM min^{-1} (less active, open form) and ~0.026 mM min^{-1}(more active, closed form). The 'on' state afforded the product (**89**) with 80% ee.

An evolution of the switching mechanism was used in the design and operation of a triple-layer switchable catalyst, **103** (Scheme 28).[52] This system consists of a Al(III)–salen motif—a catalyst for the living polymerization of lactones—linked to two P,S-bidentate ligands coordinated to Rh(I) square-planar complexes. The Rh(I) centres are also coordinated to P,N-bidentate ligands whose N atoms were derivatised with bulky blocking groups. Chloride ions or acetonitrile displace the nitrogen donor of the bidentate ligands. As a result the system can be switched from a closed conformation (c-**103**), in which the bulky groups block the catalytic unit ('off' state), to a 'semi-open' conformation (o-**103**), in which the catalytic site is exposed and can participate in catalysis ('on' state). Removal of the chloride ions with NaBArF switches the system back to the closed state. The system was tested in the Al(III)–salen-catalyzed ring-opening living polymerization of ε-caprolactone (**104**) (Scheme 28). The 'on' state of the catalyst gave full conversion of the monomer, while the 'off' state was almost completely inactive under the same reaction conditions (7% conversion to the polymer after 100 h). The active catalyst was *in situ* switched 'off' after 10 h, and the 'on' state restored with acetonitrile causing the reaction to resume.

The 'weak link approach' is not restricted to metal catalysis. A triple-layer design (**106**), which incorporates a squaramide organocatalytic group functionalised with P,S-bidentate ligands, has been used to control the rate of the Friedel–Crafts reaction between indole (**107**) and *trans*-β-nitrostyrene (**81**) (Scheme 29).[53] The ligands are coordinated to a Pt(II) centre together with another hemilabile P,S-functionalised methyl tetrafluorobenzoate ligand. The system can be switched between the closed and semi-open forms by addition or removal of Cl^-, which is able to displace the S atom of the weaker chelating S-tetrafluorobenzoate ligand. In the flexible semi-open form (o-**106**) the squaramide unit participates in intermolecular hydrogen bonding with a carboxylate group from another molecule of catalyst, resulting in aggregation and blocking of the active site ('off' state). In the rigid closed form (c-**106**) the intermolecular hydrogen bonding interactions are disfavoured due to steric hindrance and geometric constraints, so the squaramide unit can perform its catalytic function ('on' state). Catalyst c-**106**afforded the Michael adduct **108** in ~60% conversion after 24 h, while o-**106** led to a less than 5% conversion, comparable to the background reaction (Scheme 29). Addition of two equivalents of tetrabutylammonium chloride switched the active form of the catalyst to its inactive form, stopping the reaction, which was resumed after restoring the system to the active form by addition of AgBArF.

A second system investigated consists of a tweezer structure with two urea groups (the hydrogen bond organocatalytic groups) linked by P,S-hemilabile ligands to a Pt(II) metal centre.[54]The same mechanism used in **106** allows switching between the open and closed forms. The closed form is the active species of the catalyst for the Diels–Alder reaction between cyclopentadiene and methyl vinyl ketone. The open form is a poorer catalyst due to urea–urea intermolecular hydrogen bonding that blocks the active site.

Slugovc and co-workers have described a different type of catalyst that exhibits ligand-coordination-induced switching (Scheme 30).[55] They prepared a Ru(II) olefin metathesis catalyst **109** bearing a benzylidene chelate, an N-heterocyclic carbene and two chlorides as ligands, which could catalyzed the ring-opening metathesis polymerisation of norbornene derivative **110**in >90% yield in 4 h. The system could be deactivated by replacing one of the chloride ions for pyridine followed by counterion exchange to PF_6^-. This 'off' state of the catalyst afforded only 10% conversion to polymer after 19 h under the same reaction conditions. Conditions to switch the system back to the original active state were not discussed.

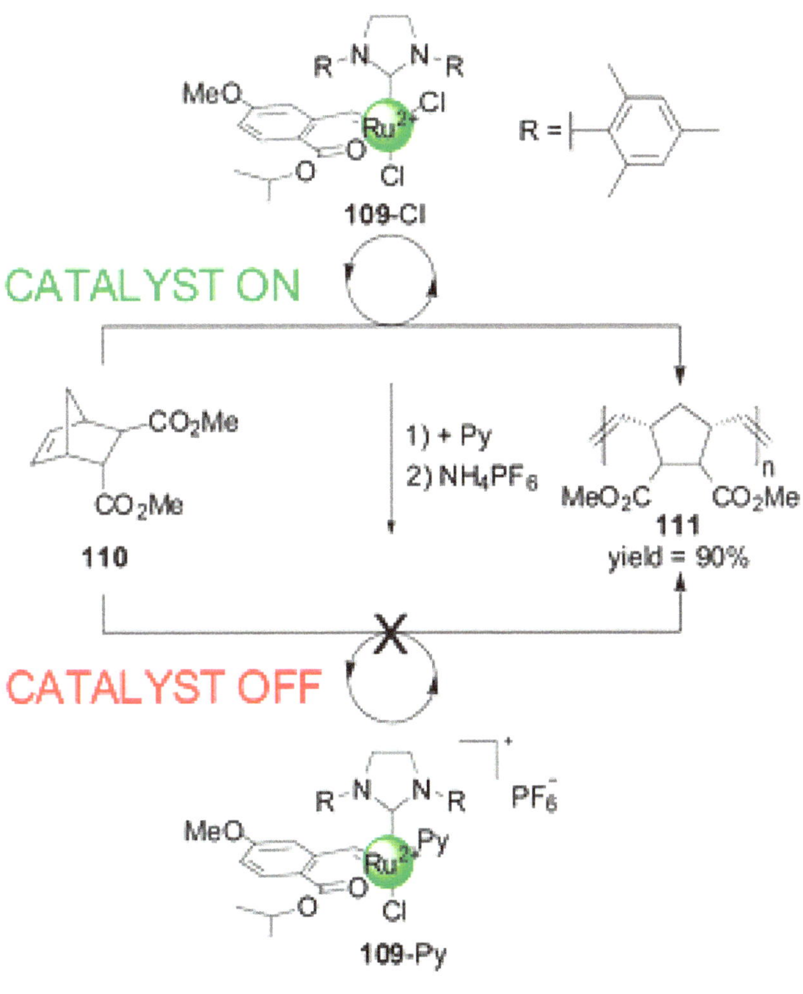

Scheme 30 Deactivation of Ru(II) metathesis catalyst **109** promoted by Cl^-–pyridine ligand exchange.[55]

Switching promoted by the addition/removal of metal cations

Following Shinkai's early example (Scheme 22), other systems were developed in which the binding of metal ions to crown-ether-like ligands modulated catalytic activity. Dervan and colleagues employed compound **112**, which contains a Fe(II)-EDTA-type complex, to cleave DNA (Fig. 2a).[56] In the absence of metal ions the system is not catalytically active, but in the presence of Ba^{2+} or Sr^{2+} the glycol chain binds to the ions adopting a conformation that bends the molecule and enables hydrogen bonding between the amide groups and the DNA bases, facilitating sequence-controlled cleavage of the strands. More recently, Kubo reported a bis(thiourea) dibenzo-diaza-30-crown-10 derivative (**113**, Fig. 2b)[57] that catalysed the cleavage of 2-hydroxypropyl-*p*-nitrophenyl phosphate (**99**) in the presence of a stoichiometric amount of K^+ 400 times faster than in the absence of the metal ion. The authors attributed this rate enhancement to the conformation adopted by the system upon coordination allowing the thioureas to bind the phosphate group, increasing the electrophility of the phosphorous centre.

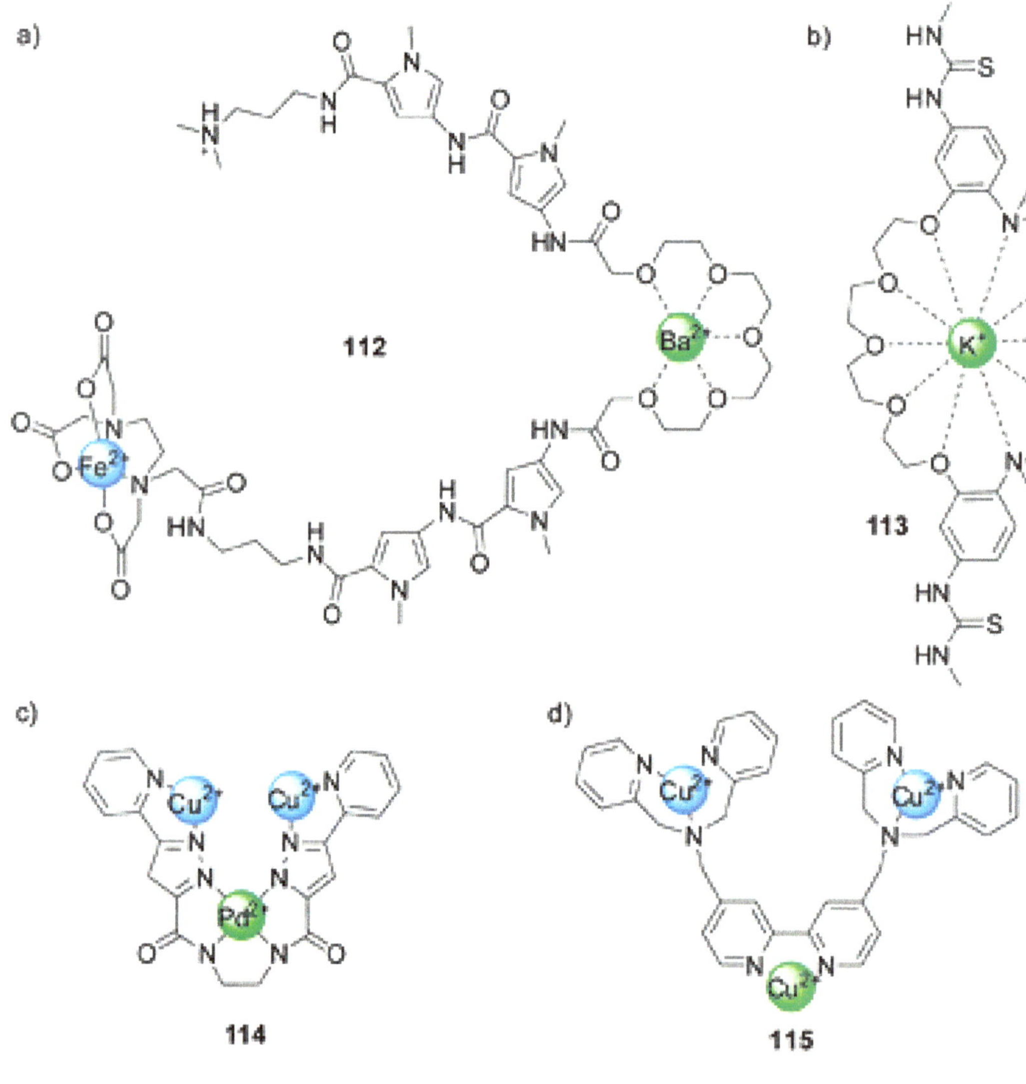

Figure 2 Catalysts activated by conformational changes induced by coordination of metal ions.[56–61]

Krämer has developed a series of systems in which the coordination of one metal ion (*e.g.* Pd(II) in compound **114**, Fig. 2c)[58] switches 'on' catalytic activity of another metal centre (Cu(II) in **114**). Complex **114** displayed activity in the hydrolysis of nitrophenyl phosphate **99** when bound to Pd(II), but was inactive when coordinated to Pt(II).[59]

Catalysts featuring multiple metal ions have been prepared that display increased activity in phosphate diester cleavage reactions due to a cooperative effect between the metal centres.[60]An example (**115**), shown in Fig. 2d, comprises a bidentate 2,2′-bipyridine ligand and two tridentate 2,2′-dipicolylamine units which bind more strongly to metal centres such as Cu(II) and Zn(II), allowing selective complexation at the different sites. When only the dipicolylamine groups are coordinated to metal centres the system displays poor phosphodiester cleavage activity. However, when the bipyridine unit is also bound catalysis was enhanced due to the coordination-induced change from the *cisoid* to the *transoid* conformation of the bidentate ligand, bringing together the Zn(II) or Cu(II) catalytic centres. A 19-fold increase of activity was achieved in the case of the Cu(II) system **115**.[61]

These systems were switched 'on' by the external stimulus, but switching back to the 'off' state was not investigated. More recently, reversible switchable catalysts that can be controlled by coordination of metal ions to crown ethers present in their structure have been reported. Miller and co-workers developed an Iridium(III) complex with a pincer ligand bearing an aza-15-crown-5 moiety (**116**, Scheme 31).[62] The activity of this complex for the activation of dihydrogen can be triggered by coordination of Li^+ or Na^+ to the crown ether. In the absence of alkali metal cations, the Ir(III) centre is hexa-coordinate with two oxygen atoms of the crown ether forming part of the coordination sphere of the metal (**116**-H). In this form, coordination of the substrate (the authors actually used deuterium in their experiments) was hampered by the crown ether (catalyst 'off') and dihydrogen activation, measured as H/D exchange at the metal centre, was slow ($k_{obs} = 1.2 \times 10^{-6}$ s^{-1} in CH_2Cl_2). In the presence of Li^+ or Na^+ cations, however, the crown ether oxygen atoms were no longer coordinated to the Ir(III) centre, allowing the formation of a penta-coordinated Ir(III) complex with a dideuterium σ-complex (**116**-H-M, catalyst 'on'). Thus, the addition of substoichiometric amounts of NaBArF (0.3 equiv.) and LiBArF (0.4 equiv.) resulted in a 20-fold and 250-fold rate enhancement of H/D exchange, respectively. The rate of the reaction could be modulated by *in situ* switching of the catalyst. Addition of chloride to the active form of the catalyst caused precipitation of the Na^+ ions, switching the system 'off'.

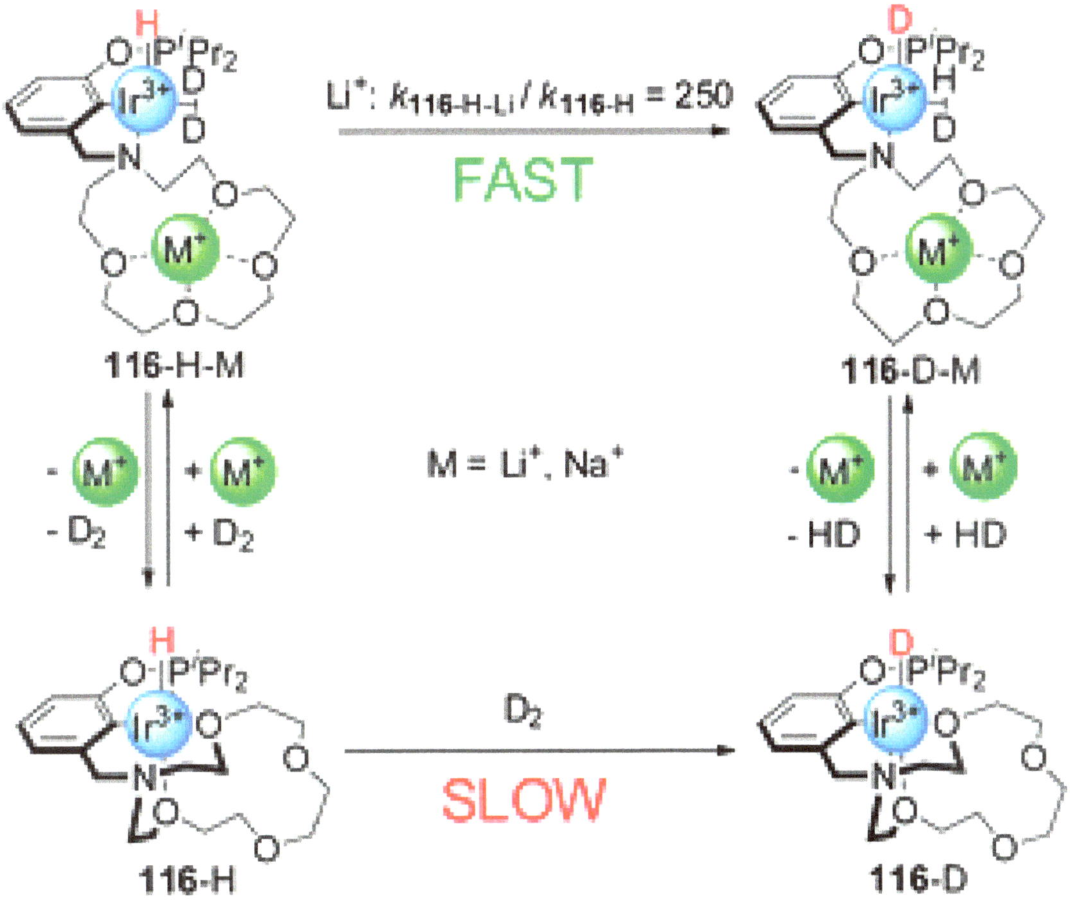

Scheme 31 D_2 activation by Ir(III) catalyst **116** triggered by reversible coordination of an alkali metal cation to a crown ether ligand.[62]

Fan and co-workers reported chiral aza-18-crown-6-phosphoramidite Rh(I) catalyst **117**, whose activity in the asymmetric hydrogenation of dehydroamino acid esters (**118**) could be switched 'on' or 'off' by the reversible coordination of alkali metal cations to pendant crown ether groups (Scheme 32).[63] In the absence of Na^+ ions the O atoms of the crown ethers coordinate to the rhodium(I) centre and the system exhibits no catalytic activity (less than 1% conversion, catalyst 'off'). Addition of two equiv. NaBArF results in the crown ethers disassociating from the Rh(I) centre (**117**-2Na). Complex **117**-2Na catalyzes the hydrogenation with full conversion and excellent enantiomeric excess (92–98% ee) (catalyst 'on'). Removal of Na^+ with cryptand[2.2.2] restored the system to the inactive state. The switching cycle could be repeated without significant loss of activity.

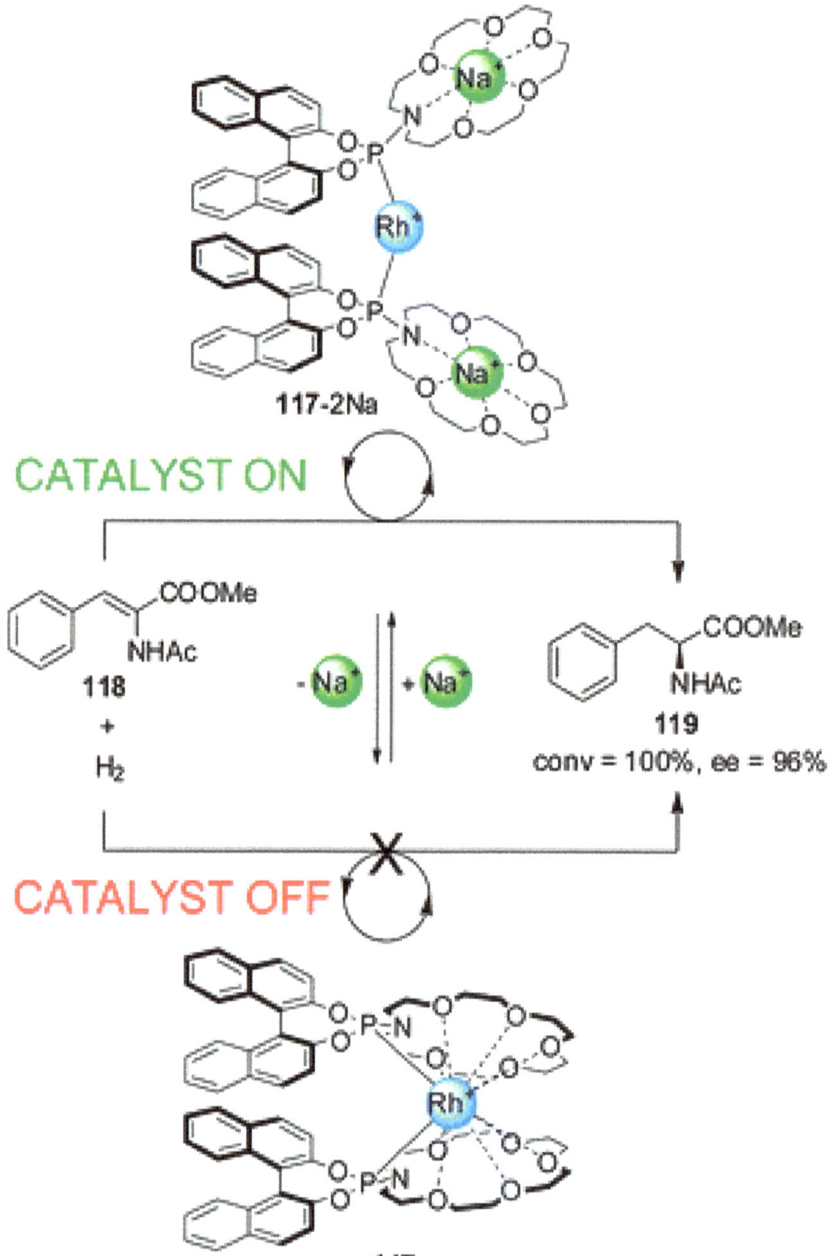

Scheme 32 'On–off' switching of Rh(I) catalyst **117** driven by addition or removal of Na^+ ions.[63]

In 2012 Schmittel and co-workers reported a switchable catalyst system triggered by the addition or removal of Cu(I) ions (Scheme 33).[64] The system consisted of three elements: a switch containing a Zn(II)–porphyrin unit and a 4-(2-pyridyl)pyrimidine moiety separated by an angular rigid oligophenylalkynyl spacer (**120**), a phenanthroline (phen) derivative (**122**) and piperidine (**121**). The latter is the active organocatalyst, able to promote the Knoevenagel reaction between 4-nitrobenzaldehyde (**41**) and diethyl malonate (**123**). In the absence of Cu(I), the pyrimidine unit coordinated to the Zn(II) porphyrin leaving the piperidine available for catalysis (**120**, catalyst 'on'). When Cu(I) is added it forms a complex with the phenanthroline and pyridyl–pyrimidine group (**120**-Cu), and the Zn(II) porphyrin binds to the piperidine, preventing it acting as a catalyst (catalyst 'off'). Removal of the Cu(I) ions with cyclam restored the system to the 'on' state. The reversibility of the system was demonstrated by three cycles of *in situ* switching.

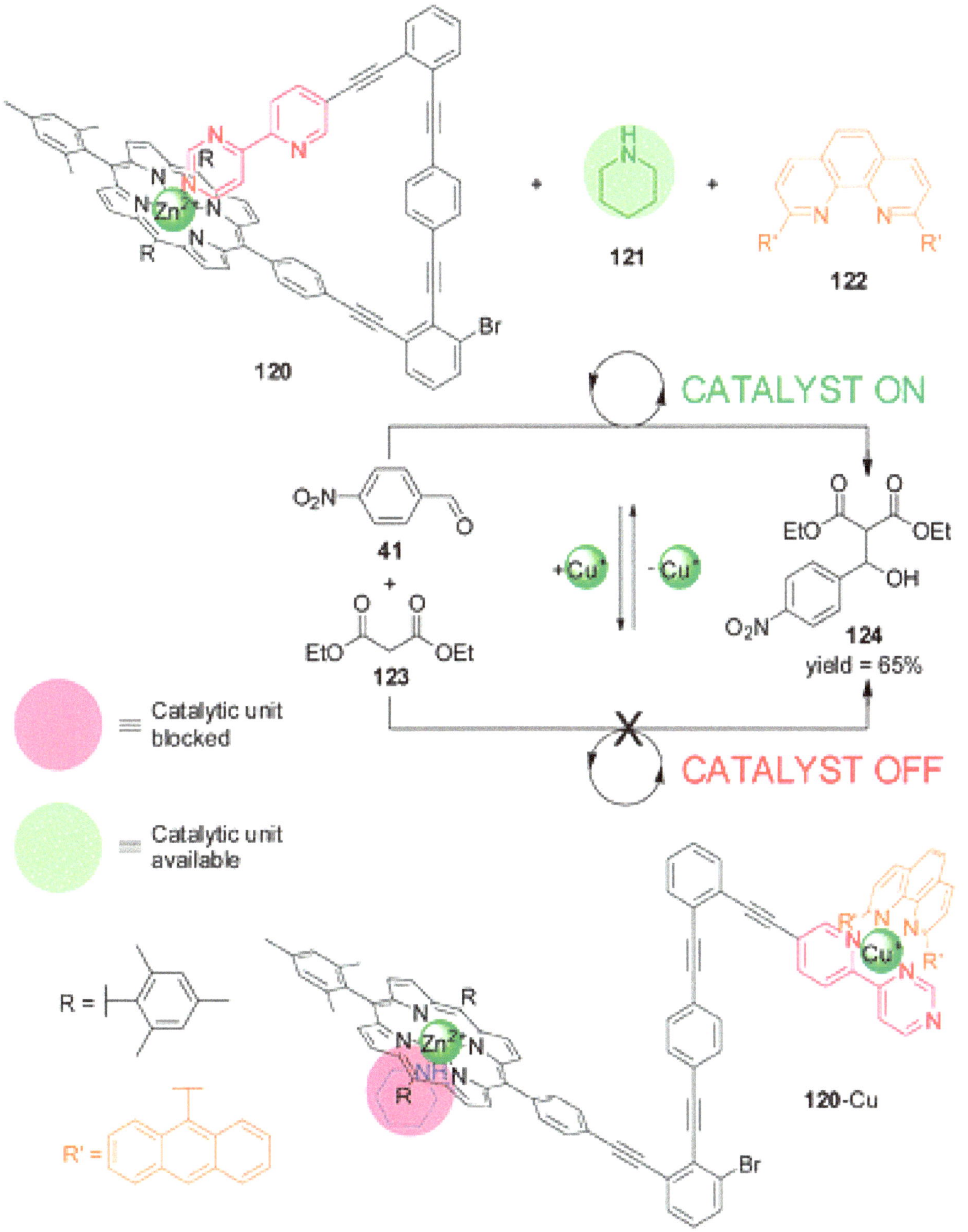

Scheme 33 Switchable catalysis of a Knoevenagel reaction based on the capture/release of a piperidine catalyst (**121**) by Zn(II)–porphyrin **120** induced by the reversible coordination of Cu(I).[64]

A related catalyst (**125**), in which all the elements required for operation are included within a single molecule, has been used to promote the *cis–trans* photoisomerization of *E*-4-styrylpyrimidine (**126**) (Scheme 34).[65] The Zn(II)–porphyrin moiety catalyzes the reaction in the presence of Cu(I) due to the metal ion forming a complex with the pyridyl–pyrimidine and the phen unit ('on' state; **125**-Cu). Addition of cyclam extracts the Cu(I) and the pyrimidine ring coordinates to Zn(II), preventing the porphyrin from binding to the substrate ('off' state; **125**). Significant differences in catalytic activity between the active and inactive states were demonstrated by *in situ* switching.

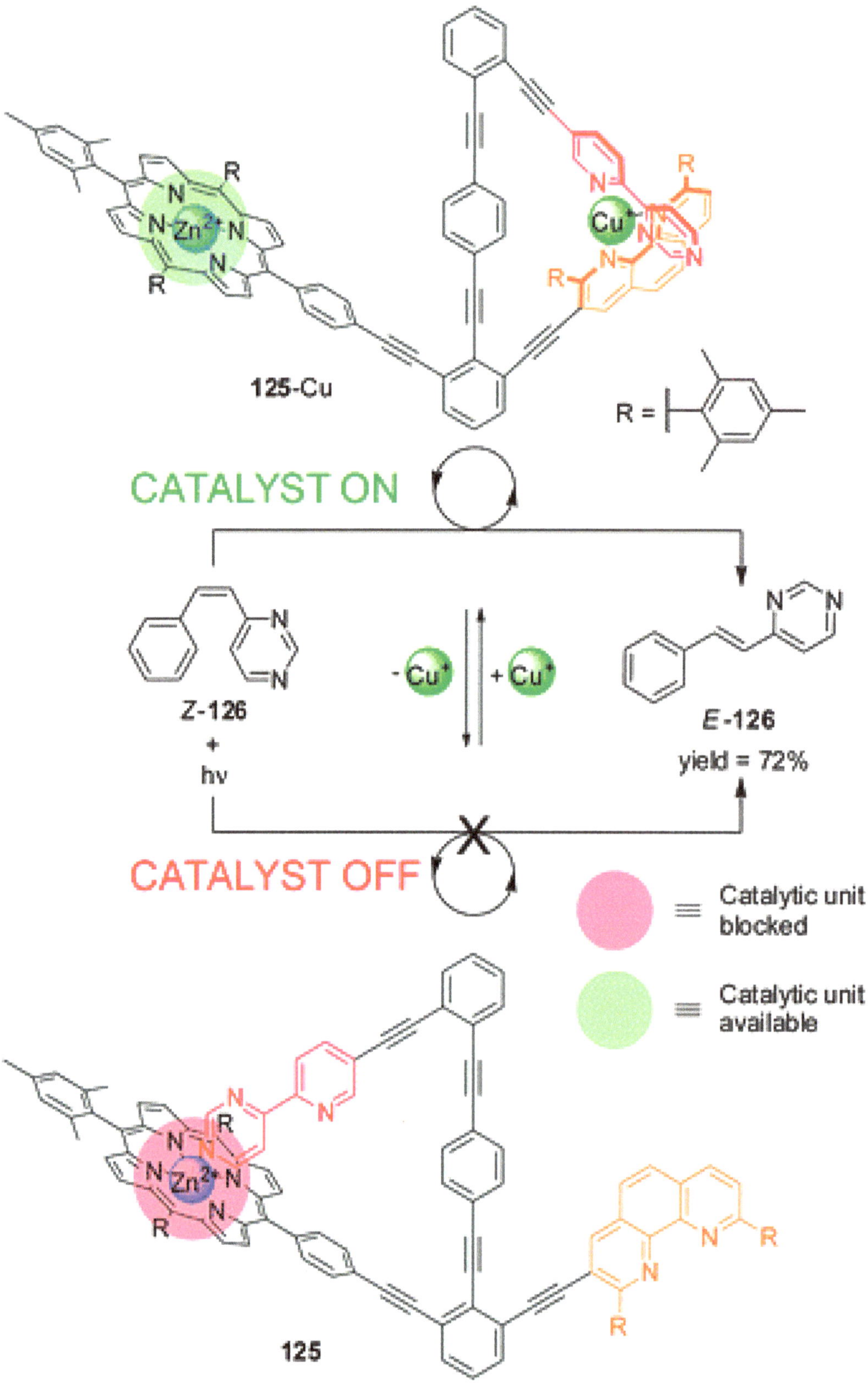

Scheme 34 Control of the activity of catalyst **125** in a *cis–trans* isomerization process triggered by the addition/removal of Cu(I).[65]

A different metal ion provided the external stimulus in a design (**127**) comprising a phenanthroline and terpyridine (terpy) ligands coordinated to a Cu(I) centre (Scheme 35).[66] Addition of Fe(II) resulted in the formation of a dimeric species (**127**$_2$-Fe) in which two terpyridine ligands are coordinated to the Fe(II) centre, leaving accessible coordination sites on the Cu(I)–phen complex available to catalyse the cyclopropanation of *Z*-cyclooctene (**68**) with ethyldiazoacetate (**17**) (catalyst 'on'; **127**$_2$-Fe). Extracting the Fe(II) with 4-*N*,*N*-dimethylaminoterpyridine restores the system to its inactive monomeric form (catalyst 'off'; **127**).

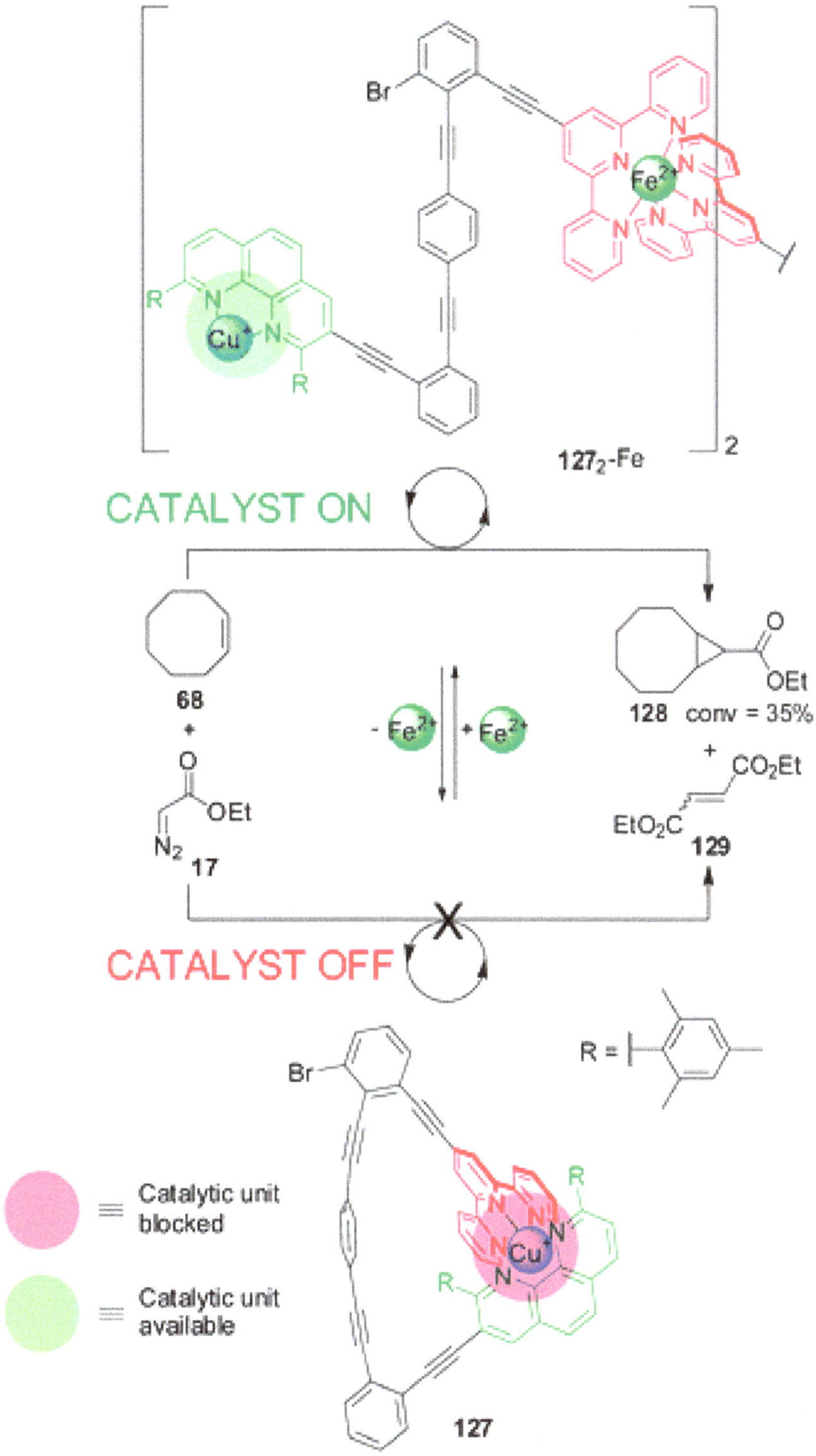

Scheme 35 Control of the activity of Cu(I) catalyst **127** by switching between a monomeric and a dimeric form promoted by the reversible coordination of Fe(II).[66]

System **125** (Scheme 34) was used in combination with piperidine (Pip, **121**) to develop a switchable catalyst (**130**-Cu-Pip), which is able to control not only the rate but also the chemoselectivity of catalysis (Scheme 36).[67] The authors sequentially applied two different chemical stimuli (Cu(I) and phen derivative **131**) to the system formed from **130** and piperidine. The system starts in an 'off' state in which the potential catalytic species, piperidine and Cu(I), are coordinated to binding sites preventing them from being catalytically active (**130**-Cu-Pip). Addition of an external phen ligand (**131**) displaces the pyridyl–pyrimidine arm from the Cu(I) complex, which then binds to the porphyrin, displacing the piperidine from the Zn(II) centre (**130**-Cu-**131**). The liberated piperidine can catalyze the Knoevenagel reaction in a similar fashion to Scheme 33 (piperidine catalysis 'on'). In this state the Cu(I) is coordinatively saturated with bidentate ligands and cannot participate in catalysis (Cu catalysis 'off'). However, if a second equivalent of Cu(I) is added, an intramolecular complex is formed and the catalytically active Cu(I)–phen complex (**131**-Cu) is released and able to promote the Cu-catalyzed azide–alkyne cycloaddition (CuAAC) 'click' reaction between benzyl azide (**132**) and 4-nitrophenylacetylene

(**133**) (Cu catalyst 'on'). As the pyridyl–pyrimidine ligand is no longer bound to the Zn porphyrin, its position is occupied by the piperidine, which is no longer available and therefore the Knoevenagel reaction does not take place (piperidine catalysis 'off'). Addition of one equivalent of **131** turned 'off' both catalytic activities sending the system back to its original state. The sequential addition of the different chemical stimuli switches the different activities with a high degree of control; only one reaction type is observed in each state.

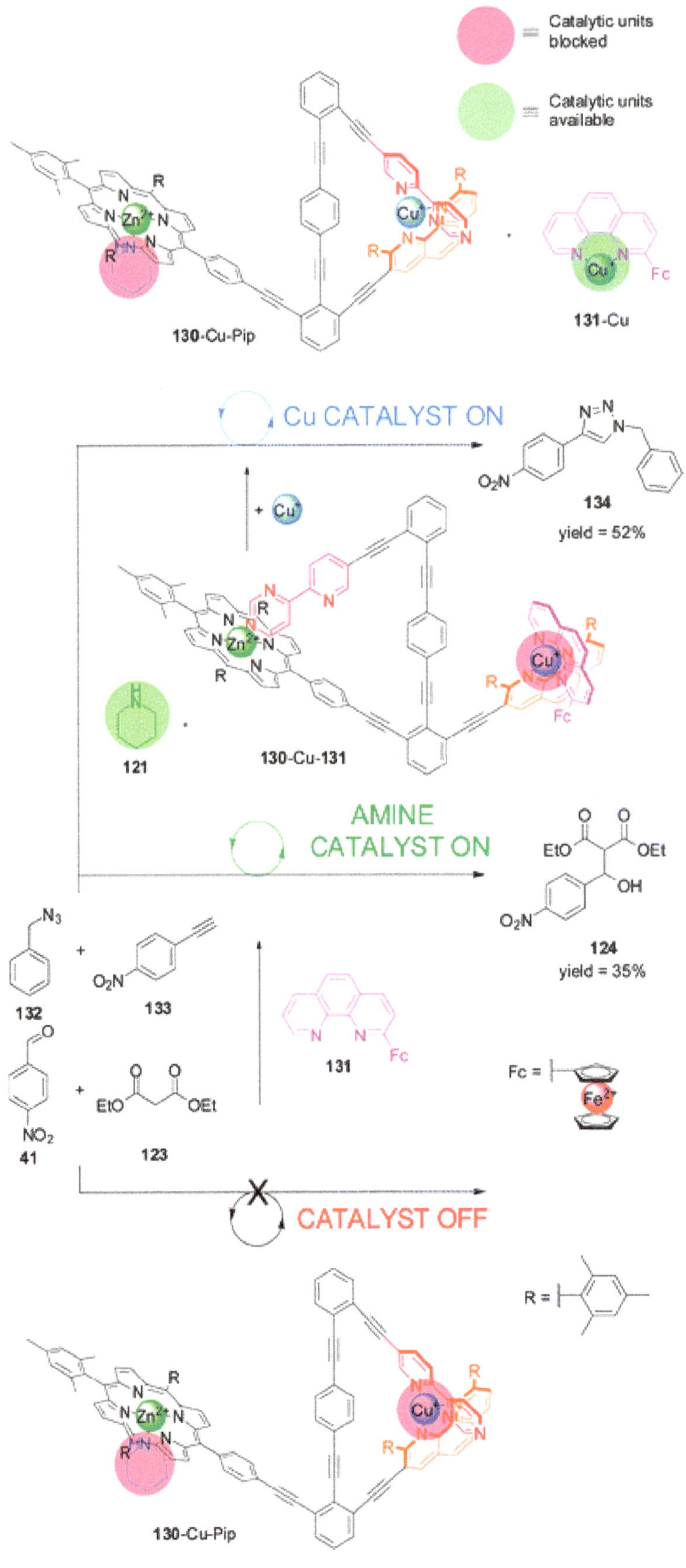

Scheme 36 Switching of catalytic system **130** between an 'off' state and two 'on' states with orthogonal catalytic processes through the sequential addition of phenanthroline derivative **131** and Cu(I).[67]

Aprahamian and co-workers recently reported a switch (**135**) that is able to liberate or extract a proton (the catalytically active species) to and from the reaction media through the reversible coordination of Zn(II) (Scheme 37).[68] The system was used to catalyze imine bond hydrolysis. In the absence of Zn(II) a proton (shown in blue in Scheme 37) is covalently bound to the hydrazone group of the switch (**135**) and cannot participate in catalysis (catalyst 'off'). Upon addition of Zn(II), the system (**135**-Zn) acts as a tridentate ligand that coordinates to the metal centre through the nitrogen atoms of the hydrazone, the quinoline and imidazol units releasing a proton. This was used to catalyse the hydrolysis of anthracen-9-yl-*N*-(4-nitrophenyl)-methanimine (**136**) into *p*-nitroaniline (**137**) and 9-anthraldehyde (**138**), which proceeded to completion (catalyst 'on'). Removal of Zn(II) with CN^- switched the system back to the 'off' state stopping the reaction. Upon addition of more Zn(II) the catalytic activity resumed and full conversion was achieved.

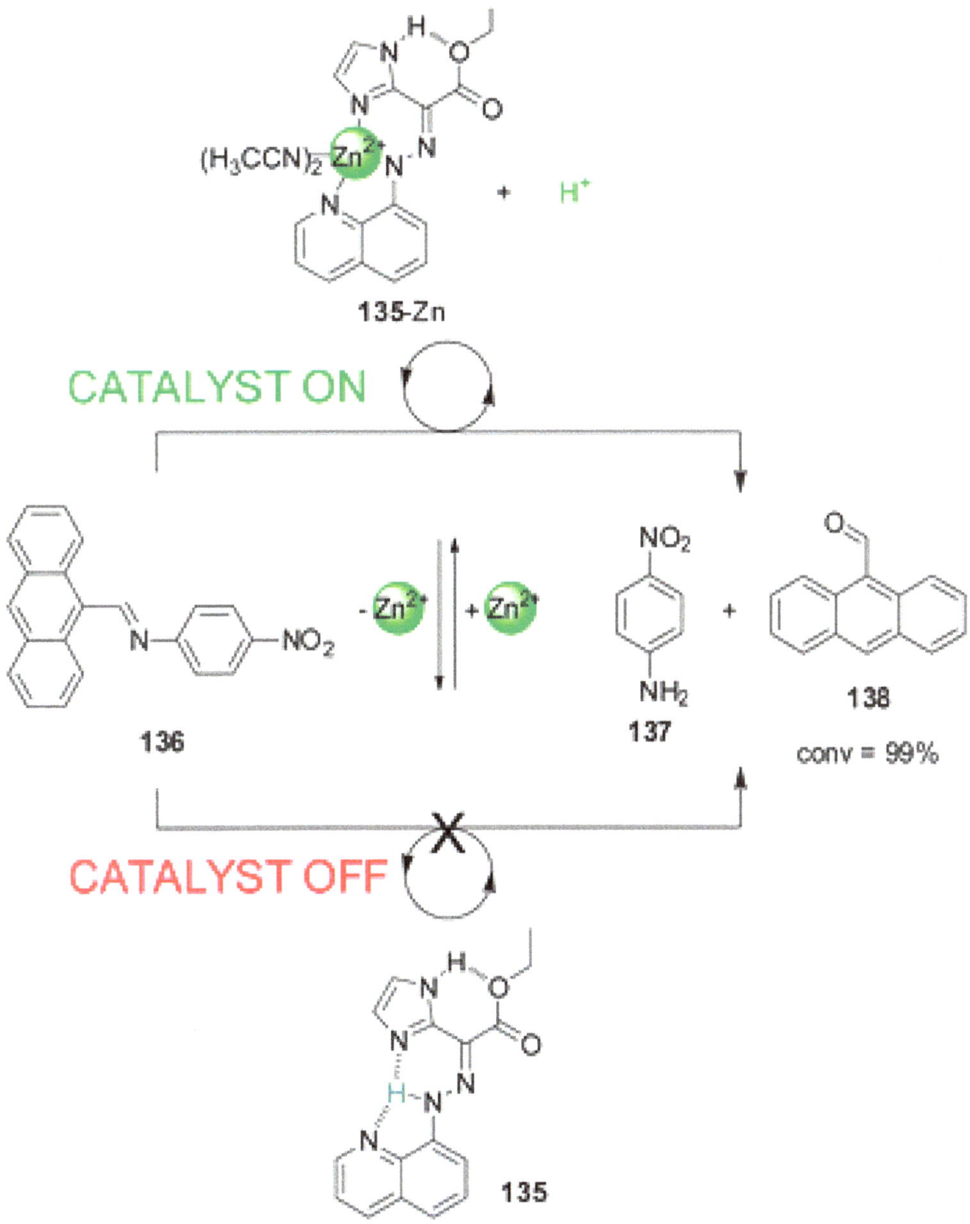

Scheme 37 Controlling the hydrolysis of imine **136** by switchable H^+ capture/release through coordination of **135** to Zn(II).[68]

Redox switching

One of the most explored redox mechanisms to influence the activity of a catalyst is the control of the oxidation state of redox-active ligands, often through incorporating ferrocene or cobaltocene groups into catalysts. In 1995 Wrighton and co-workers, developed a Rh(I) catalyst (**139**) with a 1,1′-bis(diphenylphosphino)cobaltocene (dppc) ligand in which the change of the cobalt oxidation state could be exploited to accelerate or slow down the rate of reactions catalyzed with this system (Scheme 38).[69] Interestingly, the 'on' and 'off' states of the catalyst can correspond to either form of the catalyst depending on the particular reaction. Hydrogenation of cyclohexane proceeds at 3.5 mM min^{-1} in the presence of the reduced form of the catalyst ('on' state), while with the oxidized form (**139**$_{ox}$) only a rate of 0.23 mM min^{-1} is observed. In contrast, the hydrosilylation of acetone was faster in the presence of **139**$_{ox}$ than with **139**$_{red}$. *In situ* switching was achieved by oxidation or reduction of the cobalt complex with 1 equiv. of decamethylferrocenium or cobaltocene.

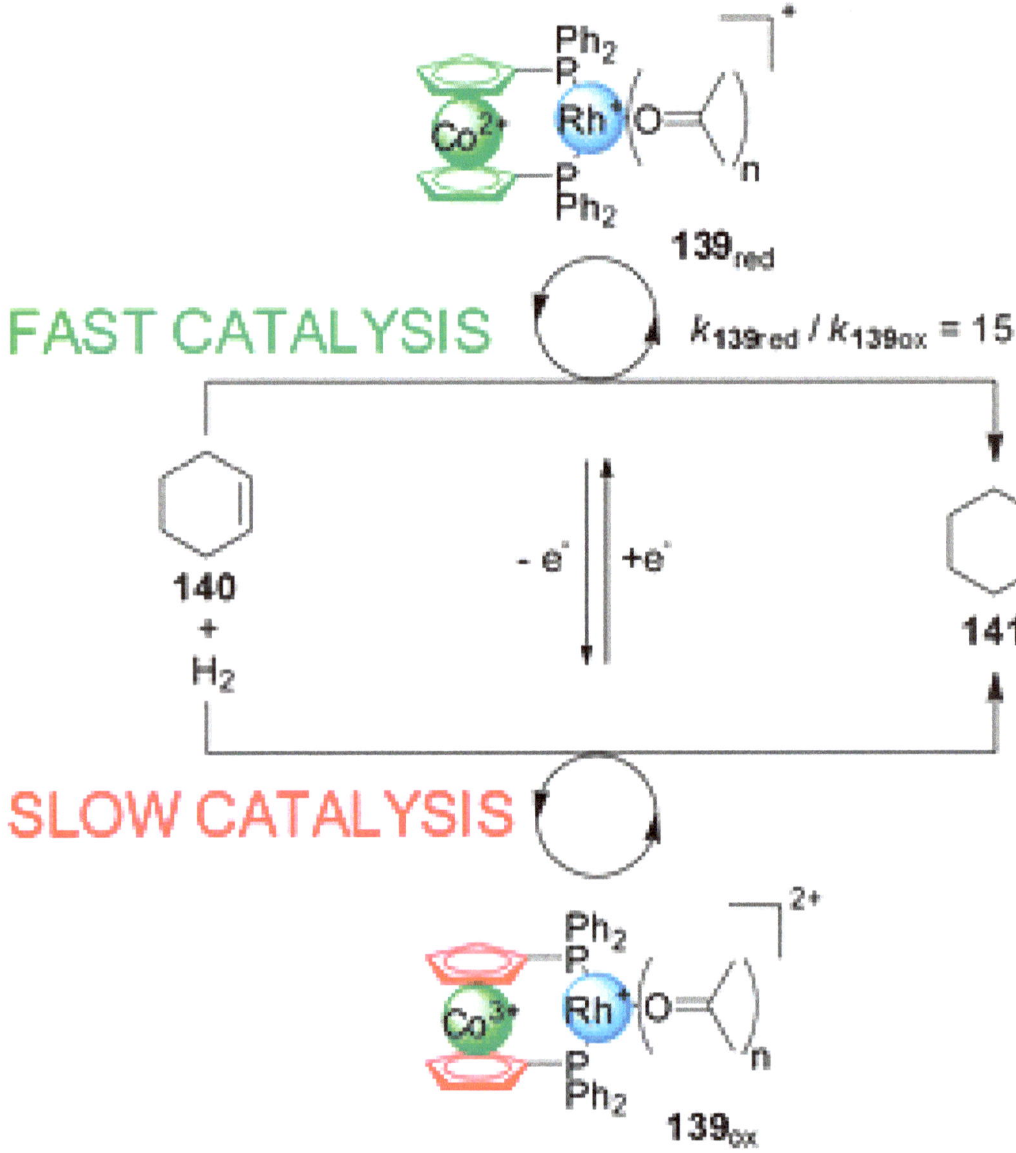

Scheme 38 Co(II)/Co(III) redox control over the activity of a Rh(I) hydrogenation catalyst **139**.[69]

Ferrocene ligands have been used to regulate the activity of metal catalysts in a similar manner. Mirkin and colleagues developed a Rh(I) catalyst with ferrocene–arene ligands that showed switchable activity in the isomerization of allyl groups to the corresponding vinyl derivatives.[70] When the Fe(II) in the ferrocene groups was oxidized to Fe(III), the initial rate of the isomerization reaction was increased 2.2–3.3-fold depending on the substrate-to-catalyst ratio. The authors attributed this difference to the increased η^6-arene ligand lability of the Fe(III) system. However, the enhanced catalytic rate was not maintained throughout the reaction.

Gibson and Long demonstrated that the activity of Ti(IV)–salen catalyst **142** could be modulated by oxidation/reduction of the redox-active ferrocene ligands (Scheme 39).[71] The reduced ferrocene-containing initiator (**142**$_{red}$, 'on' state) promoted the ring-opening polymerization of *rac*-lactide (*rac*-**143**) 30× faster than the oxidized ferrocenium-containing form (**142**$_{ox}$, 'off' state) under the same reaction conditions, resulting in 95% conversion after 70 h. The catalytic activity could be controlled by *in situ* switching of the system: catalyst **142** was switched 'off' by oxidation with AgOTf, markedly slowing the rate of polymerization; subsequent reduction with decamethylferrocene restored the 'on' state.

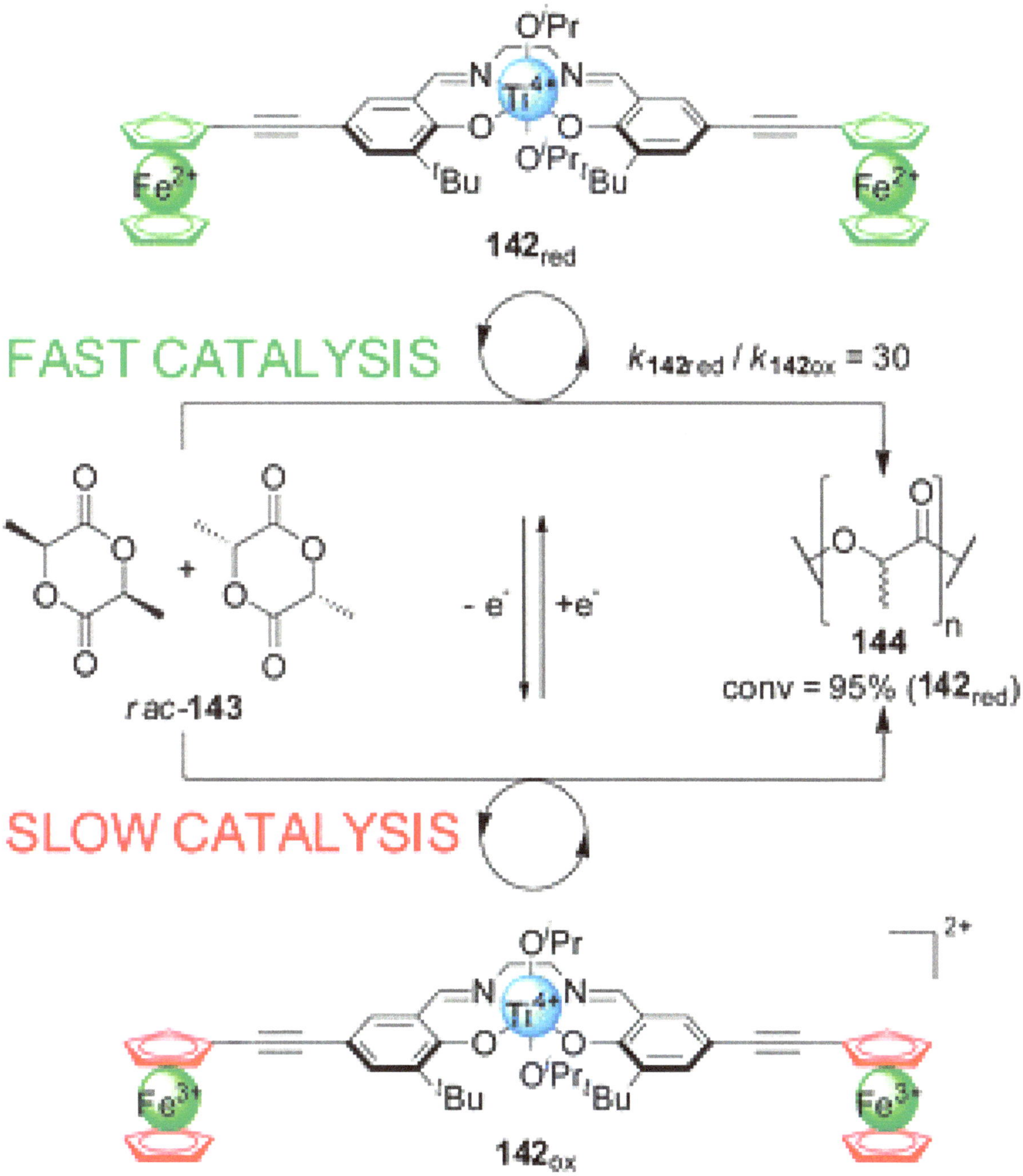

Scheme 39 Influencing the rate of polymerization of *rac*-lactide (**143**) by oxidation/reduction of ferrocene groups pendant to a Ti(IV)–salen catalyst (**142**).[71]

Following a similar strategy, Diaconescu and co-workers developed an Ytrium(III) complex, **145**, that displayed redox-controllable activity (Scheme 40).[72] In a similar way to previous examples in this Section, the reduction or oxidation of the iron centre in a phosphimine ferrocene ligand was exploited to activate or deactivate the metal catalytic centre in the polymerization of L-lactide (L-**143**). The active state of the system, **145**$_{red}$, catalyzed the reaction with 74% conversion after 3 h, while the Fe(III)-oxidized form (**145**$_{ox}$, 'off' state) did not give any conversion to the polymer. *In situ* switching allowed an excellent control over the polymerization rate with a high integrity in the 'off' state and full recovery of catalytic activity upon subsequent reduction. Replacing Y(III) with In(III) changes which of the redox states is active catalytically: the oxidized form of the catalyst gives 49% conversion of trimethylene carbonate into the corresponding polymer, while the reduced form afforded only 2% conversion under the same conditions.

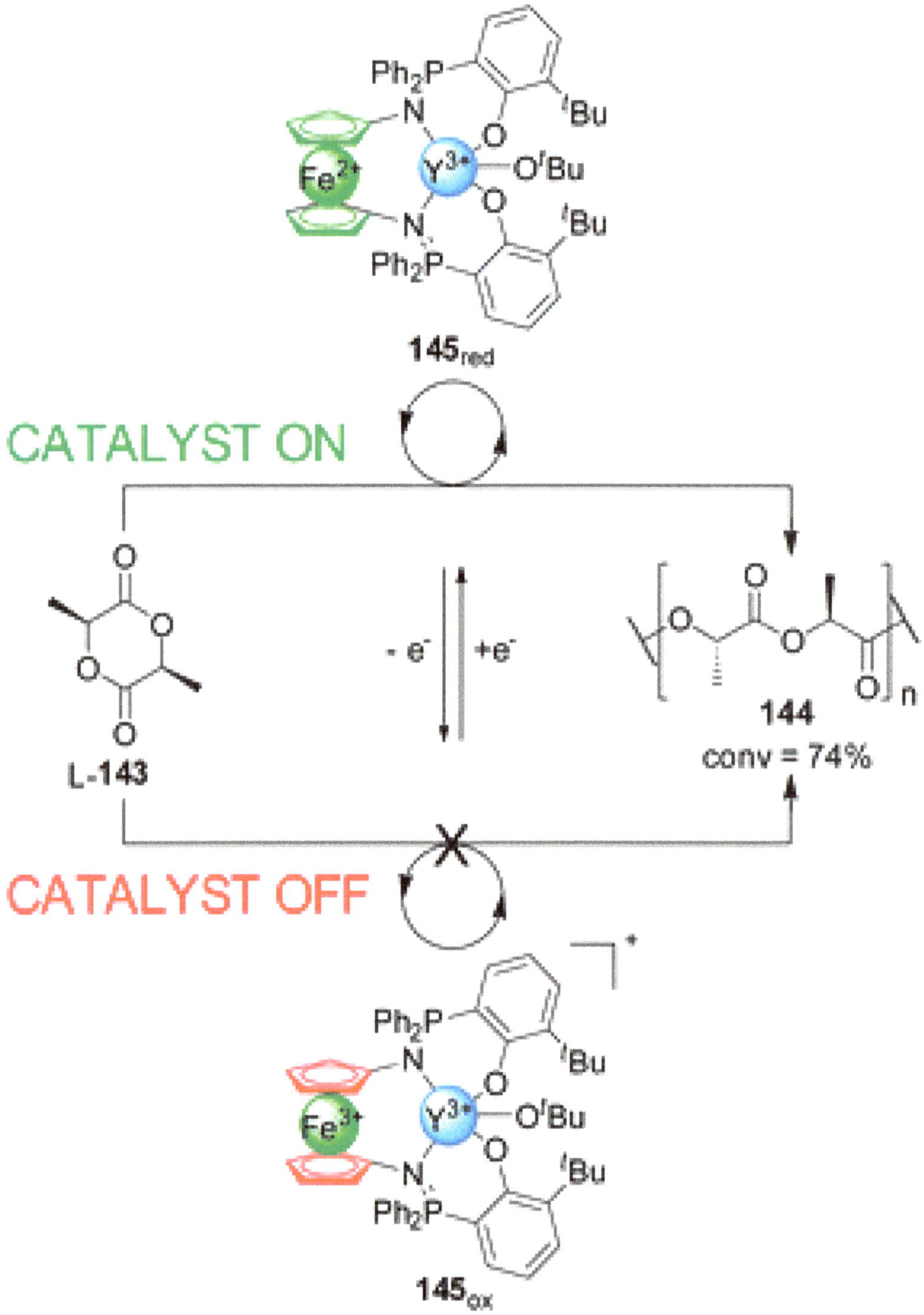

Scheme 40 'On–off' switching of catalyst **145** promoted by reversible ferrocene/ferrocenium redox reactions.[72]

Bielawski and colleagues extended this concept to ferrocene-containing monodentate ligands, in particular ferrocene NHC ligands for Ru(II) olefin metathesis catalysts (Scheme 41).[73]Ru(II) complex **146** was tested in the ring-closing olefin metathesis of diethyl diallylmalonate (**147**). In its reduced form the system catalyzed the reaction with 80% conversion within 1 h. Oxidation of ferrocene to ferrocenium reduced the activity of the catalyst, affording 7% conversion in the same period. *In situ* switching was performed but, although switching 'off' catalysis was effective, subsequent reactivation of the catalyst restored only 13% of the catalytic activity, attributed to some ruthenium oxidation. To overcome this a more robust catalyst incorporating a 1′,2′,3′,4′,5′-pentamethylferrocene-substituted NHC ligand was investigated. This system gave less than 10% of the rate enhancement in its 'on' state as the previous catalyst, but 94% of this activity was recovered after *in situ* switching.

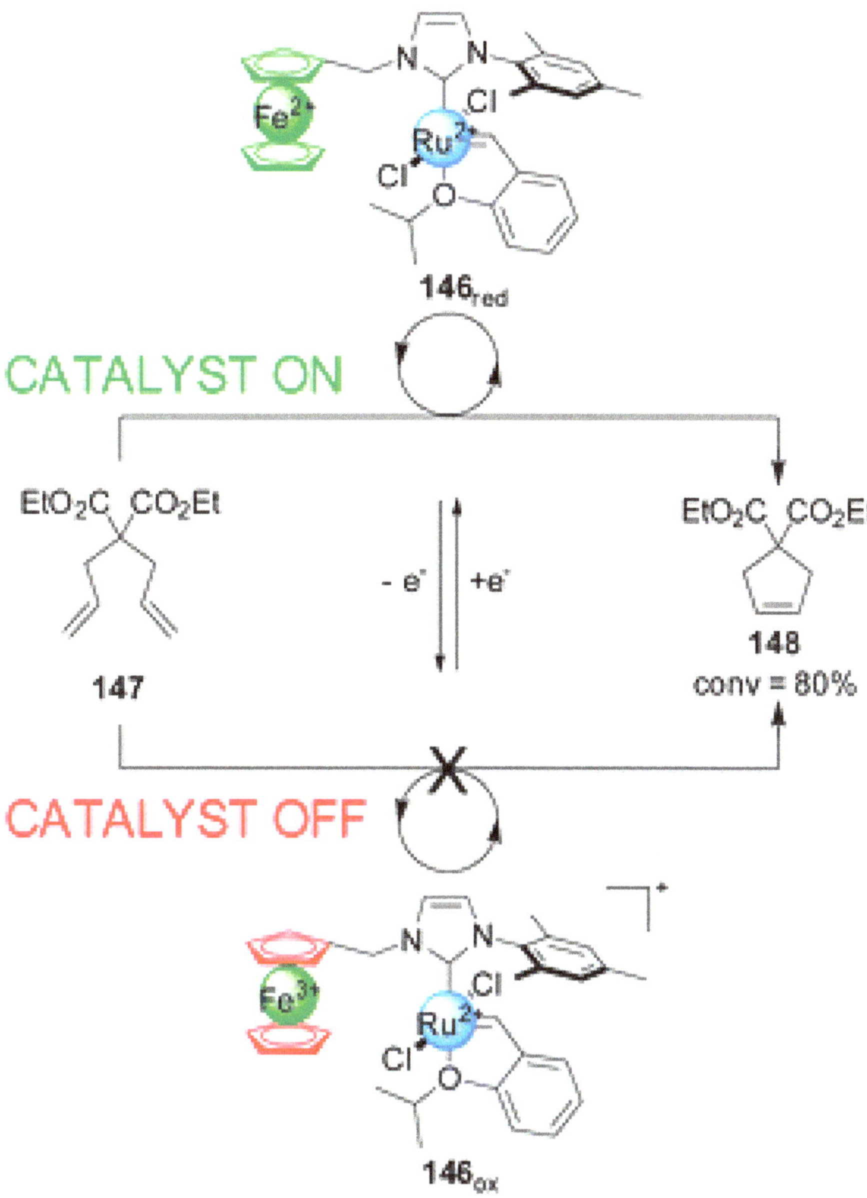

Scheme 41 Control over the activity of metathesis catalyst **146** by change of the oxidation state of a ferrocene iron centre.[73]

The rationale behind the control of reactivity of the metal catalysts discussed in this Section is that the change in the oxidation state of the metallocene results in a change in the donating ability of the ligands, increasing or decreasing the electronic density at the metal centre. However, the redox chemistry of ferrocene has also been exploited to control catalyst activity in other ways. In 2005 Plenio and colleagues fucntionalized a Grubbs–Hoveyda Ru(II) olefin metathesis catalyst with ferrocene groups.[74] The activity of this catalyst could be triggered by oxidation/reduction of the ferrocene, which caused precipitation/solubilisation of the catalyst in an apolar solvent. In the reduced state the catalyst (soluble in toluene) catalysed the ring-closing metathesis of *N*-tosyldiallylamide (catalyst 'on'). Oxidation precipitated the complex, stopping the reaction (catalyst 'off'). Reduction of the ferrocenium groups redissolved the complex restoring the catalytic activity. This behaviour allowed recycling of the catalyst by filtration of the insoluble oxidized form and reuse after reduction.

Hey-Hawkins and co-workers incorporated ferrocene-containing phosphine ligands into a dendritic structure (**149**) to take advantage of two switching mechanisms available for redox-active ferrocene ligands (Scheme 42).[75] The dendrimer comprised twelve ferrocenylphosphane units coordinated to Ru(II) centres to form (η^6-arene)RuCl$_2$(L) complexes able to catalyze the isomerization of allyl alcohols to ketones. The reduced form of the catalyst (**149**$_{red}$, 'on' state) efficiently catalyzed the isomerization of 1-octen-3-ol (**150**) (full conversion within 40 min). Interestingly, the observed activity with the dendrimer catalyst was higher than that of a monomeric ferrocene-containing Ru(II) complex at the same metal loading. When the dendritic catalyst was oxidized (**149**$_{ox}$, 'off' state) its activity dropped by almost an order of magnitude. The lower activity of the oxidised form was attributed to a combination of electronic and solubility effects. Reduction of the oxidised form restored the system to the 'on' state, recovering the catalytic activity.

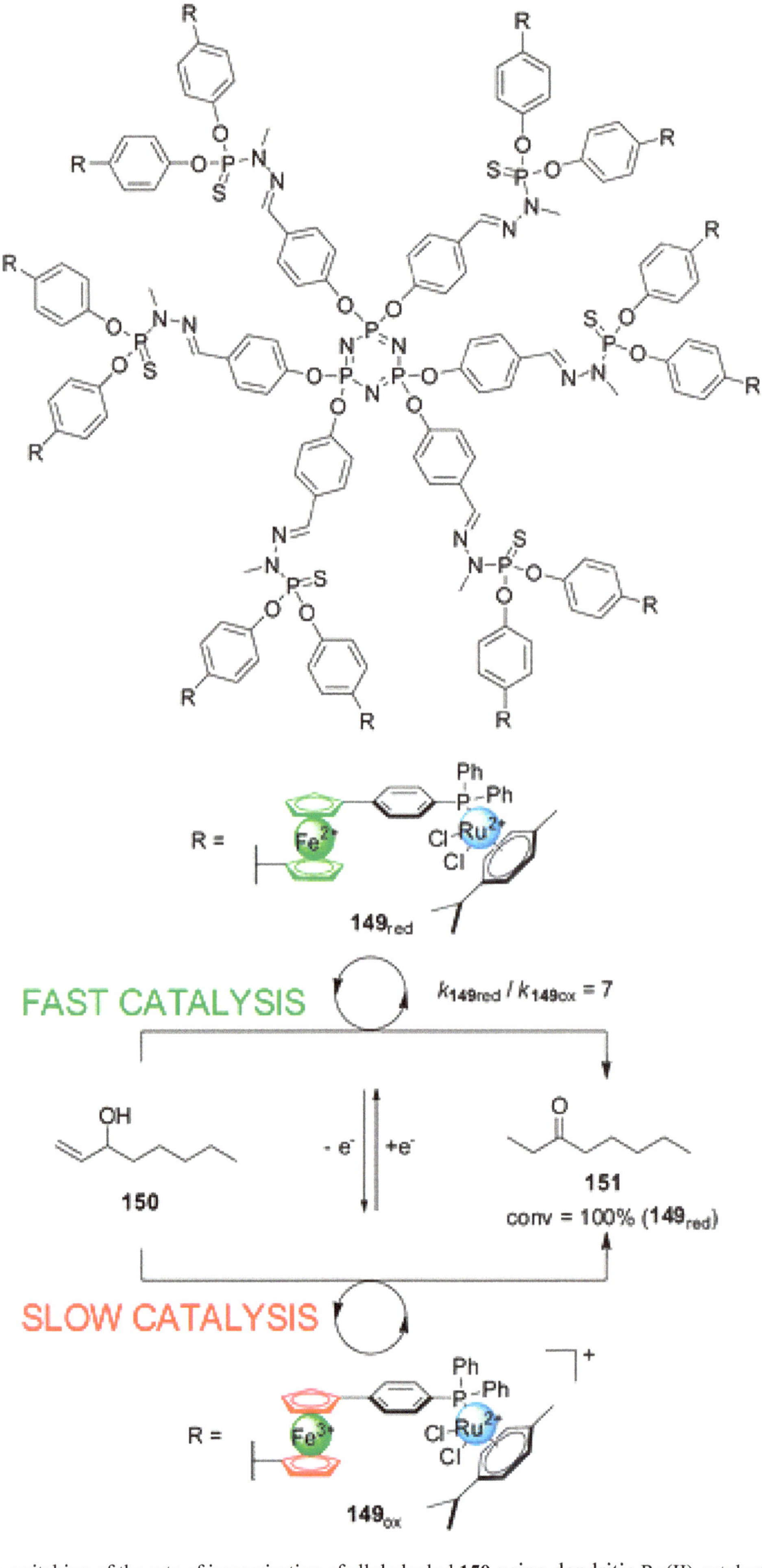

Scheme 42 Redox-switching of the rate of isomerization of allyl alcohol **150** using dendritic Ru(II)-catalyst **149**.[75]

Although the change in oxidation state of pendant metallocene ligands is the most explored mechanism for redox-switchable catalysts, control over the activity of a metal catalyst can also be achieved by a direct change in the oxidation state of the catalytic metal centre. In 2011, Matyjaszewski and co-workers described a remarkable Cu(I)/Cu(II) system for atom-transfer radical polymerization that can be controlled by electrochemical switching of the oxidation state of the copper centre (Scheme 43).[76] The catalyst (or activator, 'on' state) Cu(I)BrMe$_6$TREN (Me$_6$TREN = tris[2-(dimethylamino)ethyl]amine) reacts with alkyl bromides to form Cu(II)Br$_2$Me$_6$TREN (a deactivator, 'off' state) and a radical species that undergoes monomer addition giving rise to a polymerization reaction. However, the different species are in equilibrium as the Cu(II)Br$_2$Me$_6$TREN can be reduced by radicals reforming Cu(I)BrMe$_6$TREN and the alkyl bromide. In the absence of other factors, the reduction is the fastest process and the equilibrium favours the starting materials and so polymerization, which depends on the Cu(I)/Cu(II) ratio, is not observed. However, the equilibrium can be influenced electrochemically. By applying a potential of −0.69 V (*vs.* Ag/Ag$^+$), Cu(II)Br$_2$Me$_6$TREN formed during the reaction is reduced to Cu(I)BrMe$_6$TREN which can react again leading to a 80% conversion of methyl acrylate over 2 h. With a more positive potential (E = −0.40 V), the equilibrium is shifted towards the Cu(II) complex, which reduces the number of radicals, preventing further polymerization. The system could be switched *in situ* by changing the applied potentials. Varying the applied potential between E = −0.69 V and E = −0.40 V the polymerization reaction could be switched 'on' or 'off', respectively, as the Cu(I)/Cu(II) equilibrium is shifted towards either Cu(I) or Cu(II) species.

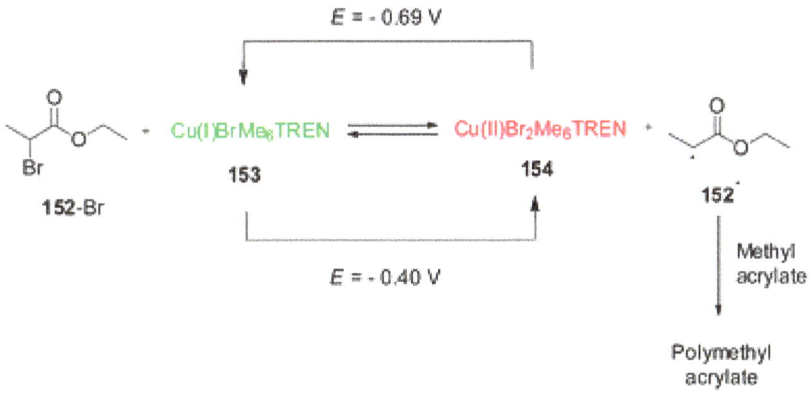

Scheme 43 Electrochemical control over the activation of a polymerization reaction by reduction/oxidation of Cu(I)/Cu(II).[76]

Diaconescu and co-workers have reported a switchable polymerization catalyst **155** based on a Ce(III)–salen complex (Scheme 44).[77] Oxidation or reduction of the cerium centre was used to activate or deactivate the catalyst. The Ce(III) complex was catalytically active in the ring-opening polymerization of L-lactide (L-**143**), resulting in 85% conversion over 2 h. In contrast, the BArF salt of the oxidized Ce(IV) complex showed no polymerization activity. The system could be switched *in situ*. After some polymer had formed, the Ce(III) complex was treated with ferrocenium BArF, forming the corresponding Ce(IV) complex and stopping further polymerization. The active state could be restored by reduction with cobaltocene, with no significant loss of catalytic activity.

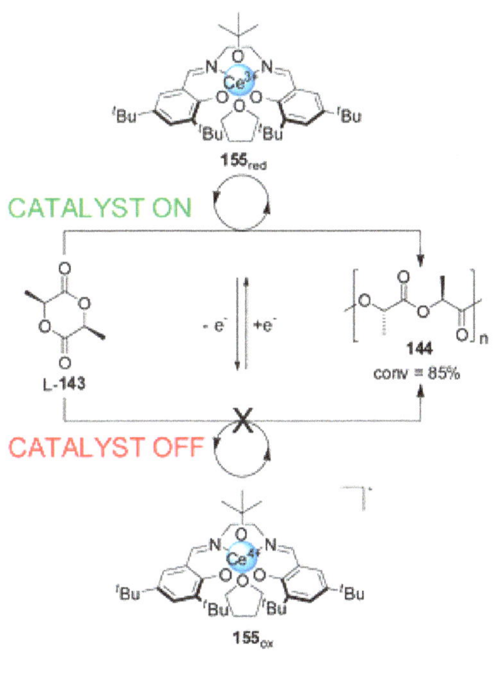

Scheme 44 'On–off' switching of lactide polymerization catalyst **155** triggered by reduction/oxidation of the cerium catalytic centre.[77]

Another switchable catalyst based on the oxidation/reduction of a cerium catalytic centre was reported by Okuda and colleagues.[78] The system comprises two tetradentate OSSO ligands coordinated to cerium. Here again, the Ce(III) complex was active (catalyst 'on') in the polymerization of *meso*-lactide, affording 78% conversion after 7 h, while the Ce(IV) system showed much lower activity (catalyst 'off') giving only 10% conversion under the same conditions.

Byers and co-workers developed a readily accessible bis(imino)pyridine iron complex (**156**) which was found to catalyze the polymerization of (*rac*)-lactide depending on the oxidation state of the iron centre (Scheme 45).[79] While the Fe(II) (**156**$_{red}$, catalyst 'on') complex catalyzed the formation of the polylactide (**144**) effectively (95% conversion over 24 h), the Fe(III) complex (**156**$_{ox}$) did not show any catalytic activity under the same reaction conditions (catalyst 'off'). The lack of activity of **156**$_{ox}$ was attributed to the electron-poor character of the complex. Switching of the catalytic activity could be performed *in situ*. After 40 min of polymerization in the presence of **156**$_{red}$, the Fe(II) centre was oxidized with ferrocenium hexafluorophosphate forming **156**$_{ox}$. The polymerization did not proceed further until cobaltocene was added to restore the active form of the catalyst.

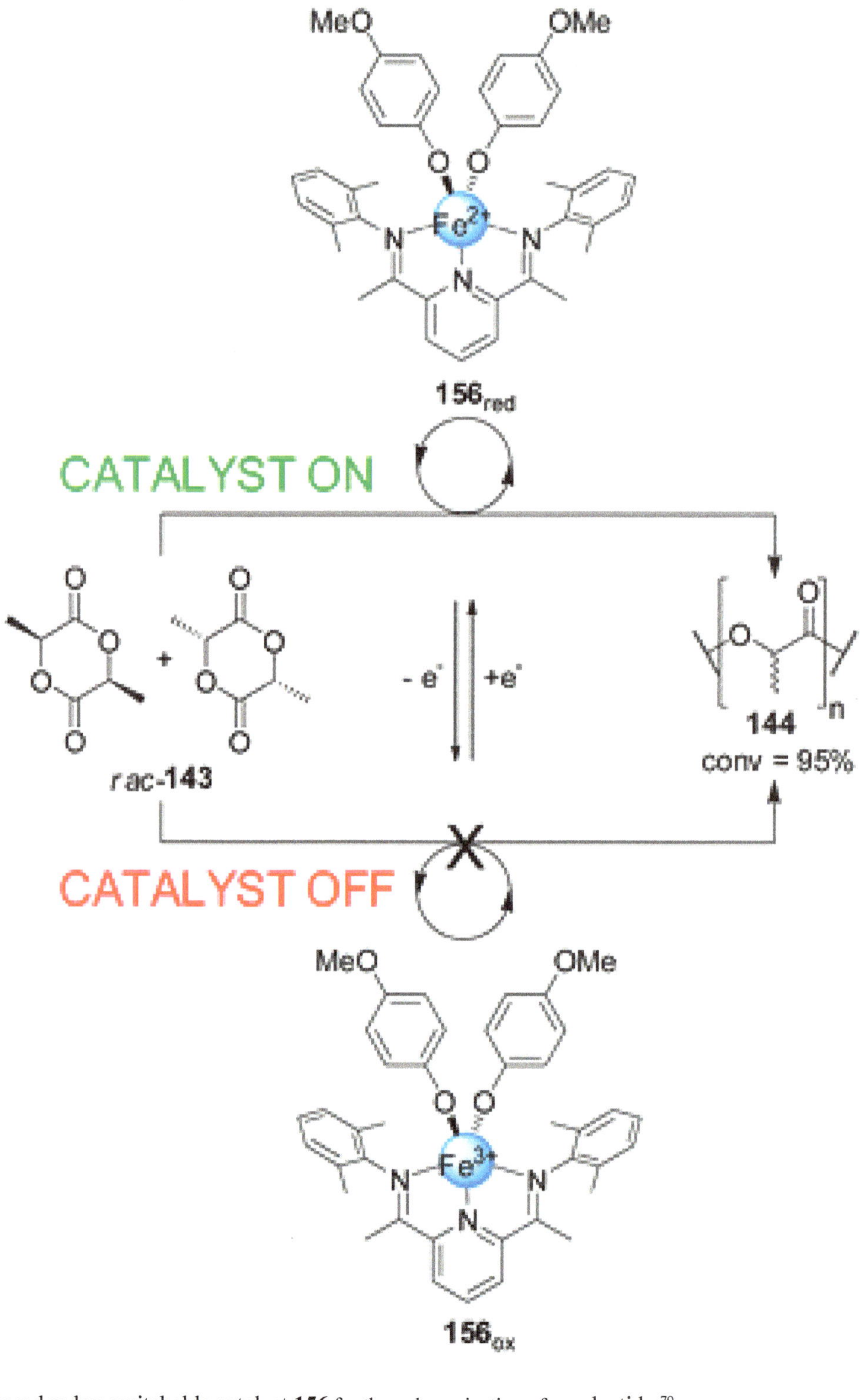

Scheme 45 Iron-based redox-switchable catalyst **156** for the polymerization of *rac*-lactide.[79]

A similar system was reported by Lang and co-workers, in which a change in the oxidation state of an iron centre bound to 2,6-di(1*H*-pyrazol-3-yl)pyridine and chloride ligands was used to switch 'on' or 'off' the catalytic activity of the complex in the ring-opening polymerization of ε-caprolactone.[80] While the Fe(III) complex can catalyse the polymerization in >90% yield in the presence of an *tert*-butanol, the reduced form of the catalyst containing an Fe(II) centre, proved to be inactive under the same reaction conditions. *In situ* switching of the catalyst could be achieved by addition of cobaltocene and ferrocenium hexafluorophosphate.

The change in oxidation state of a metal ion can also be used to control selectivity. In 2012, Canary and colleagues reported bisurea catalyst **157** incorporating a ligand motif derived from L-methionine (Scheme 46).[81] A multidentate pocket is formed that is able to complex either Cu(I) or Cu(II). The different coordination geometries of the different oxidation states induces chirality of different handedness for the two structures (**157**$_{red}$ and **157**$_{ox}$). The inversion in helicity is transferred through to the spatial arrangements of the pendant urea groups and their selectivity as organocatalysts for the asymmetric Michael addition of diethyl malonate (**123**) to *trans*-β-nitrostyrene (**81**). With the copper centre in its oxidized state (Cu(II)), **157**$_{ox}$catalyzed the formation of (*S*)-**158** in 55% yield (75% conversion) and 72% ee. In contrast complex **157**$_{red}$ afforded (*R*)-**158** in 40% yield and 70% ee. However, a definitive explanation for the helix inversion and its influence on the handedness of the products of catalysis remains elusive.

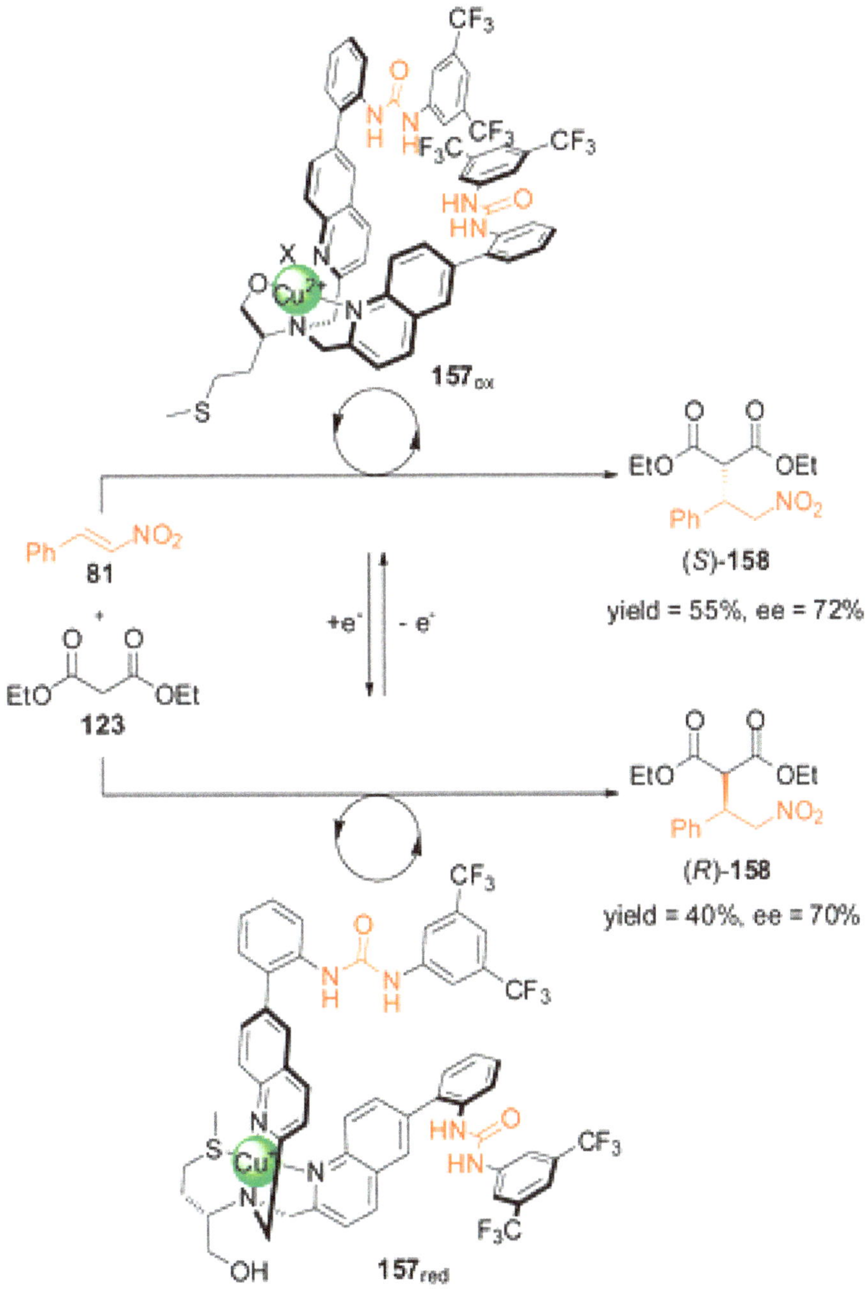

Scheme 46 Control of the stereoselectivity of the Michael addition of diethyl malonate to β-nitrostyrene promoted by redox-switchable catalyst **157**.[81]

Control over the particular substrate reacting from a mixture of building blocks has also been achieved using redox switching. Diaconescu and co-workers reported a series of Zr(IV) and Ti(IV) complexes bearing a ferrocene alkoxide ligand (Scheme 47).[82] These systems catalyse the polymerization of L-lactide (L-**143**) and ε-caprolactone (**104**) with a monomer preference dependent on the oxidation state of the ferrocene unit. The reduced form of the catalysts are selective for the polymerization of L-lactide (L-**143**), with excellent conversions (82% with $\mathbf{159}_{red}$) and only minor incorporation of the caprolactone (**105**) (<5% with $\mathbf{159}_{red}$). In contrast, the ferrocenium-containing oxidized catalysts preferentially promoted ε-caprolactone polymerization, affording high conversions to **105** (the lowest being 48% with $\mathbf{159}_{ox}$) and, in general, small amounts of incorporation of L-**143** (<5% with $\mathbf{159}_{ox}$). Each of the catalysts could be switched *in situ* when only one of the monomers (**104** or L-**143**) was present, but only the Ti(IV) complex **159** proved able to switch control of the polymerization reactions when both substrates were present (Scheme 47). By switching the oxidation state of the iron centre, the one-pot copolymerization of both monomers was achieved. Starting with $\mathbf{159}_{red}$, L-**143** was polymerized with 58% conversion after 36 h with minor incorporation of **104**. Then the selectivity of the catalyst was switched by oxidation, subsequently leading to polymerization of ε-caprolactone (18% conversion after 2 h) and affording a block copolymer poly-(**143**-minor-**104**)-(**104**-minor-**143**).

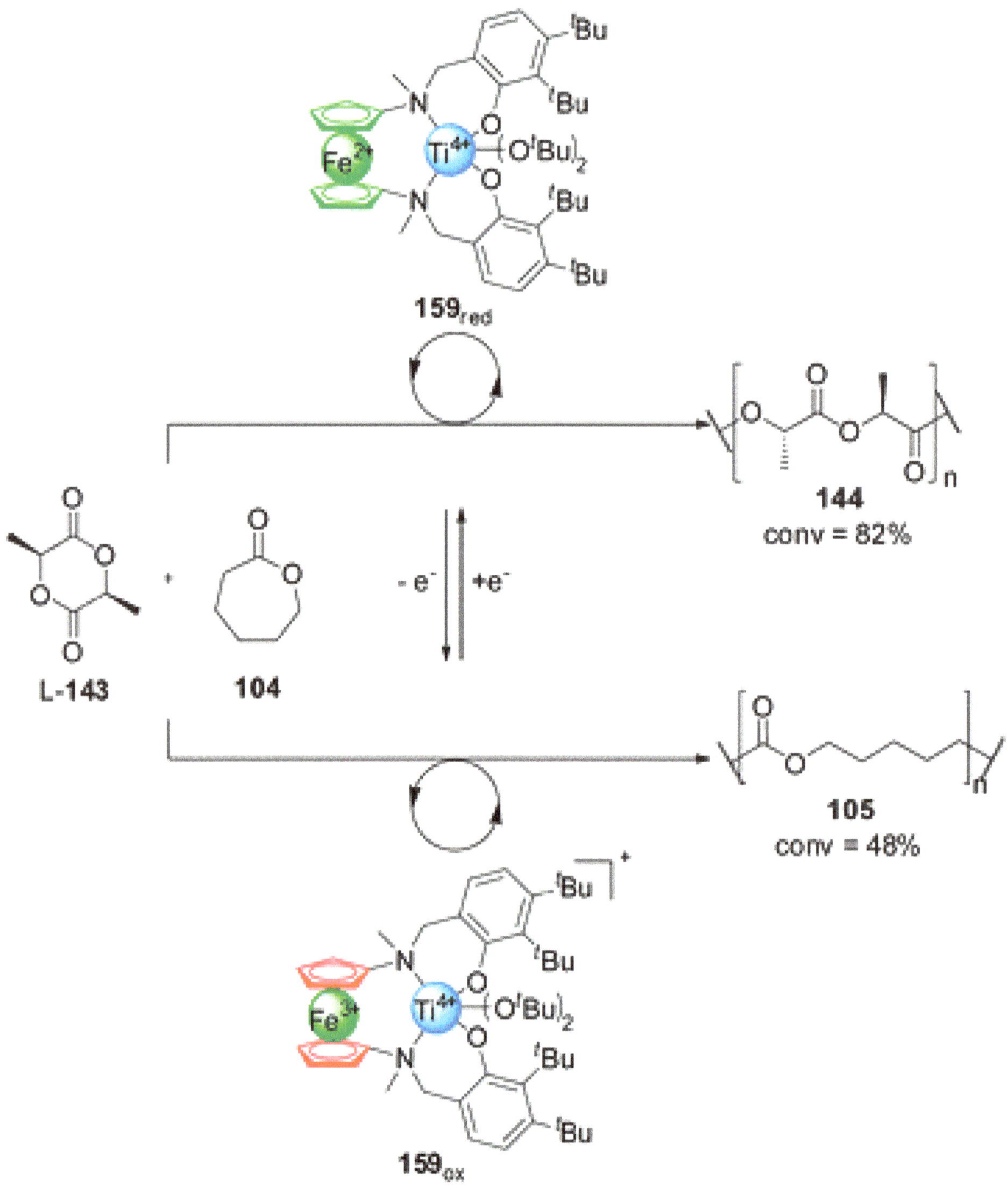

Scheme 47 Switching of monomer selectivity by polymerization catalyst **159**, triggered by oxidation/reduction of the pendant ferrocene group.[82]

Wu and co-workers developed a primary amine organocatalyst functionalized with a ferrocene unit whose activity and diastereo- and enantioselectivities in the aldol reaction between cyclohexenone and *p*-nitrobenzaldehyde depends on the oxidation state of the ferrocene. Using the reduced form of the catalyst, the reaction proceeded with low yields and poor enantio- and diastereoselectivity. However, with the oxidised Fe(III) ferrocenium moiety the catalyst afforded excellent yields (up to 95%), enantio-(up to 94% ee) and diastereoselectivities (up to 96 : 4 dr). The origin of the stereocontrol was attributed to the ability of ferrocenium to act as a Lewis acid, coordinating to the aldehyde and providing a well-expressed chiral environment for the substrate within the complex. The scope of the catalyst was extended to a range of different aldol reactions.[83]

Although most efforts on redox-triggered catalysts have focused on metal centres, reduction/oxidation of organic groups has also been used to modulate the activity of switchable catalysts:

Fujiwara and colleagues prepared an Al-functionalized mesoporous material (**160**) with a catalytically-active inner cavity able to promote the dimerization of α-methylstyrene (**161**) (Scheme 48).[84] Access to the cavity could be blocked by oxidation of thiol groups placed at the pore openings (**160**$_{ox}$; catalyst 'off'). When the disulfide groups were reduced *in situ*(**160**$_{red}$), the catalytic sites became accessible (catalyst 'on') and product **162** was formed in up to 97% yield. The material could be reused with no loss of catalytic activity.

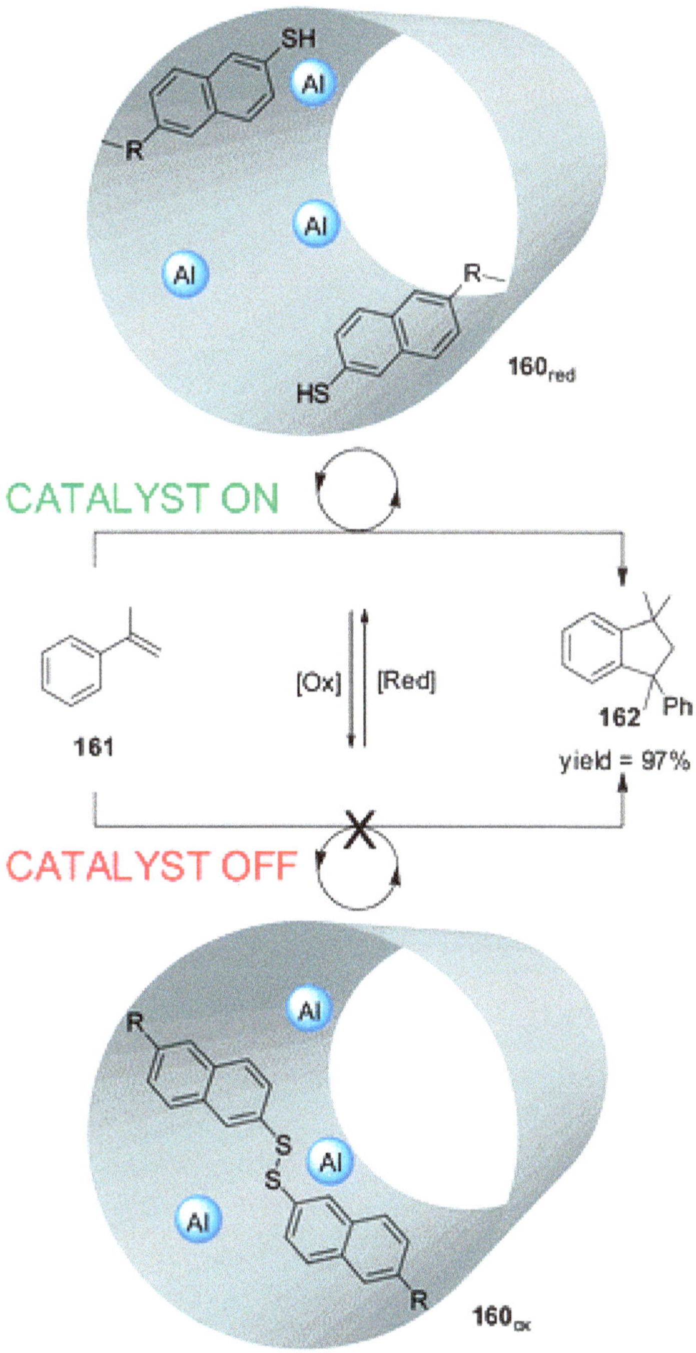

Scheme 48 Control over the activity of heterogeneous catalyst **160** through the reversible redox cleavage or formation of disulfide bonds.[84]

In 2010, Bielawski *et al.* reported a Ni(II) complex comprising two NHC ligands bearing a redox-active naphthoquinone motif.[85] With the ligand in the oxidised form (catalyst 'on') the complex efficiently catalyzed the Kumada reaction between phenylmagnesium chloride and *p*-bromoarenes, but showed lower activity in reduced form (catalyst 'off'). The rationale for this behaviour is that after oxidative-addition of the bromoarene to the Ni(0) complex (formed by reduction of the Ni(II) catalyst with the Grignard reagent), that complex could be reduced with cobaltocene, affording an electron-rich complex that undergoes the transmetallation and reductive-elimination steps at significantly slower rates.

SWITCHING DRIVEN BY MECHANICAL FORCES

Recently efforts have been made to employ mechanical force as a switching mechanism for catalysis.[86] Mechanical forces are usually applied in solution through the use of ultrasound, which induces hydrodynamic force fields around collapsing cavitation bubbles. Due to the high strain rates, polymer chains uncoil and stretch under these conditions which can cause the weakest bonds in the chain to rupture. A limitation in using this stimulus to control catalytic activity is that only sufficiently long polymer chains can be affected in this way.

Sijbesma envisaged that metal coordination bonds of complexes with polymer-functionalized ligands could be selectively dissociated with ultrasound leading to catalytically-active ligands or metal complexes.[86] In 2008 they reported a polytetrahydrofuran (pTHF) polymer with an embedded bis-N-heterocyclic carbene–silver(I) complex (Ag-(**163**)$_2$, Scheme 49, top) where the selective scission of the Ag(I)–NHC coordination bond could be achieved by sonication, exposing the active NHC–silver(I) complex **163** (catalyst 'on').[87] The system, having two pTHF ligands of ~5 kDa, was able to catalyze the transesterification of benzyl alcohol with vinyl acetate with 65% conversion over 60 min. Without applying ultrasound, the metal–ligand bond is not dissociated significantly and the metal catalyst is concealed (catalyst 'off') giving than 3% conversion under the same reaction conditions.[88] To demonstrate the generality of the approach, the authors also used a Ru(II)–benzylidene complex with NHC ligands functionalized with pTHF chains (Ru-(**163**)$_2$, Scheme 49, bottom).[88] These complexes were used to promote the ring closing olefin metathesis (RCM) of diethyl diallylmalonate. Without sonication the complexes with pTHF groups of different molecular weight did not show any catalytic activity, while the reaction proceeded efficiently when the system was sonicated. *In situ* switching of the catalyst was achieved by turning the sonication 'on' and 'off', showing no increase in the degree of polymerization when ultrasound was not applied. However, the authors pointed that the activation of the catalyst is actually irreversible as the benzylidene group does not coordinate back to the Ru(II) centre to restore the initial species when sonication is ceased. Quick decomposition of the active catalysts was proposed to explain the lack of activity during the 'off' period.

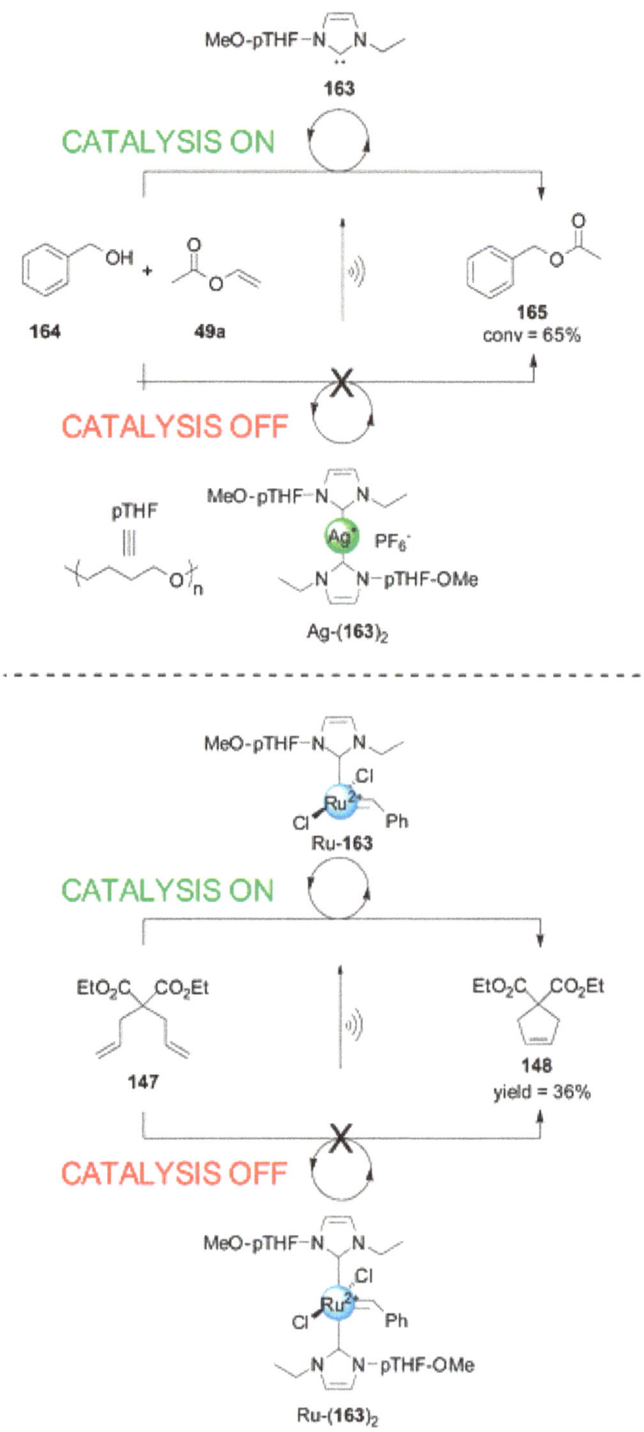

Scheme 49 Polymer-functionalized catalysts activated by ultrasound-induced breaking of metal–ligand coordination bonds.[87,88]

In both these systems the mechanical activation pathway was consistent with the higher conversions obtained using higher molecular weight polymer chains on the ligands, and through the lack of catalytic activity of similar complexes with non-polymeric ligands. Further studies with polymers containing the NHC–silver(I) complex confirmed these results.[89] Direct correlation between the conversion achieved and the concentration of the catalyst was explained by the catalytic species being deactivated by radicals thermally generated during sonication. Suppression of these species, *e.g.* by performing the reaction under a methane rather than argon atmosphere, increased the catalytic activity of both the Ag(I)- and Ru(II)-based catalysts.[90]The activity of the NHC–Ru(II) containing catalyst in ring-opening olefin metathesis polymerization (ROMP) reactions was not as clear-cut as with RCM. Although the catalysed ROMP of cyclooctene stopped when sonication was ceased under some conditions, when the ROMP was performed under optimized reaction conditions the polymerization continued for several hours after sonication ceased due to the increased stability of the active species.[91] Although useful for the application of these systems in self-healing materials,[92] this may hinder its use in other area of switchable catalysis.

SWITCHING DRIVEN BY CHANGES IN REACTION CONDITIONS

In Sections 1–6 of this Review we documented how control over the rate or selectivity of catalysis has been achieved with catalysts whose structure or reactivity is changed by a stimulus. Recently, significant changes in the chemo- or stereo-selectivity of catalyzed reactions brought about by altering reaction conditions have been reported. In this Section we describe some of the more remarkable examples of this phenomena and briefly highlight stereodivergent synthesis modulated by changes in experimental parameters that do not involve altering the chirality of the catalyst itself.[93]

Jessop and co-workers found that a 1 : 1 mixture of a strong base, such as DBU (1,8-diazobicyclo[5.4.0]undadec-7-ene), and a primary alcohol, adds to CO_2 to generate an ionic iminium carbonate, which upon purging with N_2, reverts back to its non-ionic form.[94] In 2014 Coulembier and colleagues exploited this process in developing a CO_2-switchable ROP reaction (Scheme 50).[95] The authors used DBU, TBD (triazabicyclodecene) and primary alcohol **164** (10 : 1 : 10) as the catalytic system, which is known to promote the ROP of cyclic esters such as **104**. Under a N_2 atmosphere ('on' state), the ROP polymerization of **104** to form **166** proceeded in excellent yield (96%), while simple exposure to CO_2 formed DBUH$^+$-**165** and completely turned 'off' the catalytic activity. The authors demonstrated that the catalytic activity could be switched 'on' and 'off' multiple times by alternating exposure to N_2 or CO_2, providing the ability to regulate the molar mass of the resulting polymer.

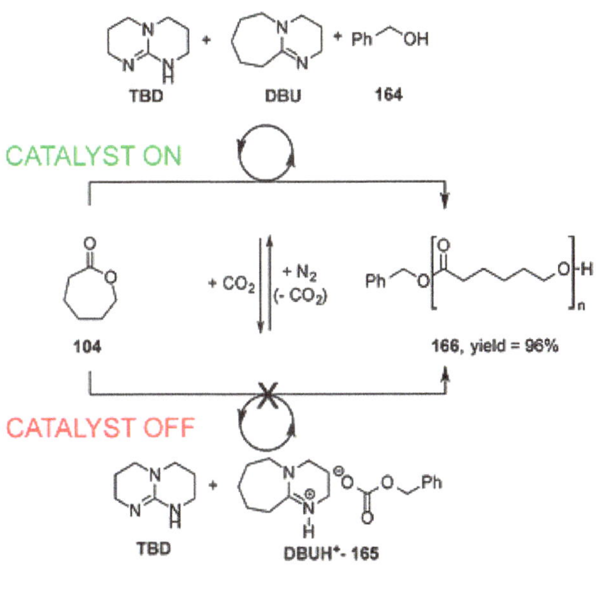

Scheme 50 CO_2-switchable catalytic system for the ROP of caprolactone (**104**).[95]

Williams and co-workers have recently developed a CO_2-switchable metal catalyst, which is able to chemoselectively control both the ring opening polymerization of lactones and the ring-opening copolymerization of epoxides by using a N_2 or CO_2 atmosphere, respectively.[96] By altering between these conditions—in a one pot procedure—the authors achieved the selective preparation of poly(carbonate-*block*-ester) copolymers.

In 2010, Sohtome and Nagasawa described an enantiodivergent organocatalyzed Mannich-type reaction using a conformationally flexible guanidine/bisthiourea catalyst (**169**) (Scheme 51).[97] When the reaction between aromatic *N*-Boc-imines (**167a–i**) and dimethylmalonate (**168**) was carried out in nonpolar solvents, such as *m*-xylene, excellent yields and enantiocontrol (99% yield, 97% ee) in favour of the (*S*)-enantiomer were achieved. In contrast, when the reaction was performed using polar aprotic solvents, such as acetonitrile, the (*R*)-enantiomer was formed ((*R*)-**170**) in similar yield and enantioselectivity. Mechanistic studies suggested that the solvent-dependent stereodiscrimination exhibited by catalyst **169** is not controlled by aggregation effects or functional group masking by the solvent. Rather, it is likely a consequence of compensating differences in the enthalpies and entropies of activation. The *S*-stereoselectivity is controlled by differences in the entropies of activation, whereas the *R*-stereoselectivity arises from differences in the enthalpies of activation.

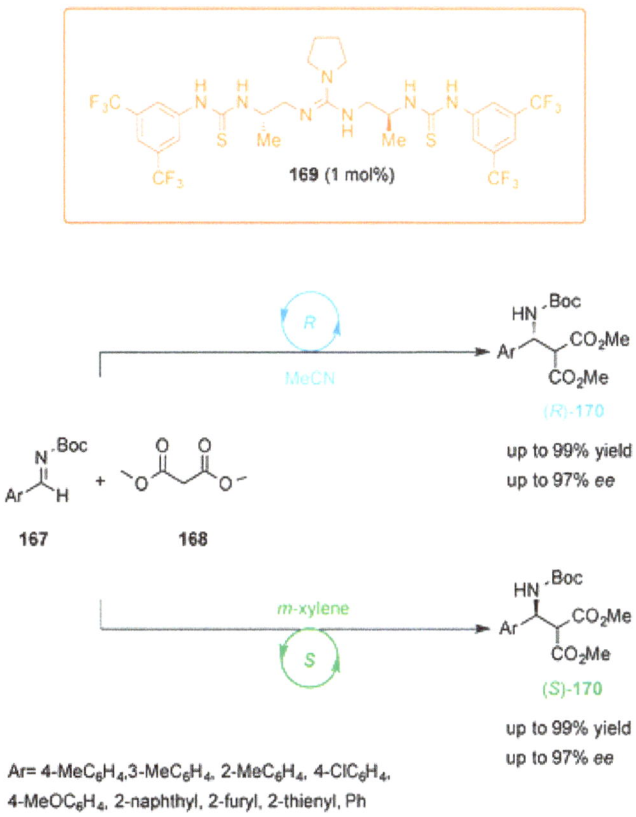

Scheme 51 A solvent-dependant enantiodivergent Mannich-type reaction using a conformationally flexible guanidine/bisthiourea catalyst **169**.[97]

In another impressive example, Melchiorre and co-workers described the diastereodivergent Michael addition of alkyl thiols (**171**) to α,β-unsaturated ketones (**172**) catalyzed by cinchona alkaloid derivative **173**, in which the sense of the diastereoselectivity could be reversed by the choice of the solvent and the acid additive employed (Scheme 52).[98] Using 2-fluorobenzoic acid in $CHCl_3$, the *syn*-Michael products (*syn*-**174**) are obtained in high yield and enantioselectivity. However, using diphenyl phosphate (DPP) as the additive in acetone, led to the formation of the *anti*-Michael adducts (*anti*-**174**) in comparable yield and enantioselectivity. The switching in diastereoselectivity is likely a consequence of a change in the structure of the iminium intermediate caused by the counter-anion (which comes from the acid additive) and the solvent polarity. *In situ* switching in the diastereoselectivity from *syn*-**174** to *anti*-**174** was also demonstrated.

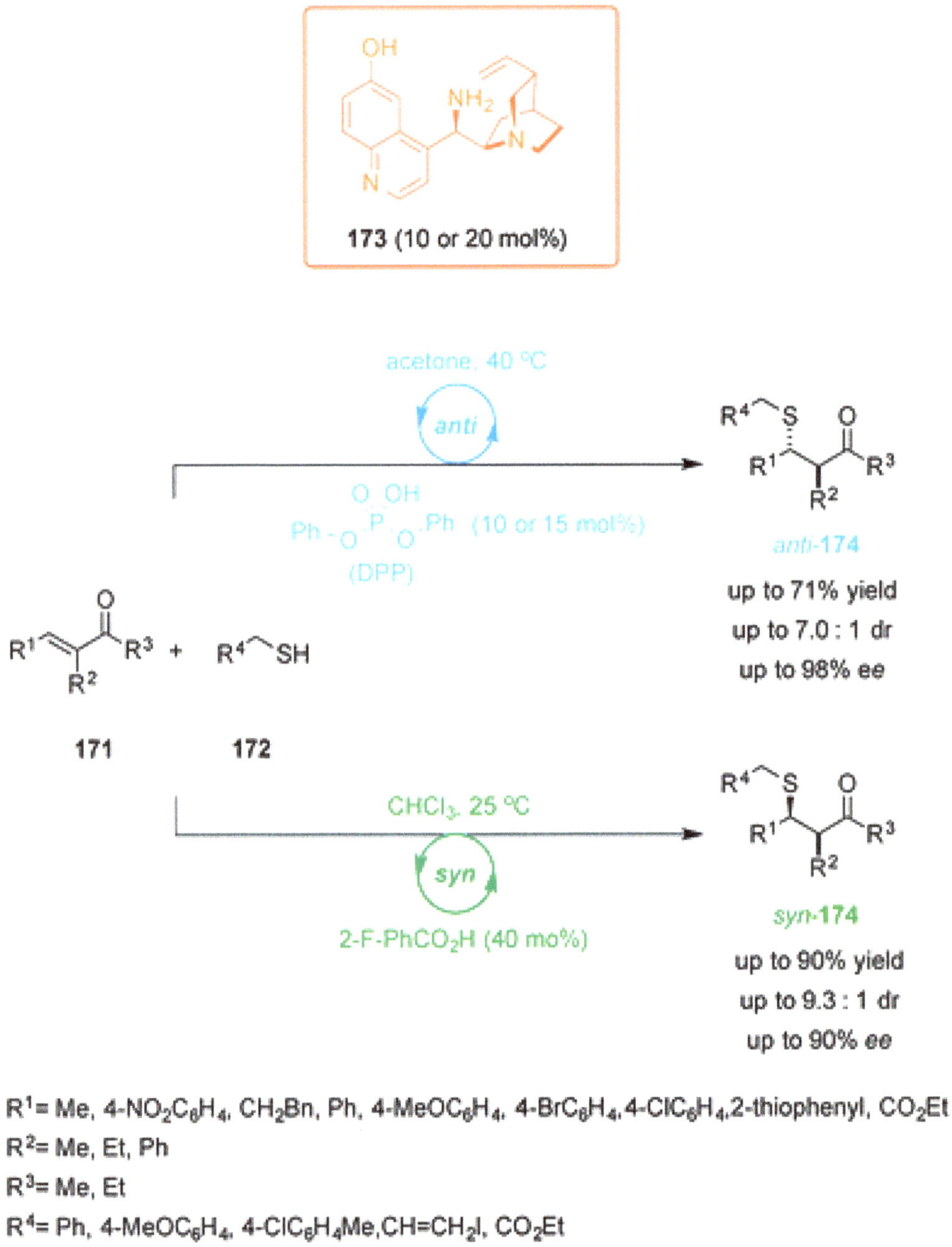

Scheme 52 Solvent- and acid-dependant diastereodivergent Michael addition of alkyl thiols to α,β-unsaturated ketones catalyzed by cinchona alkaloid derivative **173**.[98]

Solvent-dependent stereoselectivity of other catalytic systems have also been reported.[99] Changes in a variety of experimental parameters, including changes in temperature,[100] the use of different metal ions with the same chiral ligand,[101] change of the counter-anion[102] or cation,[103] the use of different acids as additives,[104] or other kind of additives[105] have all been successfully used to efficiently modulate the enantioselectivity of reactions.[106]

CONCLUSIONS AND OUTLOOK

Switchable catalysts triggered by a range of stimuli, including light, pH, coordination events, redox processes, mechanical forces and changes in reaction conditions, are starting to prove their utility in the control of a range of different types of chemical transformations. Artificial switchable catalysts are already able to alter the rate, stereochemical outcome and chemoselectivity of different types of reaction. Although the field is still in its infancy it is growing rapidly, fuelled by developments in supramolecular chemistry, metal catalysis and organocatalysis. Most efforts are currently focussed on the invention of new types of switchable catalysts with different forms of stimuli-response units incorporated within their architectures in order to achieve control over the catalytic activity and regio-, chemo- and stereocontrol of various chemical reactions. A significant limitation in the majority (but by no means all) of the systems developed to date is the modest differences in the activities of the 'on' and 'off' states of the catalysts. In order to approach the level of control exhibited by enzymatic synthesis, artificial switchable catalysts need to be developed that accelerate reactions strongly in the 'on' state and not-at-all in the 'off' state.

Two key challenges in particular remain unsolved: (i) the development of multi-responsive switchable catalysts that can control multiple catalytic processes by applying stimulus in a specific order; and (ii) the use of multiple (compatible) switchable catalysts with orthogonal responses, that can control which building blocks in a mixture react together, the order in which they react and the chemo- and regioselectivity of reactions and the outcome of one-pot tandem reaction sequences, without the need for further intervention.

This new area of catalysis research offers great potential for advanced functions and should ultimately be able to solve problems that are difficult or impossible to achieve using conventional methodologies. The next generation of switchable catalysts—featuring improved 'on'/'off' ratios, greater substrate scope and the ability to catalyze multiple classes of reaction—should bring these goals tantalisingly within reach.

ACKNOWLEDGEMENTS

We thank the EPSRC for funding our research in this area. V. B. thanks Ministerio de Economía y Competitividad (Spain) for a 'Juan de la Cierva' postdoctoral contract.

REFERENCES

1. T. Traut, Enzyme Activity: Allosteric Regulation, in *eLS*, John Wiley & Sons, 2001 .
2. T. Traut, *Allosteric Regulatory Enzymes*, Springer, New York, 2008 .
3. P. W. N. M. van Leeuwen, *Homogeneous Catalysis. Understanding the Art*, Springer, Dordrecht, 2004 ; *Catalysis for Fine Chemical Synthesis*, ed. S. M. Roberts, John Wiley & Sons, 2007 .
4. J. Barber, *Chem. Soc. Rev.*, 2009, **38**, 185–196 .
5. D. O'Brien, *Science*, 1982, **218**, 961–966 .
6. R. S. Stoll and S. Hecht, *Angew. Chem., Int. Ed.*, 2010, **49**, 5054–5075 ; B. M. Neilson and C. W. Bielawski, *ACS Catal.*, 2013, **3**, 1874–1885 ; R. Göstl, A. Senf and S. Hecht, *Chem. Soc. Rev.*, 2014, **43**, 1982–1996 ; T. Imahori and S. Kurihara, *Chem. Lett.*, 2014, **43**, 1524–1531 .
7. F. Würthner and J. Rebek, *Angew. Chem., Int. Ed. Engl.*, 1995, **34**, 446–448 .
8. F. Würthner and J. Rebek, *J. Chem. Soc., Perkin Trans. 2*, 1995, 1727–1734 .
9. R. Cacciapaglia, S. Di Stefano and L. Mandolini, *J. Am. Chem. Soc.*, 2003, **125**, 2224–2227 .
10. T. Imahori, R. Yamaguchi and S. Kurihara, *Chem. – Eur. J.*, 2012, **18**, 10802–10807 .
11. M. Samanta, V. S. Rama Krishna and S. Bandyopadhyay, *Chem. Commun.*, 2014, **50**, 10577–10579 .
12. D. Sud, T. B. Norsten and N. R. Branda, *Angew. Chem., Int. Ed.*, 2005, **44**, 2019–2021 .
13. J. Wang and B. L. Feringa, *Science*, 2011, **331**, 1429–1432 .
14. M. Vlatković, L. Bernardi, E. Otten and B. L. Feringa, *Chem. Commun.*, 2014, **50**, 7773–7775 .
15. Z. S. Kean, S. Akbulatov, Y. Tian, R. A. Widenhoefer, R. Boulatov and S. L. Craig, *Angew. Chem., Int. Ed.*, 2014, **53**, 14508–14511 .
16. D. Zhao, T. M. Neubauer and B. L. Feringa, *Nat. Commun.*, 2015, **6**, 6652 .
17. A. Ueno, K. Takahashi and T. Osa, *J. Chem. Soc., Chem. Commun.*, 1980, 837–838 ; A. Ueno, K. Takahashi and T. Osa, *J. Chem. Soc., Chem. Commun.*, 1981, 94–96 .
18. W.-S. Lee and A. Ueno, *Macromol. Rapid Commun.*, 2001, **22**, 448–450
19. L. Zhu, H. Yan, C. Y. Ang, K. T. Nguyen, M. Li and Y. Zhao, *Chem. – Eur. J.*, 2012, **18**, 13979–13983 .
20. M. Osaki, Y. Takashima, H. Yamaguchi and A. Harada, *Org. Biomol. Chem.*, 2009, **7**, 1646–1651
21. O. B. Berryman, A. C. Sather, A. Lledó and J. Rebek, *Angew. Chem., Int. Ed.*, 2011, **50**, 9400–9403 .
22. H. Sugimoto, T. Kimura and S. Inoue, *J. Am. Chem. Soc.*, 1999, **121**, 2325–2326
23. M. V. Peters, R. S. Stoll, A. Kühn and S. Hecht, *Angew. Chem., Int. Ed.*, 2008, **47**, 5968–5972 ; R. S. Stoll, M. V. Peters, A. Kühn, S. Heiles, R. Goddard, M. Bühl, C. M. Thiele and S. Hecht, *J. Am. Chem. Soc.*, 2009, **131**, 357–367 .
24. R. S. Stoll and S. Hecht, *Org. Lett.*, 2009, **11**, 4790–4793 .
25. L. Osorio-Planes, C. Rodríguez-Escrich and M. A. Pericàs, *Org. Lett.*, 2014, **16**, 1704–1707 .
26. D. Wilson and N. R. Branda, *Angew. Chem., Int. Ed.*, 2012, **51**, 5431–5434 .
27. B. M. Neilson and C. W. Bielawski, *J. Am. Chem. Soc.*, 2012, **134**, 12693–12699 .
28. B. M. Neilson and C. W. Bielawski, *Chem. Commun.*, 2013, **49**, 5453 .
29. B. M. Neilson and C. W. Bielawski, *Organometallics*, 2013, **32**, 3121–3128
30. T. Niazov, B. Shlyahovsky and I. Willner, *J. Am. Chem. Soc.*, 2007, **129**, 6374–6375 .
31. Y. Wei, S. Han, J. Kim, S. Soh and B. A. Grzybowski, *J. Am. Chem. Soc.*, 2010, **132**, 11018–11020 .
32. A. Nojiri, N. Kumagai and M. Shibasaki, *Chem. Commun.*, 2013, **49**, 4628–4630 .
33. S. L. Balof, S. J. P'Pool, N. J. Berger, E. J. Valente, A. M. Shiller and H.-J. Schanz, *Dalton Trans.*, 2008, 5791–5799 .

34. S. L. Balof, B. Yu, A. B. Lowe, Y. Ling, Y. Zhang and H.-J. Schanz, *Eur. J. Inorg. Chem.*, 2009, 1717–1722 .
35. L. H. Peeck, S. Leuthäusser and H. Plenio, *Organometallics*, 2010, **29**, 4339–4345
36. For other rotaxanes incorporating catalytic units, see: P. Thordarson, E. J. A. Bijsterveld, A. E. Rowan and R. J. M. Nolte, *Nature*, 2003, **424**, 915–918 ; Y. Tachibana, N. Kihara and T. Takata, *J. Am. Chem. Soc.*, 2004, **126**, 3438–3439 ; G. Hattori, T. Hori, Y. Miyake and Y. Nishibayashi, *J. Am. Chem. Soc.*, 2007, **129**, 12930–12931 ; Y. Li, Y. Feng, Y.-M. He, F. Chen, J. Pan and Q.-H. Fan, *Tetrahedron Lett.*, 2008, **49**, 2878–2881 ; J. Berná, M. Alajarín and R.-A. Orenes, *J. Am. Chem. Soc.*, 2010, **132**, 10741–10747 ; N. Miyagawa, M. Watanabe, T. Matsuyama, Y. Koyama, T. Moriuchi, T. Hirao, Y. Furusho and T. Takata, *Chem. Commun.*, 2010, **46**, 1920–1922 ; Y. Suzaki, K. Shimada, E. Chihara, T. Saito, Y. Tsuchido and K. Osakada, *Org. Lett.*, 2011, **13**, 3774–3777 ; B. Lewandowski, G. De Bo, J. W. Ward, M. Papmeyer, S. Kuschel, M. J. Aldegunde, P. M. E. Gramlich, D. Heckmann, S. M. Goldup, D. M. D'Souza, A. E. Fernandes and D. A. Leigh, *Science*, 2013, **339**, 189–193 .
37. U. Lüning, *Angew. Chem., Int. Ed.*, 2012, **51**, 8163–8165 ; D. A. Leigh, V. Marcos and M. R. Wilson, *ACS Catal.*, 2014, **4**, 4490–4497 .
38. V. Blanco, A. Carlone, K. D. Hänni, D. A. Leigh and B. Lewandowski, *Angew. Chem., Int. Ed.*, 2012, **51**, 5166–5169 .
39. V. Blanco, D. A. Leigh, U. Lewandowska, B. Lewandowski and V. Marcos, *J. Am. Chem. Soc.*, 2014, **136**, 15775–15780 .
40. V. Blanco, D. A. Leigh, V. Marcos, J. A. Morales-Serna and A. L. Nussbaumer, *J. Am. Chem. Soc.*, 2014, **136**, 4905–4908 .
41. J. Beswick, V. Blanco, G. De Bo, D. A. Leigh, U. Lewandowska, B. Lewandowski and K. Mishiro, *Chem. Sci.*, 2015, **6**, 140–143 .
42. S. Shinkai, Y. Ishikawa, H. Shinkai, T. Tsuno and O. Manabe, *Tetrahedron Lett.*, 1983, **24**, 1539–1542 ; S. Shinkai, Y. Ishikawa, H. Shinkai, T. Tsuno, H. Makishima, K. Ueda and O. Manabe, *J. Am. Chem. Soc.*, 1984, **106**, 1801–1808 ; S. Shinkai, K. Kameoka, K. Ueda and O. Manabe, *J. Am. Chem. Soc.*, 1987, **109**, 923–924 .
43. M. J. Wiester, P. A. Ulmann and C. A. Mirkin, *Angew. Chem., Int. Ed.*, 2011, **50**, 114–137 .
44. N. C. Gianneschi, M. S. Masar and C. A. Mirkin, *Acc. Chem. Res.*, 2005, **38**, 825–837 .
45. N. C. Gianneschi, P. A. Bertin, S. T. Nguyen, C. A. Mirkin, L. N. Zakharov and A. L. Rheingold, *J. Am. Chem. Soc.*, 2003, **125**, 10508–10509 .
46. N. C. Gianneschi, S. T. Nguyen and C. A. Mirkin, *J. Am. Chem. Soc.*, 2005, **127**, 1644–1645 .
47. H. J. Yoon and C. A. Mirkin, *J. Am. Chem. Soc.*, 2008, **130**, 11590–11591 .
48. M. S. Masar, N. C. Gianneschi, C. G. Oliveri, C. L. Stern, S. T. Nguyen and C. A. Mirkin, *J. Am. Chem. Soc.*, 2007, **129**, 10149–10158 .
49. C. G. Oliveri, N. C. Gianneschi, S. T. Nguyen, C. A. Mirkin, C. L. Stern, Z. Wawrzak and M. Pink, *J. Am. Chem. Soc.*, 2006, **128**, 16286–16296 .
50. H. J. Yoon, J. Heo and C. A. Mirkin, *J. Am. Chem. Soc.*, 2007, **129**, 14182–14183 .
51. N. C. Gianneschi, S.-H. Cho, S. T. Nguyen and C. A. Mirkin, *Angew. Chem., Int. Ed.*, 2004, **43**, 5503–5507 .
52. H. J. Yoon, J. Kuwabara, J. H. Kim and C. A. Mirkin, *Science*, 2010, **330**, 66–69 .
53. C. M. McGuirk, J. Mendez-Arroyo, A. M. Lifschitz and C. A. Mirkin, *J. Am. Chem. Soc.*, 2014, **136**, 16594–16601 .
54. C. M. McGuirk, C. L. Stern and C. A. Mirkin, *J. Am. Chem. Soc.*, 2014, **136**, 4689–4696 .
55. M. Zirngast, E. Pump, A. Leitgeb, J. H. Albering and C. Slugovc, *Chem. Commun.*, 2011, **47**, 2261–2263 .
56. J. H. Griffin and P. B. Dervan, *J. Am. Chem. Soc.*, 1987, **109**, 6840–6842
57. T. Tozawa, S. Tokita and Y. Kubo, *Tetrahedron Lett.*, 2002, **43**, 3455–3457
58. L. Kovbasyuk and R. Krämer, *Chem. Rev.*, 2004, **104**, 3161–3188 ; I. O. Fritsky, R. Ott and R. Krämer, *Angew. Chem., Int. Ed.*, 2000, **39**, 3255–3258 ; I. O. Fritsky, R. Ott, H. Pritzkow and R. Krämer, *Chem. – Eur. J.*, 2001, **7**, 1221–1231 ; I. O. Fritsky, R. Ott, H. Pritzkow and R. Krämer, *Inorg. Chim. Acta*, 2003, **346**, 111–118 ; K. P. Strotmeyer, I. O. Fritsky, R. Ott, H. Pritzkow and R. Krämer, *Supramol. Chem.*, 2003, **15**, 529–547 .
59. L. Kovbasyuk, H. Pritzkow, R. Krämer and I. O. Fritsky, *Chem. Commun.*, 2004, 880–881 .
60. A. Scarso, G. Zaupa, F. B. Houillon, L. J. Prins and P. Scrimin, *J. Org. Chem.*, 2007, **72**, 376–385 .
61. S. Takebayashi, M. Ikeda, M. Takeuchi and S. Shinkai, *Chem. Commun.*, 2004, 420–421 ; S. Takebayashi, S. Shinkai, M. Ikeda and M. Takeuchi, *Org. Biomol. Chem.*, 2008, **6**, 493–499 .
62. M. R. Kita and A. J. M. Miller, *J. Am. Chem. Soc.*, 2014, **136**, 14519–14529 .
63. G.-H. Ouyang, Y.-M. He, Y. Li, J.-F. Xiang and Q.-H. Fan, *Angew. Chem., Int. Ed.*, 2015, **54**, 4334–4337 .
64. M. Schmittel, S. De and S. Pramanik, *Angew. Chem., Int. Ed.*, 2012, **51**, 3832–3836 .
65. M. Schmittel, S. Pramanik and S. De, *Chem. Commun.*, 2012, **48**, 11730–11732 .
66. S. De, S. Pramanik and M. Schmittel, *Dalton Trans.*, 2014, **43**, 10977–10982 .
67. S. De, S. Pramanik and M. Schmittel, *Angew. Chem., Int. Ed.*, 2014, **53**, 14255–14259 .
68. J. T. Foy, D. Ray and I. Aprahamian, *Chem. Sci.*, 2015, **6**, 209–213 .
69. I. M. Lorkovic, R. R. Duff and M. S. Wrighton, *J. Am. Chem. Soc.*, 1995, **117**, 3617–3618
70. C. S. Slone, C. A. Mirkin, G. P. A. Yap, I. A. Guzei and A. L. Rheingold, *J. Am. Chem. Soc.*, 1997, **119**, 10743–10753

71. C. K. A. Gregson, V. C. Gibson, N. J. Long, E. L. Marshall, P. J. Oxford and A. J. P. White, *J. Am. Chem. Soc.*, 2006, **128**, 7410–7411 .

72. E. M. Broderick, N. Guo, C. S. Vogel, C. Xu, J. Sutter, J. T. Miller, K. Meyer, P. Mehrkhodavandi and P. L. Diaconescu, *J. Am. Chem. Soc.*, 2011, **133**, 9278–9281 .

73. K. Arumugam, C. D. Varnado, S. Sproules, V. M. Lynch and C. W. Bielawski, *Chem. – Eur. J.*, 2013, **19**, 10866–10875 .

74. M. Süßner and H. Plenio, *Angew. Chem., Int. Ed.*, 2005, **44**, 6885–6888 .

75. P. Neumann, H. Dib, A.-M. Caminade and E. Hey-Hawkins, *Angew. Chem., Int. Ed.*, 2015, **54**, 311–314 .

76. A. J. D. Magenau, N. C. Strandwitz, A. Gennaro and K. Matyjaszewski, *Science*, 2011, **332**, 81–84 .

77. E. M. Broderick, N. Guo, T. Wu, C. S. Vogel, C. Xu, J. Sutter, J. T. Miller, K. Meyer, T. Cantat and P. L. Diaconescu, *Chem. Commun.*, 2011, **47**, 9897–9899 .

78. A. Sauer, J.-C. Buffet, T. P. Spaniol, H. Nagae, K. Mashima and J. Okuda, *ChemCatChem*, 2013, **5**, 1088–1091 .

79. A. B. Biernesser, B. Li and J. A. Byers, *J. Am. Chem. Soc.*, 2013, **135**, 16553–16560 .

80. Y.-Y. Fang, W.-J. Gong, X.-J. Shang, H.-X. Li, J. Gao and J.-P. Lang, *Dalton Trans.*, 2014, **43**, 8282–8289 .

81. S. Mortezaei, N. R. Catarineu and J. W. Canary, *J. Am. Chem. Soc.*, 2012, **134**, 8054–8057 .

82. X. Wang, A. Thevenon, J. L. Brosmer, I. Yu, S. I. Khan, P. Mehrkhodavandi and P. L. Diaconescu, *J. Am. Chem. Soc.*, 2014, **136**, 11264–11267 .

83. Q. Zhang, X. Cui, L. Zhang, S. Luo, H. Wang and Y. Wu, *Angew. Chem., Int. Ed.*, 2015, **54**, 5210–5213 .

84. M. Fujiwara, S. Terashima, Y. Endo, K. Shiokawa and H. Ohue, *Chem. Commun.*, 2006, 4635–4637 .

85. A. G. Tennyson, V. M. Lynch and C. W. Bielawski, *J. Am. Chem. Soc.*, 2010, **132**, 9420–9429 .

86. R. Groote, R. T. M. Jakobs and R. P. Sijbesma, *Polym. Chem.*, 2013, **4**, 4846–4859 .

87. S. Karthikeyan, S. L. Potisek, A. Piermattei and R. P. Sijbesma, *J. Am. Chem. Soc.*, 2008, **130**, 14968–14969 .

88. A. Piermattei, S. Karthikeyan and R. P. Sijbesma, *Nat. Chem.*, 2009, **1**, 133–137 .

89. R. Groote, L. van Haandel and R. P. Sijbesma, *J. Polym. Sci., Part A: Polym. Chem.*, 2012, **50**, 4929–4935 .

90. R. Groote, R. T. M. Jakobs and R. P. Sijbesma, *ACS Macro Lett.*, 2012, **1**, 1012–1015

91. R. T. M. Jakobs and R. P. Sijbesma, *Organometallics*, 2012, **31**, 2476–2481

92. R. T. M. Jakobs, S. Ma and R. P. Sijbesma, *ACS Macro Lett.*, 2013, **2**, 613–616

93. Y. H. Kim, *Acc. Chem. Res.*, 2001, **34**, 955–962 ; M. Bartók, *Chem. Rev.*, 2010, **110**, 1663–1705 ; Z. Dai, J. Lee and W. Zhang, *Molecules*, 2012, **17**, 1247–1277 ; N. Kumagai and M. Shibasaki, *Catal. Sci. Technol.*, 2013, **3**, 41–57 .

94. P. G. Jessop, D. J. Heldebrant, X. Li, C. A. Eckert and C. L. Liotta, *Nature*, 2005, **436**, 1102 .

95. O. Coulembier, S. Moins, R. Todd and P. Dubois, *Macromolecules*, 2014, **47**, 486–491

96. C. Romain and C. K. Williams, *Angew. Chem., Int. Ed.*, 2014, **53**, 1607–1610 .

97. Y. Sohtome, S. Tanaka, K. Takada, T. Yamaguchi and K. Nagasawa, *Angew. Chem., Int. Ed.*, 2010, **49**, 9254–9257 .

98. X. Tian, C. Cassani, Y. Liu, A. Moran, A. Urakawa, P. Galzerano, E. Arceo and P. Melchiorre, *J. Am. Chem. Soc.*, 2011, **133**, 17934–17941 .

99. S. Arseniyadis, A. Valleix, A. Wagner and C. Mioskowski, *Angew. Chem., Int. Ed.*, 2004, **43**, 3314–3317 ; A. B. Northrup and D. W. C. MacMillan, *Science*, 2004, **305**, 1752–1755 ; T. Yamamoto, T. Yamada, Y. Nagata and M. Suginome, *J. Am. Chem. Soc.*, 2010, **132**, 7899–7901 ; M. Messerer and H. Wennemers, *Synlett*, 2011, 499–502 ; R. J. Chew, X.-R. Li, Y. Li, S. A. Pullarkat and P.-H. Leung, *Chem. – Eur. J.*, 2015, **21**, 4800–4804 .

100. G. Storch and O. Trapp, *Angew. Chem., Int. Ed.*, 2015, **54**, 3580–3586 .

101. A. Nojiri, N. Kumagai and M. Shibasaki, *J. Am. Chem. Soc.*, 2009, **131**, 3779–3784 ; G. Lu, T. Yoshino, H. Morimoto, S. Matsunaga and M. Shibasaki, *Angew. Chem., Int. Ed.*, 2011, **50**, 4382–4385 .

102. J. Lv, L. Zhang, Y. Zhou, Z. Nie, S. Luo and J.-P. Cheng, *Angew. Chem., Int. Ed.*, 2011, **50**, 6610–6614 .

103. P. Mazón, R. Chinchilla, C. Nájera, G. Guillena, R. Kreiter, R. J. M. Klein Gebbink and G. van Koten, *Tetrahedron: Asymmetry*, 2002, **13**, 2181–2185 .

104. J. Gao, S. Bai, Q. Gao, Y. Liu and Q. Yang, *Chem. Commun.*, 2011, **47**, 6716–6718 ; R. Doran, M. P. Carroll, R. Akula, B. F. Hogan, M. Martins, S. Fanning and P. J. Guiry, *Chem. – Eur. J.*, 2014, **20**, 15354–15359 ; R. Doran and P. J. Guiry, *J. Org. Chem.*, 2014, **79**, 9112–9124 .

105. S.-H. Chen, B.-C. Hong, C.-F. Su and S. Sarshar, *Tetrahedron Lett.*, 2005, **46**, 8899–8903 ; N. Abermil, G. Masson and J. Zhu, *Org. Lett.*, 2009, **11**, 4648–4651 .

106. M. Bartók, *Chem. Rev.*, 2010, **110**, 1663–1705 ; J. Escorihuela, M. I. Burguete and S. V. Luis, *Chem. Soc. Rev.*, 2013, **42**, 5595–5617

CHAPTER 15

MOLECULAR BIOLOGY: SELF-SUSTAINING CHEMISTRY

Paul Wrede

MOLECULAR BIOLOGY IN CHEMISTRY CENTRAL JOURNAL

At the outset, let me pose two important questions: Why is it necessary to have a molecular biology section within *Chemistry Central Journal*? Can molecular biology articles in this journal compete with contributions found in other excellent specialised molecular biology journals?

To answer these questions, let me start by first describing how the term molecular biology is understood today, before looking at the field in a chemical context.

WHAT IS MOLECULAR BIOLOGY?

Broadly, molecular biology aims to investigate the molecular genetics of all organisms, including man, by identifying mutants or genetic aberrations that are often the cause of diseases. Intense genetic studies on model organisms have cast light on the processes of morphological development, cell differentiation, and the control mechanisms governing organ (re)generation. Molecular biologists have generated their own specific methods to solve problems inherent in their field. Here of course I refer to techniques such as DNA recombination and nucleic acid amplification by PCR, whose functional basis is the manipulation of biological processes to gain deeper insight into complex molecular interactions either *in situ* or *in vitro*. The foundation of such techniques lies in the complementary nature of the genetic information contained in nucleic acids.

MOLECULAR BIOLOGY AND CHEMISTRY

The specific goals of molecular biology research are the discovery and more detailed understanding of the structure and function of genetic material, namely DNA and RNA, and in particular, how these molecules determine the smooth running of metabolic processes. Such investigations are more or less independent of the organism studied, as the conserved function of DNA and RNA in all organisms guarantees the continuation of this self-sustaining chemical system from one generation to the next.

In surveying the field of molecular biology one can only marvel at the fundamental similarities of many (bio)chemical reactions and processes – take, for example, the elaborate system of genes involved during the phase of early development. It is these similarities between complex processes that give us the possibility of categorizing molecular biology into several subdivisions. For example universal cellular processes like protein insertion and secretion or G-protein coupled signal transduction are self-dependent subdivisions of molecular biology.

THE CELL – A CHEMICAL FACTORY

It is always fascinating to consider that in a cell, thousands of different biochemical reactions can proceed under constant temperatures, such as in homoeothermic mammals and birds. In contrast, the complex metabolic networks of poikilothermic organisms, like fish or plants, also include the same metabolic chemical reactions!

In cells several anabolic and catabolic reaction chains are separated from each other by ‘compartments’ like mitochondria, the endoplasmic reticulum, lysosomes, and many more. It is noteworthy that in all these separate locations, there occur different biochemical reactions, which are not however excluded from the intracellular mass transfer. For each ‘compartment’, transport proteins exist in the membranes responsible for the uptake and discharge of small molecules. Such transport can occur actively against the concentration gradient under ATP consumption or passively through protein channels in the membrane. Other organelles, such as the

Citation: Paul Wrede, Molecular biology: Self-sustaining chemistry, doi: 10.1186/1752-153X-1-25.

Golgi apparatus, exhibit a vesicular transport system for both generating new organelles like endosomes, and transporting proteins and small molecules within these vesicles. The directed transport processes between each 'compartment' occur in a finely tuned and controlled fashion. The genetic basis for these regulatory and control processes is one major source of interest in molecular biology research.

Thus to return to our initial question, as I have outlined, it is clear that molecular biology, with its special emphasis on the self-sustaining capacity of the cell's complex chemical reaction network, has a close relationship with the field of chemistry. All (bio) chemical reactions, including their regulation, have been transferred from one generation to the next for millions of years. Selectivity has resulted in the existence of highly conserved molecular processes; whilst conversely and interestingly, some metabolic pathways cannot be traced back to any specific origin owing to their occurring in only a few recent organisms.

MAJOR DEVELOPMENTS IN MOLECULAR BIOLOGY

The formative biological research activities of the last decade have been the genome sequencing projects of uni-cellular and multi-cellular organisms, including man. Surprisingly, the sequenced genomes have only shed limited light on increasing the understanding of the relationship between a gene and its expression and regulation. For instance, even the rudimentary question, 'what is a gene?', cannot yet be answered unambiguously.

Genome sequences between several human individuals can vary drastically, so it is often impossible to predict from the DNA-sequence alone who will be susceptible to a disease and who will remain healthy. All cellular organisms have in common a high variability, which includes, for example, gene order, protein sequence, glycosylation patterns and membrane constitution. This variability is further complicated by the fact that many different organisms share conserved metabolic concepts, such as the low energy delivery glycolysis pathway, the Krebs cycle for high chemical energy production, the genetic code or the protein translation apparatus. I think it is this dialectic principle, resulting from the dichotomy of great variability coexisting with strict conservation – not only on the molecular level but also on the organism level – that makes molecular biology a particularly fascinating discipline. Almost every year, progress in the field of molecular biology provokes at least a significant modification of existing concepts, and sometimes even a paradigm change.

TECHNIQUES EMPLOYED BY MOLECULAR BIOLOGISTS

I mentioned at the outset techniques native to molecular biology, such as recombinant DNA and several other techniques like nucleic acid amplification by PCR, high resolution microscopy and very sensitive protein detection systems. These have helped accelerate the growth of our vast, but scattered, knowledge. Furthermore, significant progress in microarray techniques has delivered data on complex expression profiles, including post-translational modifications like glycosylation and phosphorylation. Here the differences in the patterns observed for normal and cancerous cells allow one to perform a classification of cellular subtypes. The expression profiles of cells from different tissues may even identify the activity status of metabolic pathways and their phylogenetic conservation. A combination of expression profiling data with confocal fluorescent microscopy and imaging processing allows one to locate protein expression activities within the cell. By combining all these data, one will be able to construct an image of the cell in time and space that will deepen our understanding of molecular interactions.

Metabolic processes can also be probed through the fine tuning of gene activities by silencing genes with small interfering RNA (siRNA) and microinterfering RNAi. Whilst the determination of the sequences and 3D structures of biological macromolecules have been significantly enhanced by many sophisticated biophysical techniques, also employed by chemists, such as MALDI and multidimensional NMR. Such techniques permit the identification and characterisation of many protein variants with unique catalytic centres and glycosylation/phosphorylation patterns that are specific for many different chemical reactions. Analytical techniques can now be performed almost quantitatively.

OPEN ACCESS AND MOLECULAR BIOLOGY

Thousands of articles on molecular biology are published each year, meaning that inevitably experts can hardly keep a detailed track of research progress, even in their particular field. I am convinced that the principal benefits of open access publishing – that is to say, free and unhindered access to scientific research, rapid publication, authors' retention of copyright for their own work, free distribution and deposition of articles – will benefit the molecular biology community, and help cope with the huge flood of research in this field.

Chemistry Central Journal's parent company, BioMed Central, already operates a number of successful journals in the field of molecular biology – notably *Algorithms for Molecular Biology*, *BMC Molecular Biology*, and the *Journal of Molecular Signaling* – and related disciplines. In addition, *BMC Bioinformatics* complements molecular biology by concentrating on data mining and computer-based simulation systems, concepts which can help significantly support molecular biologists' experimental work.

CONCLUSION

Of course a number of BioMed Central's other already-established open access titles, as well as the chemical biology section in *Chemistry Central Journal*, touch upon areas of interest to molecular biologists. Such overlap cannot be avoided, but rather serves to underline the interdisciplinary character of molecular biology.

However, that said, each of these journals or sections have their own particular focal points. For example, the chemical biology section in this journal focusses on small organic molecules and their interactions with target proteins; while the aim of the molecular biology section, however, will be to publish research highlights whose focus is the innate chemistry of the cell and its evolution, as well as the fascinating theme of gene regulation and expression in all organisms. Thus, within *Chemistry Central Journal* it is inevitable that both themolecular biology and chemical biology sections will complement and enrich each other. Of course, when submitting a manuscript, an author can select up to 3 sections by which to classify the article.

Finally, I also want to encourage molecular biologists to publish in *Chemistry Central Journal* because of the broad nature of our journal's audience – reaching from specialists in the field to those working in other fields but looking for cooperation partners to accelerate a common project – helped by the unhindered access to the journal's published research. All research presenting good science will be considered for publication and I am looking forward to receiving fresh and novel submissions.

I would like to close my commentary with two summatory statements: the first by Edmund B.Wilson: «*the key to every biological problem must finally be sought in the cell*" [1]; and the second by Theodosius Dobshanzky: "*nothing in biology makes sense except in the light of evolution*" [2].

DECLARATIONS

Acknowledgements

For helpful discussions and support I would like to thank my colleagues Ole Andersen (Roskilde), R. Stephen Berry (Chicago), Gino D'Oca (London), Jürgen Kleffe (Berlin), Gisbert Schneider (Frankfurt/M), and Reinhard Wanner (Berlin).

REFERENCES

1. Wilson EB: The Cell in Development and Heredity. 1925, New York: The Macmillan Company
2. Dobshanzky T: Nothing in biology makes sense except in the light of evolution. The American Biology Teacher. 1973, 35: 125-129.

CHAPTER 16

NUCLEIC ACID ENZYMES

Dominique Le´vesque., Francis P. Brie`re and Jean-Pierre Perreault

INTRODUCTION

Enzymes have evolved various versions of a "molecular switch" with which they control their catalytic activities. For example, regulatory proteins bind effectors that induce conformational changes that control catalytic activity [1]. Similar modes of activation have been adopted by catalytic RNA motifs such as the *glm*S riboswitch, a ribozyme that regulates gene expression in a unique fashion through autocatalytic cleavage upon its binding to glucosamine-6-phosphate [2], [3]. Mitogen-activated protein kinases (MAP kinases) have evolved a similar, but more subtle, mode of activation [4], [5]. Upon the binding of the target to a docking domain, the substrate is positioned within the catalytic site. The prodomain acts as a target-dependent module switching the kinase from an *off* state to an *on* state, a property which considerably improves the substrate specificity of the enzyme. None of the ribozymes discovered to date exhibit such a mode of activation, nor does any other known RNA. However, a Specific *On/ofF* Adaptor (SOFA) module, which includes both a blocker and a biosensor domain, has been developed for the hepatitis delta virus (HDV) ribozyme [6], [7]. The rationale behind this design was that in the absence of the target, the ribozyme should be turned *off* by the addition of a blocker domain (bl) that forms an intramolecular stem with a portion of the substrate recognition domain (stem P1; Figure 1A). Conversely, a biosensor domain (BS) must bind its complementary sequence on the substrate in order to unlock the catalytic core (turning *on*). In the SOFA module the blocker and biosensor are juxtaposed, which is not the case for the domain involved in the activation of MAP kinases. In the present work, we have adapted this mode of activation, using the techniques of molecular biology, to a wide range of catalysis types exhibited by nucleic acids.

Citation: Lévesque D, Brière FP, Perreault J-P (2007), Nucleic Acid Enzymes. PLoS ONE 2(7): e673. doi:10.1371/journal.pone.0000673.

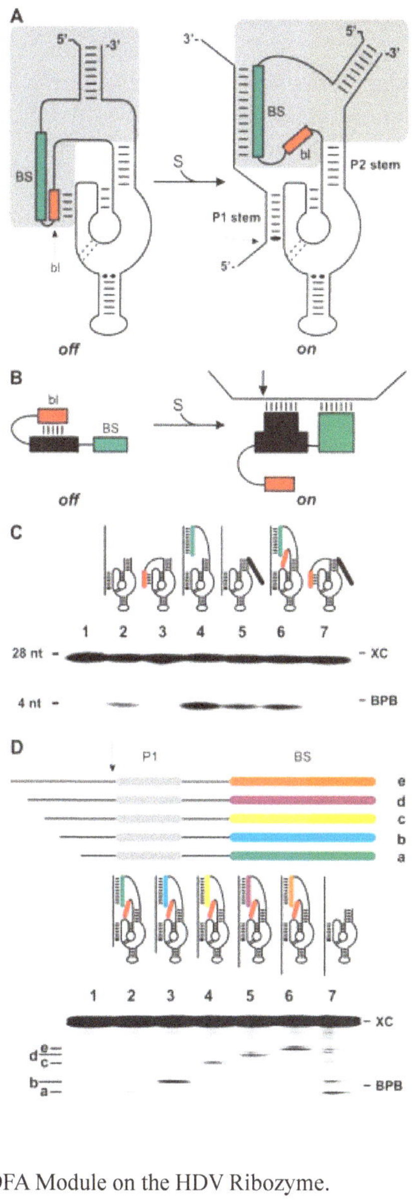

Figure 1. SOFA Module on the HDV Ribozyme.

(A) Secondary structure of both the *off* and *on* conformation of the SOFA module (in grey) engineered on the P2 stem of the HDV ribozyme. The biosensor (BS, in green) and the blocker (bl, in red) are juxtaposed. (B) Schematic representation of separated blocker and biosensor domains modulating the activity of a ribozyme in a substrate-dependent manner. (C) Autoradiogram of cleavage assays performed with various HDV ribozymes. Their structures, with both the biosensor (green) and the blocker (red), are illustrated above the lanes of the gel. The control (-) was performed in the absence of ribozyme (lane 1), while lane 2 was performed in the presence of the original version. Lane 3 is the version extended by the blocker sequence only. Lanes 4 and 5 are the versions extended by a biosensor domain either complementary, or not, to the substrate, respectively. Finally, lanes 6 and 7 are the *on* and *off* versions including separated blocker and biosensor domains, respectively. BPB and XC indicate the positions of the bromophenol blue and xylene cyanol dyes, respectively. The nucleotide sequences of each of the engineered ribozyme are shown in the. (D) Verification of the gain in substrate specificity of the *trans*-acting HDV ribozymes. The upper portion shows a schematic representation of the substrates in which the P1 binding domain is identical (grey), but the sequences of the biosensors vary (different colors). The lower portion shows the autoradiogram of a cleavage assay performed on a pool of substrates either by a specific HDV *on* ribozyme (a to e correspond to the cleavage produced by specific ribozymes, lanes 2 to 6, respectively), or by the original ribozyme (lane 7). Lane 1 is the negative control performed in the absence of ribozyme. The nucleotide sequences of each of ribozyme are shown in the.

RESULTS

Development of a new on/off Adaptor Version for HDV Ribozyme

Initially, a new design that considers the blocker and biosensor domains as distinct entities was performed by including them at each end of the P2 stem of the HDV ribozyme in a process that involved several steps (Figure 1B). First, a blocker sequence that base-pairs with the binding domain of the ribozyme was added to the 5′-end. The resulting ribozyme exhibited no cleavage activity with a model substrate derived from the internal ribosome entry site (IRES) region of the hepatitis C virus (HCV) (Figure 1C, lane 3). The blocker acts as a "safety lock" placing the ribozyme in an inactive conformation (*off* state). When the ribozyme possesses a 12 nucleotide (nt) biosensor sequence complementary to the substrate at its 3′-end, it exhibited an improved cleavage activity (Figure 1C, lane 4). The presence of an inappropriate biosensor domain (i.e. a sequence not complementary to the substrate) resulted in a cleavage level comparable to that of the original ribozyme (Figure 1C, lane 5). When an appropriate biosensor was added to a ribozyme that already possessed a blocker, a dramatic increase in the cleavage activity, as compared to that observed in the presence of an inappropriate biosensor, was observed (Figure 1C, lanes 6 and 7, respectively). Thus, the ribozyme can be switched *off* and *on* by independent domains in a way reminiscent of the MAP kinases.

Kinetic analyses were performed in order to determine the second order rate constants (k_{cat}/K_M') as an indication of the substrate specificities. The HDV *on* (i.e. the one including both a blocker and an appropriate biosensor) exhibited a cleavage activity characterized by a relative k_{cat}/K_M' 6-fold higher than that of the original ribozyme, which in turn had a value 12-fold higher than that of the HDV *off* ribozyme (i.e the one including a blocker and an inappropriate biosensor) (see Table 1). The difference between the k_{cat}/K_M' values of the *off* and *on* ribozymes is therefore greater than 60 times. This difference is smaller than the determined 1500 to 2500 times observed between the *on* and *off* forms of the original configuration of the SOFA module on other HDV ribozymes [6], [7]. The larger difference in the latter case is likely due to the fact that its configuration is probably better adapted to the architecture of this particular ribozyme.

Table 1. Kinetic parameters of four cleaving nucleic acid enzymes in single turnover conditions

nucleic acid enzymes	k_{cat} (min^{-1})	$K_{M'}$ (nM)	$k_{cat}/K_{M'}$ (min^{-1}/M^{-1})	relative $k_{cat}/K_{M'}$ values
HDV wt	0.22±0.03	128±37	1.7×10^6	12
HDV *on*	0.11±0.01	12±3	9.2×10^6	66
HDV *off*	0.034±0.004	240±82	1.4×10^5	1
hammerhead wt	0.074±0.002	48±8	1.5×10^6	1765
hammerhead *on*	0.13±0.01	59±24	2.2×10^6	2588
hammerhead *off*	$6.38\times10^{-5}\pm0.70\times10^{-5}$	75±28	8.5×10^2	1
deoxyribozyme wt	0.76±0.08	211±45	3.6×10^6	1714
deoxyribozyme *on*	1.80±0.12	312±38	5.8×10^6	2762
deoxyribozyme *off*	$1.24\times10^{-4}\pm0.04\times10^{-4}$	60±7	2.1×10^3	1
hairpin wt	0.41±0.02	163±22	2.5×10^6	2083
hairpin *on*	0.12±0.01	11±2	1.1×10^7	9167
hairpin *off*	$2.8\times10^{-4}\pm0.6\times10^{-4}$	240±155	1.2×10^3	1

As the specificity of a ribozyme is commonly defined as its ability to discriminate between similar substrates [1], a collection of five substrates possessing identical ribozyme binding domains, but different biosensor sequences, was synthesized (Figure 1D). Because these substrates harbor different 5′-ends, they can be distinguished by their electrophoretic migration (Figure 1D, lane 1). When the 5 substrates were incubated together with only one HDV *on*ribozyme, solely the substrate possessing the sequence perfectly complementary to the biosensor was cleaved (Figure 1D, lanes 2-6). Conversely, an original ribozyme (i.e. one without both a biosensor and a blocker) had the ability to cleave all of the substrates, although at different levels (Figure 1D, lane 7). Together, these data demonstrate the improvement, in terms of substrate specificity, resulting from the addition of both a blocker and a biosensor. Moreover, it shows that a ribozyme, when activated by the proper substrate, does not cleave other substrates via a *trans*-cleavage mechanism. Finally, the potential to cleave HCV-derived transcripts was tested in cell culture (data not shown). Only small amounts of transcripts were cleaved. Determination of the half-life of the new conformation of the SOFA-HDV ribozymes according to a procedure reported previously [8], revealed a relatively low molecular stability of these ribozymes (a likely by-product of the single-stranded extremities that these ribozymes possess). More important is that it has been possible to adapt the target-dependent module of activation from the MAP kinase to the HDV ribozyme.

Adaptation of this mode of Activation to Various Cleaving Nucleic Acid Enzymes

Subsequently, we tried to adapt this mode of activation to other catalytic nucleic acids. As a first attempt, we chose three cleaving molecules: the hammerhead ribozyme, the 10-23 deoxyribozyme and the hairpin ribozyme (Figure 2) [9]–[11]. All of these nucleic acid enzymes share a relatively similar architecture. The cleavage site of the substrates is located in the middle of the two stems formed between the catalytic center. The sequences of both of these ribozymes and of the deoxyribozyme have been reported to cleave various substrates that were used [12]–[15]. Each of these catalytic nucleic acids showed significant cleavage of their 5′-end labelled substrates, although at different levels (Figure 2A–C, lane 2). The addition of a blocker domain that "zips off" the binding domain dramatically decreased the cleavage activity (i.e. Figure 2A–C, lane 3). As is observed with the biosensor-blocker HDV ribozyme, the specificity of these engineered nucleic acid molecules was dramatically increased through blocker-mediated inactivation followed by reactivation via an extended complementarity with the appropriate biosensor sequence (Figure 2A–C, lanes 4 to 7). With both the biosensor and blocker domains, the presence of an inappropriate biosensor locked the catalytic core into an *off* state, while the presence of an appropriate biosensor permitted the switch into the active mode, by displacing the blocker, solely in presence of its specific substrate. The biosensors interact through more base pairs than do the blockers, thereby favoring the *on* state over the *off* state. When comparing the $k_{cat}/K_{M'}$ values for the *off* and *on* versions of these catalytic molecules, the overall improvements were of a minimum of three orders of magnitude (Table 1). Specifically, the $k_{cat}/K_{M'}$ values vary from 2588-, 2762- and 9167-folds between the *off* and *on* version of hammerhead ribozyme, hairpin ribozyme and 10-23 deoxyribozyme, respectively. This illustrates the important gain in terms of specificity. These important differences were mainly the result of dramatic decreases in the k_{cat} values between those of the original structures and those of their respective *off* states, suggesting that they lost their abilities to perform non-specific cleavage. Importantly, the addition of both a blocker and a biosensor permits the generation of cleaving nucleic acid enzymes that operate in a target-dependent fashion.

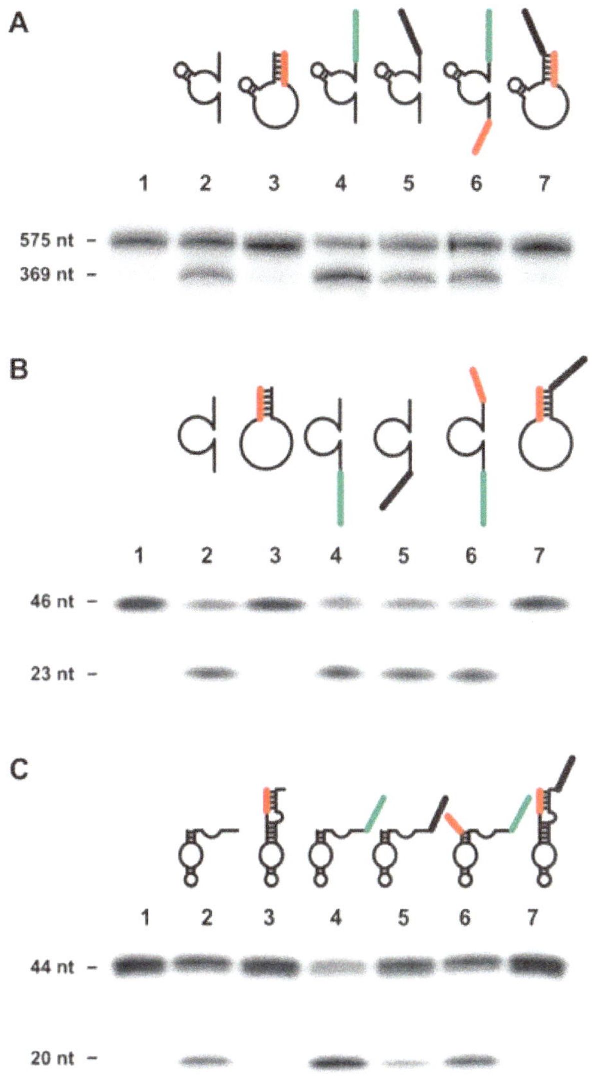

Figure 2. Autoradiograms of Cleavage Assays Performed with Controlled Cleaving Nucleic Acid Enzymes.

(A) (B) and (C) are the reactions performed with the hammerhead ribozyme, the 10-23 deoxyribozyme and the hairpin ribozyme, respectively. In each case the structures of the nucleic acid enzyme with the blocker (red) and biosensor (green) are illustrated above the appropriate lanes of the gels, and the control (-) was performed in the absence of any nucleic acid enzyme (lane 1), and, lane 2, in the presence of the unmodified version of the ribozyme. Lane 3 is the version extended by the blocker sequence. Lanes 4 and 5 are the versions extended by a biosensor that is either complementary, or not, to the substrate. Finally, lanes 6 and 7 are the *on* and *off* versions, respectively. The nucleotide sequences of each nucleic acid enzyme are depicted.

Engineering Ribozymes Catalyzing Various Reactions

Next, we investigated other ribozyme's catalyses. As a first attempt, the ligation catalyzed by the hairpin ribozyme was studied. Extending the 3′-end of the original hairpin ribozyme by 6 nt resulted in the retention of the cleavage product, and consequently favored the reverse reaction (Figure 3A) [14]. Using a 5′-^{32}P-labelled RNA strand possessing a terminal 2′-3′-cyclic phosphate and a second RNA strand possessing a 5′-hydroxyl resulted in the detection of ligation products (Figure 3B). The addition of a biosensor with a sequence complementary to that of the substrate increased the amount of ligation observed. In contrast, the addition of the blocker module almost completely abolished the ligation. Finally, an *on* version, which includes both the blocker and an appropriate biosensor, exhibited a ligation activity of up to 65%; while an *off* version (i.e. one with an inappropriate biosensor) showed less than 5% activity.

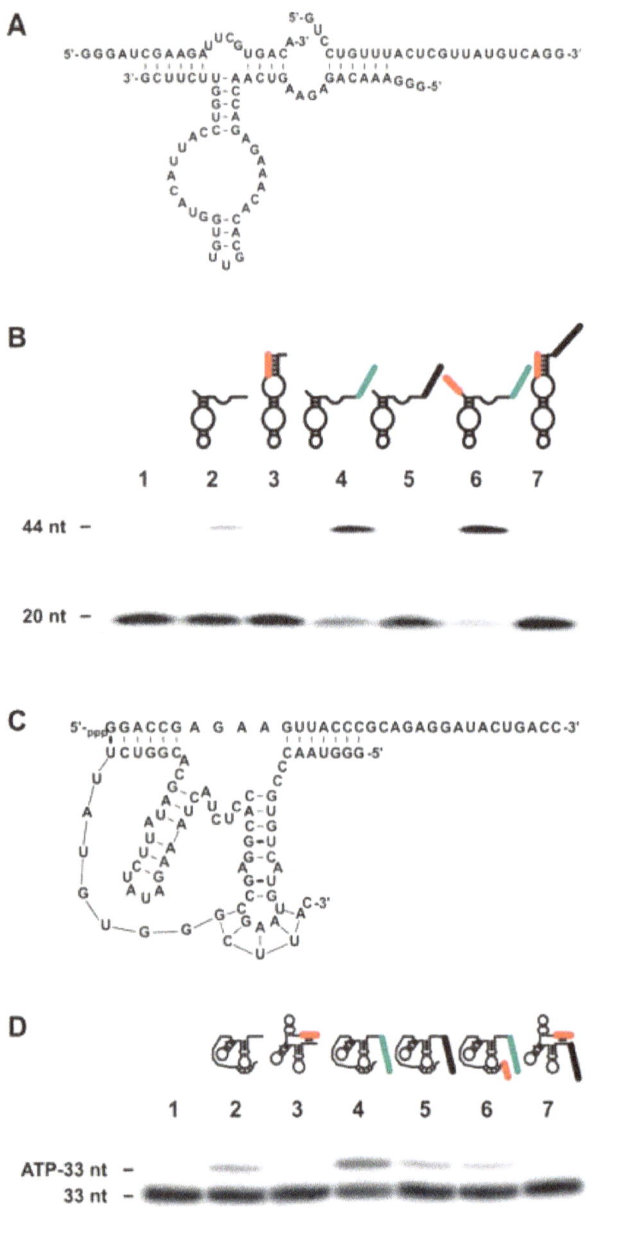

Figure 3. Catalytic activity of the engineered dependent ligating and capping ribozymes.

(A) and (C) are the nucleotide sequences and secondary structures of the ribozyme-substrate complexes of the ligating hairpin and capping ribozymes, respectively. (B) and (D) are autoradiograms of the PAGE gels used to analyse the ligation and capping reactions, respectively. The schematic structures of the nucleic acid enzymes with both the blocker (red) and biosensor (green) are illustrated above the appropriate lanes of the gels. The controls (-) was performed in the absence of ribozyme (lane 1), and, lane 2, in the presence of the unmodified versions. Lane 3 is the ribozyme extended by the blocker sequence. Lanes 4 and 5 are the versions extended by a biosensor that is either complementary, or not, to the substrate, respectively. Finally, lanes 6 and 7 are the *on* and *off* ribozymes, respectively. The nucleotide sequences of each engineered nucleic acid enzyme are depicted.

Subsequently, the concept was adapted to an artificial ribozyme that catalyzed the attack of a 5′-terminal nucleotide phosphate on the α-phosphate of the terminal triphosphate of an RNA substrate, forming a 5′-5′ phosphodiester linkage (i.e. a capping ribozyme; Figure 3C–D) [16]. A version harboring a blocker alone exhibited only residual activity for the capping of a 33-nucleotide substrate (<3%), while one possessing only an appropriate biosensor sequence was slightly more efficient than the original. The *on* version has the ability to switch from its *off* to *on* state, while the *off* version remained trapped in the inactive conformation.

DISCUSSION

Together, these experiments demonstrate the possibility of adapting the target-dependent activation concept to various types of nucleic acid enzymes. The limiting factor with such an adaptation is the fact that the suitable candidates need to have a recognition mechanism based on base pair formation in order to permit a rational design. As compared to the allosteric mode of activation that has been engineered for synthetic nucleic acid circuits based on oligonucleotide-sensing allosteric ribozymes and deoxyribozyme [17]–[19], in this case there is no need for the presence of a third partner.

As first glance it might appear somewhat surprising that any known natural catalytic RNA would resort to this mode of regulation in order to provide the appropriate substrate specificity. This might be a mode of regulation for processes that require additional selective pressures in order to ensure a greater fidelity of action, for example as it is required for the MAP kinases. The complexity of the modern cellular metabolism might have been an important selective pressure in this evolutionary process. Catalytic RNAs might not have needed to evolve such a regulation mechanism due to the fact that their recognition strategy, which is based on base pair formation, provided sufficient specificity. Moreover, these ribozymes act in *cis*, therefore there is no requirement for such mechanism. Other natural ribozymes, such as the ribonuclease P [20], ribosome [21] and spliceosome [22], catalyze their reactions with many substrates rendering high substrate specificity detrimental. Alternatively, the ability of *cis*-acting ribozymes to adopt different conformations may have contributed to their regulation. For example, viroids self-cleave their nascent transcripts through a hammerhead structure, but when the polymerase progresses further, a sequence partially complementary to the self-cleavage motif is synthesized, and, consequently, a rod-like structure is folded [23]. In this structure, the hammerhead motif is in an *off* state that prevents cleavage after viroid circularization. Thus, examples of RNA evolution that have already reached the blocker step of this mode of activation do indeed exist. Here, we show that a target-dependent activation, analogous to that of the MAP kinases, is possible for a wide range of nucleic acid catalysis types. It is therefore tempting to speculate that it already exists in nature, and that it simply remains to be found.

MATERIALS AND METHODS

DNA Templates and Constructs

Deoxyribozymes were purchased as oligonucleotides from Invitrogen. All ribozymes and small RNA substrates used were produced by PCR using oligonucleotides purchased from either Invitrogen or IDT Inc. Briefly, DNA oligonucleotides corresponding to the sequences complementary to the different templates, and possessing a T7 promoter complementary sequence, were annealed to T7 promoter primer and the primers extended in PCR reactions. The reaction products were then used as templates for run-off transcription (see below). Alternatively, some constructs were prepared by annealing of two overlapping oligonucleotides that were then filled in by PCR using Vent DNA polymerase (New England Biolab), producing a double-stranded DNA fragment. Each of these fragments was provided with a T7 promoter at the beginning of the sense strand for subsequent use as a template for run-off transcription. Hepatitis C virus (HCV) derived substrate (a 575 nt long template containing the internal ribosome entry site (IRES) and the first part of the ORF) was produced by *Kpn*I digestion of the pHCVA plasmid [6], which contains 1348 nt of the 5′ sequence of hepatitis C virus (genotype 1b) [24]. PCR products and linearized plasmid templates were purified by phenol/chloroform extraction, precipitated with ethanol and dissolved in water.

RNA Synthesis

Run-off transcriptions were performed as described previously [6]. Briefly, transcriptions were performed using purified T7 RNA polymerase (10 μg) in the presence of RNA Guard (24 U, Amersham Biosciences) and pyrophosphatase (0.01 U, Roche Diagnostics) in a buffer containing 80 mM HEPES-KOH, pH 7.5, 24 mM $MgCl_2$, 2 mM spermidine, 40 mM DTT and either linearized plasmid DNA (5 μg) or PCR product (100 pmol) as template. Each NTP (5 mM) was added to the mixture either with or without 50 μCi of [α-^{32}P]UTP (3 000 Ci/mmol, New England Nuclear) in a final volume of 100 μL, and the reaction incubated at 37°C for 2 to 4 h. Upon completion, the reaction mixtures were treated with DNase RQ1 (Promega) at 37°C for 20 min, and the RNA then purified by phenol:chloroform extraction and precipitation with ethanol. RNA products were fractionated by a denaturing (7 M urea) 6 to 20% polyacrylamide gel electrophoresis (PAGE; 19:1 ratio of acrylamide to bisacrylamide) using 45 mM Tris-borate, pH 7.5/1 mM EDTA solution as running buffer. The reaction products were visualized either by UV shadowing, or by autoradiography. The bands corresponding to the correct sizes for both the ribozymes and the substrates were excised from the gel and the transcripts eluted overnight at room temperature in a buffer containing 1 mM EDTA, 0.1% SDS and 0.5 M ammonium acetate. The transcripts were ethanol precipitated at −80°C, washed, dried and dissolved in water. The amounts retrieved were determined either by ^{32}P counting, or by absorbance at 260 nm.

In order to produce 5′-labelled RNA substrates, transcripts were dephosphorylated by adding 1 U of antartic phosphatase (New England Biolab) to 20 pmol of RNA and incubating for 30 min at 37°C in a final volume of 10 μL containing 50 mM Tris-HCl, pH 8.5, 0.1 mM EDTA and 40 U of RNAGuard (Amersham Biosciences). The enzyme was inactivated by incubating at 65°C for 5 min. Dephosphorylated RNA (5 pmol) was then radiolabelled at its 5′-end using 3 U of T4 polynucleotide kinase in the presence of 3.2 pmol of [γ-^{32}P] ATP (6000 Ci/mmol, New England Nuclear). Radiolabelled transcripts were purified by electrophoresis on a denaturing gel and recovered as described above.

Cleavage Assays and Kinetics

All reactions with the cleaving HDV, hammerhead and hairpin ribozymes, as well as those with the deoxyribozyme, were performed under single-turnover conditions ([Rz]>>[S]) using 100 nM of nucleic acid enzyme and trace amounts of 5′-^{32}P-labelled substrates at 37°C in a final volume of 10 μL containing 50 mM Tris-HCl, pH 7.5 and 10 mM $MgCl_2$. After an incubation of 1 h, the reactions were quenched by the addition of ice-cold formamide dye buffer (1 vol of 97% formamide, 10 mM EDTA, pH 8.0, 0.25% xylene cyanol, 0.25% bromophenol blue) and fractionated on 6% to 20% PAGE gels. The gels were then analyzed with a radioanalytic scanner (Storm, Molecular Dynamics). For time-course experiments, aliquots of 0.8 μL were taken at various times up to 3 h and treated as described above. Kinetic analyses were performed under single turnover conditions. Trace amounts of 5′-^{32}P-labelled substrate (<1 nM) were cleaved by various concentrations of either ribozyme or deoxyribozyme (5–400 nM). The fraction cleaved was determined, and the rate of cleavage (k_{obs}) obtained from fitting the data to the equation $A_t = A_{\square}(1-e^{-kt})$ where A_t is the percentage of cleavage at time t, $A_{\square}$ is the maximum percent cleavage (or the end-point of cleavage), and k is the rate constant (k_{obs}). Each rate constant was calculated from at least two independent measurements. The values of k_{obs} obtained were then plotted as a function of nucleic acid enzyme concentration for the determination of k_{cat}, K_M' and k_{cat}/K_M'. The values obtained from independent experiments varied by less than 15%. The experiments of substrate specificity using several substrates were performed under multiple turnover conditions using 100 nM HDV ribozyme and 1 μM substrate.

Ligation Reactions

Ligation reactions were performed as described previously [14], [15] with minor modifications. Full length 5′-^{32}P-labelled substrate (5 pmol) was cleaved in the presence of 10 μM of hairpin ribozyme in a volume of 50 μl containing 50 mM Tris-HCl, pH 7.5 and 10 mM $MgCl_2$. The mixture was incubated 2 h at 37°C, quenched by the addition of ice-cold formamide dye buffer, purified on a 20% denaturing PAGE gel and the radiolabelled 20 nt 5′-product recovered as described. This RNA strand was then used as substrate for the ligation reaction. The 3′-product used in this ligation reaction was purchased from IDT Inc. This deprotected 5′-hydroxyl RNA was purified as described above. Ligation reactions were performed using 10 μM of the hairpin ribozyme in the presence of 15 μM of the 3′-end substrate and trace amounts of 5′-^{32}P-labelled substrate mixed in a total volume of 9 μl containing 50 mM HEPES-KOH, pH 7.0. The reactions were preheated at 37°C for 10 min so as to permit the annealing of the substrate to the ribozyme, and the ligation reactions started (time=0) by adding 1 μL of 150 mM $MgCl_2$. Samples were incubated at 37°C for 1 h, fractionated on gels and visualized by autoradiography.

Capping Enzyme Reaction

The capping reaction was performed as described previously [16]. Substrate was 3′-end labelled by incubating, at 37°C, 10 pmol of transcripts with 3.2 pmoles of [^{32}P]Cp (3000 ci/mmol, New England Nuclear) and 1 U T4 RNA ligase (New England Biolab) in the buffer supplied by the manufacturer supplemented with 10% DMSO. After 3 h, the reaction was stopped and the transcripts purified by denaturing 10% denaturing PAGE and recovered as described above. Capping reactions were carried out under single turnover conditions. Both the ribozyme (2 μM) and substrate (0.2 μM) were annealed together for 5 min at 22°C in a reaction buffer containing 50 mM Tris-HCl, pH 7.5, 25 mM $MgCl_2$ and 150 mM KCl. The reactions were initiated by the addition of ATP to a final concentration of 1 mM. After 16 h of incubation at 22°C, the reactions were quenched by addition of ice-cold formamide dye buffer and analyzed on denaturing 15% PAGE gels.

ACKNOWLEDGMENTS

The authors gratefully acknowledge L. J. Bergeron and their colleagues for stimulating discussions. J.P.P. holds the Canada Research Chair of Genomics and Catalytic RNA.

REFERENCES

1. Fersht A (1999) Structure and mechanism in protein science. New York: WH Freeman and Company. pp. 289–318. ed. 3.
2. Winkler WC, Nahvi A, Roth A, Collins JA, Breaker RR (2004) Control of gene expression by a natural metabolite-responsive ribozyme. Nature 428: 281–286.
3. Jansen JA, McCarthy TJ, Soukup GA, Soukup JK (2006) Backbone and nucleobase contacts to glucosamine-6-phosphate in the glmS ribozyme. Nature Struct Mol Biol 13: 517–523.
4. Biondi RM, Nebreda AR (2003) Signalling specificity of Ser/Thr protein kinases through docking-site-mediated interactions. Biochem J 372: 1–13.
5. Tanoue T, Nishida E (2003) Molecular recognitions in the MAP kinase cascades. Cell Signalling 15: 455–462.
6. Bergeron L Jr, Perreault JP (2005) Target-dependent on/off switch increases ribozyme fidelity. Nucleic Acids Res 33: 1240–1248.
7. Bergeron L Jr, Reymond C, Perreault JP (2005) Functional characterization of the SOFA delta ribozyme. RNA 11: 1858–1868.
8. Lévesque D, Choufani S, Perreault JP (2002) Delta ribozyme benefits from a good stability in vitro that becomes outstanding in vivo. RNA 8: 464–477.
9. Blount KF, Uhlenbeck OC (2005) The structure-function dilemma of the hammerhead ribozyme. Annu Rev Biophys Biomol Struct 34: 415–440.
10. Silverman SK (2005) In vitro selection, characterization, and application of deoxyribozymes that cleave RNA. Nucleic Acids Res 33: 6151–6163.
11. Wilson TJ, Nahas M, Ha T, Lilley DM (2005) Folding and catalysis of the hairpin ribozyme. Biochem Soc Trans 33: 461–465.
12. Sakamoto N, Wu CH, Wu GY (1996) Intracellular cleavage of hepatitis C virus RNA and inhibition of viral protein translation by hammerhead ribozymes. J Clin Invest 98: 2720–2728.
13. Schubert S, Furste JP, Werk D, Grunert HP, Zeichhardt H, et al. (2004) Gaining target access for deoxyribozymes. J Mol Biol 339: 335–363.
14. Komatsu Y, Shirai M, Yamashita S, Ohtsuka E (1997) Construction of hairpin ribozymes with a three-way junction. Bioorg Med Chem 5: 1063–1069.
15. Sargueil B, Hampel KJ, Lambert D, Burke JM (2003) In vitro selection of second site revertants analysis of the hairpin ribozyme active site. J Biol Chem 278: 52783–52791.
16. Zaher HS, Watkins RA, Unrau PJ (2006) Two independently selected capping ribozymes share similar substrate requirements. RNA 12: 1949–1958.
17. Stojanovic MN, Stefanovic D (2003) A deoxyribozyme-based molecular automaton. Nature Biotech 21: 1069–1074.
18. Penchovsky R, Breaker RR (2005) Computational design and experimental validation of oligonucleotide-sensing allosteric ribozymes. Nature Biotech 23: 1424–1433.
19. Davidson EA, Ellington AD (2007) Synthetic RNA circuits. Nature Chem Biol 3: 23–28.
20. Guerrier-Takada C, Gardiner K, Marsh T, Pace N, Altman S (1983) The RNA moiety of ribonuclease P is the catalytic subunit of the enzyme. Cell 35: 849–857.
21. Noller HF, Hoffarth V, Zimniak L (1992) Unusual resistance of peptidyl transferase to protein extraction procedures. Science 256: 1416–1419.
22. Valadkhan S, Manley JL (2001) Splicing-related catalysis by protein-free snRNAs. Nature 413: 701–707.
23. Beaudry D, Bussière F, Lareau F, Lessard C, Perreault JP (1995) The RNA of both polarities of the peach latent mosaic viroid self-cleaves in vitro solely by single hammerhead structures. Nucleic Acids Res 23: 745–752.
24. Alaoui-Ismaili MH, Gervais C, Brunette S, Gouin G, Hamel M, et al. (2000) A novel high throughput screening assay for HCV NS3 helicase activity. Antiviral Res 46: 181–193.

CHAPTER 17 THE RNA–DNA HYBRID STRUCTURE DETERMINED BY EPR, CD AND RNASE H1

Olga Romainczyk, Burkhard Endeward, Thomas F. Prisner and Joachim W. Engels

INTRODUCTION

RNA–DNA hybrid helices, first reported by Rich 50 years ago, are important intermediates in biology and play an essential role in transcription of DNA into RNA, replication of double-stranded DNA and reverse transcription of viral RNA into DNA.[1] They are also key to be recognized by degrading enzymes like ribonuclease H (RNase H). The latter enzyme is an essential component of the antisense technique that degrades mRNA and reduces gene expression.[2,3] Understanding the molecular mechanism of these processes depends on more detailed knowledge of the structure of the hetero duplexes.

Electron Paramagnetic Resonance (EPR), in particular pulsed electron–electron double resonance (PELDOR),[4] can be applied to study long range distance constraints up to 7 nm of oligonucleotides.[5] The accuracy of the method is high enough to distinguish between B and A-helix forms[6,7] as present in DNA and RNA or for DNA in the presence of trifluoroethanol.[8]We successfully established the organic nitroxide TPA (2,2,5,5-tetramethyl-3-pyrrolin-*N*-oxyl-3-acetylene, Fig. 1) as a rigid paramagnetic label for DNA and RNA.[9] In addition by adding the label at a defined position on the heterocyclic base moiety we can direct it to either the major or minor groove of the double helix. Adding the spin label at the 5-position of pyrimidinesit points towards the major groove, whereas at the 2-position of purines minor grooves are addressed.

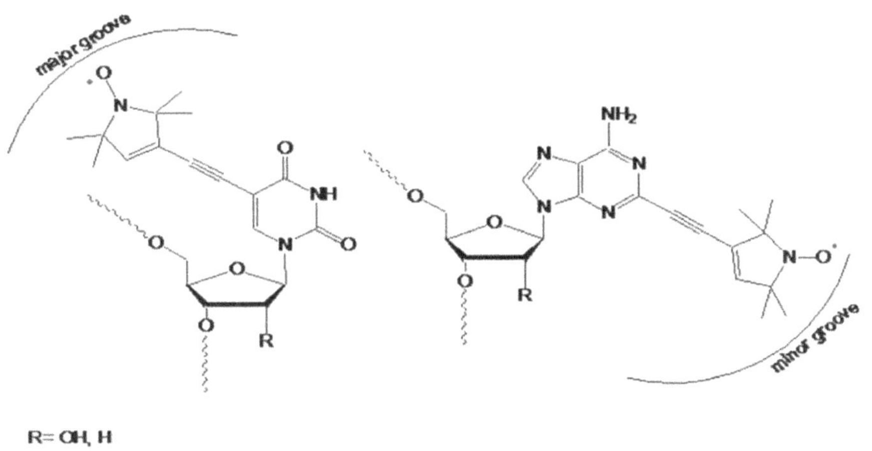

Figure 1. Left: rU/dT labelled on 5-position with TPA; right: rA/dA labelled on 2-position with TPA.

Citation: Olga Romainczyk Burkhard Endeward Thomas F. Prisner and Joachim W. Engels, The RNA–DNA hybrid structure determined by EPR, CD and RNase H1, DOI: 10.1039/c0mb00258e.

We chose non-self-complementary 15-mer sequences with identical base sequence for DNA and RNA except for T and U.

- DNA(dT) 5′-CGCTACATAGTGAGC-3′
- DNA(dA) 5′-GCTCACTATGTAGCG-3′
- RNA(rU) 5′-CGCUACAUAGUGAGC-3′
- RNA(rA) 5′-GCUCACUAUGUAGCG-3′

where the underlined bases carry the spin label. The spin label was placed in the more rigid center part, at least 3 bases apart from either end. Based on PyMol modeling the geometry of the angles and distances should show optimal discrimination between A- (2.5 nm) and B-form (2.9 nm) when hybridized. In our case an optimal distance appeared to be between 2 and 4 nm and the orientation of the two spin labels to each other was similar. From the 8 different duplexes resulting 6 were chosen: DNA(dT)/DNA(dA) to RNA(rU)/RNA(rA)*via* the two possible hybrids DNA(dT)/RNA(rA), DNA(dA)/RNA(rU) with spin label and two without DNA(dT)/RNA(rA) and DNA(dA)/RNA(rU).

RNA(rU) and (rA) were synthesized *via* "on column" solid phase synthesis as reported earlier.[7] For DNA synthesis the 2-iodo-deoxy-adenosine building block had to be prepared. This we accomplished *via* an analogous synthesis to the RNA building block but had to change the protecting groups in order to get better yields using milder conditions (see ESI†). The DNA oligomers were synthesized by phosphoramidite chemistry on a 1 μmol scale, with the same coupling time for all used phosphoramidites. The RNAs were synthesized on a 0.2 μmol scale with phosphoramidites purchased from Dharmacon (ACE chemistry).[10]

All oligonucleotides were synthesized completely without interrupting the synthesis after incorporation of the iodinated phosphoramidite. For the Sonogashira cross coupling the column was removed from the synthesizer, maintained under an argon atmosphere and acetylenic TPA was added. To achieve better yields, the Sonogashira cross coupling was performed two times with the same amount of reagents.

Finally, the DNA was cleaved from the support and completely deprotected with ammonia. In the case of RNA, the cleavage of methyl-groups of the phosphate was done first using disodium-2-carbamoyl-2-cyanoethylene-1,1-dithiolate-trihydrate. Afterwards, in one step the oligonucleotide was cleaved from the solid-support (polystyrene) and the functional groups were deprotected with methylamine and purified by anion-exchange HPLC with 1 M LiCl. Prior to use for ACE chemistry deprotection of the 2′-hydroxy group is performed by TEMED/acetic acid buffer, pH 3.8. The DNA and RNA synthesised were characterized by electron spray ionization (ESI) spectra (Table 1, ESI†). UV melting curves (T_m) of the duplexes dissolved in phosphate buffer (10 mM Na_2HPO_4, 10 mM NaH_2PO_4, 140 mM NaCl, 0.01 mM duplex, pH 7) were recorded on a UV-/VIS-600 from 5 to 95 °C with a heating rate of 0.5 °C min^{-1}. CD spectra were measured on a JASCO J-710 spectropolarimeter with a Peltier thermostat (Table 1). The duplexes for the distance measurements were dissolved in the same bufferwith addition of 20% ethylene glycol and at the 10× concentration as for the T_m-measurements.

Table 1. Comparison between different duplexes

Duplex	$T_m{}^a$/°C	Distance r^b/nm	$\lambda_1{}^c$/nm	$t_{1/2}{}^d$/min	Helix type
DNA(dT)/RNA(rA)	47.8 ± 0.2	2.8 ± 0.1	270	25.8 ± 4.6	A-like form
DNA(dA)/RNA(rU)	45.7 ± 0.2	3.0 ± 0.05	282	147.8 ± 55.9	B-like form
DNA(dT)/DNA(dA)	48.6 ± 0.1	3.0 ± 0.05	276	[—]	B-form
RNA(rA)/RNA(rU)	63.5 ± 0.4	2.7 ± 0.1	266	[—]	A-form
DNA(dT)/RNA(rA)	50.9 ± 0.2	[/]	273	110.9 ± 11.5	B-like form
DNA(dA)/RNA(rU)	49.4 ± 0.2	[/]	277	130.9 ± 40.9	B-like form

[a] In phosphate buffer, pH 7. [b] Measured by PELDOR with addition of 20% ethylene glycol. [c] The maximum in CD spectra to differentiate the helix type. [d] Treatment with RNase H at 37 °C, calculated by extrapolation of the data: [/], not spin labelled; [—], no degradation.

The dead-time free 4-pulse PELDOR sequence was used for all experiments.[11] X-band PELDOR data were collected at 40 K with a detection window of 2.5 μs. Further experimental details are given in the ESI.† The distance measured[12] for the modified RNA double helix was 2.7 nm and for the DNA double helix 3.0 nm (see Fig. 2) which nicely fits the expected distances for an A- (2.5 nm) and B-helix (2.9 nm) geometry. For the hybrids we unexpectedly obtained two different distances, depending on the labelling position. When hybridizing labelled DNA(dT) with labelled RNA(rA) we measured 2.8 nm but with labelled DNA(dA) and labelled RNA(rU) we obtained a distance of 3.0 nm for the mixed hybrid (Table 1). Therefore the first hybrid is close to the distance measured for RNA, whereas the second is identical to the distance measured for DNA.

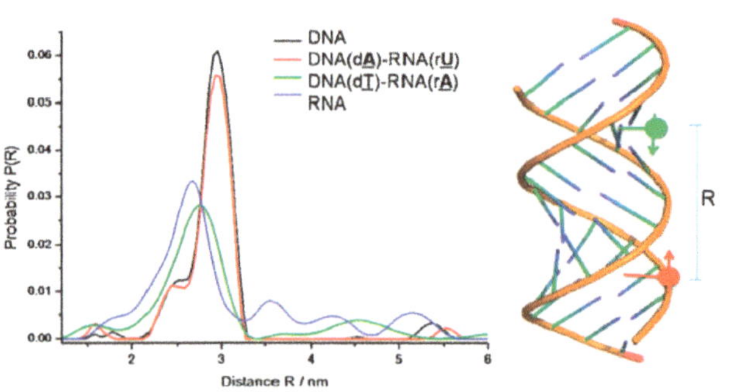

Figure 2. PELDOR determined distances distribution function $P(R)$ between the two TPA spin labels of duplexes (DNA, DNA/RNA and RNA).

To our surprise these two classes of spectra are also reflected in the UV and CD spectra of the spin labelled duplexes as well as in the reduced hyperchromicity during melting as observed by UV for DNA like structures compared to RNA (see Fig.

3 and detailed CD analysis in ESI). Therefore the distances between the spin labels measured (B- to A-like) can be grouped DNA/DNA ≈ DNA(d**A**)–RNA(r**U**) > DNA(d**T**)–RNA(r**A**) ≈ RNA/RNA. The only difference between the two hybrids is the attachment side for the spin label. Here the spin labelat the U5 position points towards the major groove whereas the label at the 2-position on A is directed to the open minor groove (Fig. 1). Considering the differences between the major and minor grooves in A- and B-helices it is obvious that a bulky spin label (as seen in the X-ray structure of TPA)[13] is less well accommodated in the narrow major groove of an A-helix than in a B-helical form with a wide open major groove (helices modeled with PyMol, see ESI†).

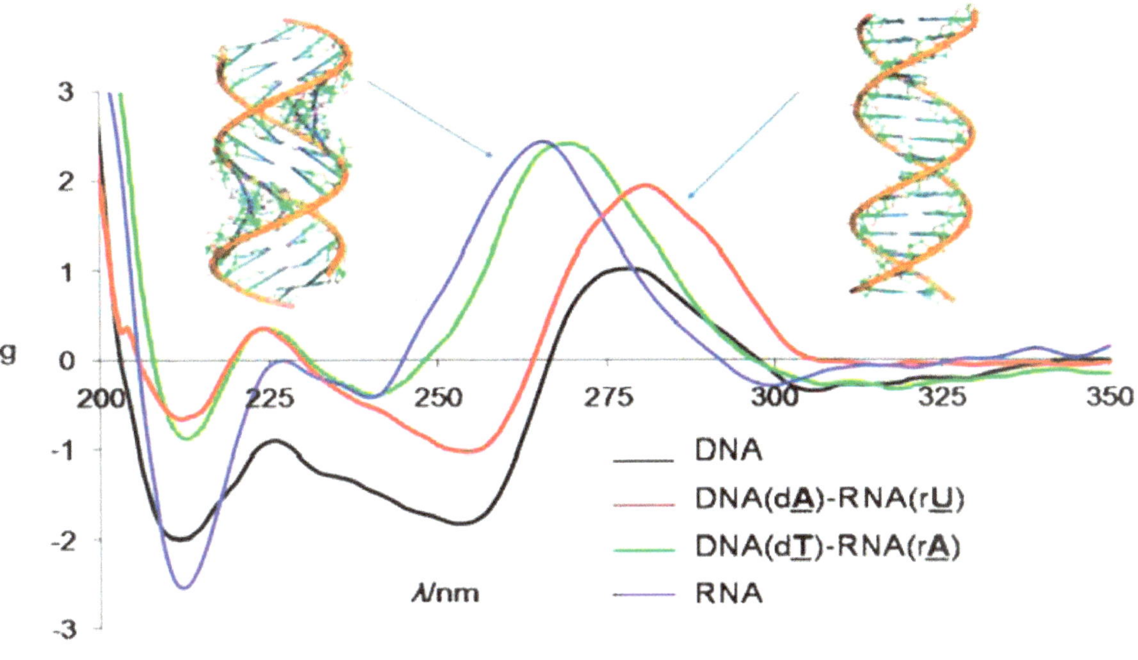

Figure 3. CD-spectra of spin labelled duplexes: left, A-; right, B-helix.

Following this observation we argue that in the hybrid DNA(dA)–RNA(rU) the RNA seems to prefer more the B-like geometry, whereas the hybrid DNA(dT)–RNA(rA) remains in the RNA-like structure. As a further proof for our assignment we checked how RNase H1 recognizes these hybrids. RNase H1 lacks a consensus sequence for cleavage and preferentially cleaves 7–10 nucleotides from the 5′-RNA/3′-DNA terminus.[14] It is known that preferred cleavage by the enzyme is a product of fitting sequential/helical geometry, predominantly A-like.[15]We incubated the 6 helices, the DNA and RNA double strands as well as the four mixed hybrids and measured the cleavage rates by HPLC-analysis. For treatment with the enzyme RNase H (Bacillus) the double helices were formed in sterile water first. RNase H-buffer (50 mM Tris-HCl, 75 mM KCl, 3 mM $MgCl_2$, 10 mM dithiothreitol, pH 8.3) was added and the probes were incubated at 37 °C. The cleavage rate was followed by anionic exchange-HPLC (1 M LiCl).

Neither DNA nor RNA was cleaved but both hybrids roughly at the same rate (see Table 1). In contrast the two spin labelled hybrid structures were cleaved at quite different rates to each other. The $t_{1/2}$ for RNA(rA)/DNA(dT) is 26 min and for RNA(rU)/DNA(dA) is 148 min, a rate difference by a factor of approx. 6. Compared to the unmodified hybrids DNA/RNA ($t_{1/2}$120 min), the first combination is about 4.5× faster and the second 1.2× slower. We interpret this rate difference as being due to the fact that RNA(rA), where the spin label points to the minor groove, has no problem to adopt the A-helix geometry and is therefore cleaved nicely. In contrast for RNA(rU) there is limited space for the spin label in the major groove of an A form, thus it is more difficult to adopt the necessary A-helix geometry. This is in full agreement with crystal data for the RNase H1 in complex with RNA/DNA, where one finds a more flexible DNA and a more rigid RNA in the A-form.[16] Furthermore for the DNA part it is known that acetylenic side chains in the 5-position of U enhance RNase H1 cleavage.[17] Thus the RNA/DNA hybrid is more easily cleaved when the RNA is predominantly A-form and the DNA can be flexible enough to adopt the A-form as observed in the X-ray structure of the RNase H.[16] This has consequences for the design of hybrid structures as in the case of antisense oligonucleotides, where strong A-form supporting nucleoside analogues inhibit RNase H cleavage.[15] Similarly this may also be applicable to RNase H like enzymes as the argonaute silencing complex Ago2 (SLICER) in the RNAi pathway.[18,19]

CONCLUSIONS

In summary we have introduced PELDOR in combination with CD for the characterization of RNA/DNA hybrid structures. We were able to show excellent agreement of PELDOR data with CD spectra. RNase H1 cleavage data support the observed structural differences. In agreement with the biological functionality of RNase H1 for faster cleaved hybrids an A-type structure should be present. This combination of PELDOR and

CD shows promise for more detailed RNA/DNA structure determination.

REFERENCES

1. (*a*) A. Rich, *J. Biol. Chem.*, 2006, **281**, 7693–7696 ; A. Rich, *Proc. Natl. Acad. Sci. U. S. A.*, 1960, **46**, 1044–1053 ; (*b*) J. T. Olimpo and J. J. DeStefano, *Nucleic Acids Res.*, 2010, **38**, 4426–4435 .
2. H. Wu, W. F. Lima, H. Zang, A. Fan, H. Sun and S. T. Crooke, *J. Biol. Chem.*, 2004, **279**, 17181–17189 .
3. J. Kurreck, *Eur. J. Biochem.*, 2003, **270**, 1628–1644 .
4. (*a*) A. D. Milov, A. B. Ponomarev and Y. D. Tsvetkov, *Chem. Phys. Lett.*, 1984, **110**, 67–72 ; Y. D. Tsvetkov, A. D. Milov and A. G. Maryasov, *Usp. Khim.*, 2008, **77**, 515–550; (*b*) O. Schiemann, *Methods Enzymol.*, 2009, **469**, 329–351 , Part B.
5. (*a*) O. Schiemann and T. Prisner, *Q. Rev. Biophys.*, 2007, **40**, 1–53 ; N. K. Kim, M. K. Bowman and V. J. DeRose, *J. Am. Chem. Soc.*, 2010, **132**, 8882–8884 ; (*b*) A. Marko, D. Margraf, P. Cekan, S. T. Sigurdsson, O. Schiemann and F. T. Prisner, *Phys. Rev.*, 2010, **E81**, 021911 ; N. A. Kuznetsov, A. D. Milov, V. V. Koval, R. I. Samoilova, Y. A. Grishin, D. G. Knorre, Y. D. Tsvetkov, O. S. Fedorova and S. A. Dzuba, *Phys. Chem. Chem. Phys.*, 2009, **11**, 6826–6832 .
6. O. Schiemann, N. Piton, Y. Mu, G. Stock, J. W. Engels and T. F. Prisner, *J. Am. Chem. Soc.*, 2004, **126**, 5722–5729 .
7. N. Piton, Y. Mu, G. Stock, O. Schiemann and J. W. Engels, *Nucleic Acids Res.*, 2007, **35**, 3128–3143 .
8. G. Sicoli, G. Mathis, O. Delalande, Y. Boulard, D. Gasparutto and S. Gambarelli, *Angew. Chem., Int. Ed.*, 2008, **47**, 735–737 .
9. (*a*) T. Strube, O. Schiemann, F. McMillan, T. Prisner and J. W. Engels, *Nucleosides, Nucleotides Nucleic Acids*, 2001, **20**(4), 1271–1274 ; (*b*) O. Schiemann, N. Piton, J. Plackmeyer, B. Bode, T. Prisner and J. W. Engels, *Nat. Protoc.*, 2007, **2**, 904–923.
10. S. A. Scaringe, F. E. Wincott and M. H. Caruthers, *J. Am. Chem. Soc.*, 1998, **120**(45), 11820–11821 .
11. R. E. Martin, M. Pannier, F. Diederich, V. Gramlich, M. Hubrich and H. W. Spiess, *Angew. Chem., Int. Ed.*, 1998, **37**, 2834–2837.
12. G. Jeschke, V. Chechik, P. Ionita, A. Godt, H. Zimmermann, J. Banham, C. R. Timmel, D. Hilger and H. Jung, *Appl. Magn. Reson.*, 2006, **30**, 473–498 .
13. O. Frolow, J. W. Bats and J. W. Engels, *Acta Crystallogr., Sect. E: Struct. Rep. Online*, 2009, **65**, o1848 .
14. P. S. Pallan and M. Egli, *Cell Cycle*, 2008, **7**, 2562–2569 .
15. W. F. Lima, J. B. Rose, J. G. Nichols, H. Wu, M. T. Migawa, T. K. Wyrzykiewicz, A. M. Siwkowski and S. T. Crooke, *Mol. Pharmacol.*, 2007, **71**, 83–91; W. F. Lima, J. B. Rose, J. G. Nichols, H. Wu, M. T. Migawa, T. K. Wyrzykiewicz, G. Vasquez, E. E. Swayze and S. T. Crooke, *Mol. Pharmacol.*, 2007, **71**, 73–82.
16. M. Nowotny, S. A. Gaidamakov, R. J. Crouch and W. Yang, *Cell (Cambridge, Mass.)*, 2005, **121**, 1005–1016 ; M. Nowotny, S. A. Gaidamakov, R. Ghirlando, S. M. Cerritelli, R. J. Crouch and W. Yang, *Mol. Cell*, 2007, **28**, 264–276 .
17. A. J. Gutierrez, M. D. Matteucci, D. Grant, S. Matsumura, R. W. Wagner and B. C. Froehler, *Biochemistry*, 1997, **36**, 743–748 .
18. Y. Wang, G. Sheng, S. Juranek, T. Tuschl and D. J. Patel, *Nature*, 2008, **456**, 209–213 .
19. W. F. Lima, H. Wu, J. G. Nichols, H. Sun, H. M. Murray and S. T. Crooke, *J. Biol. Chem.*, 2009, **284**, 26017–26028

CHAPTER 18

LIPID PEROXIDATION MODIFIES THE ASSEMBLY OF BIOLOGICAL MEMBRANES "THE LIPID WHISKER MODEL"

Ángel Catalá

Keywords: Lipid peroxidation, Lipid Whisker Model, membranes, PUFAs, Fluid Mosaic Model

INTRODUCTION

When the Fluid Mosaic Model (FMM) of biological membrane structure was introduced 42 years ago, it was visualized as a basic model for cell membranes that could explain existing data on membrane proteins and lipid structures and their dynamics. According to the (FMM), a membrane was described as a biological fluid of proteins and lipids oriented in two dimensions. The basic structure of all cell and organelle membranes is the lipid bilayer. Protein molecules are distributed in different regions of the bilayer and perform diverse functions. Cell membranes are active, fluid structures, and most of their molecules are able to travel in the plane of the membrane (Singer and Nicolson, 1972). The viscosity of a lipid membrane largely depends on whether the acyl chains attached to glycerophospholipids are grouped into a rigid state or exist in a relatively disordered, fluid state. Long chain saturated fatty acids maximize Van der Waals forces, and increase the viscosity of the membrane (Chapman and Benga, 1984). Fluidity is defined as the ease of movement and represents the reciprocal of the viscosity of the membrane (Lee, 1991). The fluid properties of biological membranes are critical for various cell functions. Still slight changes in membrane fluidity may cause unusual function and pathological processes (Garcia et al., 2005). After more than 40 years, the (FMM), described by Singer and Nicolson is still relevant to recognize the structure, function, and dynamics of biological membranes (Nicolson, 2014). However, several biological processes cannot be explained on the basis of this typical phospholipid orientation and utilize other phospholipid conformations (Catala, 2012) that are described in this article.

THE STRUCTURAL AND FUNCTIONAL PROPERTIES OF MEMBRANES ARE DETERMINED BY POLYUNSATURATED FATTY ACIDS

The fluid properties in biological membranes are recognized principally by the presence of polyunsaturated fatty acids (PUFAs) in phospholipids molecules located in both sites of the lipid bilayer. The nature and saturation of the attached fatty acid of the phospholipids generate spectacular effects on membrane packing and fluidity (Janmey and Kinnunen, 2006). The unsaturated fatty acids give a high degree of conformational flexibility to the unsaturated hydrocarbon chains in the membranes because they occupy a small wedge-shaped space. This generally results in looser packing and a more fluid membrane. In contrast, saturated fatty acids confer rigidity that results in a less fluid or more arranged membrane. The rigidity permits saturated fatty acids to group together tightly and forms a solid at inferior temperatures. When a new double bond is open in the fatty acid produces a "kink" in the molecule. For this reason unsaturated fatty acids, such as docosahexaenoic acid (22:6 n-3), assume numerous patterns since this fatty acid can turn around in the area of C–C bonds but not in the region of the rigid C=C bonds (Feller et al., 2002). Adjustments in the lipid composition modify the fluidity of membranes. Lipid composition is modified during regulation of de novo synthesis at selected cellular places, distribution or transfer to new sites, and by specific modifying reactions. Stearyl CoA desaturase is in charge for the formation of monoethylenic fatty acids from saturated fatty acids by catalyzing the addition of a double bond into the ninth carbon of saturated C16:0 and C18:0 substrates (Enoch et al., 1976).

Citation: Catalá Á (2015) Lipid peroxidation modifies the assembly of biological membranes "The Lipid Whisker Model". Front. Physiol. 5:520. doi: 10.3389/fphys.2014.00520.

THE TRANSPORT AND DIFFUSION OF OX YGEN IN MEMBRANES IS REGULATED BY CHOLESTEROL AND FATTY ACIDS

Lipid peroxidation and formation of reactive oxygen species are important chemical reactions that use oxygen and occur in cell membranes. The diffusion of oxygen into synthetic membranes prepared with phosphatidylcholine and cholesterol have been investigated by Subczynski et al. (1991). Oxygen permeability through the membrane in all the membranes studied in this work was limited by the presence of cholesterol. Thus, they showed an increase in oxygen transport in the central part of the synthetic membranes because cholesterol decreases oxygen transport in and around the head group regions, where the main obstacles to the oxygen permeability exist. Based on this explanation, it can be assumed that non-raft areas rich in PUFA-phospholipids and vitamin E will be more available by oxygen than lipid rafts areas containing sphingolipids and cholesterol. This condition will make several micro areas more vulnerable to lipid peroxidation than others.

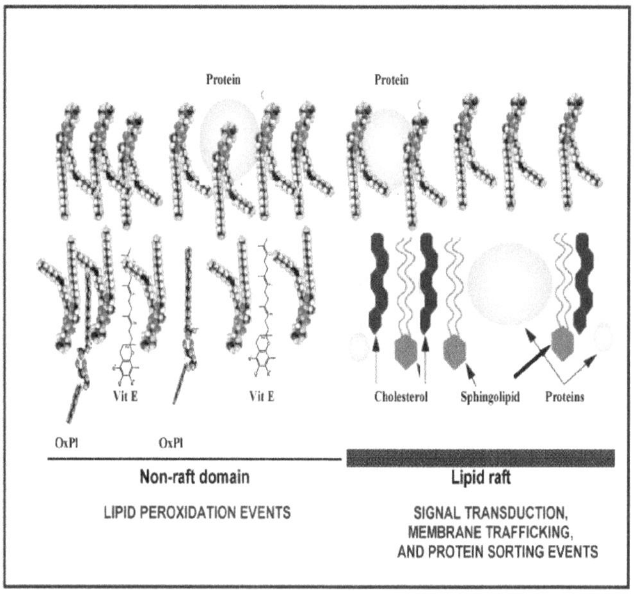

Figure 1. Hypothetical model of the plasma membrane. In terms of lipids, the heterogeneous membrane is believed to consist of a mixture of a dispersed "lipid raft" phase, enriched in cholesterol, raft-associated proteins, and saturated lipids (such as sphingolipids), and the "non-raft" matrix phase enriched with phospholipids containing PUFAs and vitamin E. Vitamin E, partition into domains that are enriched in polyunsaturated phospholipids increasing the concentration of the vitamin in the place where take place the lipid peroxidation process and oxidized phospholipids (oxPl) are formed. Reproduced from Catala (2012). Copyright © 2012 Elsevier Masson SAS. All rights reserved.

THE "LIPID WHISKER MODEL" IS SUITABLE TO EXPLAIN THE STRUCTURE OF OXIDIZED CELL MEMBRANES

The Lipid Whisker Model (LWM) (Greenberg et al., 2008) is an extension of the Fluid Mosaic Model suggested by Singer and Nicolson (1972). Recent studies into the conformation of oxidized phospholipid (oxPL) species recognized by CD36 within model membranes (Li et al., 2007) have led to the development of the LWM. A primary attribute of the (FMM) is that amphipathic phospholipids are oriented in an organization lamellar mesophase with hydrophobic fatty acyl chains embedded within the interior of the membrane and the hydrophilic polar groups facing the aqueous environment. This lipid organization accepts fast lateral circulation of lipids and membrane proteins similarly within the planar membrane surface. It also causes the water-resistant character of cell membranes to hydrophilic molecules. But, recent information proposes that in peroxidized cell membranes, numerous of the oxPL classes adopt a particular conformation. Lipid peroxidation is achieved by addition of numerous polar molecules on fatty acid chains (Catala, 2009). Consequently, when cell membranes undergo oxidation, if not adapted by the action of phospholipases, they may "produce whiskers" including a variety of oxidized sn-2 fatty acids of diverse structures. In the (LWM), the assembly of many oxPL within cell membranes is different compared with one observed in non-oxPL described in the (FMM). Biophysical studies have shown that the addition of an oxygen atom to the acyl chain produces a significant change that prevent its immersion in the interior of the membrane, however, the modified acyl chain is projected on the aqueous medium (Figure 1) (Subczynski et al., 1991; Greenberg et al., 2008).

As cell membranes are peroxidized "grow whiskers" because phospholipids are peroxidized, and several of its oxidized fatty acids are projected onto the surface (Figure 1). This change produced by the oxidation of a fatty acid chain in the membrane may be a trigger event to numerous biological activities.

THE EFFECT OF OTHER MEMBRANE MODIFICATIONS, TRIGGERED BY LIPID PEROXIDATION PRODUCTS AND REDUCING SUGARS INTRODUCES CHANGES IN CELL MEMBRANE PHYSICO-CHEMICAL AND BIOLOGICAL PROPERTIES

During lipid peroxidation, biomolecules such as proteins or amino lipids can be covalently modified by lipid decomposition products (Catala, 2009). For the case of aliphatic aldehydes (alkanals) such as 1-hexanal or 1-nonanal, the Nε-amino groups of the lysine residues in proteins can be modified through the formation of a Schiff base. α,β-Unsaturated aldehydes (alkenals) such as acrolein or 4-hydroxy-2-nonenal react with lysine, cysteine, and histidine through a Michael-type addition. Conversely, lipid hydroperoxide might covalently react with protein without serious decomposition of its structure. However, the mechanism of lipid hydroperoxide-derived protein modification is not so clear. Evidence for in situ ethanolamine phospholipid adducts with hydroxy-alkenals has been recently described.

Non-enzymatic modification of aminophospholipids by lipid peroxidation-derived aldehydes and reducing sugars through carbonyl-amine reactions are thought to contribute to the age-related deterioration of cellular membranes (Naudí et al., 2013). Much evidence demonstrates the modification of aminophospholipids by glycation, glycoxidation and lipoxidation reactions.

LIPID PEROXIDATION IN MEMBRANES: THE PEROXYL RADICAL DOES NOT "FLOAT"

Some key microscopic aspects of the lipid peroxidation reaction in cell membranes are still poorly studied. In particular, it is usually accepted that the propagation of the radical reaction in lipid bilayers is hampered by the rapid diffusion of peroxyl intermediates toward the water interface, that is, out of the reaction region. Contrary to what is expected by the LWM, Garrec et al. (2014) have recently investigated the validity of this "floating peroxyl radical" hypothesis by means of molecular modeling. Combining quantum calculations of model systems and atomistic simulations of lipid bilayers containing lipid oxidation products, these authors predict that the peroxyl radical does not "float" at the surface of the membrane. Instead, it remains located quite deep inside the bilayer.

THE CHANGES INDUCED BY LIPID PEROXIDATION IN MEMBRANE STRUCTURE AND DYNAMICS, BOTH AT THE LIPID AND PROTEIN LEVEL

The mechanisms and principles establishing the thousands of proteins and lipids that make up membrane bilayers in cells are still unclear. Mueller et al. reviewed the basic properties of biological membranes and the most common theories for lateral segregation of membrane components before discussing an emerging model of a self-organized, multi-domain membrane or "patchwork membrane" (Mueller et al., 2012).

The oxidation of phospholipids has become a recent topic of interest within the field of membrane biophysics. Still, the exact mechanism of membrane injury by oxidized lipids is uncertain (Wong-Ekkabut et al., 2007).

Membrane rafts remain one of the most controversial issues in biophysics, because the methods for their detection are still far from perfect. Optical techniques rely heavily on fluorescent markers. In biological membranes, these probes are connected with lipids and thus define only the lipid part of the membrane. Therefore, they are "blind" to the presence of membrane proteins, unless FRET techniques between proteins and membrane probes are used. Therefore, the imperative future challenge for the membrane probes will be to design molecules capable of monitoring lipid surrounding specifically around a given protein of interest (Klymchenko and Kreder, 2014).

Proteins and lipids in membrane processes are reciprocally dependent on each other. While the lipid domains as organizers of proteins has attracted wide-spread consideration, the question whether proteins control lipids or lipids control proteins in cell membranes is not a simple problem to solve. The fact that proteins exploit the biophysical properties of lipids including lipid charge and phase separation for membrane function is one of the reasons why membrane biology is such an attractive area of research (Rossy et al., 2014).

To guarantee coordination of cellular activities, cells use membrane contact sites (MCSs) between the membranes of diverse organelles. MCSs are domains where two membranes come to close proximity, typically less than 30 nm, and create microdomains that favor exchange between two organelles. Since the endoplasmic reticulum (ER) is the most widespread cellular membrane network, it is thus not surprising to find the ER involved in most MCSs within the cell. The ER contacts diverse compartments such as mitochondria, lysosomes, lipid droplets, the Golgi apparatus, endosomes, and the plasma membrane. Taken into account these observations, several levels of complexity have to be considered when analyzing the changes induced by lipid peroxidation in membrane structure and dynamics, both at the lipid and protein level.

In my opinion and in the light of new investigations, several critical aspects of lipid peroxidation in biological membranes, such as their assembly and structural organization need to be revisited.

ACKNOWLEDGMENT

Studies in the author laboratory were supported by PIP-0157 National Research Council (CONICET).

REFERENCES

1. Catala, A. (2009). Lipid peroxidation of membrane phospholipids generates hydroxyl-alkenals and oxidized phospholipids active in physiological and/or pathological conditions. *Chem. Phys. Lipids* 157, 1–11. doi: 10.1016/j.chemphyslip.2008.09.004
2. Catala, A. (2010). A synopsis of the process of lipid peroxidation since the discovery of the essential fatty acids. *Biochem. Biophys. Res. Commun.* 399, 318–323. doi: 10.1016/j.bbrc.2010.07.087
3. Catala, A. (2012). Lipid peroxidation modifies the picture of membranes from the "Fluid Mosaic Model" to the "Lipid Whisker Model." *Biochimie* 94, 101–109. (This article belongs to a special issue: Lipids in all their states Edited by Erick J. Dufourc.). doi: 10.1016/j.bbrc.2010.07.087
4. Chapman, D., and Benga, G. (1984). "Biomembrane fluidity – studies of model and natural membranes," in *Biological Membranes*, Vol. 5, ed D. Chapman (London, UK: Academic Press), 1–56.
5. Davies, S. S., and Guo, L. (2014). Lipid peroxidation generates biologically active phospholipids including oxidatively N-modified phospholipids. *Chem. Phys. Lipids* 181, 1–33. doi: 10.1016/j.chemphyslip.2014.03.002
6. Enoch, H. G., Catalá, A., and Strittmatter, P. (1976). Mechanism of rat liver microsomal stearyl-CoA desaturase. Studies of the substrate specificity, enzyme-substrate interactions, and the function of lipid. *J. Biol. Chem.* 251, 5095–5103.
7. Feller, S. E., Gawrisch, K., and MacKerell, A. D. Jr. (2002). Polyunsaturated fatty acids in lipid bilayers:

intrinsic and environmental contributions to their unique physical properties. *J. Am. Chem. Soc.* 124, 318–326. doi: 10.1021/ja0118340

8. Garcia, J. J., Martínez-Ballarín, E., Millán-Plano, S., Allué, J. L., Albendea, C., Fuentes, L., et al. (2005). Effects of trace elements on membrane fluidity. *J. Trace Elem. Med. Biol.* 19, 19–22. doi: 10.1016/j.jtemb.2005.07.007
9. Garrec, J., Monari, A., Assfeld, X., Mir, L. M., and Tarek, M. (2014). Lipid peroxidation in membranes: the peroxyl radical does not "float." *J. Phys. Chem. Lett.* 5, 1653–1658. doi: 10.1021/jz500502q
10. Greenberg, M. E., Li, X. M., Gugiu, B. G., Gu, X., Qin, J., Salomon, R. G., et al. (2008). The lipid whisker model of the structure of oxidized cell membranes. *J. Biol. Chem.* 283, 2385–2396. doi: 10.1074/jbc.M707348200
11. Janmey, P. A., and Kinnunen, P. K. J. (2006). Biophysical properties of lipids and dynamic membranes. *Trends Cell Biol.* 16, 538–546. doi: 10.1016/j.tcb.2006.08.009
12. Klymchenko, A. S., and Kreder, R. (2014). Fluorescent probes for lipid rafts: from model membranes to living cells. *Chem. Biol.* 21, 97–113. doi: 10.1016/j.chembiol.2013.11.009
13. Lee, A. G. (1991). Lipids and their effects on membrane proteins: evidence against a role for fluidity. *Prog. Lipid Res.* 30, 323–348. doi: 10.1016/0163-7827(91)90002-M
14. Li, X. M., Salomon, R. G., Qin, J., and Hazen, S. L. (2007). Conformation of an endogenous ligand in a membrane bilayer for the macrophage scavenger receptor CD36. *Biochemistry* 46, 5009–5017. doi: 10.1021/bi700163y
15. Mueller, N. S., Wedlich-Söldner, R., and Spira, F. (2012). From mosaic to patchwork: matching lipids and proteins in membrane organization. *Mol. Membr. Biol.* 29, 186–196. doi: 10.3109/09687688
16. Naudí, A., Jové, M., Ayala, V., Cabré, R., Portero-Otín, M., and Pamplona, R. (2013). Non-enzymatic modification of aminophospholipids by carbonyl-amine reactions, *Int. J. Mol. Sci.* 14, 3285–3313. doi: 10.3390/ijms14023285
17. Nicolson, G. L. (2014). The fluid-mosaic model of membrane structure: still relevant to understanding the structure, function and dynamics of biological membranes after more than 40 years. *Biochim. Biophys. Acta* 838, 1451–1466. doi: 10.1016/j.bbamem.2013.10.019
18. Pinchuk, I., and Lichtenberg, D. (2014). Analysis of the kinetics of lipid peroxidation in terms of characteristic time-points. *Chem. Phys. Lipids* 178, 63–76. doi: 10.1016/j.chemphyslip.2013.12.001
19. Rossy, J., Ma, Y., and Gaus, K. (2014). The organisation of the cell membrane: do proteins rule lipids? *Curr. Opin. Chem. Biol.* 20, 54–59. doi: 10.1016/j.cbpa.2014.04.009
20. Singer, S. J., and Nicolson, G. L. (1972). The Fluid Mosaic Model of the structure of cell membranes. *Science* 175, 720–731. doi: 10.1126/science.175.4023.720
21. Subczynski, W. K., Hyde, J. S., and Kusumi, A. (1991). Effect of alkyl chain unsaturation and cholesterol intercalation on oxygen transport in membranes: a pulse ESR spin labeling study. *Biochemistry* 30, 8578–8590. doi: 10.1021/bi00099a013
22. Vigor, C., Bertrand-Michel, J., Pinot, E., Oger, C., Vercauteren, J., Le Faouder, P., et al. (2014). Non-enzymatic lipid oxidation products in biological systems: assessment of the metabolites from polyunsaturated fatty acids. *J. Chromatogr. B Analyt. Technol. Biomed. Life Sci.* 964, 65–78. doi: 10.1016/j.jchromb.2014.04.042
23. Wong-Ekkabut, J., Xu, Z., Triampo, W., Tang, I. M., Tieleman, D. P., and Monticelli, L. (2007). Effect of lipid peroxidation on the properties of lipid bilayers: a molecular dynamics study. *Biophys. J.* 93, 4225–4236. doi: 10.1529/biophysj.107.112565
24. Yin, H., Xu, L., and Porter, N. A. (2011). Free radical lipid peroxidation: mechanisms and analysis. *Chem. Rev.* 111, 5944–5972. doi: 10.1021/cr200084z

CHAPTER 19 | COMPLEX CARBOHYDRATES

Brandi L. Cantarel, Vincent Lombard, Bernard Henrissat

INTRODUCTION

Carbohydrates in the form of glycoconjugates, oligo- and polysaccharides represent one of the most diverse sets of molecules on Earth. An astounding diversity of carbohydrates is made possible by the different monosaccharide structures that can be assembled in a number of fashions to other sugar molecules, or to virtually any molecule of life, such as lipids, nucleic acids, proteins, antibiotics [1], [2], [3]. The resulting structural diversity is exploited by living organisms to perform with a high specificity a multitude of biological roles that can be assigned to broad categories such as structure and reserve, and intrinsic vs. extrinsic recognition [4]. Plants have evolved a particularly elaborate carbohydrate metabolism to synthesize a thick protective cell wall made of numerous polysaccharides and whose structure is resistant to enzymatic conversion. Carbohydrates also play an immense role in the host:microbiome interactions, such as providing nutrients to both host and flora, or as mediators that control the complex relationships between the two partners [5].

The selective assembly of glycoconjugates and complex carbohydrates is catalyzed by glycosyltransferases (GTs) while the deconstruction of the resulting structures is achieved by specific glycoside hydrolases (GHs) and polysaccharide lyases (PLs). Collectively these enzymes have been termed "Carbohydrate-active enzymes" (CAZymes), which are classified in a number of sequence-based families in the CAZy database (www.cazy.org) [6], [7], [8], [9], [10]. Because the number of protein folds is much smaller than the number of carbohydrate structures to build or break down, the sequence (hence structural) based families of CAZymes most frequently group together enzymes of differing substrate specificity, i.e. enzymes with different EC numbers [11]. One consequence is that it is difficult to predict the exact specificity of a CAZyme based on family membership. Despite this limitation, an inspection of the content of the families shows that a broad substrate category can often be associated to a CAZy family even if the precise specificity of each protein in the family is usually hard to predict reliably. Examination of the human genome reveals that it encodes 97 GHs, with only 17 enzymes to breakdown carbohydrate nutrients, nine of which are not yet fully characterized. This is a tiny number compared to some of the gut bacteria such as *Bacteroides thetaiotaomicron* [12], which, alone, encodes over 260 GHs (see www.cazy.org/b135.html).

It has been estimated that the human body is inhabited by 10^{14} microbes, with the human gut thought to contain about 7000 different strains [13]. The Human Microbiome Project (HMP) consortium has collected and analyzed Whole Genome Shotgun (WGS) sequence information from microbial communities isolated from five major body sites inside and on the surface of the human body (Gut, Airways, Oral, Urogenital and Skin) [14]. As a collaborative analysis with the HMP consortium, we have compared the prevalence and abundances of CAZymes from 148 subjects collected in these 5 major body sites, yielding 520 samples (www.hmpdacc.org/HMPGOI). To determine the predictive power of taxonomic community structure to reconstruct the functional repertoire of a microbial community, we have built functional profiles for these samples, using their 16 S taxonomic profile and gene content based on reference genomes. Lastly, we compared the gene content of 493 reference genomes isolated from the human body in order to determine whether human-associated bacterial genomes, some of which being part of the consortium project [15], inhabiting a particular habitat had gene content more similar than microbes in the same taxonomic family. Examination of the family prevalence and of the broad carbohydrate categories shows that the CAZyme profiles of each body site are different, and adapted to the particular carbohydrate composition of the body site.

Citation: Cantarel BL, Lombard V, Henrissat B (2012) Complex Carbohydrates. PLoS ONE 7(6): e28742. doi:10.1371/journal.pone.0028742.

RESULTS

Comparison of Reference Genomes Indicates Gene Content by Taxonomic Family

Microbes living inside and on the surface of the human body have employed horizontal gene transfer and gene duplication to gain functions [16], [17], [18], [19], particularly in CAZymes [20], [21] to adapt to constraints of their particular environment. Therefore, we wanted to assess the similarity of the CAZyme repertoire in human associated bacterial genomes in the same taxonomic family compared to that in the same body site. We have decided to compare the complex carbohydrate utilization abilities of 493 human associated bacterial genomes, by determining their profiles in GH and PL enzymes which both cleave glycosidic bonds, even though they use different chemical mechanisms. When we compare a Bray-Curtis distance between samples, the distribution of distances is lower among most bacterial families (Figure 1A) compared body sites (Figure 1B). However, some bacterial families, such as *Coribacteriaceae*, *Ruminococcaceae*, and *Veillonellaceae*, are more diverse in their CAZyme profiles and have mean distances similar to the genomes of bacteria inhabiting the same body site.

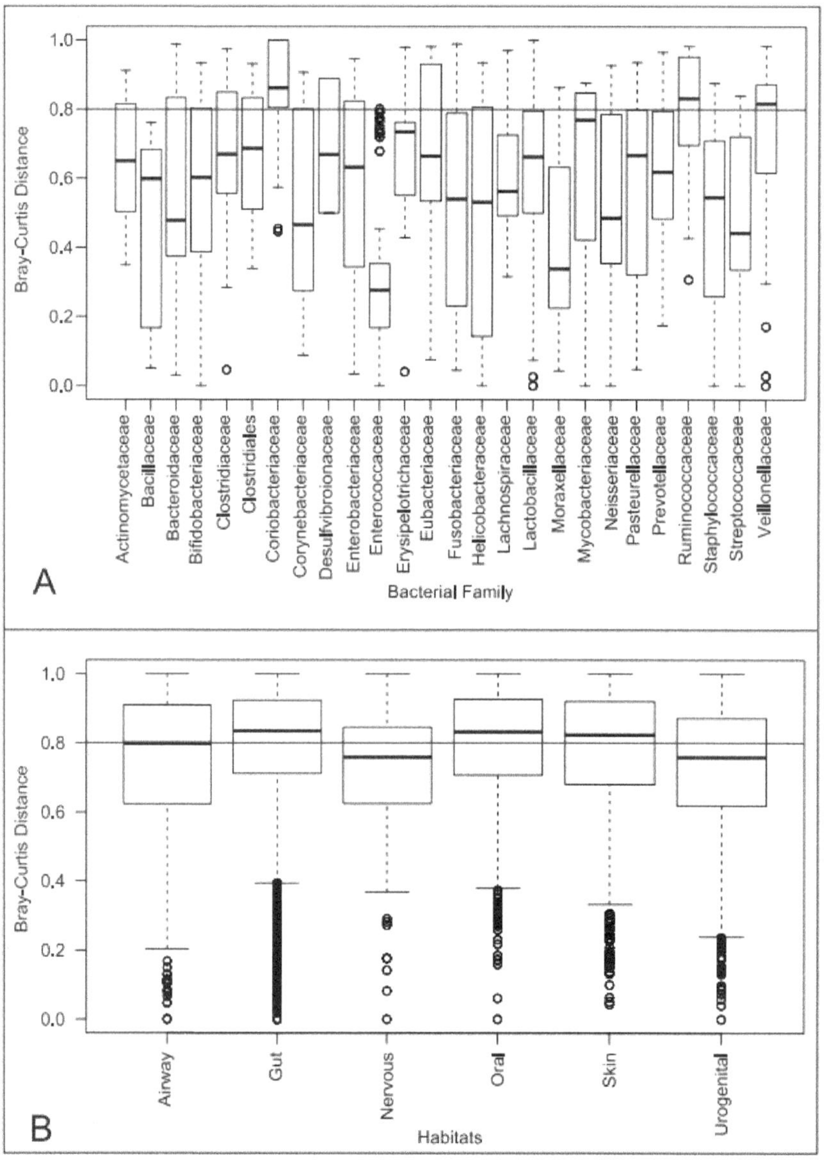

Figure 1. Comparison of Reference Genomes.

Bray-Curtis distances were calculated, using the relative abundance of gene families normalized by housekeeping genes GT28 and GT51, between the 493 human associated bacterial genomes using the ecodist library in R and compared within genomes in the same bacterial family (A) or the same general body site (B).

When we compare the mean number of sugar-cleaving enzymes, GHs and PLs, encoded by genomes within a bacterial family, the abundance of these enzymes varies greatly. For example in the family *Bacillaceae* the mean number of GHs is 25, with a standard deviation of 3.3. In contrast, *Clostridiaceae* has on average 56 GHs, yet the range of that is much broader as the standard deviation is about 79. Therefore, even within a bacterial family, the copy number of GHs and PLs can have a wide distribution, making the prediction of the gene content in an unknown genome difficult.

Relative Abundance of CAZymes in Major Body Sites

To examine microbial communities in the human body, we annotated CAZymes in the reads and contigs publically available as part of the HMP Consortium project, HMIWGS dataset [14], which represent the largest collection of WGS sequence from the human microbiome. In order to assess the CAZyme potential of these 520 microbial communities collected from five major sites inside and on the surface of the human body, we calculated the relative abundance of CAZymes on the metagenomic reads, in the 113 GHs and 19 PLs families that were known at the time of this analysis, normalized on the housekeeping genes GT51 and GT28, which have more stable copy numbers in reference genomes. Analysis of CAZyme profiles using the Bray-Curtis distance metric (Figure 2A) showed that samples in the same body habitat (white boxes) are more similar to each other than those from distinct habitats (grey box). The body site with the highest total abundance of CAZymes is the gastrointestinal tract followed by the various oral samples (Figure 2B). These digestive body sites, on average, also have greater abundances of individual CAZy families represented (Figure 2C). Samples in the buccal mucosa and anterior nares at first glance appear to be poor in these enzymes, but upon further examination, these samples along with retroauricular crease are those with the highest amount of human sequence contamination, which translates into much lower sequence coverage [14].We identified 81 protein families that exhibited statistically significant (p-value <0.05) abundances in pair-wise comparisons among samples in each major body site. Two enzyme families, GH94 (cellobiose, cellodextrin and chitobiose phosphorylases) and GH30 (β-1,6-glucanase, β-xylosidase, β-D-fucosidase, β-glucosidase and β-1,6-galactanase), are over-abundant in stool compared to the other four major body sites. Family GH19, which comprises enzymes involved in the breakdown of animal carbohydrates and fungal cell walls, is under-represented in vaginal samples compared to other sites. When we compare digestive body sites (Oral and GI) to non-digestive sites (Vaginal, Skin and Airways), we identify six families over-represented in digestive sites (Table 1), four of which are involved in plant (GH53 and GH94) and algal (GH117 and GH86) cell wall degradation.

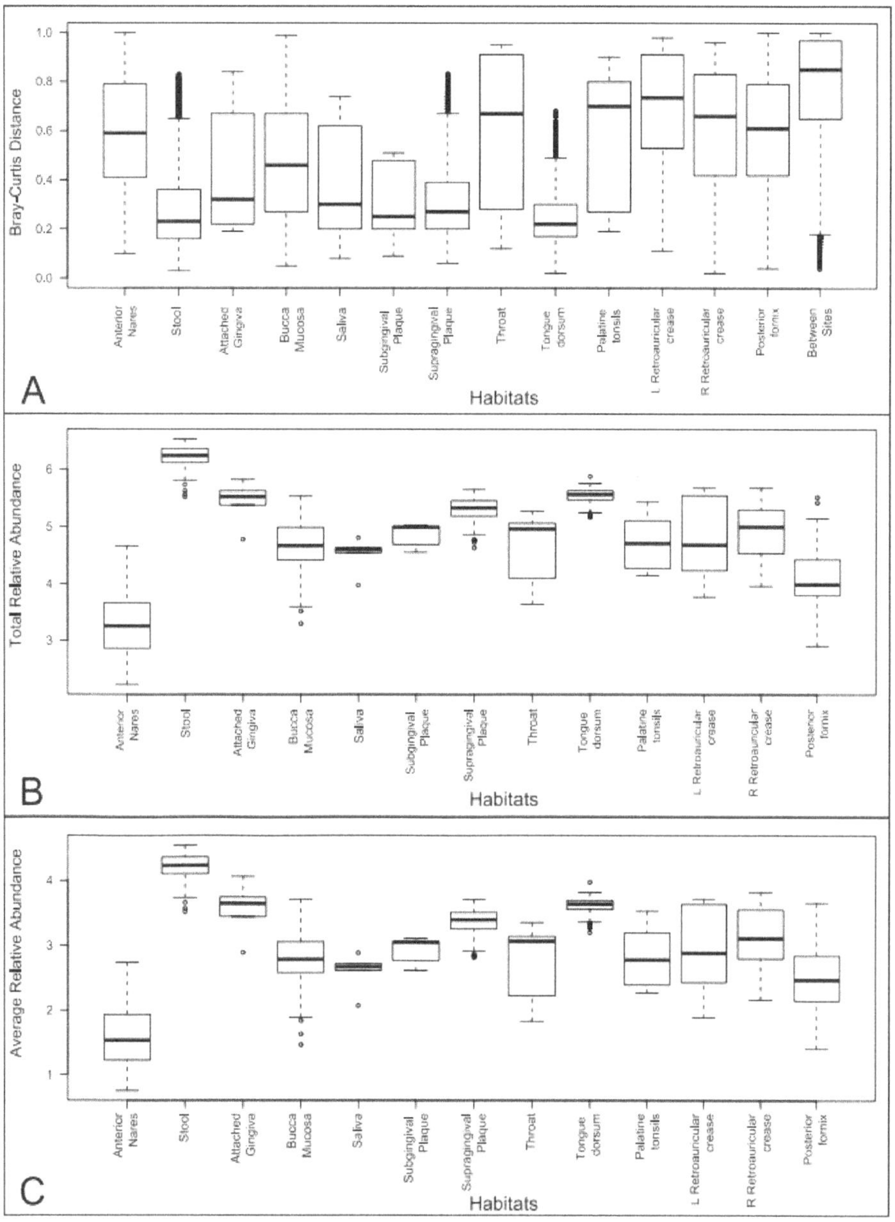

Figure 2. Comparison of Relative Abundance of Healthy Human Microbiome Samples.

(A) Bray-Curtis distances were calculated on relative gene abundances between samples and grouped by same body habitat (white boxes) or distinct habitat (grey box). (B) Total relative abundances are the results of adding the normalized relative abundances of each sample. (C) Average relative abundances are the total relative abundance divided by the total number of CAZy families in each sample.

Table 1. CAZymes statistically significant with differences in relative abundance between digestive site samples compared to non-digestive site samples

	Digestive Site Samples		Other Site Samples		
Family	**Mean Relative Abundance**	**Std Error**	**Mean Relative Abundance**	**Std Error**	**P-Value**
GH53	1327.7	143.7	947.8	209.3	0.04
GH116	634.2	84.3	401.6	102.2	0.04
GH86	2.5	0.8	0.3	0.1	0.05
GH94	1354.6	173.3	640.2	167.7	0.01
GH117	83.9	21.2	38.9	21.1	0.02
GH35	5975.9	396.5	4790.4	642.5	0.05
GH82	0.7	0.3	0.0	0.0	0.00

Differential Prevalence of CAZymes

In addition to comparing the relative abundance of enzymes in body sites, we also compared the presence of GHs and PLs on metagenomic assemblies (www.hmpdacc.org/HMGI) [14]. The prevalence of each gene family was calculated as the fraction of the samples containing the family per body site. Hierarchical clustering by gene family prevalence per body site shows similarities among habitats regardless of differences in sequence coverage seen between sites with high and low human genomic contamination. For example, by examining the enzymes present in oral samples, we discovered that these habitats (Buccal mucosa, supra- and super- gingival plaque, tonsils and tongue) are much more similar to each other than gastrointestinal, urogenital and nasal and skin habitats (Figure 3A). Core families, represented in >95% of samples, are involved in energy production (GH32, GH31, GH38, GH1, GH13) and peptidoglycan breakdown (GH73, GH25, GH23). There are a number of families that are uniquely prevalent to stool which are most likely involved in the digestion of animal (GH79, GH99, PL6, PL13), fungal (GH55, GH64, GH113), plant (GH39, GH74, GH91, GH93, GH94, PL10, PL11) and algal (GH117, PL17, PL15) polysaccharides. The latter results shows that algal polysaccharide breakdown is present in the individuals that were investigated by the HMP consortium, an observation that contrasts with a report that porphyranases and agarases are frequent in the Japanese population and are absent in metagenome data from North American individuals [22]. Our finding of families GH50 (agarases), GH86 (agarases) and GH117 (neoagarooligosaccharide hydrolases) in most oral and stool samples suggests that agarose digestion is largely present in the North American population. Additionally, we compared the gene content of samples within each body site using the Sørensen similarity index, which is a calculation of the number of genes present in two samples compared to the total number of gene families. The genes present in stools samples are the most conserved, whereas vaginal habitats exhibit higher diversity (Figure 3B). These results are consistent with other studies of this dataset based on taxonomy [23] and KEGG Orthologs (KOs) groups [24]; in these studies and in a publication by Ravel and colleagues, the posterior fornix is shown to have the most diverse repertoire of functional capability [23], [24] and a few distinct community types [23], [25].

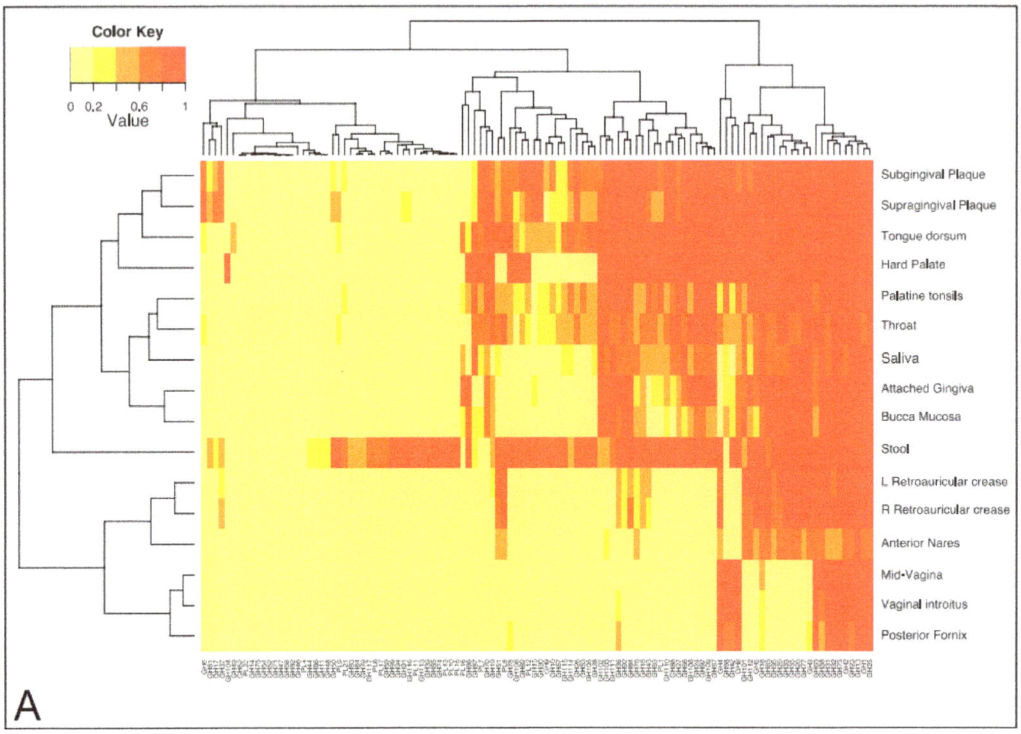

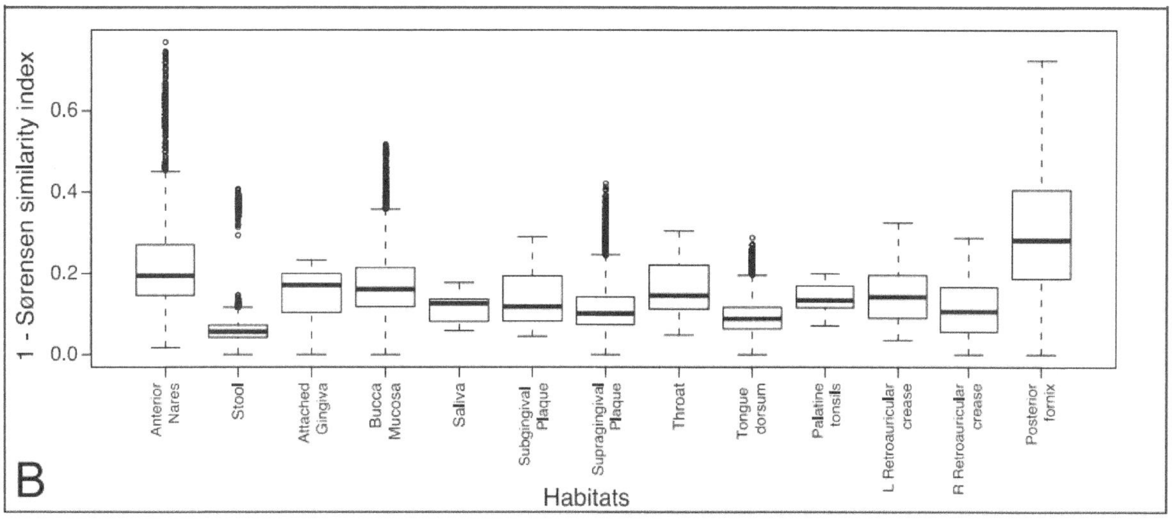

Figure 3. Comparison of Prevalence of Healthy Human Microbiome Samples.

(A) Heatmap of genes prevalence per body site using a heat color scheme (yellow to red), indicating low to high prevalence. (B) Gene repertoire distance as calculated by 1 - Sørensen's similarity coefficient, between samples originating from the same body habitat. Higher distances indicate a lower number of proportionally shared genes between any two samples from the same body site.

Using the publically available taxonomic annotations of these genes [14] (HMGI), we were able to compare the CAZyme abundance from a bacterial family by body site to the reference genome results. Because of the taxonomic diversity of most human habitats, metagenomic sequences, even assembled, are often short fragments, since we are unable to reach sequence saturation. Consequently, our ability to detect CAZymes from particular taxa is reduced compared to the single genomes, which will be a diminishing concern as sequence generation increases. For some bacterial families such as Corynebacteriaceae relative abundance of GHs and PLs is very similar (7.9±1.5 in vaginal samples vs. 9.8 in oral samples). These taxa are estimated to have 17±7.8 GHs and PLs per genome, suggesting similar copies of GHs and PLs per genome. However for other families like Enterobacteriaceae (44±18.9 GH+PLs in reference genomes), there is a 10 fold difference between the number of GHs and PLs in stool compared to oral site (12.3±2.7 vs 1±0.36), possibly reflecting either a selection of genomes with a high number of GHs and PLs in stool.

In order to get an estimate of the enzymatic capabilities and possible pathways for each body site, we have assigned each CAZy family to their broad substrate categories, namely animal, plant cell wall and fungal carbohydrates, starch and glycogen, dextran, peptidoglycan and sucrose and fructans. Although these broad substrate categories do not allow identification of the precise targeted carbohydrates, except for family GH68, which is involved in converting sucrose into biofilms made of fructose polymers (fructan), they allowed us to compare the broad sugar utilization of each body site based on the average of samples within that body site. Four major carbohydrate utilization profiles emerge (Figure 4): (a) as expected, anterior nares and retroauricular crease environments have relatively low capacity for the degradation of the sugars in these categories (b) the microbes in the vagina contain a higher proportion of enzymes involved in sucrose cleavage and polymerization to fructans than the other body sites, potentially for biofilm formation, (c) oral communities have a balanced utilization of these nutrient derived carbohydrates in each subsite, including dextran, which appears to be unique to oral site, suggesting GH70, GH66 and GH87 as potential markers for plaque formation and (d) stool communities have the highest capacity and variety of plant cell wall polysaccharide-cleaving enzymes. CAZy families GH4, GH68, GH42 and GH8, appear to be highly prevalent in the oral, the supra and subgingival plaque sites in particular, and the vagina (Figure 3A). Interestingly the GH8 and GH68 enzymes are probably involved in polysaccharidic biofilm synthesis.

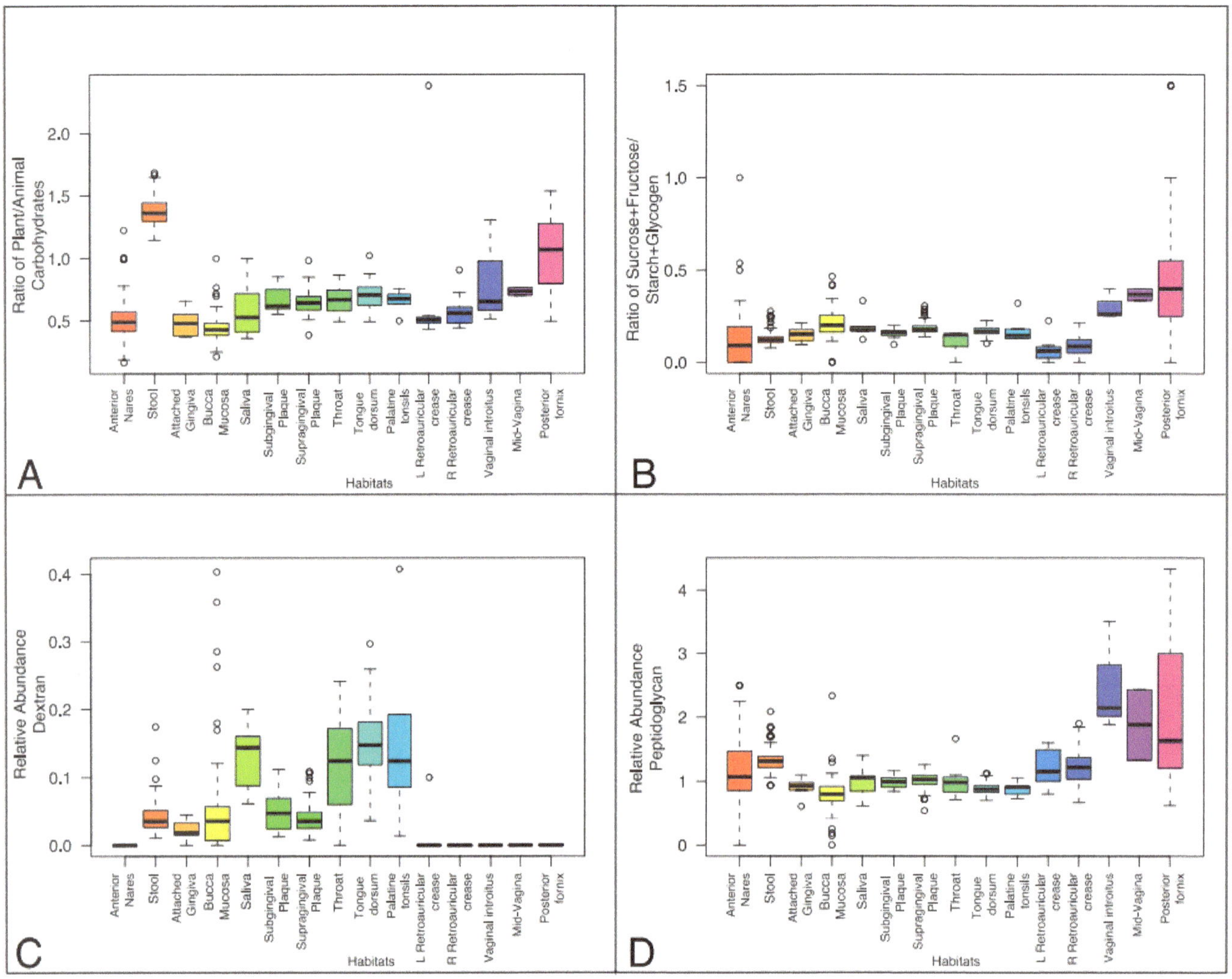

Figure 4. Sugar Utilization Potential of Microbiome Samples.

(A) The ratio of the number of proteins that hydrolyze plant cell wall carbohydrates to proteins that hydrolyze animal carbohydrates. (B) The ratio of the number of proteins that hydrolyze sucrose or fructan to proteins that hydrolyze starch or glycogen. (C) Relative abundance of the proteins that hydrolyze dextran. (D) Relative abundance of the proteins that hydrolyze peptidoglycan.

The human gut microbiota encodes a huge diversity of enzymes for the digestion of all components of plant cell wall polysaccharides including cellulose. Turnbaugh et al [26] have shown that the distal gut microbiota of humans also encodes dockerin-containing cellulolytic enzymes that indicate the presence of cellulosomes (multienzymatic complexes of plant cell wall digesting enzymes assembled on a large scaffolding protein) [27], [28]. The present HMP analysis confirms the presence of dockerin-containing proteins in the GI and identifies several in the oral, nasal and vaginal samples, which is not surprising given their presence in human-associated bacterial reference genomes. In non-digestive sites, however, these dockerin domains may have another role than attaching cellulases to form a cellulosome.

Predicting the CAZome from Taxonomic Profiling

While the cost of next-generation sequencing continues to decrease, studies of hundreds of samples with metagenomic sequencing might be prohibitory for the average researcher whom might be tempted to economize by using 16 S rRNA sequencing and reference genome gene content to predict a functional profile. However, it is unclear whether variation in bacterial communities translates to diversity in functional capability and whether these distinctions can be predicted by community taxonomic profiling. For example, at the genus-level assignment, the posterior fornix, appears to be the least diverse because of dominance by Lactobacillus, however across the population, there appears to be several distinct community types [25]. Our CAZyme analysis suggests that these distinct communities in the posterior fornix might translate into distinct functional types, a result consistent with pathway analysis of these same communities [24]. In contrast, stool has many more genera and therefore is the more taxonomically diverse, yet is the most functionally conserved. Using these 16 S taxonomic profiles and the publicly available reference genomes, we tried to reconstruct the CAZy profile of a select group of samples where 16 S and WGS data were available from the same specimen. In general, we found that the predicted profiles do not resemble very well the actual metagenomic profiles. First, based on relative abundances, we found that the discrepancy between the 16 S predicted and the actual WGS CAZy profile is greater in stool than in oral or vaginal body sites. Similarly, the gene repertoires in oral and stool communities are also underestimated using 16 S rRNA and reference genomes; red, blue and green circles). Interestingly, however, the posterior fornix shows the opposite trend, likely due to the higher coverage by reference genomes; purple circles). These results could indicate (i) there are differences in gene content between organisms in the same genus, which cannot be seen in 16 S because of gene diversity within a species or strain, and/or (ii) we lack reference genomes for some genera. For example, in stool, on average 13% (60% in some samples) of the 16 S sequences classify to a genus without a sequenced reference genome in that genus, representing about 17% of all surveyed genera. Additionally, some of the genes that we identified as differentially abundant between body sites would not have been predicted as being present or under-abundant, such as GH6 in the oral sites and GH94 or GH48 in the GI. In addition, the importance of some proteins would have been over-predicted, such as GH32 in the anterior nares and posterior fornix. Therefore, while 16 S sequencing is a good survey of taxonomic diversity, without more reference genomes, it does not serve as a good proxy for determining the precise carbohydrate utilization capabilities of digestive microbial communities.

DISCUSSION

The human genome encodes less than 20 enzymes for the digestion of complex carbohydrates, mostly plant reserve carbohydrates (sucrose and starch) and lactose. However, the cell walls of plants represent an enormous nutrient source, yet highly variable in terms of amount, diversity and botanical source [29], [30]. These carbohydrates are chemically and structurally highly complex, and are arranged in a three-dimensional network that has evolved to be intrinsically resistant to enzymatic breakdown [31]. Thus high molecular weight crystalline cellulose microfibrils are inter-twinned with hemicelluloses and pectins, which are a whole range of homo- and heteropolysaccharides composed of dozens of different monosaccharide units linked in a multitude of ways. Ester substituents or non-carbohydrate polymers, such as lignin, proteins, cutin and suberin, add a further layer of complexity. As a result, a single vegetable contains hundreds of different bonds that need to be cleaved in order to unlock the assimilable carbon of the cell wall constituents. Considering the considerable variations (in composition and in microscopic structure) in the cell walls of the vegetables and fruits in the human diet, the digestive enzymes face a huge number of different substrates. Since these enzymes are absent from the human genome, humans rely on the microbiota inhabiting the digestive track to utilize these complex plant polysaccharides [13]. The microbiota must adapt rapidly to environmental cues to determine which enzymes are necessary to metabolize the plant cell wall structures in each meal.

Digestion starts in the oral cavity, where this study suggests that microbes have a hitherto underestimated large range of enzymes to initiate plant polysaccharide breakdown as indicated by the presence of cellulases (GH6), hemicellulases (GH26) and pectin hydrolases (GH28 and GH43). Additionally, microbes in the oral cavity also initiate the processing of 'easy' plant carbohydrates, such as sucrose and starch, which can be converted into biofilms (dextran, fructans) that secure long-term residence to the bacteria in the oral sphere. When we compare the oral sites to the gut, the starch and glycogen utilization appears much reduced in stool, suggesting specialization in digestion, whereby most of these sugars are likely degraded by human salivary amylase and oral flora and taken up in the small intestine by the host [32]. The starch molecules that do reach the distal gut are particularly difficult to hydrolyze and are known as "resistant" starch [33] and are considered a component of dietary fibers.

Upon examination of the mechanisms important for plant cell-wall degradation, we found that non-reducing end acting cellobiohydrolases (CAZy family GH6) appear specific to the oral cavity while reducing end acting cellobiohydrolases (family GH48) are specific to the gut. The two types of cellobiohydrolases have been found to digest cellulose in a synergistic manner when acting together or when acting with endoglucanases [34]. The current paradigm is that cellobiohydrolases are the essential cellulases because they can deliver soluble cellobiose from polymeric cellulose in a single step, and the role of the endoglucanases is to provide chain ends to the cellobiohydrolases [31]. The GH6 genes found in the oral cavity, were highly prevalent in supragingival plaques samples and are primarily attributed to the genera *Capnocytophaga*. The GH48 genes, found in stool, are mostly in unknown taxon, although likely in some species of the *Firmicutes* genus *Ruminococcus*. However, there is no reference genome in this genus with annotated GH48 genes. Thus far, the GH6 family of enzymes has not been identified in any animal gut sample [35], [36], [37], [38], [39], suggesting that this enzyme specificity is perhaps driven by environmental factors.

Based on this study, digestion in the gut appears highly specialized for the digestion of complex carbohydrates. Since the other body sites are unlikely to be exposed to plant carbohydrates for a significant length/amount, the plant carbohydrate utilization is likely the most prominent factor to explain the great divide observed in the WSG metagenomic data between the digestive tract and the other body sites. In the gut, the proportion of genes that hydrolyze plant cell wall is greater than the genes that hydrolyze animal carbohydrates, probably reflecting the greater carbohydrate diversity elaborated by plants compared to that of animals. In almost every carbohydrate category, the gut microbiota has the highest ability to degrade these carbohydrates. The distal gut appears to have all the necessary enzymes for plant polysaccharide digestion with the puzzling exception of GH6 cellulases, suggesting that these enzymes have not been selected to breakdown cellulose substrates by anaerobic animal gut bacteria, while they are common in soil bacteria and fungi that decay plant cell walls. In conclusion, two major trends emerge. First the functional profile of the collective microbial community is more similar within a body site than between sites, despite variation of taxonomic profiles. This means that there is a specialization of the flora at each body site and that it is clearly detectable by metagenomic sequencing, suggesting that metagenomic sequencing is able to trace the functional adaptation to the carbohydrates that prevail in a given body site. Second, while broad predictions of the global number of

CAZymes could possibly be made due to the overwhelming number of GHs and PLs encoded by Bacteroidetes compared to Firmicutes, the present results show that we are unable to predict the actual CAZyme profile at each body site due to an insufficient number of reference genomes. However, without a better knowledge of the precise substrate specificity of the enzyme families showing expansion/reduction, there seems to be little correlation between the functional capability and taxonomic family. These results suggest the exact functional profile of CAZymes by body site is not currently predictable given genera abundances. Whilst the current efforts aimed at sequencing more reference genomes will sooner or later allow a finer prediction, the precise functional CAZyme profiling will also require coupling metagenomic analyses to structural genomics initiatives and to high-throughput biochemical and other functional assays of CAZymes [40].

METHODS

Reference Genome Annotations

In order to compare the gene content of reference genomes isolated from human body sites, protein sequences files from human associated genomes were downloaded from Genbank [41]. CAZyme protein family assignments were determined for each reference genome using a semi-automated pipeline [6] and normalized by the number of glycosyltransferases from families GT51 and GT28, as these are found to be stable in bacterial genomes. Bray-Curtis distances were calculated between samples using the ecodist library in R. Distances were compared within each bacterial family and within each body site as separate analyses.

Metagenomic Read Genome Annotations

Pre-human screened, trimmed and quality-filtered Illumina shotgun metagenomic reads (HMIWGS build 1.0) [14] were downloaded from http://hmpdacc.org/HMIWGS. Carbohydrate active enzyme annotations were performed on each read using MBLASTX (http://www.multicorewareinc.com/), with default parameters and against proteins in the CAZy database [6]. Relative abundances were calculated based on the matches and sequence length based on methods developed by Abubucker et al. [24] and normalized by dividing the relative count by the number of GT28 and GT51 relative counts. Bray-Curtis distances were calculated between samples using the ecodist library in R. Distances were compared between all samples and divided into same sample and distinct comparisons. Total relative abundances are the results of adding the normalized relative abundances of each sample. Average relative abundances are the total relative abundance divided by the total number of CAZy families in each sample. Gene families with statistically significant differential abundance between body sites were determined using METASTATS [42].

Metagenomic Contig Genome Annotations

Assemblies from pre-human screened, trimmed and quality-filtered Illumina shotgun metagenomic reads, HMASM build 1.0 [14], were downloaded from http://hmpdacc.org/HMASM. Functional annotations were performed by BLASTX searches against proteins in the CAZy database [6], using thresholds: E-value $<1e{-6}$ and bits/position >1, and normalized by the abundance of reads assigned to families GT51 and GT28. Hierarchical clustering was performed for these genomes, using the complete method, by the matrix of Euclidean distances between genomes based on the normalized counts of GH and PL protein families. The prevalence of a gene was determined by calculating the fraction of the samples within each body site where that gene was present, at any abundance level. Sørensen's similarity coefficient [43] was used to determine the similarity of the GH/PL profile between the samples in each body site. Based on the substrates for each GH and PL family, the sugar utilization of each sample was estimated by summing the relative abundance of genes, which act on each substrate. The sugar utilization of each body site is an average of sugar utilization of each sample.

Inferred Metagenomic CAZy Profiles

Phylotype information for samples, with corresponding 16 S rRNA profiles, was downloaded from http://hmpdacc.org/HMPOC [14]. The inferred metagenomic CAZy profiles were determined by using these taxonomic relative abundance tables for each sample. For each CAZy family in a sample, the profile was calculated by (i) determining the average number of that family for all reference genomes for each genera, (ii) multiplying that average by the fraction of the sample in each genera, (iii) summing each genera's inferred contribution to each family and (iv) normalizing each sum per family by housekeeping glycosyltransferases from families GT51 and GT28. Hierarchical clustering is calculated for these genomes, using the complete method, by the matrix of Euclidean distances between genomes based on the normalized counts of GH and PL protein families, since a Bray-Curtis distance calculation is less sensitive to differences of scale. These profiles were then compared to abundances based on reads. Fraction of families conserved was calculated by the number of families presented in the WGS metagenomic or 16 S inferred profile divided by the total number of families observed whether they be inferred or observed

ACKNOWLEDGMENTS

The authors would like to thank the Human Microbiome project research network consortium for all of the discussion on the human microbiome datasets and analysis strategies, in particular Owen White, Curtis Huttenhower and Jacque Izard for their discussion of this dataset in the scope of other analyses. The authors wish to also thank Pedro Coutinho (AFMB, Marseille) for numerous discussions about carbohydrate-active enzymes.

REFERENCES

1. Hu Y, Walker S (2002) Remarkable structural similarities between diverse glycosyltransferases. Chem Biol 9: 1287–1296.
2. Laine RA (1994) A calculation of all possible oligosaccharide isomers both branched and linear yields 1.05 x 10(12) structures for a reducing hexasaccharide: the Isomer Barrier to development of single-method saccharide sequencing or synthesis systems. Glycobiology 4: 759–767.

3. Varki A, Sharon N (2009) Historical Background and Overview.
4. Varki A (1999) Essentials of glycobiology. Cold Spring Harbor, NY: Cold Spring Harbor Laboratory Press. xvii, 653 p. p.
5. Hooper LV, Midtvedt T, Gordon JI (2002) How host-microbial interactions shape the nutrient environment of the mammalian intestine. Annu Rev Nutr 22: 283–307.
6. Cantarel BL, Coutinho PM, Rancurel C, Bernard T, Lombard V, et al. (2009) The Carbohydrate-Active EnZymes database (CAZy): an expert resource for Glycogenomics. Nucleic Acids Res 37: D233–238.
7. Henrissat B (1991) A classification of glycosyl hydrolases based on amino acid sequence similarities. Biochem J 280 (Pt 2): 309–316.
8. Henrissat B, Callebaut I, Fabrega S, Lehn P, Mornon JP, et al. (1995) Conserved catalytic machinery and the prediction of a common fold for several families of glycosyl hydrolases. Proc Natl Acad Sci U S A 92: 7090–7094.
9. Henrissat B, Davies G (1997) Structural and sequence-based classification of glycoside hydrolases. Curr Opin Struct Biol 7: 637–644.
10. Wang TC, Madiraju MV, Templin A, Clark AJ (1991) Cloning and preliminary characterization of srf-2020 and srf-801, the recF partial suppressor mutations which map in recA of Escherichia coli K-12. Biochimie 73: 335–340.
11. Nomenclature Committee of the International Union of Biochemistry and Molecular Biology (NC-IUBMB) (1999) European Journal of Biochemistry 264: 610–650.
12. Xu J, Bjursell MK, Himrod J, Deng S, Carmichael LK, et al. (2003) A genomic view of the human-Bacteroides thetaiotaomicron symbiosis. Science 299: 2074–2076.
13. Backhed F, Ley RE, Sonnenburg JL, Peterson DA, Gordon JI (2005) Host-bacterial mutualism in the human intestine. Science 307: 1915–1920.
14. Consortium HMP (2011) The Resource generated from the Human Microbiome Project Initiative. Nature *in Preparation*.
15. Nelson KE, Weinstock GM, Highlander SK, Worley KC, Creasy HH, et al. (2010) A catalog of reference genomes from the human microbiome. Science 328: 994–999.
16. Jones BV, Marchesi JR (2007) Transposon-aided capture (TRACA) of plasmids resident in the human gut mobile metagenome. Nat Methods 4: 55–61.
17. Roberts AP, Chandler M, Courvalin P, Guedon G, Mullany P, et al. (2008) Revised nomenclature for transposable genetic elements. Plasmid 60: 167–173.
18. Tamames J, Moya A (2008) Estimating the extent of horizontal gene transfer in metagenomic sequences. BMC Genomics 9: 136.
19. Xu J, Mahowald MA, Ley RE, Lozupone CA, Hamady M, et al. (2007) Evolution of symbiotic bacteria in the distal human intestine. PLoS Biol 5: e156.
20. Lozupone CA, Hamady M, Cantarel BL, Coutinho PM, Henrissat B, et al. (2008) The convergence of carbohydrate active gene repertoires in human gut microbes. Proc Natl Acad Sci U S A 105: 15076–15081.
21. Tasse L, Bercovici J, Pizzut-Serin S, Robe P, Tap J, et al. (2010) Functional metagenomics to mine the human gut microbiome for dietary fiber catabolic enzymes. Genome Res 20: 1605–1612.
22. Hehemann JH, Correc G, Barbeyron T, Helbert W, Czjzek M, et al. (2010) Transfer of carbohydrate-active enzymes from marine bacteria to Japanese gut microbiota. Nature 464: 908–912.
23. Consortium HMP (2011) Structure, Function and Diversity of the Human Microbiome in an Adult Reference Population. Nature *in Preparation*.
24. Abubucker S, Segata N, Goll J, Schubert AM, Izard J, et al. (2011) Metabolic reconstruction for metagenomic data and its application to the human microbiome. PLoS Comp Biol *in Review*.
25. Ravel J, Gajer P, Abdo Z, Schneider GM, Koenig SS, et al. (2011) Vaginal microbiome of reproductive-age women. Proc Natl Acad Sci U S A 108: 4680–4687.
26. Turnbaugh PJ, Quince C, Faith JJ, McHardy AC, Yatsunenko T, et al. (2010) Organismal, genetic, and transcriptional variation in the deeply sequenced gut microbiomes of identical twins. Proc Natl Acad Sci U S A 107: 7503–7508.
27. .Bayer EA, Belaich JP, Shoham Y, Lamed R (2004) The cellulosomes: multienzyme machines for degradation of plant cell wall polysaccharides. Annu Rev Microbiol 58: 521–554.
28. Fontes CM, Gilbert HJ (2010) Cellulosomes: highly efficient nanomachines designed to deconstruct plant cell wall complex carbohydrates. Annu Rev Biochem 79: 655–681.
29. Carpita NC (1996) Structure and Biogenesis of the Cell Walls of Grasses. Annu Rev Plant Physiol Plant Mol Biol 47: 445–476.
30. .Gorshkova TA, Wyatt SE, Salnikov VV, Gibeaut DM, Ibragimov MR, et al. (1996) Cell-Wall Polysaccharides of Developing Flax Plants. Plant Physiol 110: 721–729.
31. Gilbert HJ (2010) The biochemistry and structural biology of plant cell wall deconstruction. Plant Physiol 153: 444–455.
32. Arumugam M, Raes J, Pelletier E, Le Paslier D, Yamada T, et al. (2011) Enterotypes of the human gut microbiome. Nature.
33. Topping DL, Clifton PM (2001) Short-chain fatty acids and human colonic function: roles of resistant starch and nonstarch polysaccharides. Physiol Rev 81: 1031–1064.
34. Henrissat B, Driguez H, Viet C, Schulein M (1985) Synergism of Cellulases from Trichoderma reesei in the Degradation of Cellulose. Nat Biotech 3: 722–726.
35. Brulc JM, Antonopoulos DA, Miller ME, Wilson MK, Yannarell AC, et al. (2009) Gene-centric metagenomics of the fiber-adherent bovine rumen microbiome reveals forage specific glycoside hydrolases. Proc Natl Acad Sci U S A 106: 1948–1953.
36. Hess M, Sczyrba A, Egan R, Kim TW, Chokhawala H, et al. (2011) Metagenomic discovery of biomass-degrading

genes and genomes from cow rumen. Science 331: 463–467.

37. Turnbaugh PJ, Hamady M, Yatsunenko T, Cantarel BL, Duncan A, et al. (2009) A core gut microbiome in obese and lean twins. Nature 457: 480–484.
38. Warnecke F, Luginbuhl P, Ivanova N, Ghassemian M, Richardson TH, et al. (2007) Metagenomic and functional analysis of hindgut microbiota of a wood-feeding higher termite. Nature 450: 560–565.
39. Suen G, Scott JJ, Aylward FO, Adams SM, Tringe SG, et al. (2010) An insect herbivore microbiome with high plant biomass-degrading capacity. PLoS Genet 6.
40. Turnbaugh PJ, Henrissat B, Gordon JI (2010) Viewing the human microbiome through three-dimensional glasses: integrating structural and functional studies to better define the properties of myriad carbohydrate-active enzymes. Acta Crystallogr Sect F Struct Biol Cryst Commun 66: 1261–1264.
41. Benson DA, Karsch-Mizrachi I, Lipman DJ, Ostell J, Wheeler DL (2008) GenBank. Nucleic Acids Res 36: D25–30.
42. White JR, Nagarajan N, Pop M (2009) Statistical methods for detecting differentially abundant features in clinical metagenomic samples. PLoS Comput Biol 5: e1000352.
43. Sørensen T (1957) A method of establishing groups of equal amplitude in plant sociology based on similarity of species and its application to analyses of the vegetation on Danish commons. Biologiske Skrifter 5: 1–34.

INDEX